Phytochemicals in Nutrition and Health

Phytochemicals in Nutrition and Health

Zoey Hughes
Shawn Wallace

Kruger Brentt
Publishers
2026

Kruger Brentt Publishers UK. LTD.
Company Number 9728962

Regd. Office: 68 St Margarets Road, Edgware, Middlesex HA8 9UU

ISBN: 978-1-78715-667-8

For information on all our publications visit our website at http://krugerbrentt.com/

PREFACE

Phytochemicals in Nutrition and Health provides a comprehensive exploration of the role of plant-derived bioactive compounds in promoting human health and preventing disease. Phytochemicals, naturally occurring in fruits, vegetables, grains, and other plant-based foods, have gained increasing attention for their antioxidant, anti-inflammatory, and disease-preventing properties. This book is designed for students, researchers, nutritionists, and health professionals who seek a deep understanding of how these compounds contribute to overall wellness and disease prevention.

The book begins with an introduction to phytochemicals, their chemical structures, and classification, including polyphenols, flavonoids, carotenoids, alkaloids, and other bioactive compounds. It then explores their mechanisms of action, including antioxidant activity, modulation of metabolic pathways, and interactions with cellular signaling systems.

A significant focus is placed on the health benefits of phytochemicals, including their roles in cardiovascular health, cancer prevention, immune system support, metabolic health, and neuroprotection. The book also examines dietary sources, bioavailability, and factors influencing the absorption and effectiveness of phytochemicals in the human body.

Practical aspects of incorporating phytochemicals into nutrition and diet planning are highlighted, along with discussions on safety, potential interactions with medications, and recommendations for research and public health interventions. Case studies and real-world examples illustrate the application of phytochemicals in functional foods, nutraceuticals, and dietary strategies.

We hope that Phytochemicals in Nutrition and Health will serve as an essential reference for learners and professionals, providing both theoretical foundations and practical insights for promoting health through plant-based nutrition.

Zoey Hughes

Shawn Wallace

CONTENTS

1

Chapter

PHYTOCHEMICALS: EXTRACTION TECHNIQUES, FUNDAMENTAL STRUCTURES, AND MECHANISMS OF ACTION AS POTENTIAL CHEMOTHERAPEUTIC AGENTS

1. INTRODUCTION: AN OVERVIEW

Medicinal plants have constituted the foundation of traditional herbal medicine among rural populations globally from antiquity to the present. The medicinal application of plants dates back to the Sumerian and Akkadian civilizations about the third millennium BC.. Hippocrates (ca. 460–377 BC), one of the ancient authors who described medicinal natural products of plant and animal origins, listed approximately 400 different plant species for medicinal purposes. Natural products have been an integral part of the ancient traditional medicine systems, e.g. Chinese, Ayurvedic and Egyptian (Sarker & Nahar, 2007). Over the years they have assumed a very central stage in modern civilization as natural source of chemotherapy as well as amongst scientist in search for alternative sources of drugs. About 3.4 billion people in the developing world depend on plant-based traditional medicines. This represents about 88 per cent of the world's inhabitants, who rely mainly on traditional medicine for their primary health care. According to the World Health Organization, a medicinal plant is any plant which, in one or more of its organs, contains substances that can be used for therapeutic purposes, or which are precursors for chemo-pharmaceutical semi synthesis. Such a plant will have its parts including leaves, roots, rhizomes, stems, barks, flowers, fruits, grains or seeds, employed in the control or treatment of a disease condition and therefore contains chemical components that are medically active. These non-nutrient plant chemical compounds or bioactive components are often referred to as phytochemicals ('phyto-' from Greek - *phyto* meaning 'plant') or phytoconstituents and are responsible for protecting the plant against microbial infections or infestations by pests (Abo *et al.*, 1991; Liu, 2004; Nweze *et al.*, 2004; Doughari *et al.*, 2009). The study of natural products on the other hand is called phytochemistry. Phytochemicals have been isolated and characterized from fruits such as grapes and apples, vegetables such as broccoli and onion, spices such as turmeric, beverages such as green tea and red wine, as well as many other sources (Doughari & Obidah, 2008; Doughari *et al.*, 2009).

The science of application of these indigenous or local medicinal remedies including plants for treatment of diseases is currently called ethno pharmacology but the practice dates back since antiquity. Ethno pharmacology has been the mainstay of traditional medicines the entire world and currently is being integrated into mainstream medicine. Different catalogues including *De Materia Medica, Historia Plantarum, Species Plantarum* have been variously published in attempt to provide scientific information on the medicinal uses of plants. The types of plants and methods of application vary from locality to locality with 80% of rural dwellers relying on them as means of treating various diseases. For example, the use of bearberry (*Arctostaphylos uva-ursi*) and cranberry juice (*Vaccinium macrocarpon*) to treat urinary tract infections is reported in different manuals of phytotherapy, while species such as lemon balm (*Melissa officinalis*), garlic (*Allium sativum*) and tee tree (*Melaleuca alternifolia*) are described as broad-spectrum antimicrobial agents (Heinrich *et al.*, 2004). A single plant may be used for the treatment of various disease conditions depending on the community. Several ailments including fever, asthma, constipation, esophageal cancer and hypertension have been treated with traditional medicinal plants (Cousins & Huffman, 2002; Saganuwan, 2010). The plants are applied in different forms such as poultices, concoctions of different plant mixtures, infusions as teas or tinctures or as component mixtures in porridges and soups administered in different ways including oral, nasal (smoking, snoffing or steaming), topical (lotions, oils or creams), bathing or rectal (enemas). Different plant parts and components (roots, leaves, stem barks, flowers or their combinations, essential oils) have been employed in the treatment of infectious pathologies in the respiratory system, urinary tract, gastrointestinal and biliary systems, as well as on the skin (Rojas *et al.*, 2001; R´□os & Recio, 2005; Adekunle & Adekunle, 2009).

Medicinal plants are increasingly gaining acceptance even among the literates in urban settlements, probably due to the increasing inefficacy of many modern drugs used for the control of many infections such as typhoid fever, gonorrhoea, and tuberculosis as well as increase in resistance by several bacteria to various antibiotics and the increasing cost of prescription drugs, for the maintenance of personal health (Levy, 1998; Van den Bogaard *et al.*, 2000; Smolinski *et al.*,

2003). Unfortunately, rapid explosion in human population has made it almost impossible for modern health facilities to meet health demands all over the world, thus putting more demands on the use of natural herbal health remedies. Current problems associated with the use of antibiotics, increased prevalence of multiple-drug resistant (MDR) strains of a number of pathogenic bacteria such as methicillin resistant *Staphylococcus aureus, Helicobacter pylori,* and MDR *Klebsiela pneumonia* has revived the interest in plants with antimicrobial properties (Voravuthikunchai & Kitpipit, 2003). The rise in opportunistic infections, along with the emergence of Acquired Immune Deficiency Syndrome (AIDS) patients and those undergoing immunosuppressive chemotherapy, coupled with the toxicity of numerous antifungal and antiviral medications, has compelled the scientific community and pharmaceutical companies to explore alternative and innovative drug sources.

2. CLASSES OF PHYTOCHEMICALS

2.1 Alkaloids

These are the largest group of secondary chemical constituents made largely of ammonia compounds comprising basically of nitrogen bases synthesized from amino acid building blocks with various radicals replacing one or more of the hydrogen atoms in the peptide ring, most containing oxygen. The compounds have basic properties and are alkaline in reaction, turning red litmus paper blue. In fact, one or more nitrogen atoms that are present in an alkaloid, typically as 1°, 2° or 3° amines, contribute to the basicity of the alkaloid. The degree of basicity varies considerably, depending on the structure of the molecule, and presence and location of the functional groups (Sarker & Nahar, 2007). They react with acids to form crystalline salts without the production of water (Firn, 2010). Majority of alkaloids exist in solid such as atropine, some as liquids containing carbon, hydrogen, and nitrogen. Most alkaloids are readily soluble in alcohol and though they are sparingly soluble in water, their salts of are usually soluble. The solutions of alkaloids are intensely bitter. These nitrogenous compounds function in the defence of plants against herbivores and pathogens, and are widely exploited as pharmaceuticals, stimulants, narcotics, and poisons due to their potent biological activities. In nature, the alkaloids exist in large proportions in the seeds and roots of plants and often in combination with vegetable acids. Alkaloids have pharmacological applications as anesthetics and CNS stimulants (Madziga *et al.,* 2010). More than 12,000-alkaloids are known to exist in about 20% of plant species and only few have been exploited for medicinal purposes. The name alkaloid ends with the suffix *-ine* and plant-derived alkaloids in clinical use include the analgesics morphine and codeine, the muscle relaxant (+)-tubocurarine, the antibiotics sanguinafine and berberine, the anticancer agent vinblastine, the antiarrythmic ajmaline, the pupil dilator atropine, and the sedative scopolamine. Other important alkaloids of plant origin include the addictive stimulants caffeine, nicotine, codeine, atropine, morphine, ergotamine, cocaine, nicotine and ephedrine (Fig. 1). Amino acids act as precursors for biosynthesis of alkaloids with ornithine and lysine commonly used as starting materials. Some screening methods for the detection of alkaloids are summarized in Table 1.

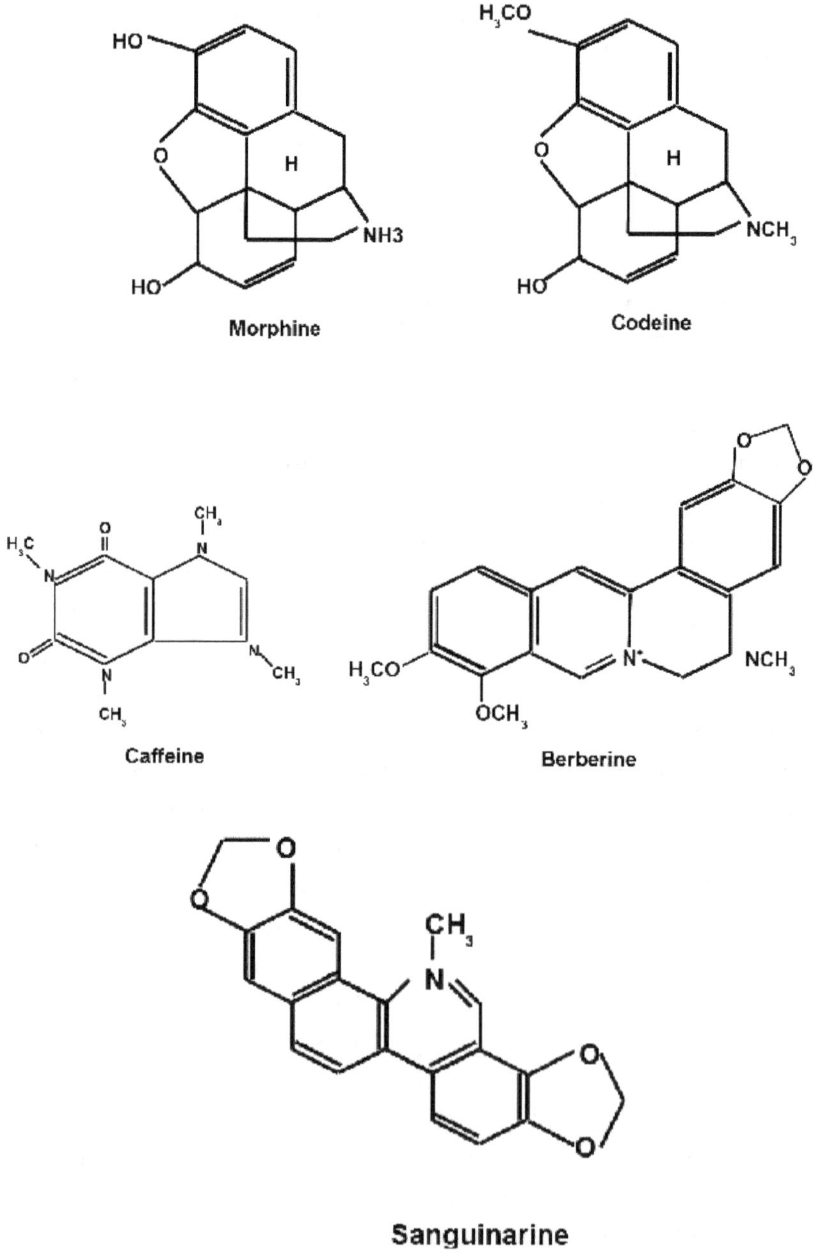

Fig. 1. Basic structures of some pharmacologically important plant derived alkaloids

Table 1. Methods for detection of alkaloids

Reagent/test	Composition of the reagent	Result
Meyer's reagent	Potassiomercuric iodide solution	Cream precipitate
Wagner's reagent	Iodine in potassium iodide	Reddish-brown precipitate
Tannic acid	Tannic acid	Precipitation
Hager's reagent	A saturated solution of picric acid	Yellow precipitate
Dragendorff's reagent	Solution of potassium bismuth iodide potassium chlorate, a drop of hydrochloric acid, evaporated to dryness, and the resulting	Orange or reddish-brown precipitate (except with caffeine and a few other alkaloids)
Murexide test for caffeine	residue is exposed to ammonia vapour	Purine alkaloids produce pink colour

2.2 Glycosides

Glycosides in general, are defined as the condensation products of sugars (including polysaccharides) with a host of different varieties of organic hydroxy (occasionally thiol) compounds (invariably monohydrate in character), in such a manner that the hemiacetal entity of the carbohydrate must essentially take part in the condensation. Glycosides are colorless, crystalline carbon, hydrogen and oxygen-containing (some contain nitrogen and sulfur) water-soluble phytoconstituents, found in the cell sap. Chemically, glycosides contain a carbohydrate (glucose) and a non-carbohydrate part (aglycone or genin) (Kar, 2007; Firn, 2010). Alcohol, glycerol or phenol represents aglycones. Glycosides are neutral in reaction and can be readily hydrolyzed into its components with ferments or mineral acids. Glycosides are classified on the basis of type of sugar component, chemical nature of aglycone or pharmacological action. The rather older or trivial names of glycosides usually has a suffix 'in' and the names essentially included the source of the glycoside, for instance: strophanthidin from *Strophanthus*, digitoxin from *Digitalis*, barbaloin from *Aloes*, salicin from *Salix*, cantharidin from *Cantharides*, and prunasin from *Prunus.* However, the systematic names are invariably coined by replacing the "ose" suffix of the parent sugar with "oside". This group of drugs are usually administered in order to promote appetite and aid digestion. Glycosides are purely bitter principles that are commonly found in plants of the Genitiaceae family and though they are chemically unrelated but possess the common property of an intensely bitter taste. The bitters act on gustatory nerves, which results in increased flow of saliva and gastric juices. Chemically, the bitter principles contain the lactone group that may be diterpene lactones (e.g. *andrographolide*) or triterpenoids (e.g. *amarogentin*).

Some of the bitter principles are either used as astringents due to the presence of tannic acid, as antiprotozoan, or to reduce thyroxine and metabolism. Examples include cardiac glycosides (acts on the heart), anthracene glycosides (purgative, and for treatment of skin diseases), chalcone glycoside (anticancer), amarogentin, gentiopicrin, andrographolide, ailanthone and polygalin (Fig. 2). Sarker & Nahar (2007) reported that extracts of plants that contain cyanogenic glycosides are used as flavouring agents in many pharmaceutical preparations. Amygdalin has been used in the treatment of cancer (HCN liberated in stomach kills malignant cells), and also as a cough suppressant in various preparations. Excessive ingestion of cyanogenic glycosides can be fatal. Some foodstuffs containing cyanogenic glycosides can cause poisoning (severe gastric irritations and damage) if not properly handled (Sarker & Nahar, 2007). To test for O-glycosides, the plant samples are boiled with HCl/H_2O to hydrolyse the anthraquinone glycosides to respective aglycones, and an aqueous base, e.g. NaOH or NH_4OH solution, is added to it. For C-glycosides, the plant samples are hydrolysed using $FeCl_3/HCl$, and and an aqueous base, e.g. NaOH or NH_4OH solution, is added to it. In both cases a pink or violet colour in the base layer after addition of the aqueous base indicates the presence of glycosides in the plant sample.

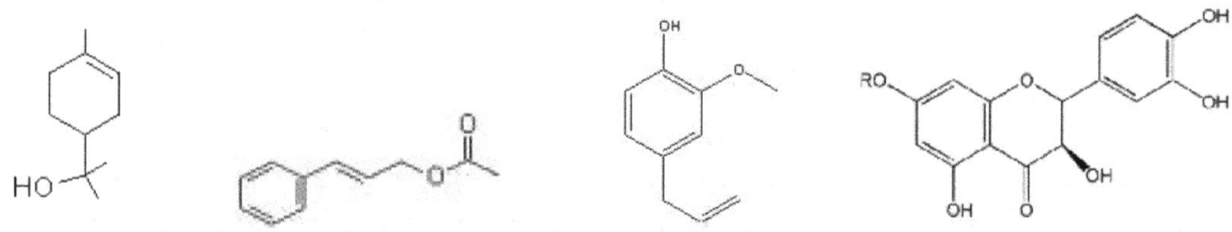

α-Terpineol Cinnamyl acetate Eugenol Taxifolin-7-O- β-glucosid

Fig. 2. Basic structures of some pharmacologically important plant derived glycossides

2.3 Flavonoids

Flavonoids re important group of polyphenols widely distributed among the plant flora. Stucturally, they are made of more than one benzene ring in its structure (a range of C15 aromatic compounds) and numerous reports support their use as antioxidants or free radical scavengers (Kar, 2007). The compounds are derived from parent compounds known as flavans. Over four thousand flavonoids are known to exist and some of them are pigments in higher plants. Quercetin, kaempferol and quercitrin are common flavonoids present in nearly 70% of plants. Other group of flavonoids include flavones, dihydroflavons, flavans, flavonols, anthocyanidins (Fig. 3), proanthocyanidins, calchones and catechin and leucoanthocyanidins.

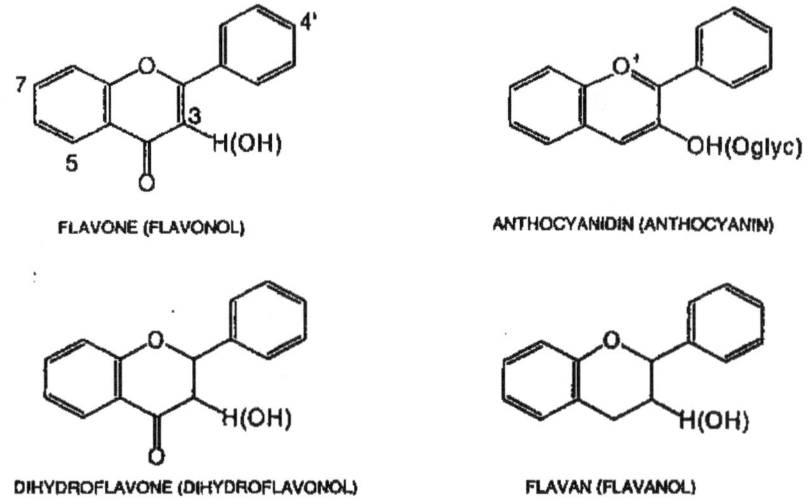

Fig. 3. Basic structures of some pharmacologically important plant derived flavonoids

2.4 Phenolics

Phenolics, phenols or polyphenolics (or polyphenol extracts) are chemical components that occur ubiquitously as natural colour pigments responsible for the colour of fruits of plants. Phenolics in plants are mostly synthesized from phenylalanine via the action of phenylalanine ammonia lyase (PAL). They are very important to plants and have multiple functions. The most important role may be in plant defence against pathogens and herbivore predators, and thus are applied in the control of human pathogenic infections (Puupponen- Pimiä *et al.*, 2008). They are classified into (i) phenolic acids and (ii) flavonoid polyphenolics (flavonones, flavones, xanthones and catechins) and (iii) non-flavonoid polyphenolies. Caffeic acid is regarded as the most common of phenolic compounds distributed in the plant flora followed by chlorogenic acid known to cause allergic dermatitis among humans (Kar, 2007). Phenolics essentially represent a host of natural antioxidants, used as nutraceuticals, and found in apples, green-tea, and red-wine for their enormous ability to combat cancer and are also thought to prevent heart ailments to an appreciable degree and sometimes are anti-inflammatory agents. Other examples include flavones, rutin, naringin , hesperidin and chlorogenic (Fig. 4).

2.5 Saponins

The term saponin is derived from *Saponaria vaccaria* (*Quillaja saponaria*), a plant, which abounds in saponins and was once used as soap. Saponins therefore possess 'soaplike' behaviour in water, i.e. they produce foam. On hydrolysis, an aglycone is produced, which is called sapogenin. There are two types of sapogenin: steroidal and triterpenoidal. Usually, the sugar is attached at C-3 in saponins, because in most sapogenins there is a hydroxyl group at C-3. *Quillaja saponaria* is known to contain toxic glycosides quillajic acid and the sapogenin senegin. Quillajic acid is strenutatory and senegin is toxic. Senegin is also present in *Polygala senega*. Saponins are regarded as high molecular weight compounds in which, a

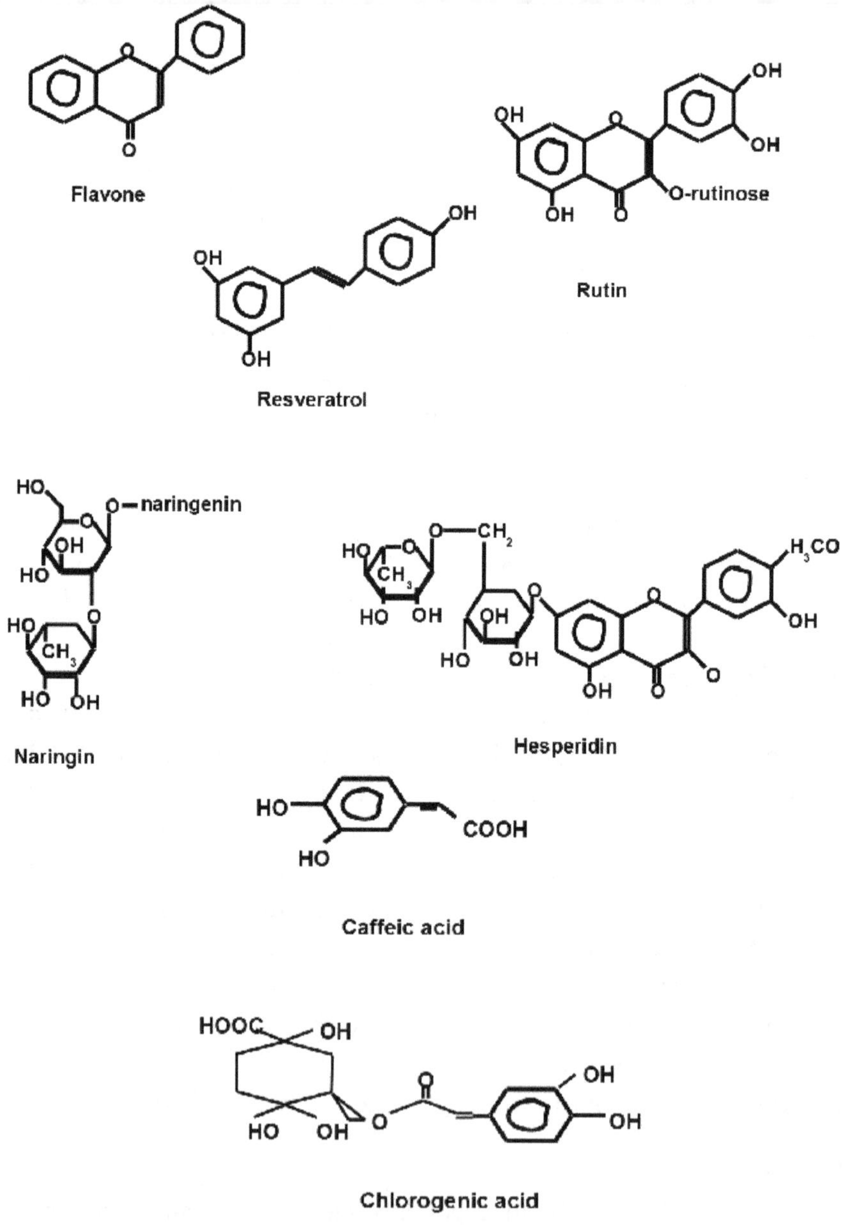

Fig. 4. Basic structures of some pharmacologically important plant derived phenolics

sugar molecule is combined with triterpene or steroid aglycone. There are two major groups of saponins and these include: steroid saponins and triterpene saponins. Saponins are soluble in water and insoluble in ether, and like glycosides on hydrolysis, they give aglycones. Saponins are extremely poisonous, as they cause heamolysis of blood and are known to cause cattle poisoning (Kar, 2007). They possess a bitter and acrid taste, besides causing irritation to mucous membranes. They are mostly amorphous in nature, soluble in alcohol and water, but insoluble in non-polar organic solvents like benzene and n-hexane.

Saponins are also important therapeutically as they are shown to have hypolipidemic and anticancer activity. Saponins are also necessary for activity of cardiac glycosides. The two major types of steroidal sapogenin are diosgenin and hecogenin. Steroidal saponins are used in the commercial production of sex hormones for clinical use. For example, progesterone is derived from diosgenin. The most abundant starting material for the synthesis of progesterone is diosgenin isolated from *Dioscorea* species, formerly supplied from Mexico, and now from China (Sarker & Nahar, 2007). Other steroidal hormones, e.g. cortisone and hydrocortisone, can be prepared from the starting material hecogenin, which can be isolated from Sisal leaves found extensively in East Africa (Sarker & Nahar, 2007).

2.6 Tannins

These are widely distributed in plant flora. They are phenolic compounds of high molecular weight. Tannins are soluble in water and alcohol and are found in the root, bark, stem and outer layers of plant tissue. Tannins have a characteristic feature to tan, i.e. to convert things into leather.

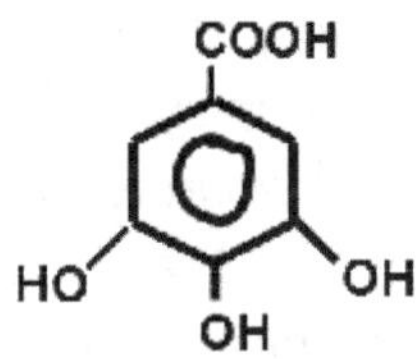

Gallic acid

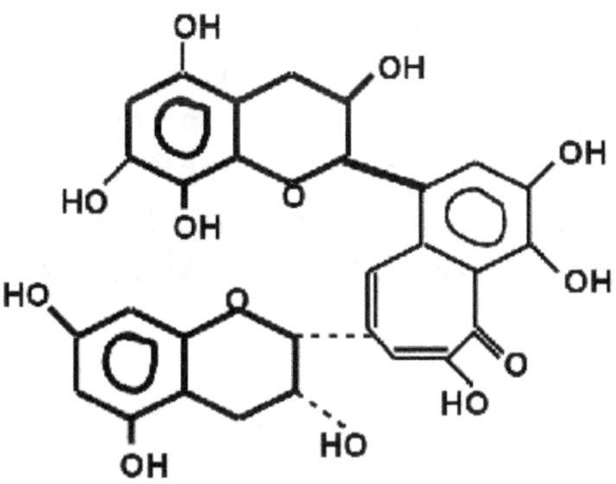

Theaflavin

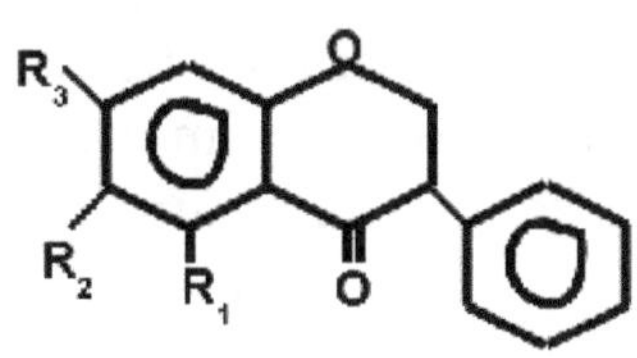

Daidzein: $R_1=R_2=H$; $R_3=OH$
Genistein: $R_1=R_3=OH$; $R_3=H$
Glycitein: $R_1=H$; $R_2=OCH_3$; $R_3=OH$

Fig. 5. Basic structures of some pharmacologically important plant derived tannins

They are acidic in reaction and the acidic reaction is attributed to the presence of phenolics or carboxylic group (Kar, 2007). They form complexes with proteins, carbohydrates, gelatin and alkaloids. Tannins are divided into hydrolysable tannins and condensed tannins. Hydrolysable tannins, upon hydrolysis, produce gallic acid and ellagic acid and depending on the type of acid produced, the hydrolysable tannins are called gallotannins or egallitannins. On heating, they form pyrogallic acid. Tannins are used as antiseptic and this activity is due to presence of the phenolic group. Common examples of hydrolysable tannins include theaflavins (from tea), daidezein, genistein and glycitein (Fig. 5). Tannin- rich medicinal plants are used as healing agents in a number of diseases. In Ayurveda, formulations based on tannin-rich plants have been used for the treatment of diseases like leucorrhoea, rhinnorhoea and diarrhea.

2.7 Terpenes

Terpenes are among the most widespread and chemically diverse groups of natural products. They are flammable unsaturated hydrocarbons, existing in liquid form commonly found in essential oils, resins or oleoresins (Firn, 2010). Terpenoids includes hydrocarbons of plant origin of general formula $(C_5H_8)_n$ and are classified as mono-, di-, tri- and sesquiterpenoids depending on the number of carbon atoms. Examples of commonly important monterpenes include terpinen-4-ol, thujone, camphor, eugenol and menthol. *Diterpenes (C20)* are classically considered to be resins and taxol, the anticancer agent, is the common example. The *triterpenes (C30)* include steroids, sterols, and cardiac glycosides with anti-inflammatory, sedative, insecticidal or cytotoxic activity. Common triterpenes: amyrins, ursolic acid and oleanic acid *sesquiterpene (C15)* like monoterpenes, are major components of many essential oils (Martinez *et al.*, 2008). The sesquiterpene acts as irritants when applied externally and when consumed internally their action resembles that of gastrointestinal tract irritant. A number of sesquiterpene lactones have been isolated and broadly they have antimicrobial (particularly antiprotozoal) and neurotoxic action. The sesquiterpene lactone, palasonin, isolated from *Butea monosperma* has anthelmintic activity, inhibits glucose uptake and depletes the glycogen content in *Ascaridia galli* (Fig. 6).

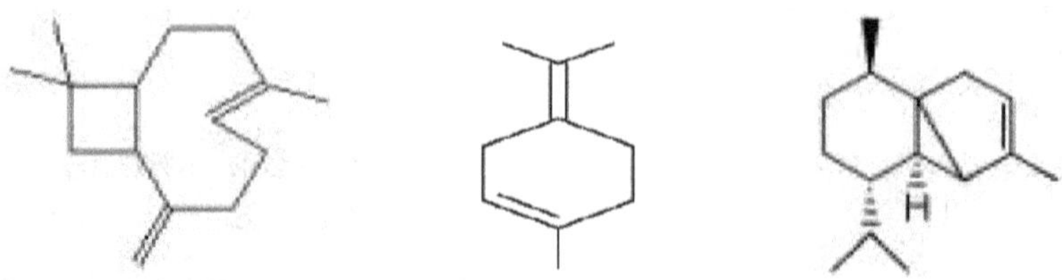

β-caryophyllene Terpenolen α-Cubebene

Fig. 6. Basic structures of some pharmacologically important plant derived tarpenes

Terpenoids are classified according to the number of isoprene units involved in the formation of these compounds. The major groups are shown in Table 2.

Table 2. Types of terpenoids according to the number of isopropene units

Type of terpenoids	Number of carbon atoms	Number of isoprene units	Example
Monoterpene	10	2	Limonene
Sesquiterpene	15	3	Artemisinin
Diterpene	20	4	Forskolin
Triterpene	30	6	a-amyrin
Tetraterpene	40	8	b-carotene
Polymeric terpenoid	several	several	Rubber

2.8 Anthraquinones

These are derivatives of phenolic and glycosidic compounds. They are solely derived from anthracene giving variable oxidized derivatives such as anthrones and anthranols (Maurya *et al.*, 2008; Firn, 2010). Other derivatives such as chrysophanol, aloe-emodin, rhein, salinos poramide, luteolin (Fig. 7) and emodin have in common a double hydroxylation at positions C-1 and C-8. To test for free anthraquinones, powdered plant material is mixed with organic solvent and filtered, and an aqueous base, e.g. NaOH or NH_4OH solution, is added to it. A pink or violet colour in the base layer indicates the presence of anthraquinones in the plant sample (Sarker & Nahar, 2007).

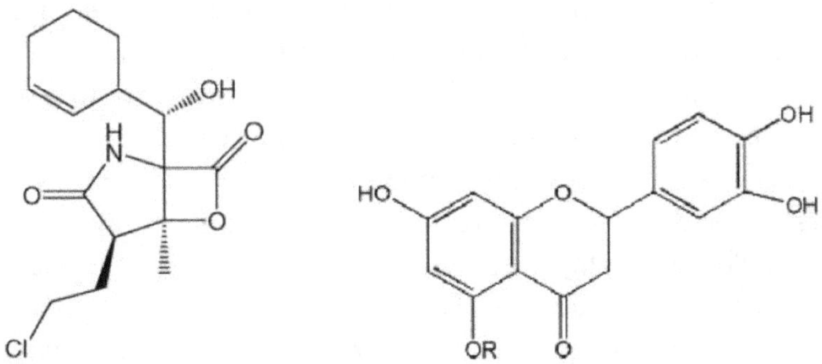

Salinos poramide Luteolin

Fig. 7. Basic structures of some pharmacologically important plant derived anthraquinones

2.9 Essential oils

Essential oils are the odorous and volatile products of various plant and animal species. Essential oils have a tendency evaporate on exposure to air even at ambient conditions and are therefore also referred to as volatile oils or ethereal oils. They mostly contribute to the odoriferous constituents or *'essences'* of the aromatic plants that are used abundantly in enhancing the aroma of some spices (Martinez *et al.*, 2008). Essential oils are either secreted either directly by the plant protoplasm or by the hydrolysis of some glycosides and structures such as directly Plant structures associated with the secretion of essential oils include: Glandular hairs (Lamiaceae e.g. *Lavandula angustifolia*), Oil tubes (or vittae) (Apiaceae eg. *Foeniculum vulgare*, and Pimpinella anisum (Aniseed), modified parenchymal cells (Piperaceae e.g. *Piper nigrum* - Black pepper), Schizogenous or lysigenum passages (Rutaceae e.g. *Pinus palustris* - Pine oil. Essential oils have been associated with different plant parts including leaves, stems, flowers, roots or rhizomes. Chemically, a single volatile oil comprises of more than 200 different chemical components, and mostly the trace constituents are solely responsible for attributing its characteristic flavour and odour (Firn, 2010).

Essential oils can be prepared from various plant sources either by direct steam distillation, expression, extraction or by enzymatic hydrolysis. Direct steam distillation involves the boiling of plant part in a distillation flask and passing the generated steam and volatile oil through a water condenser and subsequently collecting the oil in florentine flasks. Depending on the nature of the plant source the distillation process can be either water distillation, water and steam distillation or direct distillation. Expression or extrusion of volatile oils is accomplished by either by sponge method, scarification, rasping or by a mechanical process. In the sponge method, the washed plant part e.g. citrous fruit (*e.g.*, orange, lemon, grape fruit, bergamot) is cut into halves to remove the juice completely, rind turned inside out by hand and squeezed when the secretary glands rupture. The oozed volatile oil is collected by means of the sponge and subsequently squeezed in a vessel. The oil floating on the surface is separated. For the the scarification process the apparatus Ecuelle a Piquer (a large bowl meant for pricking the outer surface of citrus fruits) is used. It is a large funnel made of copper having its inner layer tinned properly. The inner layer has numerous pointed metal needles just long enough to penetrate the epidermis. The lower stem of the apparatus serve two purposes; *first*, as a receiver for the oil; and *secondly*, as a handle. Now, the freshly washed lemons are placed in the bowl and rotated repeatedly when the oil glands are punctured (scarified) thereby discharging the oil right into the handle. The liquid, thus collected, is transferred to another vessel, where on keeping the clear oil may be decanted and filtered. For the rasping process, the outer surface of the peel of citrus fruits containing the oil gland is skilfully removed by a grater. The 'raspings' are now placed in horsehair bags and pressed strongly so as to ooze out the oil stored in the oil glands. Initially, the liquid has a turbid appearance but on allowing it to stand the oil separates out which may be decanted and filtered subsequently. The mechanical process involves the use of heavy duty centrifugal devices so as to ease the separation of oil/water emulsions invariably formed and with the advent of modern mechanical devices the oil output has increased impressively. The extraction processes can be carried out with either volatile solvents (e.g hexane, petroleum ether or benzene) resulting into the production of 'floral concretes'- oils with solid consistency and partly soluble in 95% alcohol, or non volatile solvents (tallow, lard or olive oil) which results in the production of perfumes. Examples of volatile oils include amygdaline (volatile oil of bitter almond), sinigrin (volatile oil of black mustard), and eugenol occurring as gein (volatile oil of *Geum urbanum*) (Fig. 8).

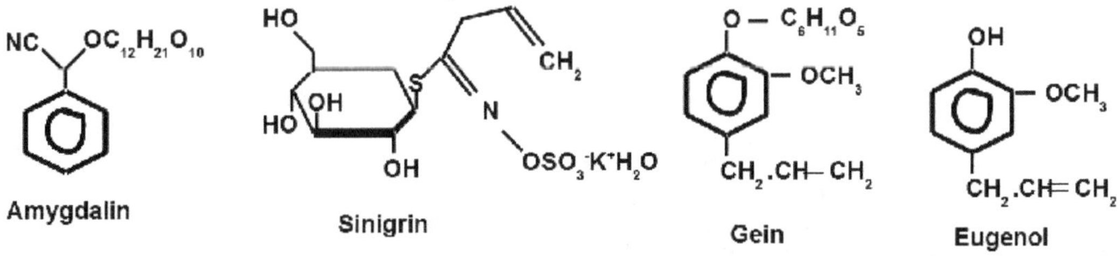

Fig. 8. Basic structures of some pharmacologically important plant derived essential oils

2.10 Steroids

Plant steroids (or steroid glycosides) also referred to as 'cardiac glycosides' are one of the most naturally occurring plant phytoconstituents that have found therapeutic applications as arrow poisons or cardiac drugs (Firn, 2010). The cardiac glycosides are basically steroids with an inherent ability to afford a very specific and powerful action mainly on the cardiac muscle when administered through injection into man or animal. Steroids (anabolic steroids) have been observed to promote nitrogen retention in osteoporosis and in animals with wasting illness (Maurya *et al.,* 2008; Madziga *et al.,* 2010). Caution should be taken when using steroidal glycosides as small amounts would exhibit the much needed stimulation on a diseased heart, whereas excessive dose may cause even death. Diosgenin and cevadine (from *Veratrum veride*) are examples of plant steroids (Fig. 9).

3. MECHANISM OF ACTION OF PHYTOCHEMICALS

Different mechanisms of action of phytochemicals have been suggested. They may inhibit microorganisms, interfere with some metabolic processes or may modulate gene expression and signal transduction pathways (Kris-Etherton *et al.,* 2002; Manson 2003; Surh 2003). Phytochemicals may either be used as chemotherapeutic or chemo preventive agents with chemoprevention referring to the use of agents to inhibit, reverse, or retard tumorigenesis. In this sense chemo preventive phytochemicals are applicable to cancer therapy, since molecular mechanisms may be common to both chemoprevention and cancer therapy (D'Incalci *et al.*, 2005; Sarkar & Li, 2006). Plant extracts and essential oils may exhibit different modes of action against bacterial strains, such as interference with the phospholipids bilayer of the cell membrane which has as a consequence a permeability increase and loss of cellular constituents, damage of the enzymes involved in the production of cellular energy and synthesis of structural components, and destruction or inactivation of genetic material. In general, the mechanism of action is considered to be the disturbance of the cytoplasmic membrane, disrupting the proton motive force, electron flow, active transport, and coagulation of cell contents (Kotzekidou *et al.*, 2008). Some specific modes of actions are discussed below.

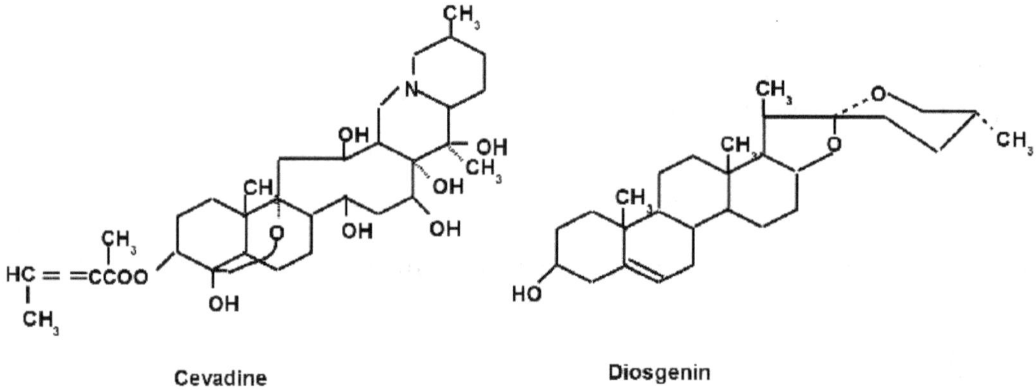

Fig. 9. Basic structures of some pharmacologically important plant derived steroids

3.1 Antioxidants

Antioxidants protect cells against the damaging effects of reactive oxygen species otherwise called, free radicals such as singlet oxygen, super oxide, peroxyl radicals, hydroxyl radicals and peroxynite which results in oxidative stress leading to cellular damage (Mattson & Cheng, 2006). Natural antioxidants play a key role in health maintenance and prevention of the chronic and degenerative diseases, such as atherosclerosis, cardiac and cerebral ischema, carcinogenesis, neurodegenerative disorders, diabetic pregnancy, rheumatic disorder, DNA damage and ageing (Uddin *et al.*, 2008; Jayasri *et al.*, 2009). Antioxidants exert their activity by scavenging the 'free-oxygen radicals' thereby giving rise to a fairly 'stable radical'. The free radicals are metastable chemical species, which tend to trap electrons from the molecules in the immediate surroundings. These radicals if not scavenged effectively in time, they may damage crucial bio molecules like lipids, proteins including those present in all membranes, mitochondria and, the DNA resulting in abnormalities leading to disease conditions (Uddin et al. 2008). Thus, free radicals are involved in a number of diseases including: tumour inflammation, hemorrhagic shock, atherosclerosis, diabetes, infertility, gastrointestinal ulcerogenesis, asthma, rheumatoid arthritis, cardiovascular disorders, cystic fibrosis, neurodegenerative diseases (e.g. parkinsonism, Alzheimer's diseases), AIDS and even early senescence (Chen *et al.*, 2006; Uddin *et al.*, 2008). The human body produces insufficient amount of antioxidants which are essential for preventing oxidative stress. Free radicals generated in the body can be removed by the body's own natural antioxidant defences such as glutathione or catalases (Sen, 1995). Therefore this deficiency had to be compensated by making use of natural exogenous antioxidants, such as vitamin C, vitamin E, flavones, β-carotene and natural products in plants (Madsen & Bertelsen, 1995; Rice- Evans *et al.*, 1997; Diplock *et al.*, 1998).

Plants contain a wide variety of free radicals scavenging molecules including phenols, flavonoids, vitamins, terpenoids hat are rich in antioxidant activity (Madsen & Bertelsen, 1995; Cai & Sun, 2003). Many plants, citrus fruits and leafy vegetables are the source of ascorbic acid, vitamin E, caratenoids, flavanols and phenolics which possess the ability to scavenge the free radicals in human body. Significant antioxidant properties have been recorded in phytochemicals that are necessary for the reduction in the occurrence of many diseases (Hertog & Feskens, 1993; Anderson & Teuber, 2001). Many dietary polyphenolic constituents derived from plants are more effective antioxidants *in vitro* than vitamins E or C, and thus might contribute significantly to protective effects *in vivo* (Rice-Evans & Miller, 1997; Jayasri *et al.*, 2009). Methanol extract of *Cinnamon* contains a number of antioxidant compounds which can effectively scavenge reactive oxygen species including superoxide anions and hydroxyl radicals as well as other free radicals *in vitro.* The fruit of *Cinnamon*, an under-utilized and unconventional part of the plant, contains a good amount of phenolic antioxidants to counteract the damaging effects of free radicals and may protect against mutagenesis.

Antioxidants are often added to foods to prevent the radical chain reactions of oxidation, and they act by inhibiting the initiation and propagation step leading to the termination of the reaction and delay the oxidation process. Due to safety concerns of synthetic compounds, food industries have focused on finding natural antioxidants to replace synthetic compounds. In addition, there is growing trend in consumer preferences for natural antioxidants, all of which has given more impetus to explore natural sources of antioxidants.

3.2 Anticacinogenesis

Polyphenols particularly are among the diverse phytochemicals that have the potential in the inhibition of carcinogenesis (Liu, 2004). Phenolics acids usually significantly minimize the formation of the specific cancer-promoting nitrosamines from the dietary nitrites and nitrates. Glucosinolates from various vegetable sources as broccoli, cabbage, cauliflower, and Brussel sprouts exert a substantial protective support against the colon cancer. Regular consumption of Brussel sprouts by human subjects (up to 300 g.day-1) miraculously causes a very fast (say within a span of 3 weeks) an appreciable enhancement in the glutathione-S- transferase, and a subsequent noticeable reduction in the urinary concentration of a specific purine meltabolite that serves as a marker of DNA-degradation in cancer. Isothiocyanates and the indole-3-carbinols do interfere categorically in the metabolism of carcinogens thus causing inhibition of procarcinogen activation, and thereby inducing the 'phase-II' enzymes, namely: NAD(P)H quinone reductase or glutathione S-transferase, that specifically detoxify the selected electrophilic metabolites which are capable of changing the structure of nucleic acids. Sulforaphane (rich in broccoli) has been proved to be an extremely potent phase-2 enzyme inducer. It predominantly causes specific cell-cycle arrest and also the apoptosis of the neoplasm (cancer) cells. Sulforaphane categorically produces d-D-gluconolactone which has been established to be a significant inhibitor of breast cancer. Indole-3-carbinol (most vital and important indole present in broccoli) specifically inhibits the Human Papilloma Virus (HPV) that may cause uterine cancer. It blocks the estrogen receptors specifically present in the breast cancer cells as well as down regulates CDK6, and up regulates p21 and p27 in prostate cancer cells. It affords G1 cell-cycle arrest and apoptosis of breast and prostate cancer cells significantly and enhances the p 53 expression in cells treated with benzopyrene. It also depresses Akt, NF-kappaB, MAPK, and Bel-2 signaling pathways to a reasonably good extent. Phytosterols block the development of tumors (neoplasms) in colon, breast, and prostate glands. Although the precise and exact mechanisms whereby the said blockade actually takes place are not yet well understood, yet they seem to change drastically the ensuing cell-membrane transfer in the phenomenon of neoplasm growth and thereby reduce the inflammation significantly.

3.3 Antimicrobial activity

Phytoconstituents employed by plants to protect them against pathogenic insects, bacteria, fungi or protozoa have found applications in human medicine (Nascimento *et al.*, 2000). Some phytochemicals such as phenolic acids act essentially by helping in the reduction of particular adherence of organisms to the cells lining the bladder, and the teeth, which ultimately lowers the incidence of urinary-tract infections (UTI) and the usual dental caries. Plants can also exert either bacteriostatic or bactericidal activity ob microbes. The volatile gas phase of combinations of *Cinnamon* oil and clove oil showed good potential to inhibit growth of spoilage fungi, yeast and bacteria normally found on IMF (Intermediate Moisture Foods) when combined with a modified atmosphere comprising a high concentration of CO_2 (40%) and low concentration of O_2 (<0.05%) (Jakhetia *et al.*, 2010). *A. flavus*, which is known to produce toxins, was found to be the most resistant microorganism. It is worthy of note that antimicrobial activity results of the same plant part tested most of the time varied from researcher to researcher. This is possible because concentration of plant constituents of the same plant organ can vary from one geographical location to another depending on the age of the plant, differences in topographical factors, the nutrient concentrations of the soil, extraction method as well as method used for antimicrobial study. It is therefore important that scientific protocols be clearly identified and adequately followed and reported.

3.4 Anti-ulcer

Plants extracts have been reported to inhibit both growth of *H. pylori in-vitro* as well as its urease activity (Jakhetia *et al.,* 2010). The efficiency of some extracts in liquid medium and at low pH levels enhances their potency even in the human stomach. Their inhibitory effect on the intestinal and kidney Na+/K+ ATPase activity and on alanine transport in rat jejunum has also been reported (Jakhetia *et al.,* 2010).

3.5 Anti-diabetic

Cinnamaldehyde, a phytoconstituent extracts have been reported to exhibit significant antihyperglycemic effect resulting in the lowering of both total cholesterol and triglyceride levels and, at the same time, increasing HDL-cholesterol in STZ-induced diabetic rats. This investigation reveals the potential of cinnamaldehyde for use as a natural oral agent, with both hypoglycaemic and hypolipidemic effects. Recent reports indicate that *Cinnamon* extract and polyphenols with procyanidin type-A polymers exhibit the potential to increase the amount of TTP (Thrombotic Thrombocytopenic Purpura), IR (Insulin Resistance), and GLUT4 (Glucose Transporter-4) in 3T3-L1 Adipocytes. It was suggested that the mechanism of *Cinnamon*'s insulin-like activity may be in part due to increase in the amounts of TTP, IRβ, and GLUT4 and that *Cinnamon* polyphenols may have additional roles as anti-inflammatory and/or anti-angiogenesis agents (Jakhetia *et al.,* 2010).

3.6 Anti-inflammatory

Essential oil of *C. osmophloeum* twigs has excellent anti- inflammatory activities and cytotoxicity against HepG2 (Human Hepatocellular Liver Carcinoma Cell Line) cells. Previous reports also indicated that the constituents of *C. osmophloeum* twig exhibited excellent anti-inflammatory activities in suppressing nitric oxide production by LPS (Lipopolysaccharide)-stimulated macrophages (Jakhetia *et al.,* 2010).

3.7 Multifunctional targets

Multiple molecular targets of dietary phytochemicals have been identified, from pro- and anti-apoptotic proteins, cell cycle proteins, cell adhesion molecules, protein kinases, transcription factors to metastasis and cell growth pathways (Awad

& Bradford, 2005; Aggarwal & Shishodia, 2006; Choi & Friso, 2006). Phytochemicals such as epigallocatechin- 3-gallate (EGCG) from green tea, curcumin from turmeric, and resveratrol from red wine tend to aim at a multitude of molecular targets. It is because of these characteristics that definitive mechanisms of action are not available despite decades of research (Francis *et al.*, 2002). The multi-target nature of phytochemicals may be beneficial in overcoming cancer drug resistance. This multi-faceted mode of action probably hinders the cancer cell's ability to develop resistance to the phytochemicals. It has also been demonstrated that EGCG has inhibitory effects on the extracellular release of verotoxin (VT) from *E. coli* 0157: H7 (Voravuthikunchai & Kitpipit, 2003). Ethanol pericarp extracts from *Punica granatum* was also reported to inhibited VT production in periplasmic space and cell supernatant. Mechanisms responsible for this are yet to be understood, however the active compounds from the plant are thought to interfere with the transcriptional and translational processes of the bacterial cell (Voravuthikunchai & Kitpipit, 2003). More work is needed to be done in order to establish this assumption. Phytochemicals may also modulate transcription factors (Andreadi *et al.*, 2006), redox-sensitive transcription factors (Surh *et al.*, 2005), redox signalling, and inflammation. As an example, nitric oxide (NO), a signalling molecule of importance in inflammation, is modulated by plant polyphenols and other botanical extracts (Chan & Fong, 1999; Shanmugam *et al.*, 2008). Many phytochemicals have been classified as phytoestrogens, with health-promoting effects resulting in the phytochemicals to be marketed as nutraceuticals (Moutsatsou, 2007).

4. METHODS OF STUDYING PHYTOCHEMICALS

No single method is sufficient to study the bioactivity of phytochemicals from a given plant. An appropriate assay is required to first screen for the presence of the source material, to purify and subsequently identify the compounds therein. Assay methods vary depending on what bioactivity is targeted and these may include antimicrobial, anti-malarial, anticancer, seed germination, and mammalian toxicity activities. The assay method however should be as simple, specific, and rapid as possible. An *in vitro* test is more desirable than a bioassay using small laboratory animals, which, in turn, is more desirable than feeding large amounts of valuable and hard to obtain extract to larger domestic or laboratory animals. In addition, *in vivo* tests in mammals are often variable and are highly constrained by ethical considerations of animal welfare. Extraction from the plant is an empirical exercise in which different solvents are utilized under a variety of conditions such as time and temperature of extraction. The success or failure of the extraction process depends on the most appropriate assay. Once extracted from the plant, the bioactive component then has to be separated from the co extractives. Further purification steps may involve simple crystallization of the compound from the crude extract, further solvent partition of the co extractives or chromatographic methods in order to fractionate the compounds based on their acidity, polarity or molecular size. Final purification, to provide compounds of suitable purity for such structural analysis, may be accomplished by appropriate techniques such as recrystallization, sublimation, or distillation.

4.1 Extraction of phytochemicals

4.1.1 Solvent extraction

Various solvents have been used to extract different phytoconstituents. The plant parts are dried immediately either in an artificial environment at low temperature (50-60°C) or dried preferably in shade so as to bring down the initial large moisture content to enable its prolonged storage life and . The dried berries are pulverised by mechanical grinders and the oil is removed by solvent extraction. The defatted material is then extracted in a soxhlet apparatus or by soaking in water or alcohol (95% v/v). The resulting alcoholic extract is filtered, concentrated in vacuo or by evaporation, treated with HCl (12N) and refluxed for at least six hours. This can then be concentrated and used to determine the presence of phytoconstituents.

Generally, the saponins do have high molecular weight and hence their isolation in the purest form poses some practical difficulties. The plant parts (tubers, roots, stems, leave etc) are washed sliced and extracted with hot water or ethanol (95% v/v) for several hours. The resulting extract is filtered, concentrated *in vacuo* and the desired constituent is precipitated with ether.

Exhaustive extraction (EE) is usually carried out with different solvents of increasing polarity in order to extract as much as possible the most active components with highest biological activity.

4.1.2 Supercritical fluid extraction (SFE)

This is the most technologically advanced extraction system (Patil & Shettigar, 2010). Super Critical Fluid Extraction (SFE) involves use of gases, usually CO_2, and compressing them into a dense liquid. This liquid is then pumped through a cylinder containing the material to be extracted. From there, the extract-laden liquid is pumped into a separation chamber where the extract is separated from the gas and the gas is recovered for re-use. Solvent properties of CO_2 can be manipulated and adjusted by varying the pressure and temperature that one works at. The advantages of SFE are, the versatility it offers in pinpointing the constituents you want to extract from a given material and the fact that your end product has virtually no solvent residues left in it (CO_2 evaporates completely). The downside is that this technology is quite expensive. There are many other gases and liquids that are highly efficient as extraction solvents when put under pressure (Patil &

Shettigar, 2010).

a. **Coupled SFE-SFC** System in which a sample is extracted with a supercritical fluid which then places the extracted material in the inlet part of a supercritical fluid chromatographic system. The extract is than chromatographed directly using supercritical fluid.
b. **Coupled SFE-GC and SFE-LC** System in which a sample is extracted using a supercritical fluid which is then depressurized to deposit the extracted material in the inlet part or a column of gas or liquid chromatographic system respectively. SFE is characterized by robustness of sample preparation, reliability, less time consuming, high yield and also has potential for coupling with number of chromatographic methods.

4.1.3 Microwave-Assisted extraction

Patil & Shettigar (2010) reported an innovative, microwave-assisted solvent-extraction technology known as Microwave-Assisted Processing (MAP). MAP applications include the extraction of high-value compounds from natural sources including phytonutrients, nutraceutical and functional food ingredients and pharmaceutical actives from biomass. Compared to conventional solvent extraction methods, MAP technology offers some combination of the following advantages: 1. Improved products, increased purity of crude extracts, improved stability of marker compounds, possibility to use less toxic solvents; 2. Reduced processing costs, increased recovery and purity of marker compounds, very fast extraction rates, reduced energy and solvent usage. With microwave-derived extraction as opposed to diffusion, very fast extraction rates and greater solvent flexibility can be achieved. Many variables, including the microwave power and energy density, can be tuned to deliver desired product attributes and optimize process economics. The process can be customized to optimize for commercial/cost reasons and excellent extracts are produced from widely varying substrates. Examples include, but are not limited to, antioxidants from dried herbs, carotenoids from single cells and plant sources, taxanes from taxus biomass, essential fatty acids from microalgae and oilseeds, phytosterols from medicinal plants, polyphenols from green tea, flavor constituents from vanilla and black pepper, essential oils from various sources, and many more (Patil & Shettigar, 2010).

4.1.4 Solid phase extraction

This involves sorption of solutes from a liquid medium onto a solid adsorbent by the same mechanisms by which molecules are retained on chromatographic stationary phases. These adsorbents, like chromatographic media, come in the form of beads or resins that can be used in column or in batch form. They are often used in the commercially available form of syringes packed with medium (typically a few hundred milligrams to a few grams) through which the sample can be gently forced with the plunger or by vacuum. Solid phase extraction media include reverse phase, normal phase, and ion-exchange media. This is method for sample purification that separates and concentrates the analyte from solution of crude extracts by adsorption onto a disposable solid-phase cartridge. The analyte is normally retained on the stationary phase, washed and then evaluated with different mobile phase. If an aqueous extract is passed down a column containing reverse phase packing material, everything that is fairly nonpolar will bind, whereas everything polar will pass through (Patil & Shettigar, 2010).

4.1.5 Chromatographic fingerprinting and marker compound analysis

Chromatographic fingerprint of an Herbal Medicine (HM) is a chromatographic pattern of the extract of some common chemical components of pharmacologically active and or chemical characteristics (Patil & Shettigar, 2010). This chromatographic profile should be featured by the fundamental attributions of "integrity" and "fuzziness" or "sameness" and "differences" so as to chemically represent the HM investigated. It is suggested that with the help of chromatographic fingerprints obtained, the authentication and identification of herbal medicines can be accurately conducted (integrity) even if the amount and/or concentration of the chemically characteristic constituents are not exactly the same for different samples of this HM (hence, "fuzziness") or, the chromatographic fingerprints could demonstrate both the "sameness" and "differences" between various samples successfully. Thus, we should globally consider multiple constituents in the HM extracts, and not individually consider only one and/or two marker components for evaluating the quality of the HM products. However, in any HM and its extract, there are hundreds of unknown components and many of them are in low amount. Moreover, there usually exists variability within the same herbal materials. Hence it is very important to obtain reliable chromatographic fingerprints that represent pharmacologically active and chemically characteristic components of the HM. In the phytochemical evaluation of herbal drugs, TLC is being employed extensively for the following reasons: (1) it enables rapid analysis of herbal extracts with minimum sample clean-up requirement, (2) it provides qualitative and semi quantitative information of the resolved compounds and (3) it enables the quantification of chemical constituents. Fingerprinting using HPLC and GLC is also carried out in specific cases. In TLC fingerprinting, the data that can be recorded using a high- performance TLC (HPTLC) scanner includes the chromatogram, retardation factor (Rf) values, the colour of the separated bands, their absorption spectra, λmax and shoulder inflection/s of all the resolved bands. All of these, together with the profiles on derivatization with different reagents, represent the TLC fingerprint profile of the sample. The information so generated has a potential application in the identification of an authentic drug, in excluding the adulterants and in maintaining the quality and consistency of the drug. HPLC fingerprinting includes recording of the chromatograms, retention time of individual peaks and the absorption spectra (recorded with a photodiode array detector) with different mobile phases. Similarly, GLC is used for generating the fingerprint profiles of volatile oils and fixed oils of herbal drugs. Furthermore, the recent approaches of applying hyphenated chromatography and spectrometry such as High-Performance Liquid Chromatography–Diode Array Detection (HPLC–DAD), Gas Chromatography–Mass Spectroscopy (GC–MS), Capillary Electrophoresis-Diode Array Detection (CE-DAD), High- Performance Liquid Chromatography–Mass Spectroscopy (HPLC–MS) and High- Performance Liquid Chromatography–Nuclear Magnetic Resonance Spectroscopy (HPLC– NMR) could provide the additional spectral information, which will be very helpful for the qualitative analysis and even for the on-line structural elucidation.

4.1.6 Advances in chromatographic techniques

4.1.6.1 Liquid chromatography

a. Preparative high performance liquid chromatography

There are basically two types of preparative HPLC. One is low pressure (typically under 5 bar) traditional PLC, based on the use of glass or plastic columns filled with low efficiency packing materials of large particles and large size distribution. A more recent form PLC, Preparative High Performance Liquid Chromatography (Prep.HPLC) has been gaining popularity in pharmaceutical industry. In preparative HPLC (pressure >20 bar), larger stainless steel columns and packing materials (particle size 10-30 μm are needed. The examples of normal phase silica columns are Kromasil 10 μm, Kromasil 16 μm, Chiralcel AS 20 μm whereas for reverse phase are Chromasil C18, Chromasil C8,YMC C18. The aim is to isolate or purify compounds, whereas in analytical work the goal is to get information about the sample. Preparative HPLC is closer to analytical HPLC than traditional PLC, because its higher column efficiencies and faster solvent velocities permit more difficult separation to be conducted more quickly (Oleszek & Marston, 2000; Philipson, 2007). In analytical HPLC, the important parameters are resolution, sensitivity and fast analysis time whereas in preparative HPLC, both the degree of solute purity as well as the amount of compound that can be produced per unit time i.e. throughput or recovery are important. This is very important in pharmaceutical industry of today because new products (Natural, Synthetic) have to be introduced to the market as quickly as possible. Having available such a powerful purification technique makes it possible to spend less time on the synthesis conditions.

b. Liquid Chromatography- Mass Spectroscopy (LC-MS)

In Pharmaceutical industry LC-MS has become method of choice in many stages of drug development. Recent advances includes electro spray, thermo spray, and ion spray ionization techniques which offer unique advantages of high detection sensitivity and specificity, liquid secondary ion mass spectroscopy, later laser mass spectroscopy with 600 MHz offers accurate determination of molecular weight proteins, peptides. Isotopes pattern can be detected by this technique (Oleszek & Marston, 2000; Philipson, 2007).

c. Liquid Chromatography- Nuclear Magnetic Resonance (LC-NMR)

The combination of chromatographic separation technique with NMR spectroscopy is one of the most powerful and time saving method for the separation and structural elucidation of unknown compound and mixtures, especially for the structure elucidation of light and oxygen sensitive substances. The online LC-NMR technique allows the continuous registration of time changes as they appear in the chromatographic run automated data acquisition and processing in LC-NMR improves speed and sensitivity of detection (Daffre et al., 2008). The recent introduction of pulsed field gradient technique in high resolution NMR as well as three-dimensional technique improves application in structure elucidation and molecular weight information. These new hyphenated techniques are useful in the areas of pharmacokinetics, toxicity studies, drug metabolism and drug discovery process.

4.1.6.2 Gas chromatography

A. Gas Chromatography Fourier Transform Infrared spectrometry

Coupling capillary column gas chromatographs with Fourier Transform Infrared Spectrometer provides a potent means for separating and identifying the components of different mixtures.

B. Gas Chromatography-Mass Spectroscopy

Gas chromatography equipment can be directly interfaced with rapid scan mass spectrometer of various types. The flow rate from capillary column is generally low enough that the column output can be fed directly into ionization chamber of MS. The simplest mass detector in GC is the Ion Trap Detector (ITD). In this instrument, ions are created from the eluted sample by electron impact or chemical ionization and stored in a radio frequency field; the trapped ions are then ejected from the storage area to an electron multiplier detector. The ejection is controlled so that scanning on the basis of mass-to-charge ratio is possible. The ions trap detector is remarkably compact and less expensive than quadrapole instruments.GC-MS instruments have been used for identification of hundreds of components that are present in natural and biological system (Oleszek & Marston, 2000; Philipson, 2007; Daffre *et al.*, 2008).

4.1.6.3 Supercritical Fluid Chromatography (SFC)

Supercritical fluid chromatography is a hybrid of gas and liquid chromatography that combines some of the best features of each. This technique is an important third kind of column chromatography that is beginning to find use in many industrial, regulatory and academic laboratories. SFC is important because it permits the separation and determination of a group of compounds that are not conveniently handled by either gas or liquid chromatography. These compounds are either non-volatile or thermally labile so that GC procedures are inapplicable or contain no functional group that makes possible detection by the spectroscopic or electrochemical technique employed in LC. SFC has been applied to a wide variety of materials including natural products, drugs, foods and pesticides.

4.2 Other Chromato-Spectrometric studies

The NMR techniques are employed for establishing connectivities between neighbouring protons and establishing C-H bonds. INEPT is also being used for long range heteronuclear correlations over multiple bondings. The application of Thin Layer chromatography (TLC), High Performance Chromatography (HPLC) and HPLC coupled with Ultra violate (UV) photodiode array detection, Liquid Chromatography-Ultraviolet (LC-UV), Liquid Chromatography-Mass Spectrophotometry (LCMS), electrospray (ES) and Liquid Chromatography-Nuclear Magnetic Resonance (LC-NMR) techniques for the separation and structure determination of antifungal and antibacterial plant compounds is on the increase frequently (Oleszek & Marston, 2000; Bohlin and Bruhn, 1999). Currently available chromatographic and spectroscopic techniques in new drug discovery from natural products Currently, computer modelling has also been introduced in spectrum interpretation and the generation of chemical structures meeting the spectral properties of bioactive compounds obtained from plants (Vlietinck, 2000). The computer systems utilise 1H, 13C, 2D-NMR, IR and MS spectral properties (Philipson, 2007). Libraries of spectra can be searched for comparison with complete or partial chemical structures. Hyphenated chromatographic and spectroscopic techniques are powerful analytical tools that are combined with high throughput biological screening in order to avoid re-isolation of known compounds as well as for structure determination of novel compounds. Hyphenated chromatographic and spectroscopic techniques include LC–UV–MS, LC–UV–NMR, LC–UV– ES–MS and GC–MS (Oleszek & Marston, 2000; Philipson, 2007).

4.3 Simple assay methods

4.3.1 Antimicrobial assay

Common methods used in the evaluation of the antibacterial and antifungal activities of plant extracts and essential oils, include the agar diffusion method (paper disc and well), the dilution method (agar and liquid broth) (Yagoub, 2008; Okigbo *et al.*, 2009; El-Mahmood, 2009; Aiyegoro *et al.*, 2009), and the turbidimetric and impedimetric monitoring of microbial growth (R′os & Recio, 2005). These methods are simple to carry out under laboratory conditions.

4.3.2 Antioxidant assays

Most common spectrophotometric assay method applied is the DPPH radical scavenging system in which the hydrogen or electrons donation ability of plant extracts are measured from bleaching of purple methanol solution of 2, 2'-diphenyl-1-picrylhydrazyl (DPPH) free radical (Changwei *et al.*, 2006). This spectrophotometric assay uses the stable radical DPPH as a reagent. DPPH absorbs at 517 nm, and as its concentration is reduced by existence of an antioxidant, the absorption gradually disappears with time. A 2-ml aliquot of a suspension of the ethanol extracts is mixed with 1 ml of 0.5 mM 2,2- diphyenyl-1-picrylhydrazyl (DPPH) solution and 2 ml of 0.1 M sodium acetate buffer (pH 5.5). After shaking, the mixture is incubated at ambient temperature in the dark for 30 minutes, following which the absorbance is measured at 517 nm using a UV-160A spectrometer. A solvent such as ethanol can be used as negative control. Radical scavenging activity is often expressed as percentage inhibition and is often calculated using the formula:

$$\%\ \text{radical scavenging activity} = [(A_{control} - A_{test})/A_{control}] \times 100$$

Where $A_{control}$ is the absorbance of the control (DPPH solution without test sample) and A_{test} is the absorbance of the test sample (DPPH solution plus antioxidant).

Phenolics content and reducing power of extracts is often determined using the Folin- Ciocalteu method. Equal volumes of Folin-Ciocalteu reagent and given quantity (mg) of plant extracts of different concentrations (e.g. 0.4, 0.3, 0.2, 0.1 and 0.05 mg/ml) are often mixed in different sets of test tubes shaken thoroughly, and left to stand for 1 min. Ten percent of $NaHCO_3$ is then added and the mixture once again allowed to stand for 30 minutes, after which the absorbance (725 nm) is measured spectrophotometrically. Gallic acid (0.05-0.5 mg/ml) is often used to produce standard calibration curve and the total phenolic content expressed as mg equivalent of gallic acid (mg GAE) per gram dry weight of the extract by computing with standard calibration curve (Djeridane *et al.*, 2006).

For determination of reducing power of plant extracts, the ferric reducing/ antioxidant power (FRAP) assay method can be applied. The assay is based on the reducing power of a compound (antioxidant). A potential antioxidant reduces the ferric ion (Fe^{3+}) to the ferrous ion (Fe^{2+}); the later forms a blue complex (Fe^{2+}/2, 4, 6, tripyridyl-*s*-triazine (TPTZ)), which increases the absorption at 593 nm. Stronger absorption at this wavelength indicates higher reducing power of the phytochemical, thus higher antioxidant activity. Reaction mixture containing test extract sample at different concentrations (10-100µl) in phosphate buffer (0.2 M, pH 6.6) and equal amounts of 1% (w/v) potassium ferricyanide are incubated at 50ºC for 20 minutes and then the reaction terminated by the addition of equal volumes of 10% (w/v) tricarboxyllic acid (TCA) solution and the mixture centrifuged at 3000rpm for 20 minutes. The supernatant is mixed with equal volume of distilled water and 0.1 % (w/v) ferric chloride solution and the absorbance measured at 700 nm. Increased absorbance of the mixture with concentration indicates the reducing power of extract (Jayasri, 2009).

4.3.3 Toxicological studies

These are often carried out to determine the toxicity of a plant part. Usually animal models such as mice, guinea pigs or rabbits are often employed. In these procedures, the LD_{50} of the extracts in the experimental animal is often determined via either oral or intradermal administration. The toxic response of experimental animals to the administration of plant alkaloids is usually detected by assay of the serum ALT and AST of the animal as sensitive indicators of hepatocellular damage (Chapatwala *et al.*, 1982). Any toxicity usually results in distortion of hepatocytes membrane integrity due to hepatocellular injury and plasma levels rise, as a consequence of high toxin levels present within hepatocytes.

5. SAFETY CONCERNS FOR PHYTOCHEMICALS

Plants are natural reservoir of medicinal agents almost free from the side effects normally caused by synthetic chemicals (Fennel *et al.*, 2004). The World Health Organization estimates that herbal medicine is still the main stay of about 75-80% of the world population, mainly in the developing countries for primary health care because of better cultural acceptability, better compatibility with the human body, and lesser side-effects (Kamboj, 2000; Yadav & Dixit, 2008). The over use of synthetic drugs with impurities resulting in higher incidence of adverse drug reactions, has motivated mankind to go back to nature for safer remedies. Due to varied locations where these plants grow, coupled with the problem of different vanacular names, the World Health Organization published standards for herbal safety to minimize adultartion and abuse (WHO, 1999).

A number of modern drugs have been isolated from natural sources and many of these isolations were based on the uses of the agents in traditional medicine (Rizvi *et al.*, 2009). Antimicrobial properties of crude extracts prepared from plants have been described and such reports had attracted the attention of scientists worldwide (Falodun *et al.*, 2006; El- Mahmood & Amey, 2007; El-Mahmood, 2009). Herbs have been used for food and medicinal purposes for centuries and this knowledge have been passed on from generation to generation (Adedapo *et al.*, 2005). This is particularly evident in the rural areas where infectious diseases are endemic and modern health care facilities are few and far thus, compelling the people to nurse their ailments using local herbs. Herbal treatments have been adjudged to be relatively safe (WHO, 1999). For instance, daily oral doses of epigallocatechin-3-gallate (EGCG) for 4 weeks at 800 mg/day in 40 volunteers only caused minor adverse effects (Phillipson, 2007). In a 90-day study of polyphenon E (a formulation of green tea extract with 53% EGCG), the oral no effect level (NOEL) values are 90 mg/kg/day for rats and 600 mg/kg/day for dogs (Boocock *et al.*, 2007). For curcumin given to cancer patients at 3600 mg/day for 4 months or 800 mg/day for 3 months, only minor adverse effects are seen. For resveratrol, a single oral dose at 5 g in 10 volunteers only causes minor adverse effects (Boocock *et al.*, 2007). Though herbs are relatively safe to use, their combined use with orthodox drugs should be done with extreme caution. Concomitant use of conventional and herbal medicines is reported to lead to clinically relevant herb–drug interactions (Liu et al., 2009). The two may interact either pharmacokinetically or pharmacodynamically resulting into adverse herbal-drug interactions (Izzo, 2005). St John's wort (*Hypericum perforatum*), used for the treatment of mild to moderate depression, interacts with digoxin, HIV inhibitors, theophylline and warfarin. Some medicinal herbs, when ingested, either affect cytochrome P450 isoenzymes by which drugs are metabolised, or, phosphoglycoprotein transporter systems that affect drug distribution and excretion. Concurrent use of some herbal medicines with other medicines may either lower blood plasma concentrations of medicinal drugs, possibly resulting in suboptimal therapeutic amounts, or lead to toxic concentrations in the blood, sometimes with fatal consequences (Phillipson, 2007).

Despite this observation however, it has been reported that phytochemicals act in synergy with chemotherapeutic drugs in overcoming cancer cell drug resistance and that the application of specific phytochemicals may allow the use of lower concentrations of drugs in cancer treatment with an increased efficacy (Liu, 2004).

Another advantage with phytochemicals is that, among an estimated 10,000 secondary products (natural pesticides), it has been proposed that human ancestors evolved a generalized defense mechanism against low levels of phytochemicals to enable their consumption of many different plant species containing variable levels of natural pesticides (carcinogens) without subsequent ill health (Liu, 2004). Traces of phytochemicals found in fruits and vegetables may potentiate the immune system and help to protect against cancer (Trewavas and Stewart, 2003). Phytochemicals show biphasic dose responses on mammalian cells. Though at high concentrations they can be toxic, sub-toxic doses may induce adaptive stress response (Ames & Gold, 1991). This includes the activation of signaling pathways that result in increased expression of genes encoding cytoprotective proteins. It is therefore suggested that hormetic mechanisms of action may underlie many of the health benefits of phytochemicals including their action against cancer drug resistance (Mattson, 2008).

Molecular mechanisms of herb–drug interaction occur, the most notable is the ATP-binding cassette drug transporters such as P-glycoprotein (You & Moris, 2007) and the drug metabolizing enzymes (known as phase I and phase II enzymes), especially cytochrome P450 3A4 (CYP3A4) (Pal & Mitra, 2006; Meijerman *et al.*, 2006).

6. FUTURE PROSPECTS OF PHYTOCHEMICALS AS SOURCES OF ANTIMICROBIAL CHEMOTHERAPEUTIC AGENTS

Though there are few disadvantages associated with natural products research. These include difficulties in access and supply, complexities of natural product chemistry and inherent slowness of working with natural products. In addition, there are concerns about intellectual property rights, and the hopes associated with the use of collections of compounds prepared by combinatorial chemistry methods. Despite these limitations, over a 100 natural-product-derived compounds are currently undergoing clinical trials and at least 100 similar projects are in preclinical development (Phillipson, 2007). Among these products the highest number are from plant origin (Table 3). Most are derived from plants and microbial sources. The projects based on natural products are predominantly being studied for use in cancer or as anti-infectives. There is also, a growing interest in the possibility of developing products that contain mixtures of natural compounds from traditionally used medicines (Charlish, 2008), while, a defined mixture of components extracted from green tea (Veregen TM) has been approved by the US Food and Drug Administration (FDA) and has recently come on the market.

Table 3. Drugs based on natural products at different stages of development

Development stage	Plant	Bacterial	Fungal	Animal	Semi-synthetic	Total
Preclinical	46	12	7	7	27	99
Phase I	14	5	0	3	8	30
Phase II	41	4	0	10	11	66
Phase III	5	4	0	4	13	26
Pre-registration	2	0	0	0	2	4
Total	108	25	7	24	61	225

Most of the leads from natural products that are currently in development have come from either plant or microbial sources. Earlier publications have pointed out that relatively little of the world's plant biodiversity has been extensively screened for bioactivity and that very little of the estimated microbial biodiversity has been available for screening (Doughari *et al.*, 2009). Hence, more extensive collections of plants (and microbes) could provide many novel chemicals for use in drug discovery assays. With the growing realization that the chemical diversity of natural products is a better match to that of successful drugs than the diversity of collections of synthetic compounds and with the global emergence of multidrug resistant pathogens (Feher and Schmidt, 2003) the interest in applying natural chemical diversity to drug discovery appears to be increasing once again (Galm & Shen, 2007).

With advances in fractionation techniques to isolate and purify natural products (e.g. counter-current chromatography (Doughari *et al.*, 2009) and in analytical techniques to determine structures (Singh & Barrett, 2006), screening of natural product mixtures is now more compatible with the expected timescale of high-throughput screening campaigns. Singh and Barrett (2006) point out that pure bioactive compound can be isolated from fermentation broths in less than 2 weeks and that the structures of more than 90% of new compounds can be elucidated within 2 weeks. With advances in NMR techniques, complex structures can be solved with much less than 1 mg of compound. It has recently been demonstrated that it is possible to prepare a screening library of highly diverse compounds from plants with the compounds being pre-selected from an analysis of the Dictionary of Natural Products to be drug-like in their physicochemical properties (Oleszek & Marston, 2000; Doughari *et al.*, 2009). It will be interesting to see if such a collection proves to be enriched in bioactive molecules. Several alternative approaches are also being explored in efforts to increase the speed and efficiency with which natural products can be applied to drug discovery. For instance, there is an attraction to screen the mixtures of compounds obtained from extracts of plant material or from microbial broths to select extracts from primary screens that are likely to contain novel compounds with the desired biological activity using the concept of 'differential smart screens'. This approach involves screening extracts of unknown activity against pairs of related receptor sites. By the comparison of the ratios of the binding potencies at the two receptor sites for a known selective ligand and for an extract, it is possible to predict which extract was likely to contain components with the appropriate pharmacological activity (McGaw *et al.*, 2005; Doughari *et al.*, 2009; Okigbo *et al.*, 2009). Another approach is the use of 'chemical-genetics profiling' (Doughari *et al.*, 2009). In this method, by building up a database of the effects of a wide range of known compounds, it is possible to interrogate drugs with unknown mechanisms or mixtures of compounds such as natural product mixtures. The technique highlighted unexpected similarities in molecular effects of unrelated drugs (e.g. amiodarone and tamoxifen) and also revealed potential anti-fungal activity of crude extracts. This activity was confirmed by isolation and testing of defined compounds, stichloroside and theopalauamide (Fig. 10).

Because these compounds are not structurally similar, they would not have been expected to act via the same biological target, thus providing more chances for a very versatile drug component with high efficacy against antibiotic resistant bacteria. It's been reported that despite the popularity of chemical drugs, herbal medicine in Africa and the rest of the world, continued to be practiced due to richness of certain plants in varieties of secondary metabolites such as alkaloids, flavonoids, tannins and terpenoids (Cowan, 1999; Lewis & Ausubel, 2006; Adekunle & Adekunle, 2009). Stapleton *et al.* (2004) reported that aqueous extracts of tea (*Camellia sinensis*) reversed methicillin resistance in methicillin resistant *Staphylococcus aureus* (MRSA) and also to some extent reduced penicillin resistance in beta- lactamase-producing *Staphylococcus aureus.* Also, Betoni *et al.* (2006) reported synergistic interactions between extracts of guaco (*Mikania glomerata*), guava (*Psidium guajava*), clove (*Syzyguim aromaticum*), garlic (*Allium sativum*) lemon grass (*Cymbopogon citratus*) ginger (*Zingiber officinale*) cargueja (*Baccharis trimera*), and mint (*Mentha pieria*) and some antibiotics against *S. aureus.* However, these are preliminary investigations and more works are needed to actually determine the active ingredients in these plants extracts and this may help in improving management of the different infectious diseases that are developing resistance to commonly used antibiotics and possibly to verocytotoxuic bacteria. Furthermore, toxicological studies can also be carried out to determine the reliance on these herbs without many side effects.

Researchers have also devised cluster of chemically related scaffolds which are very useful in guiding the synthesis of new compounds. In an attempt to combine the advantages of virtual screening of chemically diverse natural products and their synthetic analogues (scaffolds) with the rapid availability of physical samples for testing, an academic collaboration has established the Drug Discovery Portal (http://www.ddp.strath.ac.uk/). This brings together a wide variety of compounds from academic laboratories in many different institutions in a database that can be used for virtual screening. Academic biology groups can also propose structures as targets for virtual screening with the Portal's database (and with conventional commercially available databases). Access to the Portal is free for academic groups and the continued expansion of the chemical database means that there is a valuable and growing coverage of chemical space through many novel chemical compounds (Feher & Schmidt, 2003; Galm & Shen, 2007).

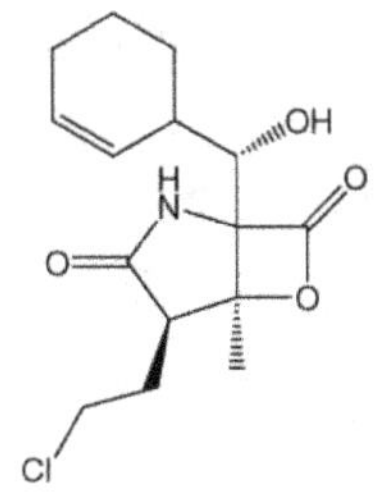

1

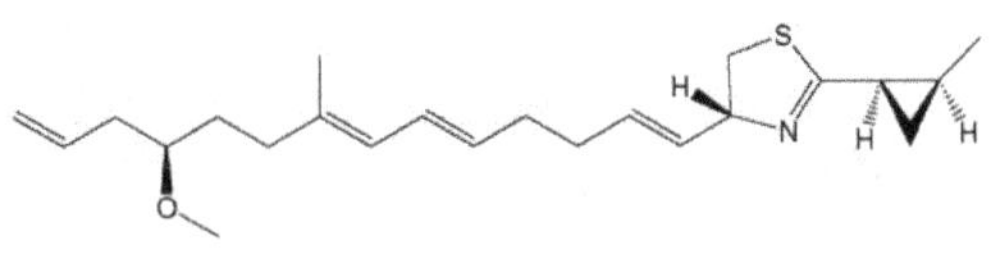

2

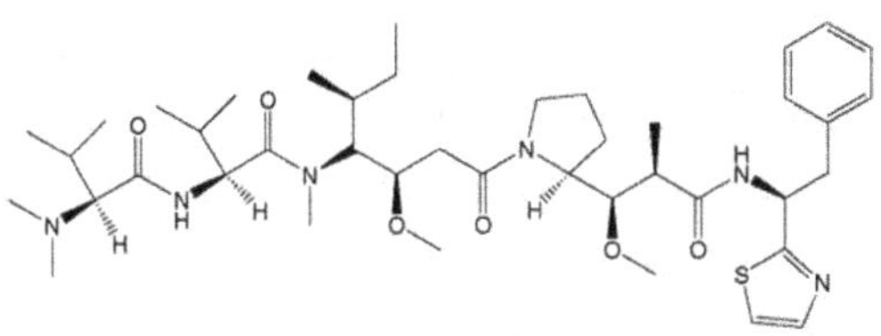

3

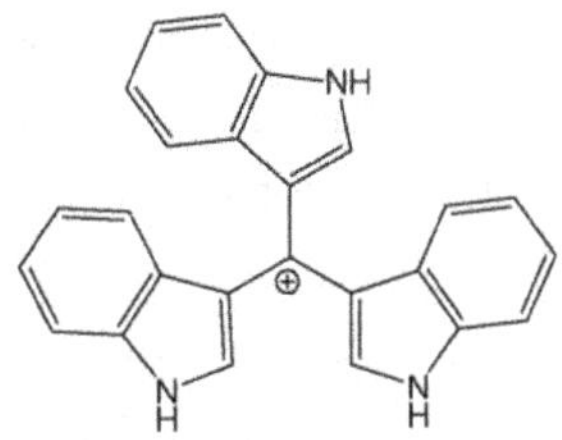

4

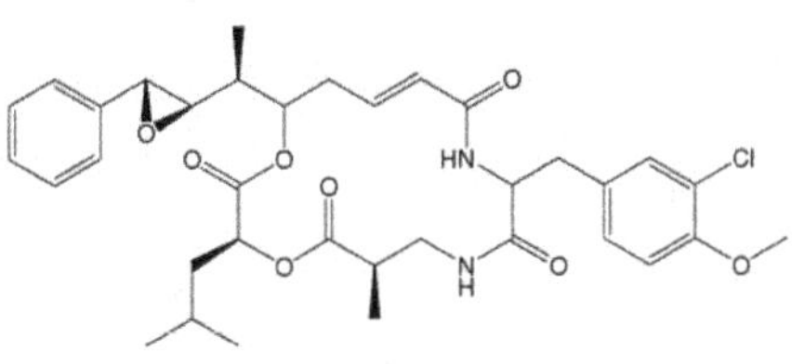

5

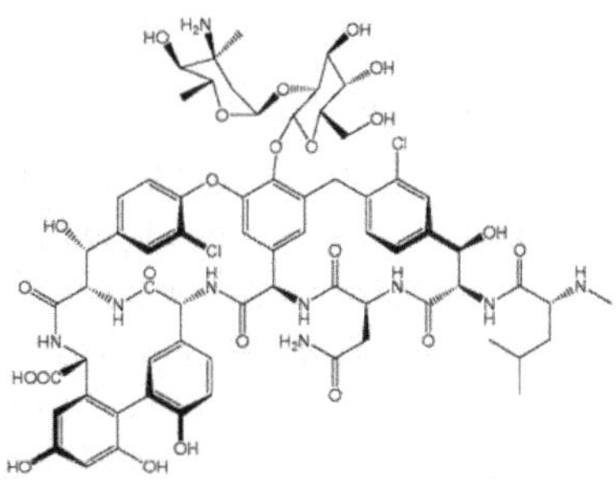

6

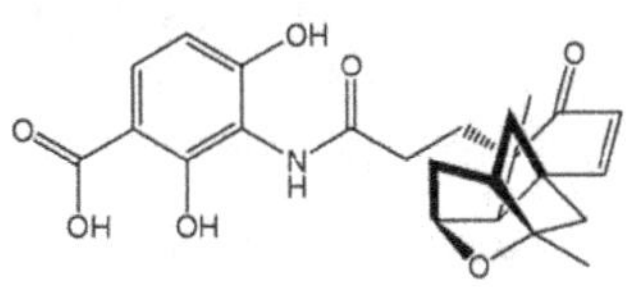

7

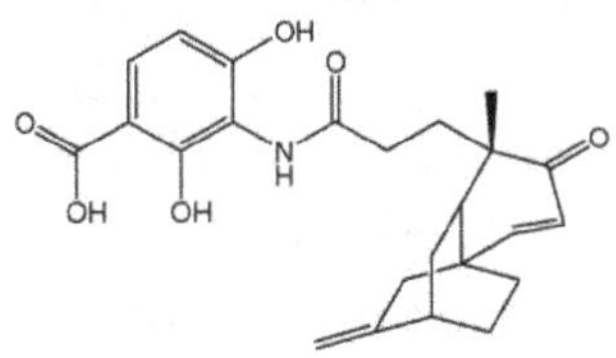

8

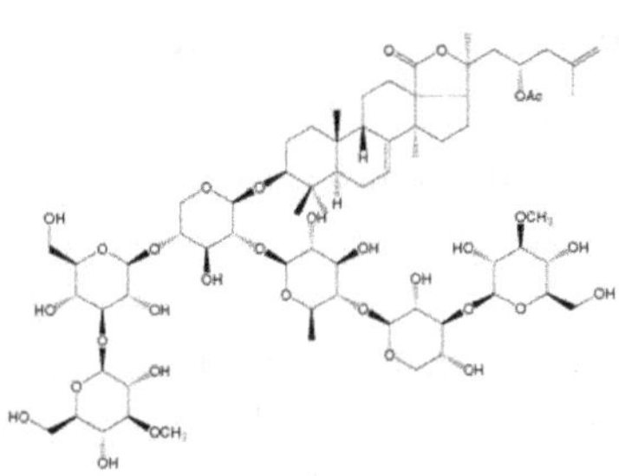

9

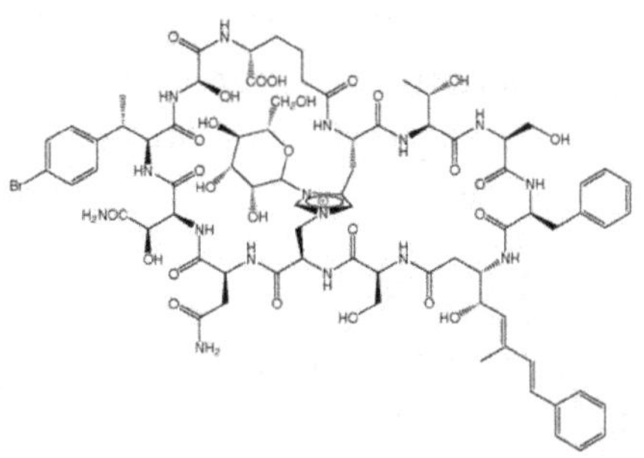

10

Fig. 10. Natural products – recently discovered and/or in development. (1) Salinosporamide A; (2) curacin A; (3) dolastatin 10; (4) turbomycin A; (5) cryptophicin; (6) vancomycin; (7) platensimycin; (8) platencin; (9) stichloroside; (10) theopalauamide (Source; Doughari *et al.*, 2009).

Despite all of the advances made by the pharmaceutical industry in the development of novel and highly effective medicines for the treatment of a wide range of diseases, there has been a marked increase in the use of herbal medicines even including the more affluent countries of the world. Germany has the largest share of the market in Europe and it was reported that the sales of herbal medicinal products (HMPs) in 1997 were US$ 1.8 billion (Barnes *et al.*, 2007). Numerous scientific medical/ pharmaceutical books have been published in recent years aiming to provide the general public and healthcare professionals with evidence of the benefits and risks of herbal medicines (Barnes *et al.*, 2007; Phillipson, 2007). The pharmaceutical industry has met the increased demand for herbal medicines by manufacturing a range of HMPs many of which contain standardized amounts of specific natural products. In the 1950s, it would not have been possible to predict that in 50 years time there would be a thriving industry producing HMPs based on the public demand for herbal medicines. To date European Pharmacopoeia has even published up to 125 monographs on specific medicinal herbs with another 84 currently in preparation (Mijajlovic *et al.*, 2006; Phillipson, 2007. The monographs are meant to provide up-to-date knowledge of phytochemistry for defining the chemical profiles of medicinal herbs and an understanding of analytical tests for identification of the herbs and for the quantitative assessment of any known active ingredients (Phillipson, 2007). Several regulatory bodies incuding Traditional Medicines Boards (TMBs, in Nigeria and other African Countries), Medicines and Healthcare products Regulatory Agency (MHRA), Herbal Medicines Advisory Committee (HMAC) (Uk) and American Herbal Products Association (AHPA) and several other pharmacopoeia (British, Chinese, German, Japanese) provide guidelines and advice on the safety, quality and utilization of the plant herbal products in several countries (Yadav & Dixit, 2008). Scientific and Research communities are currently engaged in phytochemical research, and pharmacognosy, phytomedicine or traditional medicine are various disciplines in higher institutions of learning that deals specifically with research in herbal medicines. It is estimated that >5000 individual phytochemicals have been identified in fruits, vegetables, and grains, but a large percentage still remain unknown and need to be identified before we can fully understand the health benefits of phytochemicals (Liu, 2004).

7. CONCISE SUMMARY

As interest grows and numerous promising drug candidates derived from natural sources emerge in the development pipeline, coupled with the reduction of technical limitations in natural product research, opportunities to investigate the biological activity of previously untapped natural product sources have improved. Moreover, the growing recognition that the chemical variety of natural products is ideally positioned to serve as foundational scaffolds for future pharmaceuticals will lead to advancements in the utilization of innovative natural products and chemical libraries derived from them in drug discovery initiatives.

8. REFERENCES

Abo, K.A.; Ogunleye, V.O. & Ashidi JS (1991). Antimicrobial poteintial of *Spondias mombin, Croton zambesicus* and *Zygotritonia crocea. Journal of Pharmacological Research.* 5(13): 494-497.

Adedapo, A.A.; Shabi, O.O. & Adedokun OA (2005). Antihelminthic efficacy of the aqueous extract of *Euphorbia hirta* (Linn.) in Nigerian dogs. *Veterinary Archives.* 75(1): 39-47.

Adekunle, A.S. & Adekunle, O.C. (2009). Preliminary assessment of antimicrobial properties of aqueous extract of plants against infectious diseases. *Biology and Medicine.* 1(3): 20-24.

Anderson, K.J. & Teuber, S.S. (2001). Walnut polyphenolics inhibit *in vitro* human plasma and LDL oxidation, biochemical and molecular action of nutrients. *Journal of Nutrition.* 131: 2837-2842.

Andreadi, C.K.; Howells, L.M., Atherfold, P.A. & Manson, M.M. (2006). Involvement of Nrf2, p38, B-Raf, and nuclear factor-kappaB, but not phosphatidylinositol 3-kinase, in induction of hemeoxygenase-1 by dietary polyphenols. *Molecular Pharmacology.* 69:1033–1040.

Aggarwal, B.B. & Shishodia, S. (2006). Molecular targets of dietary agents for prevention and therapy of cancer. *Biochemistry and Pharmacology.* 71:1397–1421.

Aiyegoro, O.A.; Afolayan, A,J. & Okoh, A.I. (2009). *In vitro* antibacterial activities of crude extracts of the leaves of *Helichrysum longifolium* in combination with selected antibiotics. *African Journal of Pharmacy and Pharmacology.* 3(6: 293-300.

Ames, B.N. & Gold, L.S. (1991). Endogenous mutagens and the causes of aging and cancer. *Mutation Research.* 250: 3–16.

Awad, A.B.; Bradford, P.G. & Editors. (2005). *Nutrition and Cancer Prevention.* Boca Raton: Taylor & Francis.

Barnes, J.; Anderson, L.A. & Phillipson, J.D. (2007). *Herbal Medicines. A guide for Healthcare Professionals, third ed.* Pharmaceutical Press, London. pp 1-23.

Bohlin, L. & Bruhn, J.G. (1999). *Bioassay methods in natural product research and drug development. Proceedings of the Phytochemical Society of Europe, vol. 43.* Kluwer Academic, Dordrecht. pp 288-356.

Boocock, D.J.; Faust, G.E., Patel, K.R., Schinas, A.M., Brown, V.A., Ducharme, M.P., Booth, T.D., Crowell, J.A., Perloff, M., Gescher, A.J., Steward, W.P. & Brenner, D.E. (2007). Phase I dose escalation pharmacokinetic study in healthy volunteers of resveratrol, a potential cancer chemopreventive agent. *Cancer Epidemiology Biomarkers and Prevention.* 16: 1246–1252.

Cai, Y.Z. & Sun, M. (2003). Antioxidant activity of betalins from plants of the Amaranthacea. *Journal of Agriculture and Food Chemistry.* 51: 2288-2294.

Chan, M.M. & Fong, D. (1999). *Modulation of the nitric oxide pathway by natural products.* In: *Cellular and Molecular Biology of Nitric Oxide,* eds. J.D. Laskin & Laskin, D.L. New York: Marcel Dekker, Inc. pp 333–351

Chapatwala, K.D.; Boykin, M.A. & Rajanna, B. (1982). Effects of intraperitoneally injected cadmium on renal and hepatic glycogenic enzymes in rats. *Drug Chemistry and Toxicology.* 5:305-317.

Chen, F.W.; Shieh, P., Kuo, D. & Hsieh, C. (2006). Evaluation of the antioxidant activity of *Ruellia tuberosa. Food Chemistry.* 94: 14-18.

Choi, S.W.; Friso, S. & Editors. (2006). *Nutrient-Gene Interactions in Cancer.* Boca Raton: Taylor & Francis.

Cowan, M.M. (1999): Plant products as antimicrobial agents. *Clinical Microbiology Reviews.* 12(4): 564 – 582.

Daffre, S.;Bulet, P., Spisni, A., Ehret-sabatier, L., Rodrigues, EG. &Travassos, L.R. (2008). *Bioactive natural peptides.* In: Atta-ur-Rahman (Ed.) *Studies in Natural Products Chemistry*, Vol. 35. Elsevier. pp 597- 691.

D'Incalci, M.; Steward, W.P. & Gescher, A.J. (2005). Use of cancer chemopreventive phytochemicals as antineoplastic agents. *Lancet Oncology*. 6:899–904.

Djeridane, A.; Yousfi, M., Nadjemi, B., Boutassouna, D., Stocker, P. & Vidal, N. (2006). Antioxidant Activity of Some Algerian Medicinal Plants Extracts Containing Phenolic Compounds. *Food Chemistry* 97: 654-660.

Diplock, A.T.; Charleux, J.L., Crozier-Willi, G., Kok, F.J., Rice-Evans, C., Roberfroid, M., Stahl, W. & Vina-Ribes, J. (1998). Functional food science and defense against reactive oxidative species. *Brazilian Journal of Nutrition*. 80: S77-S112.

Doughari, J.H.; Human, I.S, Bennade, S. & Ndakidemi, P.A. (2009). Phytochemicals as chemotherapeutic agents and antioxidants: Possible solution to the control of antibiotic resistant verocytotoxin producing bacteria. *Journal of Medicinal Plants Research.* 3(11): 839-848.

Doughari, J.H. & Obidah, J.S. (2008). Antibacterial potentials of stem bark extracts of *Leptadenia lancifoli.*against some pathogenic bacteria. *Pharmacologyonline* 3: 172-180.

EL-mahmood, M.A. (2009). Efficacy of crude extracts of garlic (*Allium sativum* Linn.) against nosocomial *Escherichia coli, Staphylococcus aureus, Streptococcus pneumoniea* and *Pseudomonas aeruginosa. Journal of Medicinal Plants Research.* 3(4): 179-185.

El-Mahmood, A.M. & Ameh JM. (2007). *In-vitro* antibacterial activity of *Parkia biglobosa* (Jacq) root, bark extract against some microorganisms associated with Urinary tract infections. *African Journal of Biotechnology*. 6(11): 195-200.

Falodun, A.; Okunrobe, L.O. & Uzoamaka N (2006). Phytochemical screening and anti- inflammatory evaluation of methanolic and aqueous extracts of *Euphorbia heterophylla* Linn. (Euphorbiaceae). *African Journal of Biotechnology*. 5(6):529-531

Feher, M. & Schmidt, J.M. (2003). Property distributions: differences between drugs, natural products, and molecules from combinatorial chemistry. *Journal of Chemistry and Infection and Computational Science*. 43: 218–227.

Fennell, C.W.; Lindsey, K.L., McGaw, L.J., Sparg, S.G., Stafford, G.I., Elgorashi, E.E., Grace, O.M. & van Staden, J. (2004). Assessing African medicinal plants for efficacy and safety: Pharmacological screening and toxicology. *Journal of Ethnopharmacoly*.94: 205-217.

Firn, R. (2010). *Nature's Chemicals.* Oxford University Press, Oxford. Pp 74-75.

Francis, M. S.; Wolf-Watz, H., & Forsberg, A. (2002). Regulation of type III secretion systems. *Current Opinion in Microbiology*. 5(2):166–172.

Galm, U. & Shen B (2007). Natural product drug discovery: the times have never been better. *Chemical Biology*. 14: 1098–1104.

Heinrich, M.; Barnes, J., Gibbons, S. & Williamson, E.M. (2004). *Fundamentals of Pharmacognosy and Phytotherapy*. Churchill Livingstone, Edinbrugh, pp. 245–252.

Hertog, M.G.L. & Feskens, E.J.M. (1993). Dietary antioxidant flavonoids and risk of coronary heart disease: The Zutphen Elderly study. *Lancet.* 342: 1007-1011.

Izzo, A.A. (2005). Herb-drug interactions: an overview of the clinical evidence. *Fundamental Clinical Pharmacology*. 19:1–16.

Jakhetia, V.; Patel, R., Khatri, P., Pahuja, N., Garg, S., Pandey, A. & Sharma, S.A. (2010). *Cinnamon*: a pharmacological review. *Journal of Advanced Scientific Research.* 1(2): 19- 23.

Jayasri, M.A.; Mathew, L. & Radha, A. (2009). A report on the antioxidant activities of leaves and rhizomes of *Costus pictus* D. Don. *International Journal of Integretive Biology*. 5(1): 20-26.

Kamboj, V.P. (2000). Herbal medicine. *Current Science.* 78(1): 35-39.

Kar, A. (2007). *Pharmaocgnosy and Pharmacobiotechnology (Revised-Expanded Second Edition).* New Age International LimtedPublishres New Delhi. pp 332-600

Kris-Etherton, P,M,; Hecker, K.D., Bonanome, A., Coval, S.M., Binkoski, A.E., Hilpert, K.F., Griel, A.E. & Etherton TD (2002). Bioactive compounds in foods: their role in the prevention of cardiovascular disease and cancer. *American Journal of Medicine.* 113:71S–88S.

Kotzekidou, P.; Giannakidis, P., Boulamatsis, A. (2008). Antimicrobial activity of some plant extracts and essential oils against foodborne pathogens in vitro and on the fate of inoculated pathogens in chocolate. *LWT* 41: 119–127

Levy, S.B. (1998). The challenge of antibiotic resistance. *Scientific American.* 278:32-39.

Lewis, K. & Ausubel, F.M. (2006): Prospects for plant-derived antibacterials. Nature Biotech. 24(12): 1504 – 1507.

Liu, A.G.; Volker, S.E. Jeffery, E.H. & Erdman, J.W.Jr, (2009). Feeding Tomato and Broccoli Powders Enriched with Bioactives Improves Bioactivity Markers in Rats. *Journal of Agriculture and Food Chemistry*.1: 22-28.

Liu, R.H. (2004). Potential synergy of phytochemicals in cancer prevention: mechanism of Action. *Journal of Nutrition.* 134(12 Suppl):3479S-3485S.

Madsen, H.L. & Bertelsen G (1995). Spices as antioxidants. *Trends Food Science and Technology.* 6: 271-277.

Madziga HA, Sanni S and Sandabe UK. (2010). Phytochemical and Elemental Analysis of *Acalypha wilkesiana* Leaf. *Journal of American Science.* 6(11): 510-514.

Manson, M.M. (2003). Cancer prevention – the potential for diet to modulate molecular signalling. *Trends in Molecular Medicine.* 9: 11–18.

Martinez, M.J.A.; Lazaro, R.M, del Olmo LMB. & Benito, P.B. (2008) *Anti-infectious activity in the anthemideae tribe.* In: Atta-ur- (Ed.) *Studies in Natural Products Chemistry,* Vol. 35. Elsevier. pp 445-516

Mattson, M.P. (2008). Dietary factors, hormesis and health. *Ageing Research Reviews.* 7:43–48. Mattson MP, Cheng A (2006). Neurohormetic phytochemicals: low-dose toxins that induce adaptive neuronal stress responses. *Trends in Neurosciences.* 29(11): 632-639.

Mattson, M.P. & Cheng A (2006). Neurohormetic phytochemicals: lowdose toxins that induce adaptive neuronal stress responses. *Trends in Neurosciences.* 29(11): 632-639.

Maurya, R.; Singh G. & Yadav, P.P. (2008). *Antiosteoporotic agents from Natural sources.* In: Atta-ur-Rahman (Ed.) *Studies in Natural Products Chemistry,* Vol. 35. Elsevier. pp 517-545.

McGaw, L.J.; Lager, A., Grace, O., Fennell, C. & van Staden, J. (2005). *Medicinal plants.* In: *van Niekerk, A (Ed.), Ethics in Agriculture-An African Perspective.* Springer, Dordrecht, The Netherlands. pp. 67-83.

Meijerman, I.; Beijnen, J.H. & Schellens, J.H. (2006). Herb-drug interactions in oncology: focus on mechanisms of induction. *Oncologist.* 11:742–752.

Mijajlovic, S.; Smith, J., Watson, K., Parsons, P. & Jones, G.L. (2006). Traditional Australian medicinal plants: screening for activity against human cancer cell lines. Journal of Australian Traditional-Medicine and. Sociology. 1:17-19.

Moutsatsou, P. (2007). *The spectrum of phytoestrogens in nature: our knowledge is expanding. Hormones* (Athens). 6:173–193.

Nascimento, G.G.F.; Locatelli, J., Freitas, P.C. & Silva1, G.L. (2000). Antibacterial activity of plant extracts and phytochemicals on antibioticresistant bacteria. *Brazilian Journal of Microbiology.* 31:247-256

Nweze, E.L.; Okafor, J.L. & Njoku O (2004). Antimicrobial Activityies of Methanolic extracts of *Trume guineesis* (Scchumn and Thorn) and *Morinda lucinda* used in Nigerian Herbal Medicinal practice. *Journal of Biological Research and Biotechnology.* 2(1): 34-46.

Oleszek, W. & Marston A (2000). *Saponins in food and medicinal plants.* Kluwer academic publishers. Ney York. pp 1-95.

Okigbo RN, Anuagasi CL, Amadi JE, Ukpabi UJ (2009). Potential inhibitory effects of some African tuberous plant extracts on *Escherichia coli, Staphylococcus aureus* and *Candida albicans. International Journal of. Integrative Biology.* 6(2): 91-99.

Pal, D. & Mitra, A.K. (2006). MDR- and CYP3A4-mediated drug-drug interactions. *Journal of Neuroimmune Pharmacology.* 1: 323–339.

Patil, P.S. & Shettigar, R. (2010). An advancement of analytical techniques in herbal research *J. Adv. Sci. Res.* 1(1); 08-14.

Philipson, J.D. (2007). Phytochemistry and pharamacognosy. *Phytochemistry.* 68:2960-2972.

Rice-Evans, C.; Miller, N. & Paganga, G. (1997). Antioxidant properties of phenolic compounds. *Trends in Plant Science.* 2: 152-159.

Rìos, J.L. & Recio, M.C. (2005). Medicinal plants and antimicrobial activity. *Journal of Ethnorpharmacology.* 00: 80-84

Rizvi, M.M.A.; Irshad, M., Hassadi, G.E. & Younis, S.B. (2009). Bioefficacies of *Cassia fistula:* An Indian labrum (Review). *African Journal of Pharmacy and Pharmacology.* 3(6): 287- 292.

Sarkar, F.H. & Li, Y. (2006). Using chemopreventive agents to enhance the efficacy of cancer therapy. Cancer Res. 66:3347–3350.

Sarker, S.D. & Nahar, L. (2007). Chemistry for Pharmacy Students General, Organic and Natural Product Chemistry. England: John Wiley and Sons. pp 283-359.

Sen, C.K. (1995). Oxygen toxicity and antioxidants: state of the art. *Indian Journal of Physiology and Pharmacology.* 39: 177-196.

Shanmugam, K.; Holmquist, L., Steele, M., Stuchbury, G., Berbaum, K., Schulz, O., Benavente Garc´ia, O.,Castillo, J. Burnell, J., Garcia Rivas, V., Dobson, G. & M¨unch, G. (2008). Plant-derived polyphenols attenuate lipopolysaccharide-induced nitric oxide and tumour necrosis factor production in murine microglia and macrophages. *Molecular Nutrition and Food Research.* 52:427–438.

Singh, S.B. & Barrett, J.F. (2006). Empirical antibacterial drug discovery– foundation in natural products. *Biochemistry and Pharmacology.* 71: 1006–1015.

Smolinski, M.S.; Hamburg, M.A. & Lederberg J. (eds) (2003). *Microbial threats to health: Emergence, detection, and response.* Washington, DC: Institute of Medicine, National Academies Press. pp 203-210.

Stapleton, P.D.; Shah, S., Anderson, J.C., Hara, Y., Hamilton-Miller, J.M.T. & Taylor, P.W. (2004): Modulation of B. lactam resistant *staphylococcus aureus* by catechins and gallates. *International Journal of. Antimicrobial Agents.* 23(5): 462-467.

Surh, Y.J. (2003). Cancer chemoprevention with dietary phytochemicals. *Natural Reviews in Cancer.* 3: 768–780.

Trewavas, A. & Stewart D (2003). Paradoxical effects of chemicals in the diet on health. *Current Opinions in Plant Biology.* 6: 185–190.

Uddin, S.N.; Akond, M.A., Mubassara, S. & Yesmin MN (2008). Antioxidant and Antibacterial activities of *Trema cannabina. Middle-East Journal of Scientific Research.* 3: 105-108.

Van den Bogaard, A.E. & Stobberingh, E.E. (2000). Epidemiology of resistance to antibiotics: Links between animals and humans. *International Journal of Antimicrobial Agents.*14:327-335

Vlietinck, A.J. (2000). *The future of phytochemistry in the new millennium.* In: LuijendijkTJC (Ed.), 2000 years of natural product research - past, present and future. Phytoconsult, Leiden, pp. 212–215.

Voravuthikunchai, S.P. & Kitpipit, L. (2003). Activities of crude extracts of Thai medicinal plants on methicillin- resistant *Staphylococcus aureus. Journal of Clinical Microbiology and Infection.* 9:236.

W.H.O. (1999). *WHO Monographs on Selected Medicinal Plants.* 1: 1-295.

Yadav, N.P. & Dixit, V.K. (2008). Recent approaches in herbal drug standardization. *International Journal of Intergrative Biology.*2(3): 195-203.

Yagoub, S.O. (2008). Antimicrobial activity of *Tamarindus indica* and *Adansonia digitata* extracts against *E. coli* isolated from urine and water specimens. *Research Journal of Microbiology.* 3(3): 193-197.

You, G.; Morris, M.E. & Editors (2007). *Drug Transporters.* Hoboken: John Wiley & Sons.

Chapter 2

STRUCTURAL ANALYSIS OF FLAVONOIDS AND ASSOCIATED COMPOUNDS – A REVIEW OF SPECTROSCOPIC APPLICATIONS

1. INTRODUCTION: AN OVERVIEW

In 1936, St. Rusznyák and A. Szent-Györgyi published a paper in Nature (Rusznyàk & Szent-György, 1936) detailing the alleviation of specific pathological conditions marked by heightened permeability or fragility of the capillary wall, achieved through extracts of Hungarian red pepper rich in flavonols, a category of flavonoids subsequently designated as "vitamin P." Over the subsequent five decades, there was a gradual yet consistent increase in interest regarding flavonoids and their therapeutic advantages for a wide array of diseases and ailments, such as pregnancy toxaemia, rheumatic fever, diabetes, and cancer. During the late 1980s and throughout the 1990s, flavonoids were extensively investigated for their functions as mutagenesis agents, as well as their roles as antioxidants and pro-oxidants in biological systems (for example, Aviram & Fuhrman, 1998; Lambert & Elias, 2010). In the early 90s the antioxidant activity of flavonoids was extensively studied *in vitro*, and it was assumed that such activity would be at the basis of the health promoting benefits of these compounds. However, in the late 90s and early 00s the metabolism of flavonoids was deeply scrutinized, and the results indicated that their antioxidant activity *in vivo* could not account for the overall actions attributed to them (Fraga et al., 2010). The paradigm for flavonoid action changed towards the establishment of flavonoids as inflammation modulators, and more recently their role in neuroprotection, memory and cognition has been under scrutiny (Gomes et al., 2008; Spencer et al., 2009; Spencer, 2010). However, exact mechanisms for many of the actions attributed to flavonoids have not yet been established, but the relationship between their activity and the presence of specific functional groups in the molecules is undeniable. Moreover, each role attributed to flavonoids has been linked to different structural features – for example, while antioxidant activity depends essentially on the number and location of OH groups in the molecules, their antagonist effect towards adenosine receptors depends more on the overall planarity than on the hydroxyl groups; in fact, the latter even appear to be counter-productive (González et al., 2007).

The work developed in this area strongly depends on powerful analytical techniques for quantification and structural identification, as circulating forms are usually found in the low micromolar range, and intracellular levels are even lower. This chapter will briefly review the analytical techniques employed to determine the flavonoids structure from *in vitro* and *in vivo* studies. Albeit it will be focused on the more common classes of flavonoids (flavones, flavonols, flavanones, catechins, isoflavones and anthocyanidins), it will also address recent developments in minor flavonoid classes. Two essential works must be differentiated here. The initial reference is the 1982 publication by Markham titled Techniques of Flavonoid Identification, which examines the contemporary advancements in flavonoid structural identification, primarily emphasizing UV spectroscopy (Markham, 1982). In 2005, Markham and Andersen edited "Flavonoids: Chemistry, Biochemistry and Applications," a reference book for researchers examining the structure and activity of flavonoids. It includes extensive tables of NMR data and covers numerous contemporary research topics (Fossen & Andersen, 2005). This chapter will mostly concentrate on advancements in the field from 2005 onward.

2. FLAVONOID STRUCTURAL CLASSIFICATION

Flavonoids are polyphenols of plant origin that are among the most important compounds in human diet due to their widespread distribution in foods and beverages. They can occur both in the free form (aglycones) and as glycosides, and differ in their substituents (type, number and position) and in their insaturation. As mentioned, the most common classes are the flavones, flavonols, flavanones, catechins, isoflavones and anthocyanidins, which account for around 80 % of flavonoids. Figure 1 shows the basic structure of the different flavonoid classes addressed in this chapter.

All flavonoids share a basic C6-C3-C6 phenyl-benzopyran backbone. The position of the phenyl ring relative to the benzopyran moiety allows a broad separation of these compounds into flavonoids (2-phenyl-benzopyrans), isoflavonoids (3-phenyl-benzopyrans) and neoflavonoids (4-phenyl-benzopyrans) (Figure 1). Division into further groups is made

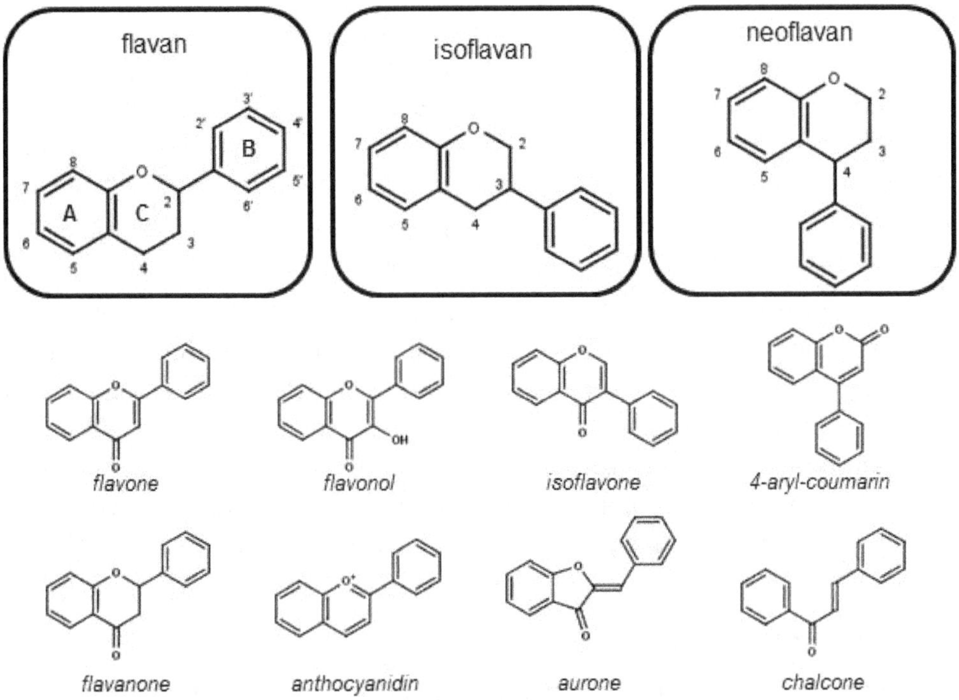

Fig. 1. Structure of the structural backbones of the main flavonoid groups (flavan, isoflavan and neoflavan) and of relevant flavonoid classes. Atom numbering and ring nomenclature are also included.

on the basis of the central ring oxidation and on the presence of specific hydroxyl groups. Most common flavonoids are flavones (with a C2-C3 double bond and a C4-oxo function), flavonols (flavones with a 3-OH group) and flavanones (flavone analogues but with a C2-C3 single bond), and abundant isoflavonoids include isoflavones (the analogue of flavones). 4- arylcoumarin (a neoflavonoid with a C3-C4 double bond) and its reduced form, 3,4- dihydro-4-arylcoumarin, are the major neoflavonoids. Other natural compounds, such as chalcones and aurones also possess the C6-C3-C6 backbone, and are henceforth included in the general group of flavonoids.

3. MASS SPECTROMETRY

Mass Spectrometry (MS) has proved to be one of the most effective techniques in biomedical research, in special when complex matrixes of biological samples must be analysed. The main advantages of MS are its high sensitivity, which allows analysis of compounds present in the µg scale, and high specificity, as it is able to separate molecules of the same molecular weight but different atom composition, and sometimes even to differentiate stereoisomeric compounds. Its easy coupling to separation techniques such as liquid and gas chromatography is also an excellent advantage. A review of separation methods, applied to flavonols, isoflavones and anthocyanidins, has been recently published (Valls et al., 2009). Sample preparation may also be critical, but that varies from sample to sample; nevertheless, some general guidelines have been reviewed (Prasain et al., 2004). For more detail on mass spectrometry equipment and experiments, an excellent book has been published in 2007 (de Hoffmann & Stroobant, 2007). Specifically for electrospray and MALDI mass spectrometry applications in the biology area, the work edited by Cole (Cole, 2010) is highly adequate.

Gas Chromatography (GC) is one of the key techniques for the separation of organics and, coupled to MS, one of the most common techniques of structural identification. However, flavonoids are largely nonvolatile, and need be derivatized; also, they are usually thermally unstable. Both these characteristics have led to the establishment of Liquid Chromatography, in particular High Performance (HPLC), as the fundamental separation technique for flavonoids. Consequently, LC-MS coupling is routinely used for the overall structure elucidation of flavonoids (Fossen & Andersen, 2005).

3.1 Ionization techniques

Various ionization techniques are available, and each has their own specificities which make them more or less useful depending on the molecules under study and on the aim of such study. The ones most applied to flavonoid research are hereafter presented.

3.1.1 FAB-MS and LSIMS

Fast-Atom Bombardment (FAB) and Liquid Secondary Ion Mass Spectrometry (LSIMS) are ionization techniques used in Secondary Mass Ion Spectrometry (SIMS) in which secondary ions, emitted by sample irradiation with a beam of energetic (primary) ions, are analyzed. Typically, these techniques are able to produce ions from polar compounds with molecular weights up to 10 kDa, but they require the analyte to be dissolved in a matrix, which may lead to the formation of more complex spectra. Also, being soft techniques, ion abundance is low, and it has been used essentially to identify flavonoid glycosides and for molecular weight determination (Stobiecki, 2000; Prasain et al., 2004).

3.1.2 ESI and APCI

Electrospray Ionization (ESI) is a technique in which ions are generated by solvent evaporation under a high voltage potential, and can be applied directly, by infusion of the sample with a flow-controlled syringe, or coupled to separation techniques such as LC or capillary electrophoresis. In both cases, a steady liquid stream enters the system, allowing multiple analyses to be performed over a relatively large period of time. ESI interfaces are mostly coupled to quadrupole mass spectrometers; both are simple and robust equipments, able to produce either positive or negative ions, and their main limitation is the relatively limited m/z range, usually below 2 kDa. In Atmospheric Pressure Chemical Ionization (APCI) sources ionization occurs via a corona discharge on a heated solvent spray, which produces solvent-derived primary ions that will, in turn, ionize the solute (Prasain et al., 2004; de Hoffmann & Stroobant, 2007). Both ESI and APCI use atmospheric pressure and high collision frequency, and thus generate large amounts of ions; as they involve solvent evaporation, the decomposition of the analytes is reduced, and full scans show limited fragmentation. The main disadvantage of both these techniques is that some HPLC solvents interfere with the ionization process, and thus chromatographic separations need to be specifically designed (Prasain et al., 2004).

3.1.3 MALDI

Matrix-Assisted Laser Desorption/Ionization (MALDI) is a soft ionization technique in which the analytes are co-crystallized with a matrix; this mixture is deposited on a plate upon which a laser beam is aimed. The laser discharge ultimately leads to analyte ionization and projection from the matrix and onto the analyser. Typical matrixes are derivatives of 4- hydroxycinnamic acid, and also 2,5-dihydroxybenzoic acid (2,5DHB) (de Hoffmann & Stroobant, 2007). Due to the structural similarity between these matrixes and the flavonoids, only recently has MALDI been applied to flavonoid structural elucidation, using a FT-ICR spectrometer, and a significant, although still informative, number of flavonoid-matrix clusters are observable (Madeira & Florêncio, 2009).

3.2 Mass analysis

Mass analysis is the second step in a MS experiment. Following ion generation, mass analysers measure the mass-to-charge ratio, *m/z*, of the ions, by using a combination of electromagnetic fields. There are many types of mass analysers, as there are of ion sources and detectors. The most common are of the ion type and of the quadrupole type, which analyse *m/z* ratios by the resonance frequency and by the trajectory stability, respectively, and time-of-flight (TOF) analysers, which measure ion velocity (or flight time). More recent resonance frequency analysers, namely Fourier Transform (FT) ion cyclotron resonance (FT-ICR) and FT orbitraps, are now starting to be applied to flavonoids (de Hoffmann & Stroobant, 2007). A recent work has compared the performance of different mass analysers coupled to the same ion source and it was concluded that fragmentation patterns are transferable among different mass analysers, only the relative abundances are changed; although applied to cyano dyes (Volná et al., 2007), the conclusions also apply to other classes of compounds, such as flavonoids, where different mass analyzers lead to similar fragmentation patterns.

3.3 Tandem mass spectrometry

Tandem mass spectrometry, usually abbreviated MS/MS, or MS^n for *n*th order fragmentation, is any method that involves at least two stages of mass analysis, in conjunction with a fragmentation process, either dissociation of reaction, which causes a change in the *m/z* ratio on an ion. Most commonly, a mass analyser is used to isolate a precursor ion, which is then fragment to yield product ions (and, eventually, neutral fragments) that will be detected in the second mass analysis – a typical MS^2 experiment. This can, at least conceptually, be expanded with further successive modification and detection steps, giving rise to MS^3,... , MS^n. However, as only a very small fraction of ions detected in one analyser makes it to the following analyzers, MS^3 is usually the highest order achieved. This spatial arrangement of equipment, analyser-modified-analyser, corresponds to tandem MS in space, where ions are treated in different regions of space. Alternatively, tandem MS can be performed in time, with analysers such as ion traps, orbitraps or FT-ICR, where the same analyser performs different tasks successively (de Hoffmann & Stroobant, 2007).

m/z modification can be achieved by various techniques, but the most common is Collision Induced (or Activated) Dissociation (CID or CAD), where precursor ions undergo collisional activation with neutral atoms or molecules (such as inert gases) in the gas phase. CID is an example of a post-source fragmentation, in which energy is added to the already vibrationally excited ions. An alternative to CID is ECD (Electron Capture Dissociation), in which multiply charged positive ions are submitted to a beam of low energy electrons, producing radical cations. In opposition to post-source fragmentation, in in-source fragmentation, ions already possess sufficient internal energy and fragment spontaneously within the mass spectrometer. Although usually this is an undesired effect, because it leads to lower abundance of precursor ions, it may in some cases become useful (de Hoffmann & Stroobant, 2007; Abrankó et al., 2011).

3.3.1 Scan modes

Four general types of tandem MS scans are possible, and all may generate valuable information. A *product ion scan* analyses all the fragment ions resulting from a single selected precursor ion (these are usually called *MS^2 spectra*). Conversely, a *precursor ion scan* will identify all the precursors of a selected product ion; a *neutral loss scan* is performed from a selected neutral fragment and will identify the fragmentations leading to the loss of that neutral fragment; these two techniques cannot be performed in time-based analysers (de Hoffmann & Stroobant, 2007). A particular application of neutral loss scans is in the identification of phase II conjugation metabolites, that can be identified by specific neutral losses (Table 1) (Prasain & Barnes, 2007). Neutral loss scans are widely used to detect phase II conjugation metabolites, such as glucuronides (loss of 176 Da) and sulfates (loss of 80 Da), as well as for the detection of glutathione adducts (loss of 129 Da). Scanning for neutral losses of 162 and 132 Da has also been used to separate flavonoids with hexose residues from those with pentose residues, respectively. More selective than these three techniques, *selected reaction monitoring* (SRM) will analyse if a specific product ion comes from the fragmentation of a specific precursor ion; although more sensitive, it is much more specific.

Table 1. Mass shift associated with possible metabolic reactions of flavonoids and correspondent detection by neutral loss scanning. Adapted from Prasain & Barnes, 2007.

Metabolic Reaction	Mass change (Neutral loss)	Metabolic Reaction	Mass change (Neutral loss)
Glucuronidation	+176 (176)	Methylation	+14 (-)
Glycosylation, hexose	+162 (162)	Hydroxylation	+16 (-)
Glycosylation, deoxyhexose	+146 (146)	Acetylation	+42 (60)
Glycosylation, pentose	+132 (132)	Carboxylation	+44 (44)
Sulfation	+80 (80)	Decarboxylation	-44 (44)
Glutathionylation	+129 (129)	Demethylation	-14 (-)

The following subsections will address the various applications of mass spectrometry to flavonoid structural elucidation, and are organized as a potential guide to explore the structure of novel compounds.

3.4 Flavonoid glycosides – differentiation and characterization

In plants, flavonoids are often found to be glycosylated; the glycoside residues can be attached to O and C atoms of the flavonoids, giving rise to *O*-glycosides, *C*-glycosides and *O*-*C*-glycosides. These can be differentiated by soft ionization techniques, with low fragmentation energy, usually by FAB-MS, in which glycoside loss from *O*-glycosides undergo heterolysis of their hemi-acetal O-C bonds, gives rise to Y_i^+ ions; at low energy, *C*- glycosides only produce $[M+H]^+$ ions, and, at higher energies, intraglycosidic cleaveages rive rise to $^{i,j}X$ fragments and water loss gives rise to characteristic ions (Cuyckens et al., 2000; Li & Claeys, 1994; Vukics & Guttman, 2010). Higher fragmentation energies lead to intraglycosidic cleavage in *O*-glycosides, to the generation of Y_i fragments in *C*-glycosides, and to both $^{i,j}X$ and Y_i fragments in *O*-*C*-glycosides, all of which correspond to complex, often misleading, mass spectra. (Li & Claeys, 1994). This nomenclature, proposed by Domon & Costello (1988) for the MS study of glycoconjugates, is presented in Figure 2.

The sugar type can be easily determined by the characteristic *m/z* values of the A_i, B_i and C_i fragments arising from hexoses, deoxyhexoses and pentoses, which are not directly observable in the mass spectra but can be computed from the *m/z* differences of the parent ions and corresponding X_i, Y_i and Z_i fragments (Vukics & Guttman, 2010; Ferreres et al., 2007; Li & Claeys, 1994). The *m/z* values for these fragments are presented in Table 2.

Flavonoid glycosides usually contain one or two glycoside residues, but molecules with more residues have been identified in nature. By definition, diglycosides can have the residues attached at different positions (di-*O*-glycosides and di-*C*,*O*-glycosides) or at the same position (*O*-diglycosides and *C*,*O*-diglycosides). The differentiation of the different types can be made from the product ions identified in the spectra, particularly from the Z_i ions – for example, Z_1 fragments do not occur in monoglycosides. Also, Yi fragments undergo well-known losses to form radical ions, which can be used to characterize the distribution of sugar residues. It must be noted that in some *O*-glycosides an internal sugar loss may take place, in which the aglycone-bound sugar is released first and, simultaneously, the other residue attaches itself to the aglycone; although this may lead to more complex spectra, this is well described in the literature, and can be overcome by the study of flavonoid-sodium adducts (Vukics & Guttman, 2010; Ma et al., 2000; Brüll et al., 1998; Cuyckens et al., 2001).

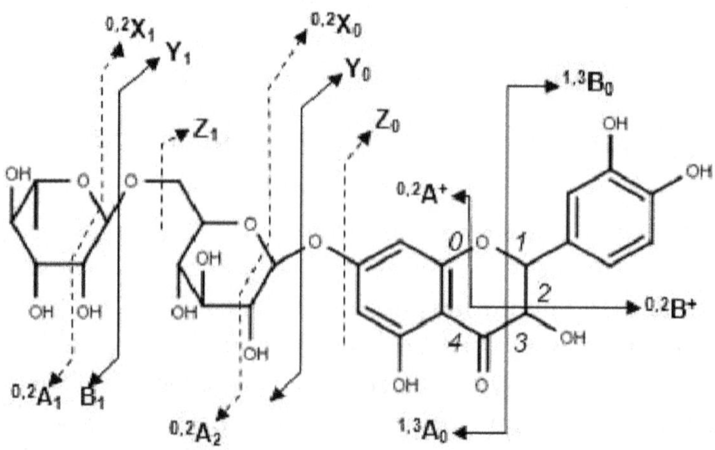

Fig. 2. Ion nomenclature used for flavonoid glycosides. Fragmentation of quercetin-7-O- rutinoside is depicted. Glycoside ions are named according to Domon & Costello, 1988; aglycone ions follow the nomenclature of Ma et al., 1997.

Table 2. Characteristic mass losses of flavonoid glycosides and characteristic cross-ring cleavages in the sugar rings of hexoses and pentoses (Vukic & Guttman, 2010; Cuyckens & Claeys, 2004).

Fragments	- $^{0,1}X$	- $^{0,2}X$ (- H_2O) (- $2H_2O$)	- $^{0,3}X$	- $^{1,5}X$	$^{2,3}X$	- $^{0,4}X$	- Y	- X
Hexose	150	120 (138) (156)	90	134	66 (84)	96	162	180
Deoxy-hexose	134	104 (122) (140)	74	120	66 (84)	80	146	164
Pentose	120	90 (108) (126)	60	104	– (–)	66	132	150

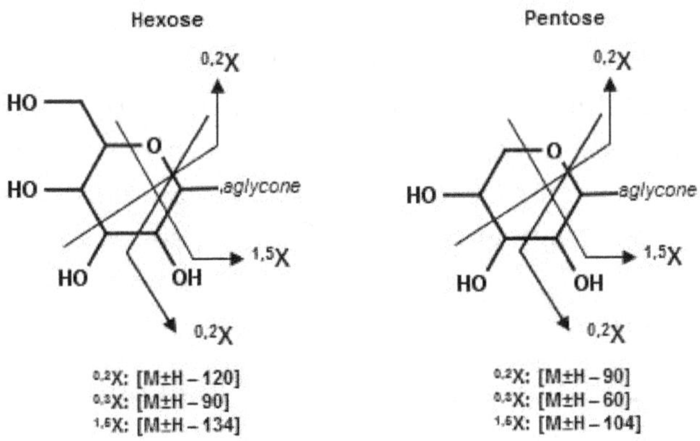

Concerning the glycoside sequence of di-, tri- and tetra-glycosides, there is no general procedure to apply for the ordering of the different residues, and it must be done rationally case by base. However, it has been shown that the interglycosidic linkage is easily accessible by MS; although these bonds can be of different types, 1→2 and 1→6 bonds are the preferred ones. For these two bond types, the Y_0^+/Y_1^+ is always larger for a 1→2 bond than for a 1→6 bond – the Y_0^+ ion is more abundant than the Y_1^+ ion for glycosides with a 1→2 bond, and the relative abundance is reversed in the case of a 1→6 bond. The negative ionisation mode may be particularly important in this case as the Z^- appears to be exclusive of 1→2 bonded glycosides. In the case of C,O-diglycosides, it is usually necessary to obtain MS^3 spectra to confirm the various interglycosidic linkages, because while 1→2 glycosides yield, by internal cleavage of the C-glycosyl moiety, $^{0,2}X^-$ ions, all other possibilities (1→3, 1→4 and 1→6) do not have characteristic fragmentation products (Ma et al., 2000; Cuyckens et al., 2000; Vukics & Guttman, 2010).

In theory, flavonoids could be glycosylated in any position, but *O*-glycosylation occurs mainly at position 7, as in flavones, isoflavones, flavanones and flavonols; *O*-glycosylation in positions 3 and 5 is also frequent, albeit less than in position 7. *C*-glycosyl flavonoids are usually flavones, and the glycosyl moieties are attached at positions 6 or 8; the literature has two reports of 3-*C*-glycosyl-flavones. Positions can be identified with basis of the product ions of the compounds because each glycosylation site appears to yield specific fragmentations. The $[Y_0\text{-CO}]^-$ is specific of 7-*O*-monoglycosides, while 3-*O*-monoglycosides are characterized by the $[Y_0\text{-2H-CO}]^-$ ion. Also, higher energy fragmentations yield product ions that contain both the glycoside moiety and fragments of the aglycone, particularly B- ring derived fragments. In what concerns *C*-glycosides, 6-*C*-glycosides usually undergo more extent fragmentations than 8-*C*-glycosides, and the former are typically associated with a ${}^{2,3}X^{\cdot-}\text{-}2H_2O$ fragment and an abundant ${}^{0,3}X^-$ ion, while for the latter no typical product ions have been out forward and the ${}^{0,3}X^-$ ion usually has a low abundance (Li & Claeys, 1994; Vukics & Guttman, 2010; Ferreres et al., 2007; Cuyckens & Claeys, 2005).

Many flavonoid glycosides possess acylated glycosyl moieties. Various acyl groups have been reported in the literature, and they are usually identifiable by characteristic mass losses, which are shown in Table 3. However, the position of glycoside acylation is not easily accessible by mass spectrometry, except for the cases in which a ${}^{0,4}X$ fragment is present in the spectra of a hexose-containg flavonoid as it inequivocally establishes that the hexosyl moiety is acylated in position 6 (Cuyckens & Claeys, 2004).

3.5 Aglycones – identification of flavonoid classes

The identification of the diverse flavonoid classes is often achieved by the MS2 spectra of the various compounds, because, as a general rule, each class of flavonoids is characterized by specific fragmentation patterns. This is especially true in the case of flavan-derived flavonoids, and, to a lesser extent, in the case of some isoflavan-derived flavonoids. For the remaining classes, qualitative MS spectra are still scarce, and, in particular, systematic studies of compounds of the same class are still to be performed.

3.5.1 Flavones, flavonols, flavanones and flavanonols

The most useful fragmentations of flavonoids, in terms of structural elucidation, are those that involve breaking the C-ring bonds, which are termed retro Diels-Alder (RDA) by analogy with the Diels-Alder cycloaddition. These fragmentations (included in Figure 2) give rise to product ions containing the A or B ring and part of the C ring; for example, the 1,3A+ ion derived from a flavone, formed by a 1/3 fragmentation (cleavage of bonds 1 and 3), will contain the whole A ring plus the O1, C4 and O4 atoms.

Table 3. Characteristic acyl groups found in acylated glycosyl flavonoids and corresponding characteristic fragmentations. ± stands for either + or –. Adapted from Cuyckens & Claeys, 2004.

Acyl group	Characteristic fragments (mass change, amu)	
Acetyl	$[M+H\text{-acetylhexose}]^+$	(204)
	[M±H–CO2]	(44)
Malonyl	$[M+H\text{-malonyl}]^+$	(86)
	$[M+H\text{-malonylhexose}]^+$	(248)
	$[M+H\text{-benzoylhexose}]^+$	(266)
Benzoyl	$[\text{benzoyl}]^+$	*m/z*=105
	$[\text{benzoylhexose}]^+$	*m/z*=267
	[M±H-galloyl]	(152)
Galloyl	$[M-H\text{-gallic acid}]^-$	(170)
	$[M+H\text{-galloylhexose}]^+$	(314)
	$[\text{Gallic acid} - H]^+$	*m/z*=169
Coumaroyl	$[M+H\text{-coumaroyl}]^+$ $[M+H\text{-coumaroylhexose}]^+$	(146) (308)
	[M±H-feruloyl]	(176)
	[M±H-feruloylhexose]	(338)
Feruloyl	$[\text{feruloyl}]^+$	m/z=177
	$[\text{ferulic acid}]^{\bullet+}$	m/z=194
	$[\text{feruloylhexose}]^+$	m/z=339
	[M±H-sinapoyl]	(206)
	[M±H-sinapoylhexose]	(368)
	$[\text{sinapoyl}]^+$	m/z=207
	$[\text{sinapic acid}]^+$	m/z=224
	$[\text{sinapoylhexose}]^+$	m/z=369
	$[\text{sinapoylhexose}]^-$	m/z=367

These RDA fragmentations allow the establishment of diagnostic product ions for the various types of flavonoids. These are valid for a large of experimental MS conditions, but it must be kept in mind that the higher fragmentation energies will lead to increased fragmentation and a higher abundance of ions with lower masses, deviating from these diagnostic ions and fragmentation pathways.

The most important RDA fragmentations in flavones and flavonols (3-hydroxyl-flavones) are the 0/2, the 0/4, the 1/3. While the 0/4 pathway appears to be exclusive of flavones, according to a low-energy CID study, where it leads to the $^{0,4}B^{+}$ ion, which looses water to form the $^{0,4}B^{+} - H_2O$, the $^{0,2}A^{+}$ is an exclusive of flavonols; also, the $^{1,3}B^{+}$ ion appears to be exclusive of flavones. In common, flavonoids from both classes give rise to the product ions $^{1,3}A^{+}$, and $^{0,2}B^{+}$ (Cuyckens et al., 2005). This information is essential when analysing, for example, flavonoid metabolites by MS. Quercetin metabolism leads to the formation of, among others, conjugates with sulphate and glucuronic acid, as well as to methoxy- quercetin. Considering a full scan MS of quercetin (molecular weight of 302 g/mol) metabolites collected from rat plasma, an ion at m/z = 479 corresponds to a protonate quercetin glucoronide; the 1/4 RDA fragmentation produces a $^{1,4}A^{+}$ at m/z = 303, that bears the mass increase, meaning the glucuronosyl moiety is present in this ion; therefore, it must be located at either position 5 or 7 (Justino et al., 2004). Similar rationales for other ions lead to the identification of the positions of metabolisation in these compounds.

More recently, a mixed ESI-MS and quantum chemical approach has analysed the fragmentation pathways of flavones and flavanols, and has proposed structures for other ions, other than RDA-derived, that are also informative in terms of structural elucidation. In particular, losses of one and two C_2H_2O moieties from the precursor ion, involving all the rings, and the formation of $^{1,3}A^{+} - C_2H_2O$ ion, are the most useful (Justino et al., 2009). Similar methodology has been applied in other cases (Madeira et al; 2010).

Flavones and flavonols constitute the vast majority of the flavan-based flavonoids studies. Flavanones, which lack the C ring 2-3 double bond, are less abundant than those two and have received less attention. In terms of MS fragmentations, in particular RDA pathways, flavanone itself undergoes cleavages to yield the $^{1,3}A^{+}$ and $^{1,4}B^{+}$ ions; these same ions are also observed for other flavanones, such as 5,7,4'-tri-hydroxy-flavanone (Nikolic & van Breemen, 2004). Studies on flavanone derivatives have evidenced that the positive mode CID spectra of isoprenylated flavanones is dominated by isoprenyl loss and $^{1,3}A^{+}$ ion formation, while the negative mode is dominated by $^{1,4}A^{-}$, [$^{1,3}A^{-}$ - isoprenyl] and [$^{1,4}A^{-}$ - isoprenyl] ions. (Zhang et al., 2008) Derived ions, such as $[^{1,3}A + H]^{+}$ and $[^{1,4}B - H_2 + H]^{+}$, have also been identified. (Zhou et al., 2007). Negative mode ESI has also allowed dividing flavanones into two groups, one with a 2'-OH group, with a spectra dominated by the $^{1,4}A^-$ ion (for 5-OH- containing molecules) or the 1,4B- (otherwise), and the other with no 2' substituents, for which the spectra were dominated by an intense $^{1,3}A^-$ ion and a low abundant $^{1,3}B^-$ ion. (Zhang et al., 2008) a 0/3 fragmentation, origination a $^{0,3}B^-$ ion, has also been observed (Xu et al., 2009); an equivalent fragmentation has also been observed in the positive mode , to produce a $^{0,3}B^+$ derived ion (Kéki et al., 2007). Flavanols (3-hydroxy-flavanones) show fragmentation patterns similar to those of flavanones; the main difference is that positive ESI spectra of flavanols are dominated by loss of water and homolytic H loss from the 3- hydroxy group to generate the $[M-H]^{\cdot+}$ ion (Zhang et al., 2008).

3.5.2 Isoflavones and Isoflavonols

Unlike flavan derived flavonoids, the fragmentation pathways of isoflavan derived flavonoids have only been explored more recently, but a good set of systematic studies are published. Besides common neutral losses, isoflavones undergo RDA fragmentation to yield the $^{0,4}B^+$-H_2O and the $^{1,3}A^+$ ions; a full B ring loss, yielding the [M-*B ring*-CO]+ ion, is also frequently reported. Also characteristic is the loss of a CO group from Ring C oxo group, leading to ring contraction (Madeira et al., 2010; Heinonen et al., 2003; Borges et al., 2001; Simons et al., 2011).

In the negative mode, the 1/3 and 0/3 RDA fragmentations are predominant, and the 0/4 fragmentation is also observed sometimes, accompanied by extensive losses of CO, CO_2 and C_3O_2 moieties. C ring expansion is also commonly observed (Vessecchi et al., 2011; Kang et al., 2007; March et al., 2004; Ablajan, 2011). For isoflavones, rare RDA fragmentations have been described in the literature, such as a 2/3 fragmentation that has also been described for various acylated 7,2'-hydroxy-3',4'-dimethoxyisoflavan glycosides (Qi et al., 2008) and a 2/4 fragmentation of daidzein (Wei et al., 2000).

The main difference in the fragmentation pathways of flavan- and isoflavan-derived flavonoids appears to be that in the latter the 0/3 RDA fragmentation occurs frequently while it appears not to occur for flavan-compounds; also, the 1/3 pathway appears to be much more important in isoflavan-compounds, in particular in those without a 2-OH group. Differentiation of isomeric aglycones of flavones and isoflavones is also possible based on a double neutral loss of CO (Kuhn et al., 2003).

3.5.3 Chalcones

Chalcones exist in nature with a variety of substituents. The MS spectra of these compounds are characterized by substituent loss, fragmentations of the substituents and chalcone fragmentations. Chalcone fragmentations are dominated by cleavage of a single bond, yielding a B ring derived ion with the attached C=O group in the charged form (C≡O+), from which a CO loss yield the free B ring (Nowakowska & Pankiewicz, 2008). Besides that, 2'-OH-chalcones, the most common ones, with an OH group in ring B adjacent to the propenone chain, are known to be converted to the corresponding flavanones by various processes, and that has also been observed to occur in ESI MS; the patter of chalcone fragmentation will then be the same of flavanone fragmentation (Zhang et al., 2008).

3.5.4 Other classes

Many other flavonoid classes exist. Many of those have been isolated and characterized, but no useful fragmentation pathways have been established.

For aurones, a positive mode of two aurone glycosides was identified as main product ions the ones formed by the loss of the glycosyl moiety, followed by a CO loss involving the heterocyclic O atom in the 5-member ring, and, from the [M-glycosyl+H]$^{\cdot+}$ ion, the $^{1,3}A^{+}$ ion, which may lose a further CO group, and the $^{1,3}B^{+}$ ion (Kesari et al., 2004).

Coumarins show a wide diversity of substituents, and many even have fused rings attached. This variability does not allow overall fragmentation patterns, but many studies observe that in many cases the heterocyclic ring undergoes a contraction by CO loss (from the C=O group) to yield an ion containing a five membered ring (for example, Nowakowska & Pankiewicz, 2008; Zhang et al., 2007; Zhang et al., 2008).

Pterocarpans, isoflavonoid derivatives, show much more complex fragmentation pathways than the above analysed classes of flavonoids; MS studies of different deuterated peterocarpan derivatives, as well as of pterocarpan glycosides, points out that these are dominated by various and successive ring openings and/or contractions (Tóth et al., 2000; Zhang et al., 2007). A rare 2/4 RDA fragmentation, like that of isoflavan-derived flavonoids, has also been observed for pterocarpans in the negative mode (Simons et al., 2011).

Neoflavonoids, in which the B ring is attached in position 4, and include 4-aryl-coumarins, have been poorly studied by MS. The few reports available indicate, however, that neoflavonoid fragmentation is dominated by loss of the CH_3 radical in methoxylated compounds and by B ring loss, to yield the [M+H-B ring] fragment (Charles et al., 2005; Hulme et al., 2005; Liu et al., 2005). Figure 3 summarizes the characteristic RDA fragmentation pathways for flavonoid classes for which there is reliable information.

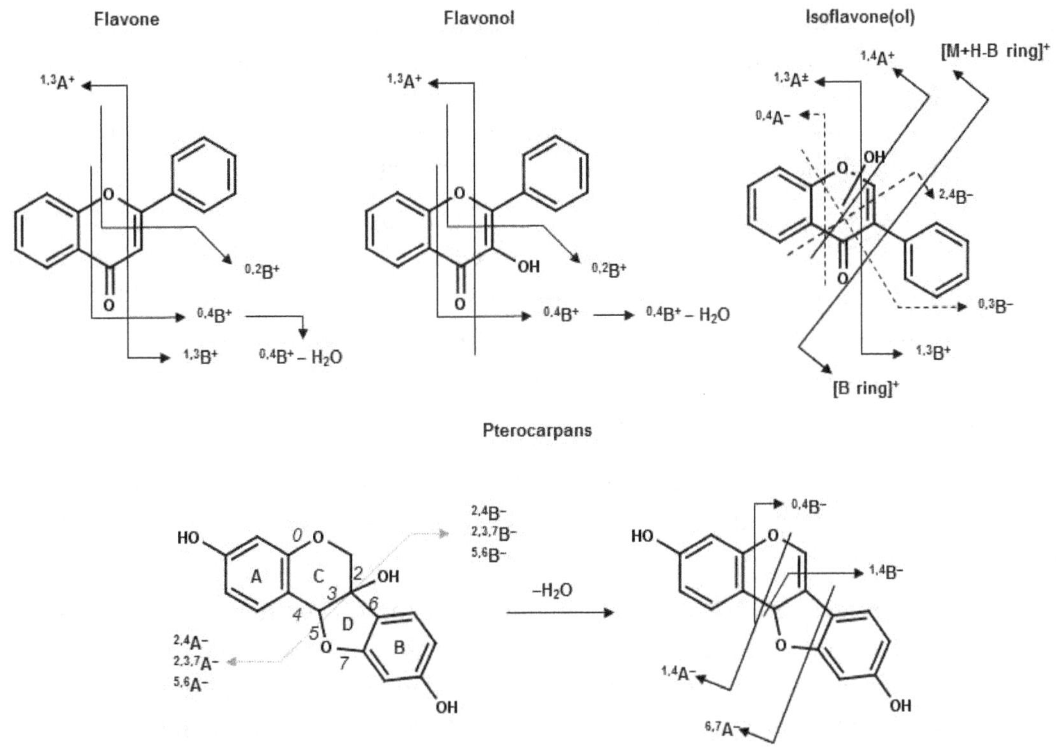

Fig. 3. Diagnostic products of flavones, flavonols, isoflavones, isoflavonols and pterocarpans. ± stands for either + or –. Adapted from Cuyckens et al, 2000; Kuhn et al., 2003; Madeira et al, 2010; Wei et al., 2000; Kang et al., 2007 and Simons et al., 2011.

3.6 Flavonoid-metal complexes

Flavonoids are good chelating agents towards metal ions and, in the case of iron and copper, the favoured places of chelation are catechol groups, hydroxyl groups adjacent to oxo groups, and 1-oxo-3-hydroxyl-containing moieties (Ren et al., 2008; Fernandez et al., 2002). This ability to chelate metals has been used to enhance the capabilities of MS; it has been used to assist the elucidation of flavonoid glucuronides (Davis et al., 2006; Davis & Brodbelt, 2007) and various diglycosides (Pikulski et al., 2007). This is a result of the spectral changes observed when flavonoids are complexed with metals, giving rise to simpler yet more intense spectra (Satterfield & Brodbelt, 2000).

4. NUCLEAR MAGNETIC RESONANCE SPECTROSCOPY

Nuclear Magnetic Resonance spectroscopy, hereafter simply designated by NMR, is one the most powerful research techniques used to investigate the structure and some properties of molecules. One of the main applications of NMR in flavonoid research is the structural elucidation of novel compounds, for which nothing is known; although NMR traditionally requires large amounts of sample, which is not easy to obtain when analysing novel compounds, the technical developments in the last decade, both in NMR instrumentation, pulse programs and in computing power, have allowed the complete assignment of all proton and carbon signals using amounts in the order of 1 mg (Fossen & Andersen, 2005).

The major goal of this section is to summarize the applications of the various NMR experiments to flavonoid research, together with the information (other than atom connectivity) that can be taken from these experiments. Flavonoid NMR data are not presented here, as these are easily accessible, for a vast number of compounds, from the literature (reviewed in Fossen & Andersen, 2005).

4.1 One Dimensional NMR: ^1H and ^{13}C

The two most basic NMR experiments are the 1H and the 13C NMR experiments, which are aimed at the determination of the resonance frequency of each 1H or 13C nucleus in the molecule.

^1H NMR experiments register the chemical shifts (δ) and spin-spin couplings, the latter described by the coupling constants (*J*). This provides valuable information about the relative number of hydrogens and also their type, by comparison of the recorded chemical shifts with compiled data. This is particularly useful in establishing the aglycone type and the acyl groups attached to it, as well as in identifying the number and the anomeric configuration of the glycoside moieties attached to the aglycone. ^{13}C NMR data is used to complement ^1H NMR data, and is particularly useful at establishing the type of groups present in the samples' molecules by comparison with compiled data; however, it must be noted that ^{13}C NMR is much less sensitive due to the abundance of ^{13}C (1.1 %) when compared to ^1H (99.9 %) (Claridge, 1999).

Together, these two 1D experiments are used primarily to identify aglycone types and substituent groups, but a definite structural elucidation, which the accurate location of the various groups, requires various 2D experiments.

4.2 Homonuclear 2D NMR

2D NMR experiments generate contour maps that show the correlations between different nuclei in the molecules, and can be either homonuclear or heteronuclear, depending on whether the interacting nuclei are of the same or different elements (Claridge, 1999). COSY (COrrelation Spectroscop Y) was one of the first multidimensional systems. COSY crosspeaks are between protons that are coupled to each other, usually two bonds apart ($^2J_{HH}$), but sometimes also three and four bonds apart ($^3J_{HH}$ and $^4J_{HH}$); the intensity of coupling affects the intensity of the peak. DQF-COSY (Double Quantum Filter COSY) is an improvement of the COSY experiment in which non-coupled proton signals, such as those from solvent, are eliminated as they may overlap signals from the analyte (Claridge, 1999). A further improvement is the TOCSY (TOtal Correlation Spectroscopy) experiment, which creates correlations between *all* protons in a given spin system, as long as there are couplings between every intervening protons; this is extremely useful to identify protons on sugar rings – every proton from one sugar ring will have a correlation with all other protons from the same ring but not with those of other rings. Magnetization is transferred over up to 5 or 6 bonds, and is interrupted by small or null ^{1}H-^{1}H couplings and hetero-atoms; also, the number of transfer steps can be adjusted by changing the spin-lock time (Fossen & Andersen, 2005). A good reference for the TOCSY transfer in various sugars is Gheysen's work (Gheysen et al., 2008). Selective 1D TOCSY (also known as HOHAHA, homonuclear Hartman-Hahn) is particularly useful in compounds with more than one sugar moiety, in which overlap occurs; in this experiment, one peak is selected and that magnetization is transferred stepwisely to the protons in the same spin system; instead of crosspeaks, transfer is shown by increased multiplet intensity (Fossen & Andersen, 2005).

4.3 Heteronuclear 2D NMR

Heteronuclear 2D NMR experiments correlate nuclei of different elements. The most powerful techniques of all are undoubtedly the 2D proton–carbon experiments HMQC/HSQC (Heteronuclear Multiple Quantum Coherence/Heteronuclear Single Quantum Coherence) and HMBC (Heteronuclear Multiple Bond Correlation) as they provide an opportunity to dovetail proton and carbon NMR data directly.

HMQC and HSQC establish one bond correlations between the protons of a molecule and the carbons to which they are attached ($^1J_{CH}$). Both these are much more sensitive than the correspondent 1D ^{13}C experiments; while in 1D experiment the low abundance of the isotope leads to a low signal-to-noise ratio, in the heteronuclear 2D experiment the initial magnetization occurs on the highly sensitive ^{1}H nuclei and is then transferred to the ^{13}C atoms that are connected to each proton. A similar NMR experiment is ^{1}H-^{13}C HMBC (Heteronuclear Multiple Bond Correlation), in which long-range interactions (typically $^2J_{CH}$ and $^3J_{CH}$) are analyzed; HMBC is usually more sensitive to 3-bond correlations than to 2-bond correlations, but this depends on the overall signal-to-noise ratios and on the adjustable parameters of each experiment. A newer experiment, $^2J_{CH}$,$^3J_{CH}$-HBMC, has been designed to differentiate these two types of correlations (Claridge, 1999; Krishnamurthy, 2000; Fossen & Andersen, 2005).

HBMC application to flavonoids usually addresses assignment on nonprotonated C atoms, from both the aglycones and acyl groups. Unlike TOCSY, HMBC transfer is not stopped by heteroatoms, and so it can also be used to determine the linkage points of heteroatom- containing groups such as sugar residues. HMBC is also useful to distinguish some classes of flavonoids, as flavones from aurones, which have similar ^{1}H and ^{13}C NMR spectra but very different HMBC spectra. Currently, only enhanced variants of the HSQC and HMBC experiments, namely gradient enhanced (*ge*) ones, are used, due to their higher sensitivity and capacity. These have been used to establish strong intramolecular H bonding between the 4-oxo and 5-hydroxy groups in flavonoids (Exarchou et al., 2002; Kozerski et al., 2003).

Further developments in NMR experiments, using new 2D and 3D techniques, have been developed in recent years, and are starting to be used for flavonoid analysis. In particular, 2D and 3D HSQC-TOCSY experiments are capable of assigning all ^{13}C signals of individual glycosides in polyglycosilated flavonoids (Fossen & Andersen, 2005).

4.4 Connectivity through space – the nuclear overhauser effect

While the above mentioned 2D techniques are useful to establish the connectivities between atoms through bonds, the Nuclear Overhauser Effect (NOE), which can be summarized as "A change in the intensity of an NMR signal from a nucleus, observed when a neighboring nucleus is saturated", is useful at establishing non-bonded connectivities, or connectivities through space. The crosspeaks in a ^{1}H-^{1}H NOESY (NOE SpectroscopY) spectrum correspond to correlations between protons that are close to each other in space (up to 4 Å) but not necessarily connected through bonds; these correlations may arise from both intramolecular and intermolecular proton interactions, and has been successfully used to establish rotational conformers and restrictions, establish intermolecular associations and even solve protein- ligand and DNA-ligand structures. A 2D NOE experiment, ROESY (Rotating Overhauser Effect SpectroscopY), has been used in flavonoid research mainly to establish the stereochemistry of various flavonoids (Claridge, 1999; Fossen & Andersen, 2005). Figure 4 presents sample NMR spectra of anthocyanidine glucosides (Jordheim et al., 2006).

4.5 Solid state NMR

X-ray crystallography depends on the ability to obtain flavonoid crystals, which has only been achieved for a small number of flavonoids; in alternative, the 1H CP-MAS (Cross Polarization Magic Angle Spinning) NMR techniques have been used to elucidate the solid state conformation of flavonoids, either pure or, for example, in tissues, providing enough sample is available (typically in the 10 mg scale). In particular, such an experiment had provided information on the planarity of the flavonoid rings, on the intramolecular H bonding between the 4-oxo group and the 3- and 5-hydroxy groups, and on intermolecular association. (Fossen & Andersen, 2005; Olejniczak & Potrzebowski, 2004)

5. UV-VIS SPECTROPHOTOMETRY

Ultraviolet and visible spectroscopy was one of the earliest techniques routinely used for flavonoid analysis due to the existence of two characteristic UV/Vis bands in flavonoids, band I in the 300 to 550 nm range, arising from the B ring, and band II in the 240 to 285 nm range, arising from the A ring. For examples, while the band I of flavones and flavonols lies in the 240 – 285 nm range, that of flavanone (no C ring instauration) lies in the 270 – 295 nm range; conversely, the band II of flavones and flavanones (no 3-OH group) lies around 303 – 304 nm, and that of 3-hydroxylated flavonoles is centred around 352 nm.

Shift reagents, such as sodium methoxide and aluminium chrolide, lead to shifts in the maximum wavelength of these bands due to methoxide-induced deprotonation of OH groups or Al^{3+} complexation by OH groups, were also routinely used to study flavonoid structure. Nowadays these techniques are not routinely used but still continue to be applied in some cases, in particular to HPLC eluates - a hyphenated LC-UV-MS has been developed using post-column UV shift reagents for the flavonoid analysis of crude extracts.

UV/Vis spectrophotometry is still widely used to study anthocyanidins, which change their form and colour depending on pH, concentration, metal ions and copigmentation (Giusti & Wrolstad, 2001). This multistate behaviour has been used to derive molecular machines based on flavonoids, particulary flavylium containing ones like anthocyanins (Melo et al., 2000; Moncada et al., 2004).

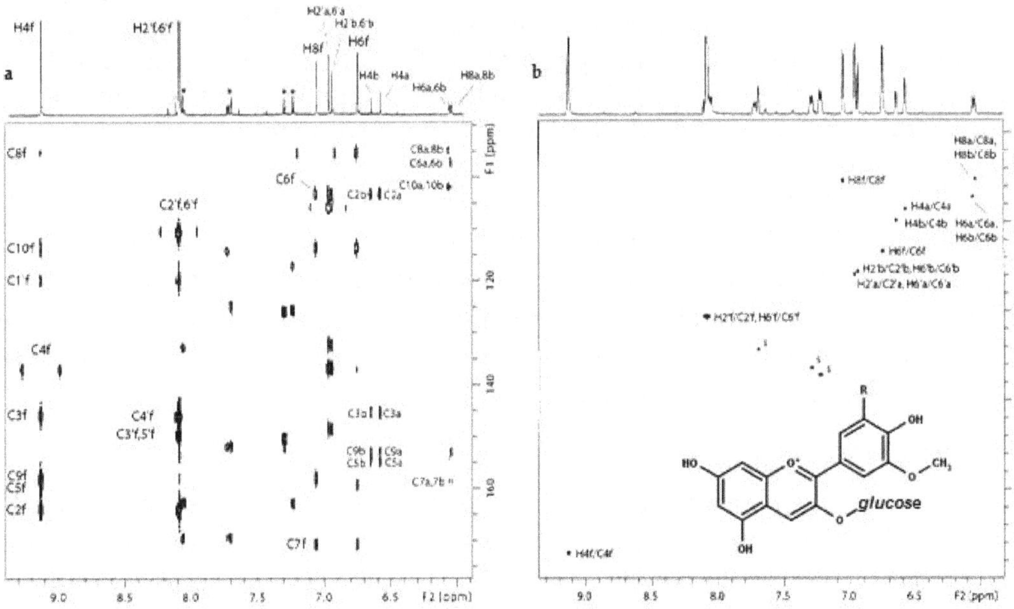

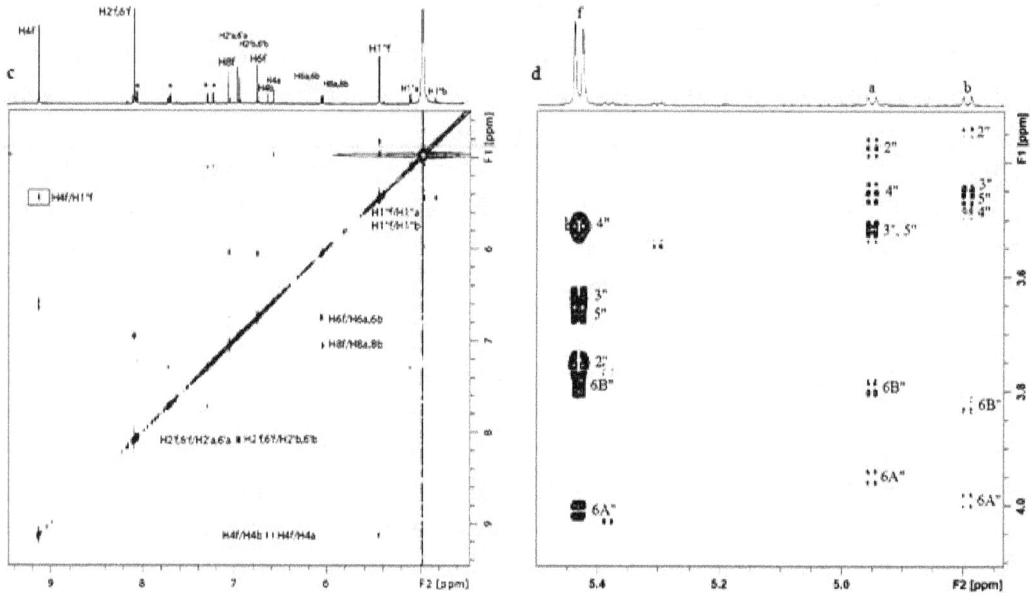

Fig. 4. Sample NMR 2D spectra of malvidin 3-*O*-β-glucopyranoside (R=OCH_3) (a, b and c) and of petunidin 3-*O*-β-glucopyranoside (R=OH) (d) obtained at 600.13 MHz, at about 11 mM at 25 ºC in CD_3OD. a) ^{1}H-^{13}C HMBC spectrum; b) ^{1}H-^{13}C HSQC spectrum; c) NOESY spectrum; A negative cross-peak due to NOE correlation between H-4 and H-1" of the flavylium cation, locating the position of the monosaccharide to the aglycone, is enclosed in a box. Other labelled cross-peaks are positive and are caused by chemical exchange; d) ^{1}H–^{1}H TOCSY NMR spectrum of petunidin 3-O-β-glucopyranoside. f = assignment for the flavylium; a = assignment for the hemiacetal a (major); b = assignment for the hemiacetal b (minor); */s = impurities. Adapted from Jordheim et al., 2006.

6. OTHER TECHNIQUES

MS, NMR and UV/Vis are the most commonly used techniques to elucidate the structure of flavonoids. Three other techniques are also used: X-ray crystallography, although with the potential to solve complete structures, is hampered by the difficulty to obtain good crystals; circular dichroism and vibrational spectroscopies are used to solve specific structural details.

6.1 X-ray crystallography

X-ray crystallography is able to detect the arrangement of atoms within a crystal by the atom-induced diffraction of X-rays. Many materials form crystals, such as salts, metals and organic and biological molecules, in particular proteins. Flavonoids, however, only form crystals in sporadic conditions, and the number of reported flavonoid crystals is very low (Fossen & Andersen, 2005).

Nevertheless, traditional X-ray crystallography has been used to identify the intermolecular π- π interactions that guide the stacking of parallel aglycones to form supramolecular layers, and to identify aminoacyl residues involved in the formation of protein-flavonoid complexes, which are critical to the circulation of flavonoids in mammals (Rolo-Naranjo et al., 2010); this is one of the strongest applications of X-ray crystallography to the flavonoid area.

More recently, X-ray powder diffraction has been used, either alone or in association with solid-state NMR, to obtain structures of flavonoids, in particular of catechins (Harper et al., 2010).

6.2 Circular dichroism spectroscopy

Circular dichroism (CD) is a spectroscopic technique that allows the analysis of the differential absorption of left and right circularly polarized light. The major advantage of CD over optical rotation measurements is that CD absorption is confined to the narrow absorption range of each individual chromophore, and so it can be used to determine the contribution of individual chromophores and to access their possible substitution patterns (Wallace & Janes, 2009).

One of the most immediate applications of electronic CD and vibrational CD is the determination of the absolute configuration of quiral flavonoid molecules, such as isoflavan- 4-ols (Kim et al., 2010). Slade et al., 2005, have reviewed the CD configuration characterization of most classes of flavonoids, and in particular of flavan-3,4-diols, which bear three quiral centers (Ferreira et al., 2004), and more recently proanthocyanidins analysis has also been reviewed, with a focus on CD results (Hümmer & Schreier, 2008).

CD is also routinely used to study the interaction of many flavonoids with biomolecules, providing valuable information on biomolecule-drug interaction, such as DNA binding of quercetin (Ahmadi et al., 2011), binding to serum albumin (di Bari et al., 2009) and hemoglobin (Chauduri et al., 2011), inhibition of β-amyloid toxicity and fibrillogenesis (Thapa et al., 2011). It must be noted that many of the CD studies of protein-flavonoid association studies are usually accompanied by UV-VIS and/or fluorescence studies, such as the probing of kaempferol interaction with human serum albumin (Matei & Hillebrand, 2010).

6.3 Vibrational spectroscopy

Vibrational spectroscopy, in its infra-red and Raman variants, is a spectroscopic technique that analyses the vibrational modes of molecules and molecular groups, allowing bond characterization, and, by comparison with known tabulated data, identification of functional groups; in the case of flavonoids, vibrational spectroscopy has been systematically used to study hydroxyl and carbonyl groups, but more recent technical developments have allowed its application to a broader set of research goals. Raman spectra are much less complex than the IR spectra of the same molecules, and for that reason Raman spectroscopy has been gradually taking over IR spectroscopy, although it is common to use both techniques as complement of each other. Vibrational spectroscopy is seldom used alone, and most studies are accompanied by other spectroscopic approaches and/or quantum chemical computations (Siebert & Hildebrandt, 2007).

Both these spectroscopies are routinely applied to study the effects of substituents on the geometry of the molecule, in particular of dihedral angles, and also on the analysis of intramolecular (Li et al., 2011; Erdogdu et al., 2010) and intermolecular H bonding, either to other flavonoid moleculer or to solvent molecules. Similarly, metal complexation by flavonoids is also routinely assessed by vibrational spectroscopy (O'Coinceanainn et al., 2004).

7. CONCISE SUMMARY

The function of flavonoids in biological systems remains inadequately defined, engaging numerous research teams globally. Notably, while numerous novel actions of flavonoids in vivo have been introduced, the previously established activities are not disregarded; rather, they are demoted to secondary mechanisms of flavonoid activity, typically seen as significant in pathological situations (Gomes et al., 2008).

Various physical-chemical approaches have been employed for the characterisation of flavonoids. The extensive research conducted since the 1940s has produced a substantial repository of structural and spectroscopic data regarding these compounds, facilitating the swift identification of new isolated species. Nonetheless, several constraints remain to be addressed. The maximum molecular size permitted in mass spectrometry (approximately 10 kDa) and in NMR spectroscopy (approximately 30 kDa) constrains the applicability of these methods in characterizing certain complex molecules and molecular complexes. Nonetheless, both techniques have facilitated significant advancements in structural elucidation that classical spectroscopic methods, such as UV/Vis and vibrational (Raman and infrared) spectroscopy or X-ray diffraction, could not achieve.

It is essential to recognize that, in some instances, information acquired using NMR or MS requires correlation with data from alternative structural analysis techniques, such as CD, to validate certain tenuous findings. Nonetheless, it has been established that MS and NMR are the most appropriate techniques for elucidating the chemical structure of flavonoids and their derivatives. NMR spectroscopy provides insights into atomic structures and interatomic bonding, whereas mass spectrometry yields data regarding molecule and ion/fragment masses, resulting in a more intricate and labor-intensive data analysis process. This issue can be resolved by the development of structural databases, facilitating more efficient and rapid data annotation for NMR spectroscopy. NMR possesses a significant advantage in biological studies: the capability to examine them in their native environments. Solid-state NMR facilitates the observation of flavonoid behavior in tissues, whereas solution NMR is advantageous for elucidating ligand-acceptor interactions through 2D-NMR techniques like NOESY, which is a premier method for investigating protein activity inhibition that underlies flavonoid biological activity.

It is reasonable to assert that, although mass spectrometry methods seldom yield complete molecular identifications, their inherent qualities render them the optimal technique for investigating flavonoid structures, complemented, where feasible, by NMR research. UV absorption data can be utilized to establish suitable methodologies for the preliminary identification of flavonoid classes in expedited, less refined screens.

8. REFERENCES

Ablajan, K. (2011) A study of characteristic fragmentation of isoflavonoids by using negative ion ESI-MSn. *Journal of Mass Spectrometry*, 46, 1, 77-84.

Ahmadi, S.M., Dehghan, G., Hosseinpurfeizi, M.A., Ezzati, J., Dolatabadi, N. & Kashanian, S. (2011) Preparation, Characterization, and DNA Binding Studies of Water-Soluble Quercetin–Molybdenum(VI) Complex. *DNA and Cell Biology*, 30, 7, 517-523.

Abrankó, L.; García-Reyes, J.F. & Molina-Diáz, A. (2011) In-source fragmentation and accurate mass analysis of multiclass flavonoid conjugates by electrospray ionization time-of- flight mass spectrometry. *Journal of Mass Spectrometry*, 46, 5, 478-488.

Aviram, M. & Fuhrman, B. (1998) LDL oxidation by arterial wall macrophages depends on the oxidative status in the lipoprotein and in the cells: role of prooxidants vs. antioxidants. *Molecular and Cellular Biochemistry*, 188, 1-2, 149-159.

Borges, C.; Martinho, P.; Martins, A.; Rauter, A.P. & Ferreira, M.A. (2001) Structural characterisation of flavonoids and flavonoid-O-glycosides extracted from *Genista tenera* by fast-atom bombardment tandem mass spectrometry. *Rapid Communications in Mass Spectrometry*, 15, 18, 1760-1767.

Brüll, L.; Huisman, M.; Schols, H.; Voragen, F.; Critchley, G.; Thomas-Oates, J. & Haverkamp, J. (1998) Rapid molecular mass and structural determination of plant cell wall-derived oligosaccharides using off-line high-performance anion-exchange chromatography/mass spectrometry. *Journal of Mass Spectrometry*, 33, 8, 713-720.

Charles, L.; Laure, F.; Raharivelomanana, P. & Bianchini, J.-P. (2005) Sheath liquid interface for the coupling of normal-phase liquid chromatography with electrospray mass spectrometry and its application to the analysis of neoflavonoids. *Journal of Mass Spectrometry*, 40, 1, 75-82.

Chauduri, S.; Chakraborty, S. & Sengupta, P.K. (2011) Probing the interactions of hemoglobin with antioxidant flavonoids via fluorescence spectroscopy and molecular modeling studies. *Biophysical Chemistry*, 154, 1, 26-34.

Claridge, T. (1999) *High-Resolution NMR Techniques in Organic Chemistry* (1st edition), Elsevier, ISBN 978-0080427997, UK.

Cole, R.B. (2010) *Electrospray and MALDI Mass Spectrometry: Fundamentals, Instrumentation, Practicalities, and Biological Applications* (2nd edition), Wiley, ISB 978-0471741077, Great Britain.

Cuyckens, F.; Ma, Y.L.; Pocsfalvi, G. & Claeys, M. (2000) Tandem mass spectral strategies for the structural characterization of flavonoid glycosides. *Analusis,* 28, 10, 888-895.

Cuyckens, F.; Rozenberg, R.; de Hoffmann, E. & Claeys, M. (2001) Structure characterization of flavonoid O-diglycosides by positive and negative nano-electrospray ionization ion trap mass spectrometry. *Journal of Mass Spectrometry,* 36, 11, 1203-1210.

Cuyckens, F. & Claeys, M. (2004) Mass spectrometry in the structural analysis of flavonoids. *Journal of Mass Spectrometry,* 39, 1, 1-15.

Cuyckens, F. & Claeys, M. (2005) Determination of the glycosylation site in flavonoid mono-O-glycosides by collision-induced dissociation of electrospray-generated deprotonated and sodiated molecules. *Journal of Mass Spectrometry,* 40, 3, 364-372.

Davis, B.D.; Needs, P.W.; Kroon, P.A. & Brodbelt, J.S. (2006) Identification of isomeric flavonoid glucuronides in urine and plasma by metal complexation and LC-ESI- MS/MS. *Journal of Mass Spectrometry,* 41, 7, 911-920.

Davis, B.D. & Brodbelt, J.S. (2007) Regioselectivity of human UDP-glucuronosyl-transferase 1A1 in the synthesis of flavonoid glucuronides determined by metal complexation and tandem mass spectrometry. *Journal of the American Society for Mass Spectrometry,* 19, 2, 246-256.

de Hoffmann, E. & Stroobant, V. (2007) *Mass Spectrometry – Principles and Applications* (3rd edition) Wiley, ISBN 978-0-470-03311-1, Great Britain.

di Bari, L., Ripoli, S., Pradhan, S., Salvadori, P. (2009) Interactions between quercetin and Warfarin for albumin binding: A new eye on food/drug interference. *Chirality,* 22, 6, 593-596.

Domon, B. & Costello, C.E. (1988) A Systematic Nomenclature for Carbohydrate Fragmentations in FAB-MS/MS of Glycoconugates. *Glycoconjugate Journal,* 5, 4, 397-409.

Erdogdu, Y.; Unsalan, O.; Sajan, D. & Gulluoglu, M.T. (2010) Structural conformations and vibrational spectral study of chloroflavone with density functional theoretical simulations. *Spectrochimica Acta Part A: Molecular and Biomolecular Spectroscopy,* 76, 2, 130-136.

Exarchou, V.; Troganis, A.; Gerothanassis, I.P.; Tsimidou, M. & Boskou, D. (2002) Do strong intramolecular hydrogen bonds persist in aqueous solution? Variable temperature gradient 1H, 1H–13C GE-HSQC and GE-HMBC NMR studies of flavonols and flavones in organic and aqueous mixtures. *Tetrahedron,* 58, 37, 7423-7429.

Fernandez, M.T.; Mira, M.L.; Florêncio, M.H. and Jennings, K.R. (2002) Iron and copper chelation by flavonoids: an electrospray mass spectrometry study. *Journal of Inorganic Biochemistry,* 92, 2, 105-111.

Ferreira, D., Marais, J.P.J., Slade, D. & Walker, L.A. (2004) Circular Dichroic Properties of Flavan-3,4-diols. *Journal of Natural Products,* 67, 2, 174-178.

Ferreres, F.; Gil-Izquierdo, A.; Andrade, P.B.; Valentão, P. & Tomás-Berberán, F.A. (2007) Characterization of C-glycosyl flavones O-glycosylated by liquid chromatography- tandem mass spectrometry. *Journal of Chromatography A,* 1161, 1-2, 214-223.

Fossen, T. & Andersen, ØM. (2005) Spectroscopic Techniques Applied to Flavonoids, In: *Flavonoids – Chemistry, Biochemistry and Applications*, Andersen ØM, Markham KR, (37-142) Taylor & Francis, 978-0-8493-2021-7, USA.

Fraga, C.G.; Galleano, M.; Verstraeten, S.V. & Oteiza, P.I. (2010) Basic biochemical mechanisms behind the health benefits of polyphenols. *Molecular Aspects of Medicine,* 31, 6, 435-445.

Gheysen, K.; Mihai, C.; Conrath, K. & Martins, J.C. (2008) Rapid Identification of Common Hexapyranose Monosaccharide Units by a Simple TOCSY Matching Approach. *Chemistry - A European Journal*, 14, 29, 8869-8878.

Giusti, M.M. & Wrolstad, R.E. (2001) Characterization and Measurement of Anthocyanins by UV-Visible Spectroscopy, In: *Current Protocols in Food Analytical Chemistry*, R.E. Wrolstad (F.1.2.1-F.1.2.13) John Wiley & Sons Inc, ISBN 0471325651, USA.

Gomes, A.; Fernandes, E.; Lima, J.L.F.C.; Mira, L. & Corvo, M.L. (2008) Molecular Mechanisms of Anti-Inflammatory Activity Mediated by Flavonoids, *Current Medicinal Chemistry*, 15, 1586-1605.

González, M.P.; Terán,C. & Teijeira, M. (2007) Search for new antagonist ligands for adenosine receptors from QSAR point of view. How close are we? *Medicinal Research Reviews*, 28, 3, 329-371.

Harper, J.K., Doebbler, J.A., Jacques, E., Grant, D.M. & von Dreele, R.B. (2010) A Combined Solid-State NMR and Synchrotron X-ray Diffraction Powder Study on the Structure of the Antioxidant (+)-Catechin 4.5-hydrate. *Journal of the American Chemical Society*, 132, 9, 2928-2937.

Heinonen, S.-M.; Hoikkalab, A.; Wähäläb, K. & Adlercreutza, H. (2003) Metabolism of the soy isoflavones daidzein, genistein and glycitein in human subjects: Identification of new metabolites having an intact isoflavonoid skeleton. *The Journal of Steroid Biochemistry and Molecular Biology*, 87, 4-5, 285-299.

Hulme, A.N.; McNab, H.; Peggie, D.A. & Quye, A. (2005) Negative ion electrospray mass spectrometry of neoflavonoids. *Phytochemistry*, 66, 23, 2766-2770.

Hümmer, W. & Schreier, P. (2008) Analysis of proanthocyanidins. *Molecular Nutrition & Food Research*, 52, 12, 1381-1398.

Jordheim, M.; Fossen, T. & Andersen, Ø.M. (2006) Characterization of Hemiacetal Forms of Anthocyanidin 3-O-β-Glycopyranosides. *Journal of Agricultural and Food Chemistry*, 54, 25, 9340-9346.

Justino, G.C.; Santos, M.R.; Canário, S.; Borges, C.; Florêncio, M.H. & Mira, L. (2004) Plasma quercetin metabolites: structure–antioxidant activity relationships. *Archives of Biochemistry and Biophysics*, 432, 1, 109-121.

Justino, G.C.; Borges, C. & Florêncio, M.H. (2009) Electrospray ionization tandem mass spectrometry fragmentation of protonated flavone and flavonol aglycones: a re- examination. *Rapid Communications in Mass Spectrometry*, 23, 2, 237-248.

Kang, J.; Hick, L.A. & Price, W.E. (2007) A fragmentation study of isoflavones in negative electrospray ionization by MSn ion trap mass spectrometry and triple quadrupole mass spectrometry. *Rapid Communications in Mass Spectrometry*, 21, 6, 857-868.

Kéki, S.; Tóth, K.; Zsuga, M.; Ferenczi, R. & Antus, S. (2007) (+)-Silybin, a pharmacologically active constituent of *Silybum marianum*: fragmentation studies by atmospheric pressure chemical ionization quadrupole time-of-flight tandem mass spectrometry. *Rapid Communications in Mass Spectrometry*, 21, 14, 2255-2262.

Kesari, A.N.; Gupta, R.K. & Watal, G. (2004) Two aurone glycosides from heartwood of Pterocarpus santalinus. *Phytochemistry*, 65, 23, 3125-3129.

Kim, M., Won, D., & Han, J. (2010) Absolute configuration determination of isoflavan-4-ol stereoisomers. *Bioorganic & Medicinal Chemistry Letters*, 20, 15, 4337-4341.

Kozerski, L.; Kamieński, B.; Kawecki, R.; Urbanczyk-Lipkowska, Z.; Bocian, W.; Bednarek, E.; Sitkowski, J.; Zakrzewska, K.; Nielsen, K.T. & Hansen, P.E. (2003) Solution and solid state ^{13}C

NMR and X-ray studies of genistein complexes with amines. Potential biological function of the C-7, C-5, and C4'-OH groups. *Organic and Biomolecular Chemistry*, 1, 20, 3578-3585.

Krishnamurthy, V. (2000) 2J,3J-HMBC: A New Long-Range Heteronuclear Shift Correlation Technique Capable of Differentiating 2JCH from 3JCH Correlations to Protonated Carbons. *Journal of Magnetic Resonance*, 146, 1, 232-239.

Kuhn, F.; Oehme, M.; Romero, F.; Abou-Mansour, E. & Tabacchi, R. (2003) Differentiation of isomeric flavone/isoflavone aglycones by MS2 ion trap mass spectrometry and a double neutral loss of CO. *Rapid Communications in Mass Spectrometry*, 17, 17, 1941-1949.

Lambert, J.D. & Elias, R.j (2010) The antioxidant and pro-oxidant activities of green tea polyphenols: a role in cancer prevention. *Archives of Biochemistry and Biophysics*, 501, 1, 65-72.

Li, Q.M. & Claeys, M. (1994) Characterization and differentiation of diglycosyl flavonoids by positive ion fast atom bombardment and tandem mass spectrometry. *Biological Mass Spectrometry*, 23, 7, 406–416.

Li, X.-H., Liu, X.-R.; Zhang, X.-Z. (2011) Molecular structure and vibrational spectra of three substituted 4-thioflavones by density functional theory and ab initio Hartree–Fock calculations. *Spectrochimica Acta Part A: Molecular and Biomolecular Spectroscopy*, 78, 1, 528-536.

Liu, R.; Ye, M.; Guo, H.; Bi, K. & Guo, D.A. (2005) Liquid chromatography/electrospray ionization mass spectrometry for the characterization of twenty-three flavonoids in the extract of Dalbergia odorifera. *Rapid Communications in Mass Spectrometry*, 19, 11, 1557-1165.

Ma, Y.L.; Li, Q.M.; van den Heuvel, H. & Claeys, M. (1997) Characterization of flavone and flavonol aglycones by collision-induced dissociation tandem mass spectrometry. *Rapid Communications in Mass Spectrometry*, 11, 12, 1357-1364.

Ma, Y.L.; Vedernikova, I.; van den Heuvel, H. & Claeys, M. (2000) Internal glucose residue loss in protonated O-diglycosyl flavonoids upon low-energy collision-induced dissociation. *Journal of the American Society of Mass Spectrometry*, 11, 2, 136-144.

Madeira, P.J.A. & Florêncio, M.H. (2009) Flavonoid-matrix cluster ions in MALDI mass spectrometry. *Journal of Mass Spectrometry*, 44, 7, 1105-1113.

Madeira, P.J.; Borges, C.M. & Florêncio, M.H. (2010) Electrospray ionization Fourier transform ion cyclotron resonance mass spectrometric and semi-empirical calculations study of five isoflavone aglycones. *Rapid Communications in Mass Spectrometry*, 24, 23, 3432-3440.

March, R. E.; Miao, X.-S.; Metcalfe, C. D.; Stobiecki, M. & Marczak, L. (2004) A fragmentation study of an isoflavone glycoside, genistein-7-O-glucoside, using electrospray quadrupole time-of-flight mass spectrometry at high mass resolution. *International Journal of Mass Spectrometry*, 232, 2, 171-183.

Markham, K. R. (1982) *Techniques of Flavonoid Identification* (1st edition), Academic Press, ISBN 978-0124726802, UK.

Matei, I. & Hillebrand, M. (2010) Interaction of kaempferol with human serum albumin: A fluorescence and circular dichroism study. *Journal of Pharmaceutical and Biomedical Analysis*, 51, 3, 768-773.

Melo, M.J.; Moura, S.; Roque, A.; Maestri, M. & Pina, F. (2000) Photochemistry of luteolinidin - ''Write-lock-read-unlock-erase'' with a natural compound. *Journal of Photochemistry and Photobiology A: Chemistry*, 135, 1, 33-39.

Moncada, M.C.; Fernandéz, D.; Lima, J.C.; Parola, J.; Lodeiro, C.; Folgosa, F.; Melo, J.M. & Pina, F. (2004) Multistate properties of 7-(N,N-diethylamino)-4'-hydroxyflavylium. An example of an unidirectional reaction cycle driven by pH. *Organic & Biomolecular Chemistry*, 2, 2802-2808.

Nikolic, D. & van Breemen, R.B. (2004) New metabolic pathways for flavanones catalyzed by rat liver microsomes. *Drug Metabolism & Disposition*, 32, 4, 387-397.

Nowakowska, Z. & Pankiewicz, P. (2008) *(E)*-4-Alkoxycarbonylalkylthiochalcones: differentiation of isomeric derivatives by electron ionization mass spectrometry. *Rapid Communications in Mass Spectrometry*, 22, 15, 2301-2306.

O'Coinceanainn, M.; Bonnely, S.; Baderschneider, B. & Hynes, M.J. (2004) Reaction of iron(III) with theaflavin : complexation and oxidative products. *Journal of Inorganic Biochemistry*, 98, 4, 657-663.

Olejniczak, S. & Potrzebowski, M.J. (2004) Solid state NMR studies and density functional theory (DFT) calculations of conformers of quercetin. *Organic and Biomolecular Chemistry*, 2, 16, 2315-2322.

Pikulski, M.; Aguilar, A. & Brodbelt, J. S. (2007) Tunable Transition Metal–Ligand Complexation for Enhanced Elucidation of Flavonoid Diglycosides by Electrospray Ionization Mass Spectrometry. *Journal of the American Society of Mass Spectrometry*, 18, 3,422-431

Prasain, J.K.; Wang, C.-C. & Barnes, S. (2004) Mass spectrometric methods for the determination of flavonoids in biological samples. *Free Radical Biology & Medicine*, 37, 9, 1324-1350.

Prasain, J.K. & Barnes, S. (2007) Metabolism and Bioavailability of Flavonoids in Chemoprevention: Current Analytical Strategies and Future Prospectus. *Molecular Pharmaceutics*, 4, 6, 846-864.

Qi, L. W.; Wen, X. D.; Cao, J.; Li, C. Y.; Li, P.; Yi, L.; Wang Y.-X.; Cheng X.-L. & Ge, X.X. (2008) Rapid and sensitive screening and characterization of phenolic acids, phthalides, saponins and isoflavonoids in Danggui Buxue Tang by rapid resolution liquid chromatography/diode-array detection coupled with time-of-flight mass spectrometry. *Rapid Communications in Mass Spectrometry*, 22, 16, 2493-2509.

Ren, J.; Meng, S.; Lekka, C. E. & Kaxiras, E. (2008) Complexation of Flavonoids with Iron: Structure and Optical Signatures. *The Journal of Physical Chemistry B*, 112, 6, 1845-1850.

Rolo-Naranjo, A., Codorniu-Hernández, E. & Ferro, N. (2010) Quantum Chemical Associations Ligand–Residue: Their Role to Predict Flavonoid Binding Sites in Proteins. *Journal of Chemical Information and Modelling*, 50, 5, 924-933.

Rusznyàk, St. & Szent-György, A. (1936) Vitamin P: Flavonols as Vitamins. *Nature*, 138, 3479, 27. Satterfield, M. & Brodbelt, J.S. (2000) Enhanced Detection of Flavonoids by Metal Complexation and Electrospray Ionization Mass Spectrometry. *Analytical Chemistry*, 2000, 72, 5898-5906.

Siebert, F. & Hildebrandt, P. (2007) *Vibrational Spectroscopy in Life Science* (1st edition), Wiley VCH, ISBN 978-3527405060, ISA.

Simons, R.; Vincken, J. P.; Bohin, M. C.; Kuijpers, T.F.; Verbruggen, M.A. & Gruppen, H. (2011) Identification of prenylated pterocarpans and other isoflavonoids in *Rhizopus* spp. elicited soya bean seedlings by electrospray ionisation mass spectrometry. *Rapid Communications in Mass Spectrometry*, 25, 1, 55.

Slade, D., Ferreira, D. & Marais, J.P. (2005) Circular dichroism, a powerful tool for the assessment of absolute configuration of flavonoids. *Phytochemistry*, 66, 18, 2177-2215.

Spencer, J.P.; Vauzour, D. & Rendeiro, C. (2009) Flavonoids and cognition: the molecular mechanisms underlying their behavioural effects. *Archives of Biochemistry and Biophyics*, 492, 1-2, 1-9.

Spencer, J.P. (2010) The impact of fruit flavonoids on memory and cognition. *British Journal of Nutrition*, 31, 6, 546-557.

Stobiecki, M. (2000) Application of mass spectrometry for identification and structural studies of flavonoid glycosides. *Phytochemistry*, 54, 3, 237-256.

Thapa, A., Woo, E.-R.; Chi, E. Y.; Sharoar, G.; Jin, H.-G.; Shin, S.Y.; Park, I.-S. (2011) Biflavonoids Are Superior to Monoflavonoids in Inhibiting Amyloid-β Toxicity and Fibrillogenesis via Accumulation of Nontoxic Oligomer-like Structures. *Biochemistry*, 50, 13, 2445-2455.

Tóth, E.; Dinya, Z. & Antus, S. (2000) Mass spectrometric studies of the pterocarpan skeleton. *Rapid Communications in Mass Spectrometry*, 14, 24, 2367-2372.

Valls, J.; Millán, S.; Martí, M.P.; Borràs, E. & Arola, L. (2009) Advanced separation methods of food anthocyanins, isoflavones and flavanols. *Journal of Chromatography A*, 1216, 43, 7143-7172.

Vessecchi, R.; Zocolo, G.J.; Gouvea, D.R.; Hübner, F.; Cramer, B.; de Marchi, M.R.R.; Humpf, H.U. & Lopes, N.P. (2011) Re-examination of the anion derivatives of isoflavones by radical fragmentation in negative electrospray ionization tandem mass spectrometry: experimental and computational studies. *Rapid Communications in Mass Spectrometry*, 25, 14, 2020-2026.

Volná, K.; Holčapek, M.; Kolářová, L.; Lemr, K.; Čáslavský, J.; Kačer, P.; Poustka, J. & Hubálek, M. (2007) Comparison of negative ion electrospray mass spectra measured by seven tandem mass analyzers towards library formation. *Rapid Communications in Mass Spectrometry*, 22, 2, 101-108.

Vukics, V. & Guttman, A. (2010) Structural Characterization of Flavonoid Glycosides by Multi-Stage Mass Spectrometry. *Mass Spectrometry Reviews*, 29, 1, 1-16.

Wallace, B.A. & Janes, R.W. (Eds.) (2009) *Modern Techniques for Circular Dichroism and Synchrotron Radiation Circular Dichroism Spectroscopy* (1st edition) IOS Press, ISBN 978-1607500001, USA.

Wei, J.; Liu, S.; Fedoreyev, S. A. & Voinov, V. G. (2000) A study of resonance electron capture ionization on a quadrupole tandem mass spectrometer. *Rapid Communications in Mass Spectrometry*, 14, 18, 1689-1694.

Xu, F.; Liu, Y.; Zhang, Z.; Yang, C. & Tian, Y. (2009) Quasi-MS^n identification of flavanone 7- glycoside isomers in *Da Chengqi Tang* by high performance liquid chromatography- tandem mass spectrometry. *Chinese Medicine*, 4, 15.

Zhang, L.; Zu, L.; Xiao, S.-S.; Liao, Q.-F.; Li, Q.; Liang, J.; Chen, X.-H- & Bi, K.-S. (2007) Characterization of flavonoids in the extract of *Sophora flavescens* Ait. by high- performance liquid chromatography coupled with diode-array detector and electrospray ionization mass spectrometry. *Journal of Pharmaceutical and Biomedical Analysis*, 44, 5, 1019-1028.

Zhang, Y.; Zhang, P. & Cheng, Y. (2008) Structural characterization of isoprenylated flavonoids from Kushen by electrospray ionization multistage tandem mass spectrometry. *Journal of Mass Spectrometry*, 43, 10, 1421-1431.

Zhou, D.y.; Xu, Q.; Xue, X.-y.; Zhang, F.-f.; Jing, Y. & Liang, X.-m. (2007) Rapid qualitative and quantitative analyses of flavanone aglycones in *Fructus aurantii* by HPLC ion- trap MS. *Journal of Separation Science*, 30, 6, 858-867.

Chapter 3

ANALYTICAL TECHNIQUES FOR THE ISOLATION, SEPARATION, AND IDENTIFICATION OF SPECIFIC FURANOCOUMARINS IN BOTANICAL SAMPLES

1. INTRODUCTION: AN OVERVIEW

Coumarins are α-pyrone derivatives produced as secondary metabolites in plants. They exist as free chemicals or glycosides within plants. They were isolated from A. Vogel in 1820 from tonka beans (Coumarouna odorata Aubl. = Dipteryx odorata Will.) and synthesized in 1868 by W. H. Perkin via the renowned Perkin process (Dewick, 2009).

Furanocoumarins are derivatives of coumarin. They can be categorized into the linear type, characterized by the attachment of the furan ring (dihydro) at C(6) and C(7), and the angular type, which features substitution at C(7) and C(8).

The predominant linear furanocoumarins are psolaren, xanthotoxin, bergapten, and isopimpinellin, whereas the angular form is primarily represented by angelicin, sphondin, and pimpinellin. Table 1 presents various structures of furanocoumarins. Numerous minor furanocoumarins, such as bergamottin (5-geranoxy-psolaren) (Stanley & Vannier, 1967), have been documented in the literature, particularly due to their role as significant grapefruit constituents that disrupt drug metabolism via intestinal CYP3A4. (Bourgaud et al., 2006; Wen et al., 2008).

2. DISTRIBUTION OF FURANOCOUMARINS IN PLANTS

Linear furanocoumarins (syn. psolarens) are principally distributed in four angiosperm families: Apiaceae (Umbelliferae), Moraceae *(Brosimum, Dorstenia,*

Fatoua and *Ficus),* Rutaceae and Leguminose (restricted to Psoralea and Coronilla generae). The angular (dihydro) furanocoumarins are less widely distributed and primarily confined to the Apiaceae and Leguminosae (Berenbaum et al., 1991; Bourgaud et al., 1995).

Moreover, furanocoumarins have been reported from Asteraceae (Compositae), Pittosporaceae, Rosaceae, Solanaceae and Thymelaeaceae (Milesi et al., 2001; Murray et al., 1982). Certain precursors to this group of compounds are found in the Cneoraceae (Murray, 1982).

Table 1. The structure of some furanocoumarins

Name of compound	Structure
Psoralen	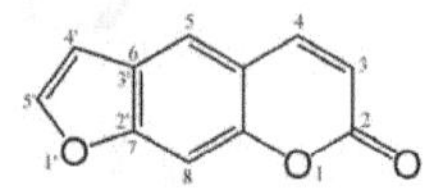
Phellopterin	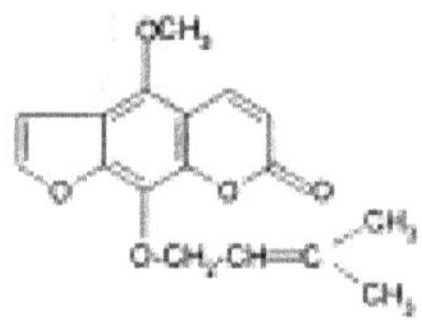
Xanthotoxin	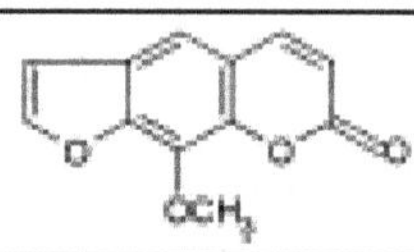
Isopimpinellin	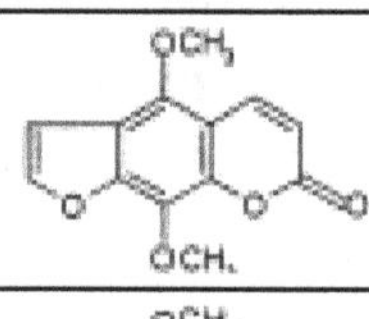
Bergapten	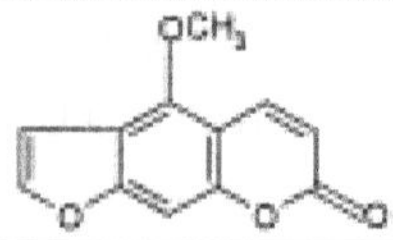
Pimpinellin	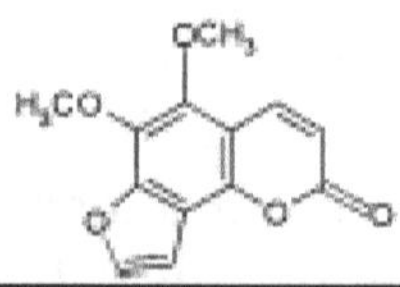
Imperatorin	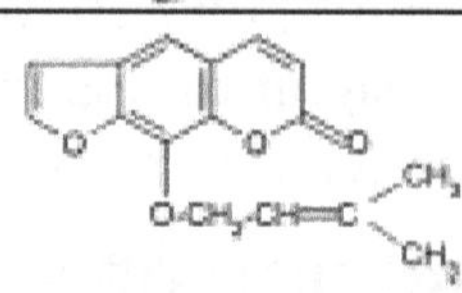
Angelicin	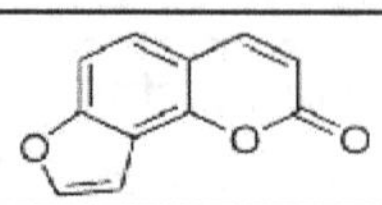

Coumarins are distributed across different parts of the plants, and they have specific histological locations in the tissues. Within the plant they are most abundant in fruits and roots. However, in flowers and leaves they are evident in fewer quantities. In some plant species coumarins were also found in the bark or stems (Głowniak, 1988).

The amount of particular furanocoumarins depends on the enzymes active in plants secondary metabolism. Plants with similar enzyme profiles contain comparable amount of secondary metabolites that are products of chemical reactions induced by these enzymes. Thus, furanocoumarins' content, in different species, varieties and forms may contribute to their better distinction, and better understanding of the taxonomy of genuses within which they are present.

Diawara et al. (1995) examined the relative distribution of furocoumarins in celery (*Apium graveolens* L. var. *dulce* Miller) plant parts and found that leaves of the outer petioles contained significantly higher levels of the three phototoxic constituents than did other plant parts, followed by leaves of the inner petioles.

On the other hand, levels of furanocoumarins observed in plants grown in the field are higher than those observed in plants grown in laboratory or greenhouse conditions and may fluctuate over the season (Trumbe et al., 1992; Diawara et al., 1995). In most studies, bergapten has been found to occur in highest concentrations, followed by xanthotoxin, but psoralen is often observed only in trace quantities (Trumbe et al., 1990; Trumbe et al., 1992; Diawara et al., 1993). However, other studies have found that xanthotoxin (Beier et al., 1983) or psoralen (Diawara et al., 1993; Trumble et al., 1990) is most abundant (Stanley-Horn, 1999).

Considering the histological location of furanocoumarins in plant tissues, they are arranged differently. For example, celery contains schizogenous canals scattered throughout the pericycle, which are secretory and are thought to extend through the stem and foliage (Maksymowych & Ledbetter, 1986). Furanocoumarins are thought to be restricted to schizogenous canals in seeds of celery (Berenbaum, 1991) and accumulate primarily in petiolar and foliar canals in cow-parsnip, *Heracleum lanatum* Michx. (Apiaceae). However, there is also evidence suggesting that this group of compounds occur in and on the surfaces of tissues as well. A study of several apiaceous and rutaoeous species by Zobel and Brown (1990) revealed that a large proportion of each furocoumarin was located on the leaf surface in most of the plants studied. Furanocoumarins of *Ruta graveolens* L. are present in the epidermal layer of both stems and leaves and in the mesophyll directly below the epidermis, while glands of leaves contain only traces of furanocoumarins. In fact, the cuticular layer contains 15.60% of the psoralens found in leaves (Zobel et al., 1989). The occurrence of bergapten and xanthotoxin in the surface wax of leaves of wild carrot, *Daucus carota* L., a plant containing only trace levels of furanocoumarins has also been reported (Ceska et al., 1986; Stanley-Horn, 1999).

The content of coumarins in plants is conditioned by the degree of the development of the plant and its vegetation stage, too. Concentrations of linear furanocoumarins increase dramatically with plant age between 8 and 18 weeks (Reitz et al., 1997) with a subsequent decline in bergapten concentrations in the last six to eight weeks before harvest (Trumble et al., 1992). Significant decreases in levels of furanocoumarins were also observed both in and on senescing leaves of *Ruta graveolens* (Zobel & Brown, 1991). The content of some furanocoumarins in *Apium graveolens* and *Petroselinum sativum* decreases in summer and in autumn increases (Kohlmünzer, 2010).

3. BIOSYNTHESIS OF FURANOCOUMARINS

The biosynthesis of linear and angular furanocoumarins is still poorly understood at the molecular level. They are produced *via* the shikimic acid biosynthetic pathway beginning with the conversion of phenyloalanine to trans-cinnamic acid by phenylalanine ammonia lyase. Orto-hydroxylation of trans-cinnamic acid yields 2'-hydroxycinnamic acid, which is converted to its cis form, the precursor to coumarin, in the presence of UV light. Alternatively, trans-cinnamic acid may undergo parahydroxylation to yield p-coumaric acid. P-coumaric acid may undergo 2'–hydroxylation followed by conversion by 4- coumarate: CoAligase to 4-coumaryl CoA. This compound is intermediate in the biosynthesis of both flavonoids and phenylpropanoids, including 7-hydroxycoumarin (umbelliferone). Umbelliferone is the precursor to both the angular and linear furanocoumarins. The production of the latter involves prenylation to form marmesin, followed by oxidative loss of the hydroxypropyl group in marmesin by 'psoralensynthase' to yield psoralen (Berenbaum, 1991; Stanley-Horn, 1999). A second cytochrome P-450- dependent monooxygenase enzyme then cleaves off the hydroxyisopropyl fragment (as acetone) from marmesin to give the furocoumarin psoralen. Psoralen can act as a precursor for the further substituted furanocoumarins bergapten, xanthotoxin and isopimpinellin. On the other hand, angular furanocoumarins, such an angelicin can arise by a similar sequence of reactions, but these involve initial dimethylallylation at the alternative position *ortho* to the phenol (Dewick, 2009).

4. BIOLOGICAL ACTIVITIES OF FURANOCOUMARINS

Due to their biological activities, furanocoumarins are very interesting compounds and widely investigated. The various biological and pharmacological activities of coumarins, have been known for a long time.

They play the role of phytoalexin in plants (Szakiel, 1991), which can be synthesized as a result of elicitation by microorganisms, insects, fungi as well as abiotic elicitors such as UV radiation, environment pollutants and mechanical breakage. Defensive activity of furanocoumarins consists in their toxicity against phytopathogens (e.g. retardation of DNA synthesis) (Waksmundzka-Hajnos et al., 2004).

Linear furocoumarins can be troublesome to humans since they can cause photosensitization towards UV light, resulting in sunburn or serious blistering. Used medicinally, this effect may be valuable in promoting skin pigmentation and treating psoriasis. Plants containing psoralens have been used internally and externally to promote skin pigmentation and suntanning. Bergamot oil obtained from the peel of *Citrus aurantium* ssp. *bergamia* (Rutaceae) can contain up to 5% bergapten and is frequently used in external suntan preparations. The psoralen absorbs in near UV light and allows this radiation to stimulate formation of melanin pigments (Dewick, 2009).

Methoxsalen (xanthotoxin; 8-methoxypsoralen), a constituent of the fruit of *Ammi majus* (Umbelliferae/Apiaceae), is used medically to facilitate skin repigmentation where severe blemishes exist (vitiligo). An oral dose of methoxsalen is followed by long – wave UV irradiation, though such treatments must be very carefully regulated to minimize the risk of burning, cataract formation, and the possibility of causing skin cancer. The treatment is often referred to as PUVA (psoralen + UV-A). PUVA is also of value in the treatment of psoriasis, a widespread condition characterized by proliferation of skin cells. Similarly, methoxsalen is taken orally, prior to UV treatment. Reaction with psoralens inhibits DNA replication and reduces the rate of cell division. Because of their planar nature, psoralens intercalate into DNA, and this enables a UV – initiated cycloaddition reaction between pyrimidine bases (primarily thymine) in DNA and the furan ring of psoralens. A second cycloaddition can then occur, this time involving the pyrone ring, leading to interstrand cross – linking of the nucleic acid (Dewick, 2009; Żołek et al., 2003).

A troublesome extension of these effects can arise from the handling of plants which contain significant levels of furocoumarins. *Apium graveolens* (= celery; Umbelliferae/Apiaceae) is normally free of such compounds, but fungal infection with the natural parasite *Sclerotinia sclerotiorum* induces the synthesis of furanocoumarins as a response to the infection. Field workers handling these infected plants may become very sensitive to UV light and suffer from a form of sunburn termed photophytodermatitis. Infected parsley (*Petroselinum crispum*) can give similar effects. Handling of rue (*Ruta graveolens;* Rutaceae) or giant hogweed (*Heracleum mantegazzianum*; Umbelliferae/Apiaceae), which naturally contain significant amounts of psoralen, bergapten, and xanthotoxin, can cause similar unpleasant reactions, or more commonly rapid blistering by direct contact with the sap. The giant hogweed can be particularly dangerous. Individuals vary in their sensitivity towards furanocoumarins, some are unaffected, whilst others tend to become sensitized by an initial exposure and then develop the allergic response on subsequent exposures (Dewick, 2009).

Methoxsalen in combination with ultraviolet light is also used for antineoplastic effects and for treating certain skin disorders, including alopecia, cutaneos T-cell lymphoma, excema, lichen planus, mycosis fungoides and psoriaris. A recent report has found that this drug inhibits the enzyme, CYP2A6, which is responsible for the metabolism of nicotine. When 8- methoxypsoralen is taken with oral nicotine, this drug can reduce the number of cigarettes smoked by about one quarter and decrease overall levels of tobacco smoke exposure by almost half in tobacco dependent individuals (Lehr et al., 2003).

Xanthotoxin is used orally or topically in combination with controlled exposure to long wavelength ultraviolet radiation (UVA) or sunlight to repigment vitiliginous skin in patients with idiopathic vitiligo. Many studies have shown that naturally occurring furocoumarins, e.g. imperatorin and isopimpinellin, inhibit P450-mediated enzyme activities *in vitro*. Imperatorin and isopimpinellin have also the potential chemopreventive effects when administered in the diet. The stimulation of melanogenesis by bergapten is related to increased tyrosinase synthesis. In addition, bergapten stimulated TRP-1 synthesis and induced a dose-dependent decrease of DCT activity without modification of protein expression. Osthole could prevent postmenopausal osteoporosis. It can also delay aging, build up strength, enhance immune function, and adjust sex hormone levels (Chen et al., 2007).

Psoralen and bergapten exert their photosensibilising effects through a covalent interaction with DNA triggered by light of a specific wavelength (320-400 nm). The resulting complex blocks the DNA interaction with transcriptases and polymerases, avoiding cell replication. This mechanism consist of three steps, i.e., (1) drug intercalation between DNA nucleotide bases, (2) drug absorption of a UVA photon and covalent bond formation between the furan ring double bond and a thymine base (T2) of the DNA molecule, (3) absorption of a second photon (UVA) and covalent bonding between the lactone ring double bond and another thymine base (T1), which, in the end, results in a psoralen cross-linked DNA (da Silva et al., 2009; Panno et al., 2010; Cardoso et al., 2002). The same effects have been alternatively utilized for the treatment of human lymphoma and of autoimmune diseases through extracorporeal photochemotherapy (Panno et al., 2010).

Panno et al. (2010) investigated the pro-apoptotic effects induced by high doses of bergapten (methoxypsoralen; 5-MOP), in the absence of UV rays, in human breast cancer cells. The same authors examined the effects of bergapten, alone and in combination with UV light, on the cellular growth of breast tumoral cells. Their study suggested that bergapten alone, or as a photoactivated product, could be used as an active molecule able to counteract effectively the survival and growth of breast hormone-responsive tumors.

Furanocoumarins isolated from fruits of *Heracleum sibiricum* L. inducing apoptosis by forming adducts with DNA. Bogucka – Kocka (2004) reported a visible influence of these compounds on the inhibition of the proliferation and on induction of apoptosis processes in the human HL-60 cell lines. Moreover, compounds isolated from *Angelica dahurica* (Apiaceae) were examined regarding their cytotoxic activity against L1210, HL-60, K562, and B16F10 tumor cells lines using the MMT cell assay. It was discovered that pangelin and oxypeucedanin hydrate acetonide exhibited the most cytotoxic activity against all selected tumor cell lines (Heinrich et al., 2004). Um and co-authors (2010) were isolated four furonocoumarins (bergapten, isopimpinellin, xanthotoxin and imperatorin) from *Glehnia littoralis* F. Schmidt ex Miquel (Apiaceae), which exhibited dose-dependent inhibitory effectson the cell proliferation. Their study demonstrated that *G. littoralis* has potent inhibitory effect on proliferation of HT-29 human colon cancer cells.

In addition, Oxypeucedanine (= prangolarin), which was isolated from *Prangos, Hippomarathrum, Angelica* and *Ferulago* (genera of Apiaceae) and *Ruta* genus of Rutaceae, has pharmacological and biological activities. It was reported to have antiarrhytmic, channel blocer and antiestrogenic activity. Razavi et al. (2010) studied phytotoxic, antibacterial, antifungal, antioxidant and cytotoxic effects of oxypeucedanin. Their results revealed that this compound exhibits considerable phytotoxic activity and might play an allelopathic role for plants. On the other hand, oxypeucedanin exhibits considerable cytotoxicity against Hela cell line (IC_{50} value of 314 μg/ml).

The ethanol extract of the *Cnidii* fructus and coumarins separated from it have growth- inhibitory effects on the tumor cells (Chen et al., 2007).

One of the major bioactive components of the fruits of *Cnidium monnieri* (L.) Cusson, bergapten, possesses antiinflammatory and analgesic activities. However, imperatorin exhibits strong cytotoxic activity on human leukemia, chemopreventive effects on hepatitis and skin tumor, and antiinflammatory activity (Li & Chen, 2004).

In addition of bergapten, this plant also contained numbers of others coumarins, such as xanthotoxin, isopimpinellin, bergapten, imperatorin and osthole. These constituents regarded for biological activity of this crude drug, which is used for treatment of pain in female genitalia, impotence and supportive (Chen et al., 2007).

Pharmacological studies have indicated that coumarins such as isoimperatorin, notopterol and bergapten possess anti-inflammatory, analgesic, anti-cancer and anti-coagulant activities (Qian et al., 2007).

Imperatorin (8-isopentenyloxypsoralen; 9-(3-methylbut-2-enyloxy)-7H-furo [3,2-g]chromen- 7-one) is a bio-active furanocoumarin isolated e.g. from roots of *Angelica dahurica* and fruits of *Angelica archangelica* (Umbelliferae) (Baek et al., 2000). Experimental evidence indicates that imperatorin irreversibly inactivates γ-aminobutyric acid (GABA)-transaminase (the enzyme responsible for the degradation of GABA) and thus, increases the GABA content in the synaptic clefts of neurons and elevates the inhibitory neurotransmitter GABA level in the brain (Choi et al., 2005; Łuszczki et al., 2007).

Quite recently, it has been documented that imperatorin in a dose-dependent manner increased the threshold for maximal electro-convulsions in mice (Łuszczki et al., 2007a). The time–course and dose–response relationship analyses revealed that the time to peak of the maximum anticonvulsant effect for imperatorin was established at 30 min after its systemie (i.p.) administration in mice (Łuszczki et al., 2007a).

Recently results indicate that imperatorin administered at subthreshold doses enhanced the anticonvulsant effects of carbamazepine, phenytoin and phenobarbital, but not those of valproate against maximal electroshock-induced seizures in mice. It is important to note that the anti-seizure effects of carbamazepine combined with imperatorin were greater than those observed for the combinations of phenobarbital and phenytoin with imperatorin.

The difference in the anti-seizure effects of carbamazepine and phenytoin or phenobarbital in the maximal electroshock seizure test may be explained through pharmacokinetic interaction between imperatorin and carbamazepine. It was found that imperatorin significantly increased total brain carbamazepine concentrations, having no impact on the total brain phenytoin and phenobarbital concentrations in experimental animals. The selectivity in the increase in total brain carbamazepine concentration one can try to explain through the fact that imperatorin probably enhances the penetration of carbamazepine into the brain by modifying the blood–brain barrier permeability. On the other hand, it may be hypothesized that the selective increase in carbamazepine content in the brain tissue resulted from imperatorin-induced inhibition of multi-drug resistance proteins or P- glycoproteins that normal physiological activity is related to the removal of drugs from the brain tissue. Thus, inhibitors of these proteins may contribute to the accumulation of antiepileptic drugs in the brain (Brandt et al., 2006; Łuszczki et al., 2007).

Considering molecular mechanisms of the action of conventional antiepileptic drugs and imperatorin, one can ascertain that imperatorin-induced irreversible inactivation of GABA- transaminase and subsequent increases in GABA content in the brain, as well as, the enhanced GABA-mediated inhibitory neurotransmitter action through the interaction of imperatorin with benzodiazepine receptors. This may exhibit complementary potentials to the anticonvulsant activity of carbamazepine, phenytoin and phenobarbital shown in experimental animals testing. Noteworthy, the main anticonvulsant mechanism of the action of carbamazepine and phenytoin is related to the blockade of Na+ channels in certain neurons (Łuszczki et al., 2007).

Considering molecular mechanisms of the action of conventional antiepileptic drugs and imperatorin, one can ascertain that imperatorin-induced irreversible inactivation of GABA- transaminase and subsequent increases in GABA content in the brain, as well as, the enhanced GABA-mediated inhibitory neurotransmitter action through the interaction of imperatorin with benzodiazepine receptors. This may exhibit complementary potentials to the anticonvulsant activity of carbamazepine, phenytoin and phenobarbital shown in experimental animals testing. Noteworthy, the main anticonvulsant mechanism of the action of carbamazepine and phenytoin is related to the blockade of Na+ channels in certain neurons (Łuszczki et al., 2007).

It is interesting to note that imperatorin did not potentiate the protective action of valproate against maximal electroshock-induced seizures. This apparent lack of effects of imperatorin on the antiseizure action of valproate, one can try to explain by the fact that valproate possesses a number of variousmechanisms of action that contribute to its anti-seizure activity in both rodents and humans (Łuszczki et al., 2007).

The evaluation of acute adverse effect potentials is exhibited within combinations of imperatorin with conventional antiepileptic drugs revealing that the combinations did not disturb long-term memory, impair motor co-ordination, or change neuromuscular grip- strength in experimental animals. Therefore, the investigated combinations seem to be secure and well tolerated by experimental animals (Łuszczki et al., 2007).

It was shown that imperatorin enhances the protective action of carbamazepine, phenytoin and phenobarbital, but not that of valproate against maxima electroshock-induced seizures in mice. The lack of any changes in total brain phenytoin and phenobarbital concentrations suggested that the observed interactions of imperatorin with phenytoin and phenobarbital were pharmacodynamic in nature and thus, they deserve more attention from a preclinical viewpoint. If the results from the study of Łuszczki and co-authors (2007) can be extrapolated to clinical settings, a novel therapeutic option in the management of epilepsy may be created for epileptic patients.

Piao et al. (2004) assayed eleven furanocoumarins, isolated from *Angelica dahuricae* to determine its antioxidant activities. 9-hydroxy-4-methoxypsoralen inhibited DPPH formation by 50% at a concentration of 6.1 µg/ml (IC_{50}), and alloisoimperatorin 9.4 µg/ml, thus the other nine furanocoumarins (oxypeucedanin hydrate, byakangelicol, pabulenol, neobyakangelicol, byakangelicin, oxypeucedanin, imperatorin, phellotorin, and isoimperatorin), with an IC_{50} values higher than 200 µg/ml, showed only a little DPPH radical-scavenging activities.

Tosun et al. (2008) evaluated the anticonvulsant activity of the furanocoumarins among others compounds, obtained from the fruits of *Heracleum crenatifolium.* This activity was estimated against maximal electroshock seizures induced in mice. Among analyzed compounds, bergapten showed significant anticonvulsant activity.

Osthole, a coumarin derivative extracted from many plants, such as *Cnidium monnieri* and *Angelica pubescens,* has been showed to exhibit estrogen-like effects and prevent postmenopausal osteoporosis in overiectomized rats. The latest research suggested that this compound can alleviate hyperglycemia and could be potentially developed into a novel drug for treatment of diabetes mellitus (Liang et al., 2009).

Tang et al. (2008) have reported that imperatorin and bergapten induce osteoblast differentiation and maturation in primary osteoblasts. These compounds increased also BMP-2 (bone morphogenetic protein type 2) expression via p38 and ERK-dependent (extracellular signal-regulated protein) pathways. Long–term administration of imperatorin and bergapten into the tibia of young rats also increased the protein level of BMP-2 and bone volume of secondary spongiosa.

However, the toxic effects of furanocoumarins are also well known. Da Silva et al. (2009) at computational analysis of psoralen, bergapten and their predicted metabolites revealed the presence of six toxicophoric groups related to carcinogenicity, mutagenicity, photoallergenicity, hepatotoxicity and skin sensitization.

Numerous studies have indicated that furanocoumarins are carcinogenic, and their ability to intercalate into DNA in the presence of long wave UV light accounts for their mutagenicity. Linear furocoumarins have been shown to exhibit varying levels of phototoxicity. It must be stated that with isopimpinellin, it results in having the least photosensitizing activity (Lehr et al., 2003).

Moreover, coumarin derivatives in high doses can produce significant side effects. They may induce headaches, nausea, vomiting, sleepiness, and in extreme cases, serious liver damage with potential hemorrhages as a result of hypoprothrominemia (Lozhkin & Sakanyan, 2006).

5. ANALYTICAL METHODS OF FURANOCOUMARINS ISOLATION

5.1 Extraction from plant material

As furanocoumarins have wide applications in biology and have many therapeutic activities, the study of isolation and identification of these compounds is very important. In this part of our work, review of possible methods of isolation of furanocoumarins will be described as follows.

Coumarins typically appear as colorless or yellow crystalline substances, well soluble in organic solvents (chloroform, diethyl ether, ethyl alcohol), as well as in fats and fatty oils. Coumarin and its derivatives exhibit sublimation on heating to 100°C (Lozhkin & Sakanyan, 2006).

In this process of quantitative analysis of plant secondary metabolites, preliminary treatment of the plant materials is one of the most time-consuming steps. The first problem is the extraction of the compounds from the plant material - usually performed by liquid – solid extraction (LSE).

In research of the content of pharmacologically active compounds in medicinal plants, the routine procedure of extraction from plant tissues is usually applied. The extraction from plant material is frequently carried out by means of "classic" solvent-based procedures, in Soxhlet apparatus, or more simply, in laboratory flask at the temperature of the solvent's boiling under reflux (de Castro & da Silva, 1997; Saim et al., 1997). The imperfection of the time and solvent-consuming methods consists of poor penetration of the tissues by the solvent and also possible destruction of thermolabile compounds. Advantages of conventional extraction methods result from basic, inexpensive and simple equipment to operate. In the Soxhlet extraction, the sample is repeatedly contacted with fresh portions of the solvent in relatively high temperature and with no filtration required after the leaching step (de Castro & da Silva, 1997; de Castro & Garcia-Ayuso, 1998). Recently, modern alternative extraction methods, applied in the environmental analysis and in phyto- chemistry, are sometimes reported: (1) ultrasonification (USAE) (maceration in ultrasonic bath at various temperatures) (de Castro & Garcia-Ayuso, 1998; Court et al. 1996; Saim et al., 1997); (2) microwave-assisted solvent extraction in closed and open systems (MASE) (de Castro & Garcia-Ayuso, 1998; Saim et al., 1997); (3) accelerated solvent extraction (ASE) (called also PLE, pressurized solvent extraction) (Boselli et al., 2001; de Castro & Garcia- Ayuso, 1998; Ong et al., 2000; Papagiannopoulos et al., 2002; Saim et al., 1997); and (4) supercritical fluid extraction (SFE) (Saim et al., 1997). The above methods give better penetration of solvents into plant tissues or other solid matrices that are rapid and solvent saving. ASE apart from this advantage is dynamic, fast and also enables automatization of extraction and analysis procedures (Waksmundzka-Hajnos et al., 2004; Waksmundzka- Hajnos et al., 2007).

Coumarins are usually isolated from plants by extraction with solvents such as ethanol, methanol, benzene, chloroform, diethyl and petroleum ethers, or their combinations (Lozhkin & Sakanyan, 2006). The most exhaustive extraction of coumarins is achieved with ethanol and its aqueous solutions, either in cold or on heating. The total dense extract obtained after the evaporation of extractant is purified by treatment with chloroform and diethyl or petroleum ethers (Lozhkin & Sakanyan, 2006).

Petroleum ether is the extractant usually used in selective extraction of furanocoumarin fraction from plant tissues (Głowniak, 1988), whereas more polar coumarins—hydroxyderivatives are extracted with methanol. Methanol, used after petroleum ether on the same plant material, extracts more hydrophylic coumarins, but also the residual of furanocoumarins.

Historically, exhaustive extraction with different solvents, which can be performed in Soxhlet apparatus, proved to be the most accurate method of isolation of these groups of compound (Głowniak, 1988; Hadacek et al., 1994). The extraction of the same plant material is usually continued with methanol. For example, peucedanin was successfully isolated using this type of extraction with methanol (Lozhkin & Sakanyan, 2006).

Waksmundzka-Hajnos et al. (2004) compared methods of extraction of furanocoumarins. Some of furanocoumarins from *Pastinaca sativa* fruits were extracted using exhaustive extraction with petroleum ether in Soxhlet apparatus, ultrasonification (USAE), accelerated solvent extraction (ASE) and microwave-assisted solvent extraction (MASE).

USAE was performed with petroleum ether in ultrasonic bath at an ambient temperature of 20°C or at a temperature of 60°C for 30 min three times.

In the ASE method, the plant material was mixed with neutral glass and placed into a stainless steel extraction cell. The application of neutral glass, playing the role of dispersion agent, is recommended to reduce the volume of the solvent used for the extraction (ASE 200, 1995). This extraction was performed with pure methanol or petroleum ether at the same pressure (60 bar).

MASE was also used in the isolation of furocoumarin fractions performing with 80% methanol in a water bath using a two-step extraction with results of 40% generator power during 1 min and by 60% generator power during 30 mins in open and closed systems.

In most cases of the Waksmundzka-Hajnos et al. (2004) experiment, exhaustive extraction in Soxhlet apparatus indicates low yields of furanocoumarins. For example, the use of ultrasonification at 60°C gives, in most cases a higher yield than the exhaustive Soxhlet method. In some cases, this method gives the highest yield of extraction (for xanthotoxin and for isopimpinellin) in comparison to all methods used in experimentation. Also, the use of ASE gives, in most cases, higher yields than the Soxhlet extraction (compare yield of extraction of isopimpinellin, bergapten, imperatorin and phellopterin). In case of bergapten, imperatorin, and phellopterin the yield of extraction by ASE was highest in comparison to all extraction methods used in experiments.

Microwave-assisted solvent extraction gives fair extraction yield for more polar furanocoumarins, probably because of the necessary use of more polar extractant (80% MeOH in water). From the gathered data, it is seen that the extraction yield of phellopterin and imperatorin in pressurized MASE is distinctly lower than in open systems. It shows that in a closed system, the extracted compounds were changed by microwaves. Hence, pressurised MASE cannot be recommended as a leaching method of furanocoumarin fraction (Waksmundzka-Hajnos et al., 2004).

These results are similar to those obtained from the same authors in previous investigations, in which they isolated furanocoumarins from *Archangelica officinalis* fruits. This study indicated the highest yield of psoralens by ASE, using methanol or petroleum ether as the extractant. It was also reported that microwave-assisted solvent extraction in the closed system probably causing the change of analytes (Waksmundzka-Hajnos et al., 2004a).

Soxhlet extraction, ultrasound-assisted extraction and microwaves-assisted extraction in the closed system have been investigated to determine the content of coumarins in flowering tops of *Melilotus officinalis*. Soxhlet extraction was performed in a Soxhlet apparatus equipped with cellulose extraction thimbles. Extraction was performed with ethanol (85°C). Ultrasound-assisted extraction was conducted with 50% (v/v) aq. ethanol, in an ultrasonic bath, and MASE with 50% (v/v) aq. ethanol was performed using a closed-vessel system (Martino et al., 2006).

Soxhlet extraction was used in the isolation of oxypeucedanin from *Prangos uloptera*. Dried and powdered leaves were extracted with n-hexane, dichloromethane and methanol (Razavi et al., 2010).

Celeghini et al. (2001) studied the extraction conditions for coumarin analysis in hydroalcoholic extracts of *Mikania glomerata* Spreng leaves. Maceration, maceration under sonication, infusion and supercritical fluid extraction (SFE) were compared. In SFE method, the solvent extraction system was pressurized in the high pressure vessel with the aid of a nitrogen cylinder. Several solvent mixtures were used including CO_2:EtOH (95:5), (90:10), (85:15) and CO_2:EtOH:H_2O (95:2.5:2.5). The experiment was conducted at the same pressure and temperature. The evaluation of these methods showed that maceration under sonication had the best results.

Kozyra & Głowniak (2006) examined the influence of using solvent in the isolation of furanocoumarins. They carried out extraction techniques with different eluents such n- heptane, dichloromethane and methanol. These extractions were performed on a water bath with boiling eluent and on an ultrasonic bath, for 12 and 24 hrs. The more efficient for bergapten was extraction with dichloromethane.

In another study, six solvents (n-hexane, chloroform, ethyl acetate, ethanol, acetonitrile and water) were used to extract *Cnidii Fructus* in order to evaluate their efficiency in extracting osthole. A comparative evaluation showed that aqueous alcoholic solvent was the most efficient solvent (100%) (Yu et al., 2002).

The furanocoumarin determination from air-dried plant material was also performed using 75% methanol in an ultrasonic ice–water bath (Yang et al., 2010), with 100% methanol (Cardoso et al., 2000; Ojala, 2000), 70% methanol (Chen & Sheu, 1995), with hot (70°C) pure methanol on a water bath (Bartnik & Głowniak, 2007), with pure ethanol in heated reflux (Yu et al., 2002), with 95% ethanol at 80°C (Wang et al., 2007; Zheng et al., 2010), with acetone at room temperature (Taniguchi et al., 2011), with ether at 40°C (Liu et al., 2004a), with dichloromethane at room temperature (Um et al., 2010), with petroleum ether at room temperature (Tosun et al., 2008), with chloroform in a sonic bath (Cardoso et al., 2000).

The extraction with all solvents was usually done 2–5 times, obtaining solutions that were filtered and evaporated under reduced pressure. Frequently, residuals after methanol/ethanol extractions were suspended in water and portioned a few times with chloroform or petroleum ether (Wang et al., 2007; Zheng et al., 2010).

5.2 Sample purification

The next step in sample preparations is the purification of the crude extract. Plant extracts contain much ballast material, both non-polar (chlorophylls, waxes) and polar such as tannins or sugars. Most often liquid-liquid extraction (LLE) is used, which takes advantages of solubility differences of hydrophobic substances, which have affinity for non-polar solvents, and hydrophobic substances, which have an affinity for aqueous solutions. Although the analyses can be easily obtained by evaporation of the solvent, the method has many disadvantages – for example emulsions can be formed and the process is time- consuming. Purification can also be achieved by solid-phase-extraction (SPE). This method uses a variety of adsorbents and ion-exchangers and is widely used for a variety of purposes (Fritz & Macha, 2000; Hennion, 1999; Nilsson, 2000; Snyder et al., 1997; Waksmundzka-Hajnos et al., 2007).

The SPE method is very often used in sample pre-treatment for HPLC. This method has been developed for the purification of furanocoumarins from *Peucedanum tautaricum* Bieb. In the first step, aqueous methanol (50%; v/v) solutions of the samples were passed through conditioned microcolumns to adsorb furanocoumarins on the adsorbent bed. The microcolumns were washed with 50% methanol (Zgórka & Głowniak, 1999), and the compounds of interest group were separated from fatty components and chlorophyll by use of SPE microcolumns (LiChrolut RP-18 E; 500 mg, 3 mL). In the next step, the absorbed furanocoumarins were eluted at a flow-rate of 0.5 mL min^{-1} with 80% methanol into vials previously calibrated with a pipette (Bartnik & Głowniak, 2007).

Sidwa-Gorycka et al. (2003) used SPE for purification furanocoumaric fractions obtained from Ammi majus L. and Ruta graveolens L. methanolic (30%) extracts. They were loaded into octadecyl-SPE microcolumns activated previously with 100% methanol, followed by the selective elution of compounds. The cartridges were washed with 20 ml of 60% methanol to elute the coumarins. The eluting solvents were passed through the sorbent beds at a flow rate of 0.5 ml min-1.

In addition, the SPE has been developed for purification of furanocoumarin fractions from creams and pomades. The obtained samples were cleaned-up using two methods. Each extracted sample was re-dissolved in chloroform and fractionated on cartridges, which were previously conditioned with chloroform and sequentially eluted with chloroform (first fraction), chloroform:methanol (90:10; v/v) (second fraction, furanocoumarins), chloroform:methanol (1:1; v/v) (third fraction) and methanol (fourth fraction). Next, each sample extracted above was re-dissolved in methanol in a sonic bath and fractionated on cartridges, which were previously conditioned with methanol and sequentially eluted with methanol (first fraction), methanol:chloroform (80:20; v/v) (second fraction, furanocoumarins), methanol:chloroform (1:1; v/v) (third fraction) and chloroform (fourth fraction). All fractions were evaporated to dryness in a stream of nitrogen (Cardoso et al., 2000).

5.3 Chromatographic methods in the analysis of furanocoumarins

5.3.1 Column Chromatography (CC)

The good results for purification, separation of the total furanocoumarins and the isolation of individual compounds give column chromatography (CC) a significant advantage of the use of various sorbents and solvent systems.

Furanocoumarins can be fractionated on an aluminum oxide column eluted with petroleum ether, petroleum ether-chloroform (2:1), chloroform, and chloroform-ethanol (9:1; 4:1; 2:1) mixtures or on silica gel column eluted sequentially with hexane-chloroform and chloroform-ethanol systems with increased proportion of a more hydrophilic component (Lozhkin & Sakanyan, 2006).

Separation of the psoralens from *Heracleum sibiricum* L. (Apiaceae) fruits was performed by gravitation column chromatography. Glass columns were filled with silica gel (230-400 mesh) and run, under UV-lamp control, in the following eluents: 1) benzene-ethyl acetate; mixtures of increasing polarity (12.5 to 77.5%); 2) benzene-ethyl acetate (17:3); 3) benzene- chloroform-ethyl acetate (1:1, v/v, 5%). Chromatographically pure clean compounds, in this study, were crystallized from 96% ethanol (Bogucka-Kocka, 1999).

In another investigation, the coumarin mixture from fruits of *Heracleum crenatifolium* was subjected to CC on silica gel and eluted successively with an n-hexane-ethyl acetate solvent system, with increasing polarity (99:1 to 80:20). The collected fractions were applied to preparative-TLC on silica gel plates and pure furanocoumarins were obtained. After chromatography with the use of n-hexane-ethyl acetate (3:1), isobergapten and pimpinellin were obtained. Fractions, which were chromatographed with n-hexane-dichloromethane- ethyl acetate (4:4:2) resulted in a production of bergapten, and fractions, after chromatography using toluene-ethyl acetate (9:1), resulted in a production of yielded isopimpinellin, sphondin and byak-angelicol (Tosun et al., 2008).

On silica-gel column chromatography was also subjected to chloroform residue from roots of *Angelica dahurica*. The furanocoumarins were eluted stepwise with petroleum ether- acetone mixtures (Wang et al., 2007).

Similar techniques were performed for furanocoumarins from the roots *Angelicae dahuricae*. The methylene chloride soluble was chromatographed using column chromatography over silica gel. In this study, a stepwise gradient solution with hexane-ethyl acetate (5:1 to 0:1) was used. Repeated column chromatography of obtained fractions produced an isoimperatorin, imperatorin, oxypeucedanin, phellotorin, byakangelicol, neobyakangelicol, alloisoimperatorin, pabulenol, byakangelicin, and 9-hydroxy-4-methoxypsoralen (Piao et al., 2004).

A vacuum liquid chromatography on silica gel was developed for the isolation of oxypeucedanin from the leaves of *Prangos uloptera*. Hexane extract was subjected, starting with 100% hexane, followed by step gradient of ethyl acetate mixtures (1:99; 5:95; 10:90; 20:80; 40:60; 60:40; 80:20; 100) and finally methanol. The obtained fractions were purified by preparative-TLC on silica gel using $(CH_3)_2CO$-$CHCl_3$, 5:95 as the mobile phase to yield oxypeucedanine (Razavi et al., 2010).

The CC technique was used to separate furanocoumarins from roots of several *Dorstenia* species. The afforded hexane residues were chromatographed on silica gel (230-400 mesh) eluting with hexane-chloroform mixtures (1:1) (gives psoralen) or with hexane-ethyl acetate mixtures of an increasing polarity to give bergapten, 4-[3-(4,5-Dihydro-5.5-dimethyl-4-oxo-2-furanyl)-butoxy]=7H-furo[3,2-g][1] benzopyran-7-one, psoralen and 7-hydroxycoumarin. The chloroform extracts were eluted using hexane-chloroform mixtures to give psoralen, 7- hydroxycoumarin and 4-[3-(4.5-Dihydro-5.5-dimethyl-4-oxo-2-furanyl)-butoxy]=7H- furo[3.2-g][1] benzopyran-7-one or hexane-ethyl acetate of increasing polarity to give psoralen, psoralen dimer and 7-hydroxycoumarin. The polar fractions from the methanol extract were acetylated using pyridine and Ac_2O to give (2'S, 1''S)-2.3-dihydro-2(2-acetoxy- 1-hydroxymethylethyl)-7H-furo [3.2-g][1]benzopyran-7-one (Rojas-Lima et al., 1999).

Another useful adsorbent for column chromatography is Florisil (100-200 mesh), which was used to fractionate furanocoumarins obtained from fruits of *Peucedanum alsaticum* L. and *P. cervaria* (L.) Lap. Concentrated petroleum ether extracts were fractionated on this sorbent with a dichloromethane-ethyl acetate (0-50%) gradient, then ethyl acetate and methanol as mobile phases. After CC separation, the fractions richest in coumarins were analyzed by preparative-TLC on silica gel. Separated zones of selected furocoumarins were eluted from the plates (Skalicka-Woźniak et al., 2009).

The Florisil was also used in an investigation performed by Suzuki et al. (1979). Bergamot oil was eluted on this column with methylene chloride and ethyl acetate. The ethyl acetate fractions were re-chromatographed with methylene chloride. The obtained residue was analyzed by preparative-TLC on silica gel using cyclohexane-tetrahydrofuran (1:1) as eluent. The bergapten zone was scraped and eluted with acetone.

Isolation of the furanocoumarins from grapefruit juice was accomplished by preparative thin layer chromatography. The obtained fractions were applied to tapered silica gel GF TLC plates with a fluorescent indicator. Resolution of compounds was accomplished by using solvent systems consisting of hexane:ethyl acetate (3:1 to 2:3; v/v), chloroform, chloroform/methanol (95:5), and benzene: acetone (9:1). The zones containing furanocoumarins were scraped and extracted with acetone (Manthey et al., 2006).

5.3.2 Thin Layer Chromatography (TLC)

The physicochemical properties of coumarins depend upon their chemical structure, specifically, the presence and position of functional hydroxy or methoxy groups, and methyl or other alkyl chains. As a result of these differences, group separation of the all groups of coumarins does not cause any difficulties (Jerzmanowska, 1967; Waksmundzka-Hajnos et al., 2006). Separation of individual compounds in each group – structural analogs, i.e. closely related compounds – is, however, a difficult task.

Several analytical methods for the quality control of furanocoumarins in plant materials, such as thin layer chromatography (TLC), high performance liquid chromatography (HPLC), high performance liquid chromatography–mass spectrometry (HPLC–MS), high-speed counter- current chromatography (HSCCC), gas chromatography (GC), gas chromatography–mass spectrometry (GC–MS), capillary electrophoresis (CE)), and pressurized capillary electrochromatography (pCEC), has been reported (Chen et al., 2009; Wang et et., 2007).

The oldest publications recommended one- or two- thin layer chromatography for separation and identification of furanocoumarins. This method provides a quite rapid separation of components in a sample mixture. Fractions obtained from column chromatography were usually checked with the use of TLC technique.

Several adsorbents have been applied for the chromatographic analysis of furanocoumarins, e.g. silica gel, C18 layers, alumina, poliamide, Florisil, etc. (Cieśla & Waksmundzka-Hajnos, 2009). Analyzed fractions of studied compounds are eluted using several solvent systems. Borkowski (1973) proposed the following eluents: 1) benzene-acetone (90:10, v/v); 2) toluene-acetone (95:5, v/v); 3) benzene-ethyl acetate (9:1, v/v); 4) benzene-ethylic ether-methanol-chloroform (20:1:1:1, v/v); 5) chloroform, and 6) ethyl acetate-hexane (25:75, v/v) for analysis of coumarins.

The spots of coumarins on thin-layer and paper chromatograms are usually revealed by UV fluorescence at certain characteristic wavelengths, before or after the treatment with an aqueous-ethanol solution of potassium hydroxide or with ammonia vapor, or using some other color reactions. The fluorescent color does not provide accurate identification of the structure of coumarins; nevertheless, sometimes it is possible to determine the type of functional groups (Celeghini et al., 2001; Lozhkin & Sakanyan, 2006).

Joint TLC – colorimetric methods based on the azo-addition reaction with TLC separation on an aluminum oxide layer eluted in the hexane – benzene – methanol (5:4:1) system were developed for the quantitative determination of peucedanin in *Peucedanum morrissonii* (Bess.) and for the analysis of beroxan, pastinacin, and psoralen preparations (Lozhkin & Sakanyan, 2006). Colorimetric determination of xanthotoxin, imperatorin, and bergapten in *Ammi majus* (L.) fruits can be performed after TLC separation on silica gel impregnated with formamide and eluted in dibutyl ether. In order to determine psoralen alone and together with bergapten in *Ficus carica* (L.) leaves, the extract was purified from ballast substances and chromatographed in a thin layer of aluminum oxide in diethyl ether (Lozhkin & Sakanyan, 2006).

Thin layer chromatographic analyses were made by Celeghini and co-authors (2001) on silica gel 60G. As eluent a mixture of toluene:ethyl ether (1:1) saturated with 10% acetic acid was used. The plates were sprayed with an ethanolic solution (5% v/v) of KOH and examined under UV light at 366 nm.

In the other investigation, purification of the eluted furanocoumarins from leaves of *Conium maculatum* was also carried out on TLC plates (Si gel 60, f-254). The chromatograms were developed at room temperature using one of the following solvent systems: chloroform:ethyl acetate (2:1), or chloroform, or toluene:ethyl acetate (1:1) (Al-Barwani & Eltayeb, 2004).

Bogucka-Kocka (1999) analyzed furanocoumarins in fruits of *Heracleum sibiricum* L., after silica gel column chromatography, using 2D-TLC. Two-dimensional thin-layer chromatography was conducted of silica gel plates and run in the following phases: 1) benzene-chloroform-acetonitrile (1:1, v/v, 5%) or 2) benzene-chloroform-ethyl acetate (1:1, v/v, 5%) (first direction); 3) benzene-ethyl acetate (1:1) (second direction). The chromatograms were analyzed using UV light and daylight, after spaying with one of the derivativers: 1) 0.5% I dissolvent in KI; 2) Dragendorf's reagent or 25% $SbCl_5$ in CCl_4.

Unfortunately, in furocoumarin' group, these substances have comparable polarity and similar chemical structures. As a result, multi-dimensional separations are required in such cases.

Thin-layer chromatography gives the possibility of performing multi-dimensional separation – two-dimensional separation with the use of the same stationary phase, with different mobile phases (Gadzikowska et al., 2005; Härmäla et al., 1990; Waksmundzka- Hajnos et al., 2006), or by using a stationary phase gradient (Glensk et al., 2002; Waksmundzka-Hajnos et al., 2006). In TLC, there are almost no limits as far as mobile phases are concerned, because they can be easily evaporated from the layer after the development in the first dimension. Both methods, use of the same layer and different mobile phases or two different layers developed with two mobile phases, make use of different selectivity to achieve complete separation in the two-dimensional process. The largest differences are obtained with a normalphase system, with an adsorption mechanism of separation, and a reversed-phase system, with a partition mechanism of separation, are applied for two-dimensional separations (Nyiredy, 2001). Two-dimensional thin-layer chromatography with adsorbent gradient is an effective method for the separation of large group of substances present in natural mixtures, e.g. plant extracts.

Silica gel is the most popular adsorbent, thus it has been widely used in different chromatographic methods. However, in case of two-dimensional separations of coumarins, it has been rarely applied as it is difficult to select solvent systems which are complementary in selectivity. Härmälä et al. (1990) proposed a very interesting method for the separation of 16 coumarins from the genus *Angelica* with the use of silica gel as an adsorbent. The application of two-dimensional over-pressured layer chromatography enabled complete resolution of the analyzed substances. The authors described a very useful procedure of choosing complementary systems that can be applied in the analysis of complex mixtures. It turned out that the systems, I direction – 100% $CHCl_3$ and II direction – AcOEt/*n*-hexane (30:70, v/v) provided excellent separation of all coumarins, although having only the fourth poorest correlation value.

Due to the possibility of the application of normal- and reversed – phase systems, polar bonded phases have been often a choice for two-dimensional separations. In the case of coumarins, the use of diol- and cyanopropyl-silica have been reported.

Waksmundzka-Hajnos et al. (2006) reported the use of diol-silica for the separation of 10 furanocoumarin standards. Firstly, the compounds were chromatographed with the use of 100% diisopropyl ether (double development), then in the perpendicular direction: 10% MeOH/H_2O (v/v) containing 1% HCOOH. The use of the first direction eluent caused the separation of analyzed substances into three main groups, which is useful for group separation of natural mixtures of coumarins. Chromatography in reversed-phase system enabled the complete resolution of all tested standards. The disadvantage of the applied reversed-phase system is the fact that it has low efficiency, and most of the substances, especially those containing hydroxyl groups are tailing. Diol-silica is similar in its properties to deactivated silica, thus the application of aqueous eluent may be responsible for tailing, which was only slightly reduced after the addition of formic acid.

Better results were obtained after the application of CN-silica. In this case, coumarin standards were firstly chromatographed with the use of normal-phase, then in reversed- phase system. The plate was triple developed in the first direction to improve separation of strongly retained polar coumarins.

The authors also investigated the use of multiphase plates for identification purposes. Coumarins were firstly chromatographed on a RP-18W strip with 55% MeOH/H_2O (v/v), and then in a perpendicular direction they were triple-developed with: 35% AcOEt/*n*- heptane (v/v). The use of reversed-phase system caused the separation of investigated coumarins into two groups: coumarins containing hydroxyl group, and furanocoumarins. The separation, according to the differences in polarity, is even greater than that observed on diol-silica. This system was then applied for separation of the furanocoumarin fraction from fruits of *H. sibiricum*, where seven compounds were identified in the extract (Cieśla & Waksmundzka-Hajnos, 2009; Waksmundzka-Hajnos et al., 2006).

The use of graft thin-layer chromatography of coumarins was also reported (Cieśla et al., 2008; Cieśla et al., 2008a; Cieśla et al., 2008b). The authors applied two combinations of adsorbents: silica + RP-18W, and CN-silica + silica gel. In the first stage of this experiment, plates pre-coated with CN-silica were developed in one dimension by unidimensional multiple development. The same mobile phase (35% ethyl acetate in n-heptane) was used, over the same distance, and the same direction of the development. Plates were triple- developed with careful drying of the plate after each run. Unidimensional multiple development (UMD) results in increased resolution of neighboring spots (Poole et al., 1989). After chromatography the plates were linearly scanned at 366 nm with slit dimensions 5 mm × 0.2 mm. This chromatographic system was not suitable for separation of structural analogs. Isopimpinellin and byacangelicol are coeluted and phellopterin and bergapten also have very similar retention behavior. The isopimpinellin and byacangelicol molecules have two medium polarity groups in positions 5 and 8, which have similar physicochemical properties. Therefore, it was also easily noticeable that different non-polar substituents did not cause significant difference in retention behavior.

Compounds with polar substituents – hydroxyl groups in simple coumarins are more strongly retained on CN-silica layer in normal-phase systems.

When other systems, for example silica with AcOEt–n-heptane and RP 18W with 55% MeOH in water, were used only partial separation of standards was achieved. This results from the similar structures and physicochemical properties of the compounds. On silica layers only polar aesculetin and umbelliferone are more strongly retained. Phellopterin with a long chain in the 8 position (with a shielding effect on neighboring oxygen) is weakly retained. These differences cause the aforementioned coumarins to be completely separated from other standards. Byacangelicol and umbelliferone, and bergapten, isopimpinellin, and xanthotoxin, with only slight differences in number and position of medium-polarity methoxy groups, are eluted together. More significant resolution of the investigated compounds was obtained on RP-18 plates, eluted with aqueous mobile phases. The differences in number, length, and position of medium-polarity and non-polar substituents cause differences in retention behavior of the analytes. These differences result in good separation of bergapten, xanthotoxin, and phellopterin by reversed-phase systems.

In the next step Cieśla et al. (2008b) investigated the search for orthogonal systems, which would ensure better separation selectivity for the coumarins, was conducted. To achieve this, graft TLC, with two distinct layers, was applied. The authors experimentally chose two pairs of orthogonal TLC systems:

- first dimension, CN-silica with 30% ACN + H_2O (three developments); second dimension, SiO_2 with 35% AcOEt + n-heptane (three developments);
- first dimension, SiO_2 with 35% AcOEt + n-heptane (three developments); second dimension, RP-18 with 55% MeOH + H_2O.

An application of multiple development technique (UMD) in the first dimension results in partly separated spots, which are transferred to the second layer with methanol. Use of methanol causes narrowing of starting bands, similarly to the effect of a preconcentrating zone. The preconcentration is responsible for symmetric and well separated spots being obtained after development of the plate in the second dimension. This makes the densitometric estimation easier.

In the last step of Cieśla and co-authors (2008b) investigations, the separation of furanocoumarin fractions from *Archangelica officinalis, Heracleum sphondylium,* and *Pastinaca sativa* fruits was performed by the use of grafted plates SiO_2 with RP-18W and CN with SiO_2, with appropriate mobile phases. The identity of the extract components was confirmed by comparing retardation factors and UV spectra with the R_f values and spectra obtained for the standards.

Graft TLC in orthogonal systems characterized by different separation selectivity enables complete separation of structural analogs such as furanocoumarins. The use of two different TLC systems enables complete separation and identification of some furanocoumarins present in extracts obtained from *Archangelica officinalis, Heracleum sphondylium*, and *Pastinaca sativa* fruits (Cieśla et al., 2008b).

The graft-TLC system silica + RP-18W were successfully applied for construction of chromatographic fingerprints of different plants from the *Heracleum* genus.

Two-dimensional chromatography has also been applied for quantitative analysis of furanocoumarins in plant extracts (Cieśla et al., 2008b). In order to obtain reproducible results, all investigated compounds should be completely separated. Graft-TLC with the use of adsorbents silica + RP-18W was proven to be the most suitable for quantitative analysis. Resolution of compounds was insufficient in case of 2D-TLC on one adsorbent (CN-silica), as the standards had to be divided into two separate groups for an accurate estimation of peak surface area.

Quantitative analysis is difficult to perform after two-dimensional chromatographic run, as densitometers are not adjusted to scan two-dimensional chromatograms. This problem may be overcome if small steps between scans are used. In the proposed method, the authors scanned the plate with the slit of a dimension 5 mm×0.2 mm, operated at λ= 366 nm, obtaining 36 tracks that were not overlapping. This wavelength was chosen to get rid of intensive baseline noise, observed at lower wavelengths. Peak areas were measured with the use of the method called "peak approximation" (Cieśla et al., 2008b; Cieśla & Waksmundzka-Hajnos, 2009).

5.3.3 High Performance Liquid Chromatography (HPLC)

Furanocoumarins are also examined by means of high performance liquid chromatography (HPLC). This technique has shown to be a very efficient system for separation of this group of compounds. HPLC methods have been reported for the determination of psoralens in callus cultures, vitro culture, serum, dermis, plants, citrus essential oils, phytomedicines, but only the most recently published methods has reported assay validation (Cardoso et al., 2000; Dugo et al., 2000; Markowski & Czapińska, 1997; Pires et al., 2004).

Linear furanocoumarins, such as psoralen, bergapten, xanthotoxin, and isopimpinellin isolated from three varieties of *Apium graveolens* were examined by normal-phase HPLC equipped with a variable wavelength detector set at 250 nm. The mobile phase consisted of a mixture of ethyl acetate (0.1%) and formic acid (0.1%) in chloroform (Waksmundzka- Hajnos & Sherma, 2011).

In most recent applications, reversed-phase HPLC is used to evaluate furanocoumarins quantitatively.

For example, the quantitative analysis of some furanocoumarins from *Pastinaca sativa* fruits was performed by RP-HPLC in system C18/methanol + water in gradient elution. The authors used the following gradient: 0-10 min, 45% MeOH; 10-20 min, 45-55% MeOH; 20-30 min, 55-70% MeOH, and 30-40 min, 70% MeOH in bidistilled water (Waksmundzka-Hajnos et al., 2004).

The determination of two furocoumarins (bergapten and bergamottin) in bergamot fruits, was carried out by the HPLC system equipped with a diode array detector. C18 column and the mobile phase consisted of methanol and 5% (v/v) acetic acid aqueous solution in the following gradient: 5-20% (0-13 min), 20-100% (13-25 min), 100-5% (20-30 min), were used in this investigation (Giannetti et al., 2010).

The optimized HPLC-UV method was used to evaluate the quality of 21 samples of Radix *Angelica dahurica* from different parts of China. Bergapten, imperatorin and cnidilin were separated on C18 column; the mobile phase was 66:34 (v/v) methanol-water (Wang et al., 2007).

The HPLC technique was ensued for analyses of psoralen and bergapten. HPLC separation of the psoralens was performed using a Shimadzu octadecyl Shim-pack CLC-ODS reversed- phase column with a small pre-column containing the same packing. Elution was carried with acetonitrile-water 55:45 (v/v) and detections of the peaks were recording at 223 nm (Cardoso et al., 2002). The same conditions were used for determination of furanocoumarins in three oral solutions by Pires et al. (2004).

A rapid and sensitive reversed-phase HPLC method has been used for the determination of furanocoumarins in methanolic extracts of *Peucedanum tautaricum* Bieb. Compounds were separated on stainless-steel column packed with 5µm particle Hypersil ODS C18. The mobile phase was methanol-water gradient used following: 0-5 mins, isocratic elution with 60% (v/v) methanol; 5–20 mins, linear gradient from 60 to 80% methanol; 20 to 30 mins, linear gradient from 80 to 60% methanol; 30–40 mins isocratic elution with 60% methanol. An acetonitrile–water mobile phase gradient was also used (0–8 mins, isocratic elution with 50% acetonitrile; 8–25 mins, linear gradient from 50 to 70% acetonitrile; 25–28 mins, linear gradient from 70 to 50% acetonitrile; 28–40 mins isocratic elution with 50% acetonitrile) (Bartnik & Głowniak, 2007).

The mobile phase consisted of water with orthophosphoric acid 1:10000 (solvent A), methanol (solvent B) and acetonitrile (solvent C) was used for analysis of coumarins from *Melilotus officinalis* (L.) Pallas. The starting mixture (80% A, 5% B and 15% C) was modified as follows: within 20 mins the mobile phase composition became 65% A, 20% B, 15% C and was kept constant for 10 mins; in the following 10 mins the mixture composition came back to the initial eluting system (Martino et al., 2006).

The search for better conditions for application of HPLC has led to development of UPLC (Ultra Performance Liquid Chromatography), a relatively new liquid chromatography technique enabling faster analysis, consumption of less solvent and better sensitivity. The UPLC method enables a reduction of analysis time by up to a factor of nine compared with conventional HPLC without loss of quality of the analytical data generated. Another very important advantage is high column efficiency which increases the possibility of compound identification and results in better quantitative analysis. UPLC is more efficient and therefore has greater resolving power than traditional HPLC (Novakova et al., 2006; Skalicka-Woźniak et al., 2009; Wren & Tchelitcheff, 2006).

The quantitative analysis by UPLC was performed for the furocoumarins in *Peucedanum alsaticum* and *P. cervia* (Skalicka-Woźniak et al., 2009). The optimalization of the RP-UPLC separation of the coumarins was achieved by the use of DryLab. The investigation was performed with an Acquity Ultra Performance LC (Waters, Milford, MA, USA) coupled with a DAD detector. Compounds were separated on a stainless-steel column packed with 1.7 µm BEH C18. Two linear mobile phase gradients from 5 to 100% of acetonitrile with gradient times of 10 and 20 min were used. Detection was at 320 nm.

A paper by Desmortreux et al. (2009) reports separation of furocoumarins of essential oils (lemon residue) by supercritical fluid chromatography (SFE). The authors studied many types of stationary phases and the effects of numerous analytical parameters. Amongst the numerous tested columns, good separation of analyzed furanocoumarins was obtained on a pentafluorophenyl (PFP) phase (Discovery HS F5), based on an aromatic ring substituted by five fluorine atoms. The mobile phase used was CO_2-EtOH 90:10 (v/v). Amongst the standard compounds, bergapten was well separated being eluted after the other furocoumarins in the lemon residue sample. The results obtained in this study show that SFC is a perfectly suited method to investigate the psoralens in essential oil composition, because of the great number of compounds separated in a reduced analysis time, and with a very short time for re-equilibration of the system at the end of the gradient analysis. Because of the absence of water in the mobile phase in SFC, the stationary phase can establish more varied interactions than in HPLC, making the stationary phase choice highly significant.

5.3.4 Hyphenated HPLC techniques

A hyphenated, HPLC-TLC procedure for the separation of couamrins, has been proposed by Hawrył et al. (2000). A mixture of 12 coumarins from *Archangelica officinalis* was completely separated as a result of the different selectivities of the two combined chromatographic techniques, RP-HPLC and NP-TLC. Firstly, the analyzed compounds were separated by means of RP-HPLC. The optimal eluent:

60% MeOH in water was chosen with the use of DryLab program. All HPLC fractions were collected, evaporated and finally developed in normal-phase system, on silica gel, with the use of a solvent mixture: 40% AcOEt (v/v) in dichloromethane/heptane (1:1). All fractions were completely separated. The combination of these methods gave successful results, although both methods, if used separately, failed to give good resolution. This procedure may be useful for micropreparative separation of coumarins (Cieśla & Waksmundzka-Hajnos, 2009).

The liquid chromatography coupled with mass spectrometry (LC-MS) technique is becoming increasingly popular, in particular, the introduction of atmospheric pressure chemical ionization (APCI) has dramatically influenced the possibilities for analyzing poorly ionizable compounds. The use of hyphenated techniques such as LC-MS provides great information about the content and nature of constituents of complex natural matrices prior to fractioning and carrying out biological assays. Moreover, MS presents a great advantage not only in its ability to measure accurate ion masses but also in its use in structure elucidation (Chaudhary et al., 1985; Dugo et al., Waksmundzka-Hajnos & Sherma, 2011).

Coumarins can be detected in both positive and negative ion modes. Whereas, the positive ion mode often generates higher yields, the noise level is lower in the negative ion mode, thus improving the quality of the signals. So, preliminary investigations regarding the polarity used are very important.

The main problem of working with LC-MS of natural products is the choice of the ionisation technique. Particle beam (PB) and thermospray (TSP) interfaces are the most commonly used for natural component analysis. Both of them exhibit many drawbacks, such as the difficulty to optimize ionisation conditions and the lack of sensitivity. Electrospray (ESI) and atmospheric pressure chemical ionisation (APCI) techniques, which operate under atmospheric pressure, seem to be very promising. These ionisations differ in the way they generate ions, but show many similarities: both operate at atmospheric pressure, giving molecular weight information and additional structural information. Many classes of compounds can be analyzed by both APCI and ESI. However, ESI is the technique of choice for polar and higher molecular weight compounds, while APCI is suitable for less polar compounds and of lower molecular weight than ESI (Dugo et al., 2000).

A sensitive, specific and rapid LC–MS method has been developed and validated for the simultaneous determination of xanthotoxin (8-methoxypsoralen), psoralen, isoimpinellin (5,8-dimethoxypsoralen) and bergapten (5-methoxypsoralen) in plasma samples from rats after oral administration of *Radix Glehniae* extract using pimpinellin as an internal standard. A chromatographic separation was performed on a C18 column with a mobile phase composed of 1mmol ammonium acetate and methanol (30:70, v/v). The detection was accomplished by multiple-reaction monitoring (MRM), scanning via electrospray ionization (ESI) source operating in the positive ionization mode. The optimized mass transition ion- pairs (m/z) for quantitation were 217.1/202.1 for xanthotoxin, 187.1/131.1 for psoralen, 247.1/217.0 for isoimpinellin, 217.1/202.1 for bergapten, and 247.1/231.1 for pimpinellin (Yang et al., 2010).

A paper by Zheng et al. (2010) reports the quantitation of eleven coumarins including furocoumarins in *Radix Angelicae dahuricae*. By using this HPLC–ESI-MS/MS method, all coumarins were separated and determined within 10 min. These compounds were detected by ESI ionization method and quantified by multiple-reaction monitoring (MRM). The mass spectral conditions were optimized in both positive- and negative-ion modes, and the positive-ion mode was found to be more sensitive. The all coumarins exhibited their quasi- molecular ions $[M^{+}H]^{+}$, $[M^{+}Na]^{+}$, $[M^{+}NH_4]^{+}$, $[M^{+}K]^{+}$ and fragment ions $[M^{+}H\text{-}CO]^{+}$, $[M^{+}H\text{-}C_5H_9O]^{+}$, $[M^{+}H\text{-}C_5H_8]^{+}$, $[M^{+}H\text{-}C_5H_8\text{-}CO]^{+}$, $[M^{+}H\text{-}C_5H_8\text{-}CO_2]^{+}$, $[M^{+}H\text{-}CH_3]^{+}$.

Yang et al. (2010a) proposed a practical method for the characterization of coumarins, i.e. linear furanocoumarins, in *Radix Glehniae* by LC–MS. They described in details over 40 derivatives of psoralens. First, 10 coumarin standards were studied, and mass spectrometry fragmentation patterns and elution time rules for the coumarins were found. Then, an extract of *Radix Glehniae* was analyzed by the combination of two scan modes, i.e., multiple ion monitoring-information-dependent acquisition-enhanced product ionmode (MIM-IDA- EPI) and precursor scan information-dependent acquisition-enhanced product ionmode (PREC-IDA-EPI) on a hybrid triple quadrupole-linear ion trap mass spectrometer. This study has demonstrated the unprecedented advantage of the combination of these two scan modes. The MIM-IDA-EPI mode is sensitive, and no pre-acquisition of MS/MS spectra of the parent ion is required due to the same precursor ion and product ion. A PREC-IDA-EPI mode was used to provide information on the parent ions, fragment ions and retention times of specified ions so the molecular weights of unknown coumarins and their glycosides could be identified. The information on the fragment ions from the MIM-IDA-EPI mode could be supplemented, and the retention time could be verified. Therefore, the characterization of trace furanocoumarins has become very easy and accurate by the combined use of the two modes and may play an important role in controlling the quality of medicinal herbs.

A high performance liquid chromatography–diode array detection–electrospray ionization tandem mass spectrometry (HPLC/DAD/ESI-MS^n) method was used for the chromatographic fingerprint analysis and characterization of furocoumarins in the roots of *Angelica dahurica* (Kang et al., 2008). The HPLC fingerprint technique has been considered as a useful method in identification and quality evaluation of herbs and their related finished products in recent years, because the HPLC fingerprint could systematically and comprehensively exhibit the types and quantification of the components in the herbal medicines (Drasar & Moravcova, 2004; Kang et al., 2008; Wang et al., 2007). Kang and co- authors (2008) showed that the samples from different batches had similar HPLC fingerprints, and the method could be applied for the quality control of the roots of *Angelica dahurica*. In addition, they identified a total of 20 furocoumarins by HPLC/DAD/ESI-MS^n technique, and their fragmentation patterns in an electrospray ion trap mass spectrometer were also summarized.

Recently, high-speed counter-current chromatography (HSCCC) equipped with a HPLC system for separation and purification of furanocoumarins from crude extracts of plant materials, was also described.

High-speed counter-current chromatography (HSCCC), which was first invented by Y. Ito (1981), is a kind of liquid–liquid partition chromatography. The stationary phase of this method is also a liquid. It is retained in the separation column by centrifugal force. Because no solid support is used in the separation column, HSCCC successfully eliminates irreversible adsorption loss of samples onto the solid support used in conventional chromatographic columns (Ito, 1986). As an advanced separation technique, it offers various advantages including high sample recovery, high-purity of fractions, and high-loading capacity (Ma et al., 1994). In the past 30 years, HSCCC has made great progress in the preparation of various reference standards for pharmacological studies and good manufacturing practice, such as coumarins, alkaloids, flavonoids, hydroxyanthraquiones (Liu et al., 2004b).

Liu and co-authors (2004b) isolated and purified psoralen and isopsoralen from *Psoralea corylifolia* using HSCCC technique. In their investigation, they utilized TBE-300A HSCCC instrument with three multilayer coil separation column connected in series. The two-phase solvent system composed of n-hexane-ethyl acetate-methanol-water was used for HSCCC separation. Each solvent was added to a separatory funnel and roughly equilibrated at room temperature. The upper phase (stationary phase) and the lower phase (mobile phase) of the two-phase solvent system were pumped into the column with the volume ratio of 60:40. When the column was totally filled with the two phases, the lower phase was pumped, and at the same time, the HSCCC apparatus was run at a revolution speed of 900 rmp. After hydrodynamic equilibrium was reached, the sample solution containing the crude extract was injected into the separation coil tube through the injection valve. Each peak fraction was collected according to the chromatogram and evaporated under reduced pressure. The results of HSCCC tests indicated that n-hexane-ethyl acetate-methanol-water (5:5:4.5:5.5, v/v) was the best solvent system for the separation of psoralen and isopsoralen (Liu et al., 2004b).

In another investigations, the same authors had used the same HSCCC technique to induce preparative isolation and purification of furanocoumarins from *Angelica dahurica* (Fisch. ex Hoffm) Benth, et Hook. f (Liu et al., 2004) and from *Cnidium monnieri* (L.) Cusson (Liu et al., 2004a). The results of the first study (Liu et al., 2004) indicated that the best separation of imperatorin, oxypeucedanin and isoimperatorin was when the lower phase of n-hexane- methanol-water (5:5:5, v/v) and n-hexane-methanol-water (5:7:3, v/v) were used in gradient elution. The following gradient was used: 0-150 min, only the lower phase of n-hexane-methanol-water (5:5:5, v/v); 150-300 min, the volume ratio of the lower phase of n-hexane- methanol-water (5:7:3, v/v) changed from 0 to 100%.

A HSCCC method for separation and purification of psoralens from *C. monnieri* was developed by using with a pair of two-phase solvent system composed of light petroleum- ethyl acetate-methanol-water at volume ratios of 5:5:5:5, 5:5:6:4 and 5:5:6.5:3.5 (Liu et al., 2004a).

In described cases, the crude extracts obtained by HSCCC technique were analyzed by HPLC method. The column used was a reversed-phase symmetry C18 column. The mobile phase was methanol-water (68:32, v/v) (analysis of extract from *A. dahurica*), methanol- water (40:60, v/v) (analysis of extract from *P. corylifolia*), or methanol-acetonitrile-water system in gradient mode as follows: 30:30:40 to 50:30:20 in 30 min (analysis of extract from *C.monnieri*) (Liu et al., 2004, 2004a, 2004b). Bergapten and imperatorin obtained by HSCCC method from *Cnidium monnieri* were also analyzed by high-performance liquid chromatography. The column used was a reversed-phase symmetry C18 column, and the mobile phase adopted was methanol (solvent A) – water (solvent B) in the gradient mode as follows: 0-5 min, 60% A; 5-14 min, 60-80% A; 14-15 min, 80-60% A (Li & Chen, 2004).

5.3.5 Capillary electrophoresis

In some cases, capillary electrophoresis was chosen to determine quantities of furanocoumarins. For example Ochocka et al. (1995) used this method for separating psoralens from roots and aerial parts of *Chrysanthemum segetum* L. The analyses were performed with electrophoresis apparatus with UV detection at 280 nm. The best overall separation was obtained on uncoated silica capillary with 7-s pneumatic injection using a buffer solution of 0.2 M boric acid-0.05 M of borax in water (11:9, v/v) (pH 8.5). In another example, micellar electrokinetic capillary chromatography (MEKC) was used in the separation of coumarins contained in *Angelicae Tuhou Radix* (Chen & Sheu, 1995). In this investigation, the electrolyte was buffer solution [20 mM sodium dodecyl sulfate (SDS) – 15 mM sodium borate – 15 mM sodium dihydrogenphosphate (pH 8.26)] – acetonitrile (24:1).

The pressurized capillary electrochromatography (pCEC) was utilized for the separation and determination of coumarins in *Fructus cnidii* extracts from 12 different regions (Chen et al., 2009). Capillary electrochromatography (CEC), as a novel microcolumn separation technology, couples the high efficiency of capillary electrophoresis with high selectivity of HPLC. The CEC analytes separation is usually achieved in capillaries containing packed stationary phases by an electroosmotic flow (EOF) generated by a high electric field. The experiments were performed in an in-house packed column with a monolithic outlet frit under the optimal conditions: pH 4.0 ammonium acetate buffer at 10 mM containing 50% acetonitrile at -6 kV applied voltage. This analytical method, with use of the novel column, gives good results in the determination of coumarins.

5.3.5 Gas chromatography

In the recent decade, tasks related to the isolation of furocoumarins and the quality control of related preparations were most frequently solved using GC techniques. Gas chromatography was predominantly used for the identification and quantitative analysis of furocoumarins in preparations and raw plant materials. Investigations of the chromatographic behavior (retention times) of substituted furocoumarins revealed the following general laws: 1) on passage from hydroxy- to methoxycoumarins, the retention time decreases (because of reduced adsorption via hydrogen bonds); 2) furocoumarins with O-alkyl substituents at C5 are eluted after 8-hydroxy isomers; 3) the logarithm of the relative retention time is a linear function of the molecular weight. This GC data can be used for determining the structure and estimating the retention time of analogous coumarins (Lozhkin & Sakanyan, 2006). A number of methods have been described for the analysis of furanocoumarins using capillary gas chromatography (GC) (Beier et al., 1994; Wawrzynowicz & Waksmundzka-Hajnos, 1990).

Gas chromatographic method was used to determine osthole content in *Cnidii Fructus* extract. The analytical conditions are the following: nitrogen as the carrier gas, the flow rate of 40 mL/min; the split ratio of 120:1. The column used was DBTM-5 (30 m × 0.53 mm I.D. 1.5 μm) equipped with a flame ionization detector (FID). The initial oven temperature was programmed to be at 135°C for 12 minutes. The temperature was then raised to 215°C at a rate of 12°C/min for 20 minutes. Caffeine anhydrous was used as the internal standard (Yu et al., 2002).

In another example, GC-FID was used to analyze of psoralen, bergapten, pimpinellin and isopimpinellin present in phytomedicines (creams and pomades) employed in the treatment of vitiligo in Brazil. The GC-FID assay method present here is rapid, sensitive and robust and can be applied to the determination of furanocoumarins in routine analysis of creams, pomades and other lipophilic phytocosmetics. These analyses were performed in a VARIAN 3400 gas chromatograph equipped with a capillary fused silica LM-5 and with a flame ionization detector (FID). H_2 was used as carrier gas at a flow rate 0.8 ml min^{-1} and the injection split ratio was 1:20. The injection temperature was 280°C. Column temperature was programmed from 150 to 240°C with a linear increase of 10°C min^{-1}, then 240–280°C with a linear increase of 5°C min^{-1} and was then held for 15 mins. The detector temperature was 280°C (Cardoso et al., 2000).

5.4 Structural analysis

For the structural identification and characterizing of the psoralen compounds, especiallyifthey are novel, instrumental techniques such as nuclear magnetic resonance (NMR) spectroscopy and infrared spectroscopy (IR) are used. NMR spectroscopy is an invaluable technique for the structural determination of all furanocoumarins. As well as providing information on the chemical environment of each proton or carbon nucleus in the molecule, the technique can be employed to determine linkages amongst nearby nuclei, often enabling a complete structure to be assembled (Rice-Evans & Packer, 2003).

Dimethyl sulfoxide (DMSO-d_6) and methanol (CD_3OD) are both suitable solvents for furanocoumarins.

The reader is referred to Rojas-Lima et al., 1999; Um et al., 2010; Taniguchi et al., 2011, and Tesso et al., 2005 publications, for details of the principles of NMR and general interpretation of NMR spectra.

6. CONCISE SUMMARY

Furanocoumarins have a lactone structure, which confers a diverse array of biological activities. Bergapten and other furanocoumarins are utilized in the treatment of dermatological conditions such as psoriasis and vitiligo. Their photosensitizing properties are consequently significant (Bhatnagar et al., 2007; Trott et al., 2008). Their capacity to covalently alter nucleic acids is employed in a procedure known as "extracorporeal photopheresis," which is clinically essential for either of the following indications: erythrodermic variants of cutaneous T-cell lymphoma (e.g., mycosis fungoides, Sezary's syndrome) or chronic graft-versus-host disease unresponsive to conventional immunosuppressive treatment (Hotlick et al., 2008; Lee et al., 2007).

The objective of this chapter was to provide a comprehensive overview of the methodologies for the isolation, separation, and identification of furanocoumarins in plant materials. Diverse analytical methodologies are available for the detection of coumarins, and these techniques must fulfill the following criteria: brevity, cost-effectiveness, high accuracy, and precision across multiple applications. This review may assist in selecting the analytical method for furanocoumarin compounds.

7. REFERENCES

Ahn, M.J., Lee. M.K., Kim, Y.C., Sung, S.H. (2008) *The simultaneous determination of coumarins in Angelica gigas root by high performance liquid chromatography-diode array detector coupled with electrospray ionization/mass spectrometry*. J Pharm Biomed Anal. 46, 258- 266.

Al-Barwani, F.M. & Eltayeb, E.A. (2004) *Antifungal compounds from indused Conium maculatum L. plants.* Biochem Syst Ecol. 32, 1097-1108.

Baek, N.I., Ahn, E.M., Kim, H.Y., Park, Y.D. (2000). *Furanocoumarins from the root of Angelica dahurica.* Arch Pharm Res. 23, 467-470.

Baetas, A.C.S., Arruda, M.S.P., Müller, A.H., Arruda, A.C. (1999) *Coumarins and Alkaloids from the Stems of Metrodorea Flavida.* J Braz Chem Soc. 10(3), 181-183.

Bartnik, M. & Głowniak, K. (2007) *Furanocoumarins from Peucedanum tauricum Bieb. and their variability in the aerial parts of the plant during development.* Acta Chromatogr. 18, 5-14.

Bartnik, M., Głowniak, K., Maciąg, A., Hajnos, M. (2005) *Use of reversed-phase and normal- phase preparative thin-layer chromatography for isolation and purification of coumarins from Peucedanum tauricum Bieb. Leaves.* J Planar Chromatogr. 18, 244-248.

Bączek, T. (2008) *Computer-assisted optimization of liquid chromatography separations of drugs and related substances.* Curr Pharm Anal. 4, 151-161.

Beier, R. C. & Oertli, E. H. (1983). *Psoralen and other linear furanocoumarins as phytoalexins in celery.* Phytochem 22, 2595-2597.

Berenbaum, M. R., Nitao, J. K., Zangerl, A. R. (1991). *Adaptive significance of furanocoumarin diversity in Pastinaca sativa (Apiaceae).* J Chem Ecol. 17, 207-215.

Bhatnagar, A., Kanwar A.J., Parsad D., De D. (2007) *Psoralen and ultraviolet A and narrow- band ultraviolet B in inducing stability in vitiligo, assessed by vitiligo disease activity score: an open prospective comparative study.* J Eur Acad Dermatol Venerol. 21, 1381- 5.

Bogucka-Kocka, A. & Kocki, J. (2002) *Influence of two furocoumarins: bergapten and xanthotoxin from meadow cow parsnip (Heracleum sibiricum L.) on apoptosis induction and apoptotic genes expression in HL-60 human leukaemic cell line.* Bull Vet Inst Puławy Suppl. 1, 111-116.

Bogucka-Kocka, A. & Krzaczek, T. (2003) *The furanocoumarins in the roots of Heracleum sibiricum L.* Acta Polon Pharm-Drug Res, 60/5, 401-403.

Bogucka-Kocka, A., Rułka, J., Kocki, J., Kubiś, P., Buzała., E. (2004) *Bergapten apoptosis induction in blood lymphocytes of cattle infected with bovine leukemia virus (BLV).* Bull Vet Inst Puławy. 48, 99-103.

Borkowski B. (Rd.) (1973) *Thin layer chromatography in the pharmaceutical analysis.* PZWL, Warsaw, Poland (in Polish).

Boselli, E., Velazco, V., Caboni, M.F., Lercker, G. (2001) *Pressurized liquid extraction of lipids for the determination of oxysterols in egg-containing food.* J Chromatogr A. 917, 239- 244.

Bourgaud, F., Nguyen, C., Guckert, A. (1995) *Psoralea species: in vitro culture and production of furanocoumarins and other secondary metabolites.* Y.P.S. Bajaj (Ed.), Medicinal and aromatic plants, vol. VIII, Springer Verlag, Berlin, 388-411.

Bourgaud, F., Hehn, A., Larbat, R., Doerper, S., Gontier, E., Kellner, S., Matern, U. (2006) *Biosynthesis of coumarins in plants: a major pathway still to be unravelled for cytochrome P450 enzymes.* Phytochem. Rev. 5/2-3, 293-308.

Brandt, C., Bethmann, K., Gastens, A.M., Löscher, W. (2006) *The multidrug transporter hypothesis of drug resistance in epilepsy: proof-of-principle in a rat model of temporal lobe epilepsy.* Neurobiol Dis. 24, 202–211.

Cardoso, C.A.L., Honda, N.K., Barison, A. (2002) *Simple and rapid determination of psoralens in topic solutions using liquid chromatography.* J Pharm Biomed Anal. 27, 217-224.

Cardoso, C.A.L., Vilegas, W., Honda, N.K. (2000) *Rapid determination of furanocoumarins in creams and pomades using SPE and GC.* J Pharm Biomed Anal. 22, 203-214.

Castro de, M.D.L. & Garcia-Ayuso, L.E. (1998) *Soxhlet extraction of solid materials: An outdated technique with a promising innovative future.* Anal Chem Acta. 369/1-2, 1.

Castro de, M.D.L. & Silva da, M.P. (1997) Trends Anal Chem. 16, 16.

Celeghini, R.M.S., Vilegas, J.H.Y., Lanças, F.M. (2001) *Extraction and Quantitative HPLC Analysis of Coumarin in Hydroalcoholic Extracts of Mikania glomerata Spreng. ("guaco") Leaves.* J Braz Chem. 12(6), 706-709.

Ceska, O., Chaudhary. S. K., Warrington, P. J., Ashwood-Smith, M. J. (1986). *Furanocoumarins in the cultivated carrot, Daucus carota.* Phytochem 25, 81-83.

Chaudhary, S.K., Ceska, O., Warrington, P.J., Ashwood-Smith, M.J. (1985*) Increased furocoumarin content of celery during storage.* J Agric Food Chem. 33/6, 1153-1157.

Chen, J.R., Dulay, M.T., Zare, R.N., Svec, F., Peters, E. (2000) *Macroporous Photopolymer Frits for Capillary Electrochromatography.* Anal Chem. 72, 1224-1227.

Chen, C.-T. & Sheu, S.-J. (1995) *Separation of coumarins by micellar electrokinetic capillary chromatography.* J Chromatogr A. 710, 323-329.

Chen, D., Wang, J., Jiang, Y., Zhou, T., Fan, G., Wu, Y. (2009) *Separation and determination of coumarins in Fructus cnidii extracts by pressurized capillary electrochromatography using a packed column with a monolithic outlet frit.* J Pharm Biomed Anal. 50, 695- 702.

Chen, Y., Fan, G., Zhang, Q., Wu, H., Wu, Y. (2007) *Fingerprint analysis of the fruits of Cnidium monnieri extract by high-performance liquid chromatography-diode array detection-electrospray ionization tandem mass spectrometry.* J Pharm Biomed Anal. 43, 926-936.

Cherng, J.-M., Chiang, W., Chiang, L.-C. (2008) *Immunomodulatory activities of common vegetables and spices of Umbelliferae and its related coumarins and flavonoids.* Food Chem. 106, 944-950.

Choi, S.Y., Ahn, E.M., Song, M.C., Kim, D.W., Kang, J.H., Kwon, O.S., Kang, T.C., Baek, N.I. (2005). *In vitro GABA-transaminase inhibitory compounds from the root of Angelica dahurica.* Phytother Res. 19, 839-845.

Cieśla, Ł., Bogucka-Kocka, A., Hajnos, M., Petruczynik, A., Waksmundzka-Hajnos, M. (2008) *Two-dimensional thin-layer chromatography with adsorbent gradient as a method of chromatographic fingerprinting of furanocoumarins for distinguishing selected varieties and forms of Heracleum spp.* J Chromatogr A 1207, 160-168.

Cieśla, Ł., Petruczynik, A., Hajnos, M., Bogucka-Kocka, A., Waksmundzka-Hajnos, M. (2008a) *Two-Dimensional Thin-Layer Chromatography of Structural Analogs. Part I: Graft TLC of Selected Coumarins.* J Planar Chromatogr 21(4), 237-241.

Cieśla, Ł., Petruczynik, A., Hajnos, M., Bogucka-Kocka, A., Waksmundzka-Hajnos, M. (2008b) *Two-Dimensional Thin-Layer Chromatography of Structural Analogs. Part II: Method for Quantitative Analysis of Selected Coumarins in Plant Material.* J Planar Chromatogr. 21(6), 447-452.

Cieśla, Ł., Skalicka - Woźniak, K., Hajnos, M., Hawrył, M., Waksmundzka - Hajnos, M. (2009) *Multidimensional TLC Procedure for Separation of Complex Natural Mixtures Spanning a Wide Polarity Range; Application for Fingerprint Construction and for Investigation of Systematic Relationships within the Peucedanum Genus.* Acta Chromatogr. 21 (4), 641-657.

Cieśla, Ł. & Waksmundzka-Hajnos, M. (2009) *Two-dimensional thin-layer chromatography in the analysis of secondary plant metabolites.* J Chromatogr A. 1216, 1035-1052.

Court, W.A., Hendel, J.G., Elmi, J. (1996) *Reversed-phase high-performance liquid chromatographic determination of ginsenosides of Panax quinquefolium.* J Chromatogr A. 755, 11-17.

Desmortreux, C., Rothaupt, M., West, C., Lesellier, E. (2009) *Improved separation of furocoumarins of essential oils by supercritical fluid chromatography.* J Chromatogr A. 1216, 7088-7095.

Dewick, P.M. (2009) *Medicinal Natural Products. A Biosynthetic Approach.* 3rd Edition. A John Willey and Sons, Ltd.. United Kingdom, ISBN: 978-0470-74168-9, 163-165.

Diawara, M.M., Trumble, J.T., White, K.K., Carson, W.G., Martinez, L.A. (1993) *Toxicity of linear furanocoumarins to Spodoptera exigua: evidence for antagonistic interactions.* J Chem Ecol. 19(11), 2473-2484.

Diawara, M. M., Trumble, J. T., Quiros, C. F., Hanson, R. (1995). *Implications of distribution of linear furanocoumarins within celery.* J Agric Food Chem. 43, 723-727.

Diwan, R. & Malpathak, N. (2009) *Furanocoumarins: Novel topoisomerase I inhibitors from Ruta graveolens L.* Bioorg Med. Chem. 17, 7052-7055.

Dolan, J.W., Lommen, D.C., Snyder, L.R. (1989) *Drylab Computer Simulation For High- Performance Liquid Chromatographic Method Development.* J Chromatogr. 485, 91- 112.

Drasar, P. & Moravcova, J. (2004) *Recent advances in analysis of Chinese medical plants and traditional medicines.* Analyt Technol Biomed Life Sci, 812, 3-21.

Dugo, P., Mondello, L., Dugo, L., Stancanelli, R., Dugo, G. (2000) *LC-MS for the identification of oxygen heterocyclic compounds in citrus essential oils.* J Pharm Biomed Anal. 24, 147- 154.

Dugo, P., Favoino, O., Luppino, R., Dugo, G., Mondello, L. (2004) *Comprehensive Two- Dimensional Normal-Phase (Adsorption)-Reversed-Phase Liquid Chromatography.* Anal Chem. 76/9, 2525-2530.

Elgamal, M.H.A., Elewa, N.H., Elkhrisy, E.A.M., Duddeck, H. (1979) *^{13}C NMR Chemical shifts and carbon-proton coupling constants of some furanocoumarins and furochromones.* Phytochem. 18, 139-143.

Fritz, J.S. & Macha, M. (2000) *Solid-phase trapping of solutes for further chromatographic or electrophoretic analysis* J. Chromatography A, 902, 137-166.

Gadzikowska, M., Petruczynik, A., Waksmundzka-Hajnos, M., Hawrył, M., Jóźwiak, G. (2005) *Two-dimensional planar chromatography of tropane alkaloids from Datura innoxia Mill.* J Planar Chromatogr 18, 127–131.

Giannetti, V., Mariani, M.B., Testani, E., D'Aiuto, V. (2010) *Evaluation of flavonoids and furocoumarins in bergamot derivatives by HPLC-DAD.* J Commodity Sci Technol Quality. 49(I), 63-72.

Glensk, M., Sawicka, U., Mażol, I., Cisowski, W. (2002) *2D TLC – graft planar chromatography in the analysis of a mixture of phenolic acids.* J Planar Chromatogr. 15, 463-465.

Głowniak, K. (1988) *Investigation and isolation of coumarin derivatives from polish plant material.* Dissertation, Medical University, Lublin, Poland. (in Polish).

Hadaček, F., Müller, C., Werner, A., Greger, H., Proksch, P. (1994) *Analysis, isolation and insecticidal activity of linear furanocoumarins and other coumarin derivatives from Peucedanum (Apiaceae: Apioideae).* J Chem Ecol. 20, 2035-2054.

Hamerski, D., Beier, R.C., Kneusel, R.E., Matern, U., Himmelspachz, K. (1990) *Accumulation of coumarins in elicitor-treated cell suspension cultures of Ammi majus.* Phytochem. 4, 1137-1142.

Han, J., Ye, M., Xu, M., Sun, J.H., Wang, D.A., Guo, D.A. (2007) *Characterization of flavonoids in the traditional Chinese herbal medicine-Huangqin by liquid chromatography coupled with electrospray ionization mass spectrometry.* J Chromatogr B. 848, 355-362.

Härmälä, P., Botz, L., Sticher, O., Hiltunen, R. (1990) *Two-dimensional planar chromatographic separation of a complex mixture of closely related coumarins from the genus Angelica.* J Planar Chromatogr. 3, 515-520.

Hawrył, M.A., Soczewiński, E., Dzido, T.H. (2000) *Separation of coumarins from Archangelica officinalis in high-performance liquid chromatography and thin-layer chromatography systems.* J Chromatogr A. 886, 75-81.

Heinrich, M., Barnes, J., Gibbons, S., Williamson, E.M. (2004) *Fundamentals of Pharmacognosy and Phytotherapy. Churchill Livingstone, Edinburgh,* UK.

Hennion, M.-C. (1999) *Solid-phase extraction: method development, sorbents, and coupling with liquid chromatography.* J Chromatogr A. 856/1-2, 3-54

Holtick, U., Marshall, S.R., Wang, X.N., Hilkens, C.M., Dickinson, A.M. (2008) *Impact of psoralen/UVA-treatment on survival, activation, and immunostimulatory capacity of monocyte-derived dendritic cells.* Transplantation 15, 757-66.

Hoult, J.R.S. & Payá, M. (1996) *Pharmacological and Biochemical Actions of Simple Coumarins: Natural Products with Therapeutic Potential.* Gen Pharmac. 27(4), 713-722.

Ito, Y. (1981) *Efficient preparative countercurrent chromatography with a coil planet centrifuge.* J Chromatogr. 214, 122-125.

Ito, Y. (1986) *A new angle rotor coil planet centrifuge for countercurrent chromatography: Part I. Analysis of acceleration.* J Chromatogr. 358, 313-323.

Izquierdo, M.E.F., Granados, J.Q., Mir, M.V., Martinez, M.C.L. (2000) *Comparison of methods for determining coumarins in distilled beverages.* Food Chem. 70, 251-258.

Jerzmanowska Z. (1967) *The plant substances. Methods of determination.* PWN, Warsaw. (in Polish)

Kang, J., Zhou, L., Sun, J., Han, J., Guo, D.-A. (2008) *Chromatographic fingerprint analysis and characterization of furocoumarins in the roots of Angelica dahurica by HPLC/DAD/ESI-* MS^n *technique.* J Pharm Biomed Anal. 47, 778-785.

Katoh, A. & Ninomiya, Y. (2010) *Relationship between content of pharmacological components and grade of Japanese Angelica radixes.* J Ethnopharm. 130, 35-42.

Kohlmünzer S. (2010) *Pharmacognosy. Textbook for students of pharmacy.* PZWL. Warsaw. ISBN: 978-83-200-4291-7. (in Polish)

Kozyra, M. & Głowniak, K. (2006) *Influence of the extraction mode and type of eluents on the isolation of coumarins from plant material.* Herba Pol. 52(4), 59-64.

Królicka, A., Kartanowicz, R., Wosiński, S.A., Szpitter, A., Kamiński, M., Łojkowska, E. (2006) *Induction of secondary metabolite production in transformed callus of Ammi majus L. grown after electromagnetic treatment of the culture medium.* Enzyme and Microbial Technology. 39, 1386-1391.

Lee, C.H., Mamela, A.J., Vonderheid, E.C. (2007) *Erythrodermic cutaneous T cell lymphoma with hypereosinophilic syndrome: Treatment with interferon alfa and extracorporeal photopheresis.* Int. J. Dermatol. 46, 1198-1204.

Lehr, G.J., Barry, T.L., Franolic, J.D., Petzinger, G., Scheiner, P. (2003) *LC determination of impurities in methoxsalen drug substance: isolation and identification of isopimpinellin as a major impurity by atmospheric pressure chemical ionization LC/MS and NMR.* J Pharm Biomed Anal. 33, 627-637.

Lesellier, E. (2008) *Overview of the retention in subcritical fluid chromatography with varied polarity stationary phases.* J Sep Sci. 31/8, 1238-51.

Lesellier, E. (2009) *Retention mechanisms in super/subcritical fluid chromatography on packed columns* J Sep Sci. 1216/10, 1881-90.

Lesellier, E. & West, C. (2007) *Combined supercritical fluid chromatographic methods for the characterization of octadecylsiloxane-bonded stationary phases.* J Chromatogr A. 1149, 345-357.

Li, B., Abliz, Z., Tang, M., Fu, G., Yu, S. (2006) *Rapid structural characterization of triterpenoid saponins in crude extract from Symplocos chinensis using liquid chromatography combined with electrospray ionization tandem mass spectrometry.* J Chromatogr A. 1101/1-2, 53- 62.

Li, H.-B. & Chen, F. (2004) *Preparative isolation and purification of bergapten and imperatorin from the medicinal plant Cnidum monnieri using high-speed counter-current chromatography by stepwise icreasing the flow-rate of the mobile phase.* J Chromatogr A. 1061, 51-54.

Liang, H.J., Suk, F.M., Wang, C.K., Hung, L.F., Liu, D.Z., Chen, N.Q., Chen, Y.C., Chang, C.C., Liang, Y.C. (2009) *Osthole, a potential antidiabetic agent, alleviates hyperglycemia in db/db mice.* Chem Biol Interact. 181/3, 309-315.

Lindholm, J., Westerlund, D., Karlsson, K., Caldwell, K. (2003) *Use of liquid chromatography- diode-array detection and mass spectrometryfor rapid product identification in biotechnological synthesis of ahydroxyprogesterone.* J Chromatogr A. 992, 85-100.

Liu, A.H., Lin, Y.H., Yang, M., Guo, H., Guan, S.H., Sun, J.H., Xu, M., Guo, D.A. (2007) *Development of the fingerprints for the quality of the roots of Salvia miltiorrhiza and its related preparations by HPLC-DAD and LC-MS^n.* J Chromatogr B. 846, 32-41.

Liu, R., Feng, L., Sun, A., Kong, L. (2004) *Preparative isolation and purification of coumarins from Cnidium monnieri (L.) Cusson by high-speed counter-current chromatography.* J Chromatogr A. 1055, 71-76.

Liu, R., Li, A., Sun, A. (2004a) *Preparative isolation and purification of coumarins from Angelica dahurica (Fisch. ex Hoffm) Benth, et Hook. f (Chinese traditional medicinal herb) by high- speed counter-current chromatography.* J Chromatogr A. 1052, 223-227.

Liu, R., Li, A., Sun, A., Kong, L. (2004b) *Preparative isolation and purification of psoralen and isopsoralen from Psoralea corylifolia by high-speed counter-current chromatography.* J Chromatogr A. 1057, 225-228.

Lozhkin, A.V. & Sakanyan, E.I. (2006) *Structure of chemical compounds, methods of analysis and process control. Natural coumarins: methods of isolation and analysis.* Pharm Chem J. 40(6), 337-346.

Łuczkiewicz, M., Migas, P., Kokotkiewicz, A., Walijewska, M., Cisowski, W. (2004) *Two- dimensional TLC with adsorbent gradient for separation of quinolizidine alkaloids in the herb and in-vitro cultures of several Genista species.* J Planar Chromatogr. 17, 89-94.

Łuszczki, J.J., Głowniak, K., Czuczwar, S.J. (2007) *Imperatorin enhances the protective activity of conventional antiepileptic drugs against maximal electroshock-induced seizures in mice.* European J Pharmacol. 574, 133-139.

Łuszczki, J.J., Głowniak, K., Czuczwar, S.J. (2007a) *Time-course and dose-response relationships of imperatorin in the mouse maximal electroshock seizure threshold model.* Neurosci Res. 59, 18-22.

Ma, Y., Ito, Y., Sokolovsky, E., Fales, H.M. (1994) *Separation of alkaloids by pH-zone-refining countercurrent chromatography.* J Chromatogr A. 685, 259-262.

Maksymowych, R. & Ledbetter, M. C. (1986) *Fine Structure of epithelial canal cells in petioles of Xanthium pensylvanicum.* Amer J Bot. 74, 65-73.

Manthey, J.A., Myung, K., Martens-Talcott, S., Derendorf, H., Butterweck, V., Wildmer, W.W. (2006) *The isolation of minor-occuring furanocoumarins in grapefruit and analysis of their inhibition of CYP3A4 and p-glycoprotein transport of talinolol from CACO-2 cells.* Proc Fla State Hort Soc. 119, 361-366.

Markowski, W. & Czapińska, K.L. (1997) *Computer-assisted separation by HPLC with diode array detection and quantitative determination of furanocoumarins from Archangelica officinalis.*Chem Anal. 42/3, 353-363.

Matysik, G., Skalska-Kamińska, A., Stefańczyk, B., Wójciak-Kosior, Rapa, D. (2008) *Application of a new technique in two-dimensional TLC separation of multicomponent mixtures.* J Planar Chromatogr. 21/4, 233-236.

Malhotra, S., Bailey, D.G., Paine, M.F., Watkins, P.B. (2000) *Seville orange juice-felodipine interaction: Comparison with dilute grapefruit juice and involvement of furocoumarins.* Clinical Pharmacology & Therapeutics. 69(1), 14-23.

Manthey, J.A., Myung, K., Martens-Talcott, S., Derendorf, H., Butterweck, V., Widmer, W.W. (2006) *The isolation of minor-occurring furanocoumarins in grapefruit and analysis of their inhibition of CYP 3A4 and P-glycoprotein transport of talinolol from CACO-2 cells.* Proc Fla State Hort Soc. 119, 361-366.

Martino, E., Ramaiola, I., Urbano, M., Bracco, F., Collina, S. (2006) *Microwave-assisted extraction of coumarin and related compounds from Melilotus officinalis (L.) Pallas as an alternative to Soxhlet and ultrasound-assisted extraction.* J Chromatogr A. 1125, 147- 151.

Marumoto S. & Miyazawa, M. (2011) *Microbial reduction of coumarin, psoralen, and xanthyletin by Glomerella cingulata.* Tetrahedron. 67, 495-500.

Milesi, S., Massot, B., Gontier, E., Bourgaud, F., Guckert, A. (2001) *Ruta graveolens L.: a promising species for the production of furanocoumarins.* Plant Science. 161, 189-199.

Mistry, K., Krull, I., Grinberg, N. (2002) *Capillary electrochromatography: An alternative to HPLC and CE.* J Sep Sci. 25/15-17, 935-958.

Molnar, J. (2002) *Computerized design of separation strategies by reversed-phase liquid chromatography: Development of DryLab software.* J Chromatogr A. 965, 175-194.

Murray, R. D. H.. Mendez, J. and Brown, S. A. [eds.]. (1982). *The Natural Coumarins: Occurrence, Chemistry and Biochemistry.* John Wiley 8 Sons Ltd., Toronto.

Nilsson, U.J. (2000) *Solid-phase extraction for combinatorial libraries.* J Chromatogr A, 885/1-2, 305-319.

Novakova, L., Matysova, L., Solich, P. (2006) *Advantages of application of UPLC in pharmaceutical analysis.* Talanta. 68, 908-918.

Nyiredy, Sz. (2001) *Planar Chromatography. A Retrospective View for the Third Millenium*, Springer, Hungary.

Obradovič, M., Strgulc Krajšek, S., Dermastia, M., Kreft, S. (2007) *A new method for the authentication of plant samples by analyzing fingerprint chromatograms.* Phytochem Anal. 18, 123-132.

Ochocka, R.J., Raizer, D., Kowalski, P., Lamparczyk, H. (1995) *Determination of coumarins from Chrysanthemum segetum L. by capillary electrophoresis.* J Chromatogr A. 709, 197- 202.

Ojala, T., Remes, S., Haansuu, P., Vuorela, H., Hiltunen, R., Haahtela, K., Vuorela, P. (2000) *Antimicrobial activity of some coumarin containing herbal plants growing in Finland.* J Ethnopharm. 73, 299-305.

Ong, E.S., Woo, S.O., Yong, Y.L. (2000) *Pressurized liquid extraction of berberine and aristolochic acids in medicinal plants.* J Chromatogr A. 313, 57-64.

Panno, M.L., Giordano, F., Mastroianni, F., Palma, M.G., Bartella, V., Carpino, A., Aquila, S., Andó, S. (2010) *Breast cancer cell survival signal is affected by bergapten combined with an ultraviolet irradiation.* FEBS Letters. 584, 2321-2326.

Papagiannopoulos, M., Zimmermann, B., Mellenthin, A., Krappe, M., Maio, G., Galensa, R. (2002) *Online coupling of pressurized liquid extraction, solid-phase extraction and high- performance liquid chromatography for automated analysis of proanthocyanidins in malt.* J Chromatogr A. 958, 9-16.

Piao, X.L., Park, I.H., Baek, S.H., Kim, H.Y., Park, M.K., Park, J.H. (2004) *Antioxidative activity of furanocoumarins isolated from Angelicae dahuricae.* J Ethnopharm. 93, 243- 246.

Pires, A.E., Honda, N.K., Cardoso, C.A.L. (2004) *A method for fast determination of psoralens in oral solutions of phytomedicines using liquid chromatography.* J Pharm Biomed Anal. 36, 415-420.

Poole, C.F., Poole, S.K., Fernando, W.P.N., Dean, T.A., Ahmed, H.D., Berndt, J.A. (1989) *Multidimensional and multimodal thin layer chromatography: pathway to the future.* J Planar Chromatogr. 2, 336–345.

Qian, G.-S., Wang, Q., Leung, K.S.-Y., Qin, Y., Zhao, Z., Jiang, Z.-H. (2007) *Quality assessment of Rhizoma et Radix Notopterygii by HPTLC and HPLC fingerprinting and HPLC quantitative analysis.* J Pharm Biomed Anal. 44, 812-817.

Razavi, S.M., Zahri, S., Motamed, Z., Ghasemi, G. (2010) *Bioscreening of Oxypeucedanin, a Known Furanocoumarin.* Iranian J Basic Med Sci. 13(3), 133-138.

Reich, E. & Schibli, A. (2006) *High-Performance Thin-layer Chromatography for the Analysis of Medicinal Plants*, Thieme, New York, 2006.

Reitz, S. R., Karowe, D. N., Diawara, M. M., Trumble, J. T. (1997). *Effects of elevated atmospheric carbon dioxide levels on the growth and linear furanocoumarin content of celery.* J Agric Food Chem. 45: 3642-3646.

Rice-Evans, C.A. & Packer, L. (Ed(s).). (2003) *Flavonoids in Health and Disease.* Marcel Dekker, Inc., ISBN: 0-8247-4234-6, 115-119.

Rocco, A. & Fanali, S. (2007) *Capillary electrochromatography without external pressure assistance: Use of packed columns with a monolithic inlet frit.* J Chromatogr A. 1191/1-2, 263-267.

Rojas-Lima, S., Santillan, R.L., Dominguez, M.-A., Gutiérrez, A. (1999) *Furocoumarins of three species of the genus Dorstenia.* Phytochem. 50, 863-868.

Saim, N., Dean, J.R., Abdullah, Md. P., Zakaria, Z. (1997) J Chromatogr A. 791, 361.

Sidwa-Gorycka, M., Królicka, A., Kozyra, M., Głowniak, K., Bourgaud, F., Łojkowska, E. (2003) *Establishment of a co-culture of Ammi majus L. and Ruta graveolens L. for the synthesis of furanocoumarins.* Plant Science. 165, 1315-1319.

Silva da, V.B., Kawano, D.F., Carvalho, I., da Conceição, E.C., de Freitas, O., da Silva, C.H.T.P. (2009) *Psoralen and Bergapten: In Silico Metabolism and Toxicophoric Analysis of Drugs Used to Treat Vitiligo.* J Pharm Pharmaceut Sci. 12(3), 378-387.

Skalicka-Woźniak, K., Markowski, W., Świeboda, R., Głowniak, K. (2009) *Computer-Assisted Searching for Coumarins in Peucedanum alsaticum L. and Peucedanum cervaria (L.) Lap.* Acta Chromatogr. 21 (4), 531-546.

Snyder, L.R., Kirkland, J.J., Glajch, J.L. (1997) *Practical HPLC Method Development*, John Wiley and Sons, New York.

Stanley, W.L. & Vannier, S.H. (1967) *Psoralens and Substituted Coumarins from Expressed Oil of Lime.* Phytochem. 6, 585-596.

Stanley-Horn, D.E. (1999) *Induction of linear furanocoumarins in celery, Apium graveolens by insect damage and their effects on Lygus lineolaris and the parasitoid Persitenus stygicus. A Thesis Presented to The Faculty of Graduated Studies of The University of Guelph.* Canada.

Steck, W. & Bailey, B.K. (1969) *Characterization of plant coumarins by combined gas chromatography, ultraviolet absorption spectroscopy, and nuclear magnetic resonance analysis.* Can J Chem. 47, 3577-3583.

Su, J., Zhang, C., Zhang, W., Shen, Y.H., Li, H.L., Liu, R.H., Hu, X.Z.X.J., Zhang, W.D. (2009) *Qualitative and quantitative determination of the major coumarins in Zushima by high performance liquid chromatography with diode array detector and mass spectrometry.* J Chromatogr A. 1216, 2111-2117.

Suzuki, H., Nakamura, K., Iwaida, M. (1979) *Detection and determination of bergapten in bergamot oil and in cosmetics.* J Soc Cosmet Chem. 30, 393-400.

Szakiel, A. (1991) *The role of phytoalexins in natural plant resistance.* Postępy Biochemii. 37/2, 104-112.

Tang, C.-H., Yang, R.-S., Chien, M.-Y., Chen, C.-C., Fu, W.-M. (2008) *Enhancement of bone morphogenetic protein-2 expression and bone formation by coumarin derivatives via p38 and ERK-dependent pathway in osteoblasts.* European J Oharmacol. 579, 40-49.

Taniguchi, M., Inoue, A., Shibano, M., Wang, N.-H., Baba K. (2011) *Five condensed furanocoumarins from the root of Heracleum candicans Wall.* J Nat Med. 65, 268-274.

Terreaux, C., Maillard, M., Stoeckli-Evans, H, Gupta, M.P., Downum, K.R., Quirke, J.M.E., Hostettmann, K. (1995) *Structure revision of a furanocoumarin from Dorstenia contrajerva.* Phytochem. 39(3), 645-647.

Tesso, H., König, W.A., Kubeczka, K.-H., Bartnik, M., Głowniak, K. (2005) *Secondary metabolites of Peucedanum tauricum fruits.* Phytochem. 66, 707-713.

Thompson, H.J. & Brown, S.A. (1984) *Separations of some coumarins of higher plants by liquid* chromatography. J Chromatogr. 314, 323-336.

Tosun, F., Kizilay, Ç.A., Erol, K., K$_1$1$_1$ç, F.S., Kürkçüoğlu, M., Başer, K.H.C. (2008) *Anticonvulsant activity of furanocoumarins and the essential oil obtained from the fruits of Heracleum crenatifolium.* Food Chem. 107, 990-993.

Trott, J., Gerber W., Hammes, S., Ockenfels, H.M. (2008) *The effectiveness of PUVA treatment in severe psoriasis is significantly increased by additional UV 308-nm excimer laser sessions.* Eur J Dermatol. 18, 55-60.

Trumble, J. T., Dercks, W., Quiros, C. F., Beier, R.C. (1990). *Host plant resistance and linear furanocoumarin content of Apium accessions.* J. Econ. Entomol. 83, 519-525.

Trumble, J. T., Millar, J. G.. Ott, D. E., Carson, W.C. (1992). *Seasonal patterns and pesticidal effects on the phototoxic linear furanocoumarins in celery. Apium graveolens* L. J. Agric.Food Chem. 40, 1501 -1506.

Um, Y.R., Kong, C.-S., Lee, J.I., Kim, Y.A., Nam, T.J., Seo, Y. (2010) *Evaluation of chemical constituents from Glehnia littoralis for antiproliferative activity against HT-29 human colon cancer cells.* Process Biochem. 45, 114-119.

Waksmundzka-Hajnos, M., Oniszczuk, A., Szewczyk, K., Wianowska, D. (2007) *Effect of sample-preparation methods on the HPLC quantitation of some phenolic acids in plant materials.* Acta Chromatogr. 19, 227-237.

Waksmundzka-Hajnos, M., Petruczynik, A., Dragan, A., Wianowska, D., Dawidowicz, A.L., Sowa, I. (2004) *Influence of the extraction mode on the yield of some furanocoumarins from Pastinaca sativa fruits.* J Chromatogr B. 800, 181-187.

Waksmundzka-Hajnos, M., Petruczynik, A., Dragan, A., Wianowska, D., Dawidowicz, A.L. (2004a) *Effect of extraction method on the yield of furanocoumarins from fruits of Archangelica officinalis Hoffm.* Phytochem Anal. 15, 313-319.

Waksmundzka-Hajnos, M., Petruczynik, A., Hajnos, M.Ł., Tuzimski, T., Hawrył, M., Bogucka-Kocka, A. (2006) *Two-dimensional thin-layer chromatography of selected coumarins.* J Chromatogr Science. 44/8, 510-517.

Waksmundzka-Hajnos, M. & Sherma, J. (Ed(s).). (2011) *High Performance Liquid Chromatography in Phytochemical Analysis.* Taylor & Francis Group, ISBN: 978-1- 4200-9260-8, 513-534.

Waksmundzka-Hajnos, M. & Wawrzynowicz, T. (1990) *The combination of thin-layer chromatography and column liquid chromatography for the separation of plant extracts.* J Planar Chromatogr. 3, 439-441.

Waksmundzka-Hajnos, M. & Wawrzynowicz, T. (1992) *On-line purifucation and micropreparative separation of contaminated fruit extacts by TLC in equilibrium sandwich chambers.* J Planar Chromatogr. 5, 169-174.

Wang, J.J., Chen, Y., Lin, M., Fan, G.R., Zhao, W.Q., Yan, C., Wang, J.M. (2007) *Development of a quality evaluation method for Fructus schisandrae by pressurized capillary electrochromatography.* J Sep Sci. 30/3, 381-390.

Wang, T.T., Jin, H., Li, Q., Cheng, W.M, Hu, Q.Q., Chen, X.H., Bi, K.S. (2007) *Isolation and Simultaneous Determination of Coumarin Compounds in Radix Angelica dahurica.* Chromatogr. 65, 477-481.

Wawrzynowicz, T. & Waksmundzka-Hajnos, M. (1990) *The Application of Systems with Different Selectivity for the Separation and Isolation of some Furocoumarins.* J Liq Chromatogr. 13/20, 3925-3940.

Wen, B., Ma, L., Nelson, S.D., Zhu, M.S. (2008) *High-Throughput Screening and Characterization of Reactive Metabolites Using Polarity Switching of Hybrid Triple Quadrupole Linear Ion Trap Mass Spectrometry.* Anal Chem. 80, 1788-1799.

West, C. & Lesellier, E.(2006) *Characterization of stationary phases in subcritical fluid chromatography by the solvation parameter model. I. Alkylsiloxane-bonded stationary phases.* J Chromatogr A. 1110/1-2, 181-190.

West, C. & Lesellier, E.(2006a) *Characterisation of stationary phases in subcritical fluid chromatography by the solvation parameter model. II. Comparison tools.* J Chromatogr A. 1110/1-2, 191-199.

West, C. & Lesellier, E.(2006b) *Characterisation of stationary phases in subcritical fluid chromatography with the solvation parameter model. III. Polar stationary phases* J Chromatogr A. 1110/1-2, 200-213.

West, C. & Lesellier, E. (2006c) *Characterisation of stationary phases in subcritical fluid chromatography with the solvation parameter model IV. Aromatic stationary phases.* J Chromatogr A. 1115, 233-245.

West, C. & Lesellier, E. (2007) *Characterisation of stationary phases in supercritical fluid chromatography with the solvation parameter model V. Elaboration of a reduced set of test solutes for rapid evaluation.* J Chromatogr A. 1169, 205-219.

West, C. & Lesellier, E. (2008) *A unified classification of stationary phases for packed column supercritical fluid chromatography.* J Chromatogr A. 1191, 21-39.

West, C. & Lesellier, E. (2008a) *Orthogonal screening system of columns for supercritical fluid chromatography.* J Chromatogr A. 1203, 105-113.

Wren, S.A.C. & Tchelitcheff, P. (2006) *Use of ultra-performance liquid chromatography in pharmaceutical development.* J Chromatogr A. 1119, 140-146.

Yang, W., Feng, C., Kong, D., Shi, X., Cui, Y., Liu, M., Wang, Q., Wang, Y., Zhang, L. (2010) *Simultaneous and sensitive determination of xanthotoxin, psoralen, isoimpinellin and bergapten in rat plasma by liquid chromatography-electrospray ionization mass spectrometry.* J Chromatogr B. 878, 575-582.

Yang, W., Ye, M., Liu, M., Kong, D., Shi, R., Shi, X., Zhang, K., Wang, Q., Lantong, Z. (2010) *A practical strategy for the characterization of coumarins in Radix Glehniae by liquid chromatography coupled with triple quadrupole-linear ion trap mass spectrometry.* J Chromatogr A. 1217, 4587-4600.

Yu, C.-C., Chen, I.-S., Cham, T.-M. (2002) *Isolation and Gas Chromatographic Method for Determination of Osthole from Cnidii Fructus.* J Food Drug Anal. 10(3), 154-158.

Zhang, H.J., Wu, Y.J., Cheng, Y.Y. (2003) *Analysis of 'SHENMAF injection by HPLC/MS/MS.* J Pharm Biomed Anal. 31, 175-183.

Zheng, X., Zhang, X., Sheng, X., Yuan, Z., Yang, W., Wang, Q., Zhang, L. (2010) *Simultaneous characterization and quantitation of 11 coumarins in Radix Angelicae Dahuricae by high performance liquid chromatography with electrospray tandem mass spectrometry.* J Pharm Biomed Anal. 51, 599-605.

Zhu, G.Y., Chen, G.Y., Li, Q.Y., Shen, X.L., Fang, H.X. (2004) *HPLC/MS/MS Method for Chemical Profiling of Radix Peucedani (Baihua Quianhu).* J Nat Med. 2, 304-308.

Zobel, A. M. & Brown, S. A. (1989) *Histological location of furanocoumarins in Ruta graveolens shoots.* Can J of Bot. 67, 915-921.

Zobel, A. M. & Brown, S. A. (1990) *Dermatitis-inducing furanocoumarins on leaf surfaces of eight species of rutaceous and umbelliferous plants.* J Chem Ecol. 16, 693-700.

Zobel, A. M., Brown, S. A., Nighswander, J.E. (1991) *Influence of acid and salt sprays on furanocoumarin concentrations on the Ruta graveolens leaf surface.* Ann Bot. 67, 213- 218.

Żołek, T., Paradowska, K., Wawer, I. (2003) *^{13}C CP MAS NMR and GIAO-CHF calculations of coumarins.* Solid State Nucl Magn Reson. 23, 77-87.

ASE 200 Accelerated Solvent Extraction Operator's Manual, Document no. 031149, revision 01, Dionex, Sunyvale, CA, 1995 (Section 3-5).

4 Chapter

ANALYTICAL TECHNIQUES AND PHYTOCHEMICAL COMPOSITION OF THE TRADITIONAL ITALIAN LIQUEUR "LIMONCELLO": A REVIEW

1. INTRODUCTION: AN OVERVIEW

The analysis of food or food-derived products that naturally occur in plants is crucial, particularly concerning food safety, composition, adulteration, and quality for Protected Geographical Indication (IGP), Controlled Origin Denomination (DOC), or Controlled and Guaranteed Origin Denomination (DOCG) designations.

A interdisciplinary strategy integrating analytical chemistry, phytochemistry, organic synthetic chemistry, and biochemistry is very advisable in this area of research.

This contest features the Italian liqueur "Limoncello," derived from the maceration of lemon peels in a hydroalcoholic saccharose solution, as a pertinent example.

Limoncello is manufactured in Southern Italy, predominantly in the Campania region (Gulf of Naples, Sorrentine Peninsula, Amalfi Coast, and the islands of Procida, Ischia, and Capri), as well as in Sicily and Sardinia.

Limoncello, adhering to an ancient tradition, is a handmade liqueur derived from the peels of Sorrento lemons. Currently, because to its widespread industrial production and the rising demand in the international market, almost all varieties of lemons are utilized to produce this liqueur. Its primary distinguishing characteristic is a vivid yellow hue. Limoncello is particularly served post-dinner as a digestif, having been refrigerated at 4 °C. It is widely popular and frequently consumed throughout Italy and various regions of the Mediterranean, but it is also gaining recognition in nations such as the United States, Canada, Australia, New Zealand, and Northern Europe.

Limoncello's popularity makes it a common addition in various cocktails, owing to its robust lemon flavor devoid of the sourness or bitterness found in lemon juice.

A standard process for the preparation of limoncello includes the following ingredients. Eight Sorrento lemons, one liter of water, one liter of pure ethanol, and eight hundred grams of sugar. Lemons are peeled to extract only the flavedo, the outer yellow layer of the skin; the resultant raw material is subsequently macerated in darkness for 15 days. The alcoholic solution obtained from maceration is subsequently combined with the sugar solution in water, and the resultant combination is filtered. The resulting liqueur is stored in darkness for 40 days, following which it is refrigerated and consumed as previously outlined.

Originating from a historic ethnic practice and utilized as a digestive in various regions globally, it is intriguing to examine the chemical composition and medicinal characteristics of Limoncello.

Council Regulation No. 1576/89 establishes a definition and description of spirit drinks and mandates an enhanced understanding of the chemical composition of limoncello. The legislation explicitly prohibits the use of nature-identical flavoring chemicals and preparations in liqueurs made from citrus fruits, in addition to outlining a number of rules regarding alcoholic beverages.

Despite several studies on lemon liquors documented in journals pertinent to the beverage industry (Bonomi et al., 2001; Moio et al., 2000; Naviglio et al., 2001, 2003, 2005a, 2005b; Romano et al., 2004), it can be asserted that the burgeoning market interest in limoncellos is not matched by the volume of scientific literature.

The primary challenge in phytochemical research is the accurate identification and quantification in natural sources using highly sensitive, selective, and validated analytical methods.

The inaugural study detailing the chemical makeup of essential oil extracted from Sicilian lemons was published in 1986 by Cotroneo and colleagues (Cotroneo et al., 1986). The same group obtained additional insights on the topic through a series of manuscripts published from the early 1990s until 2001 (Dugo G. et al., 1993, 1999, 2001; Dugo P. et al., 1998, 2000; Mondello et al., 1999). The authors employed very sensitive and selective analytical techniques, such as high-resolution gas chromatography coupled with chiral capillary columns-mass spectrometry (HRGC-MS) and two-dimensional gas chromatography (2D-GC). In 2002, they ultimately compiled a review article encompassing all their research on Citrus species (Dugo et al., 2002).

The articles by Dugo and colleagues highlight the necessity of employing advanced instrumental techniques for analyzing raw samples extracted from Citrus fruits, mostly due to their chemical complexity.

The following paragraph will review the existing literature regarding extraction methodologies and experimental setups employed in the analysis of this Italian liqueur derived from lemon peel maceration.

2. ANALYTICAL METHODS

In the last twenty years a huge progress was made concerning instrumental configuration, sensitivity and selectivity improvement (Locatelli et al., 2011) for analytical methods applied in many fields, especially for food safety.

The main problem consisted in the great complexity of extracted samples and the large number of components. For these reasons the qualitative and quantitative analyses must be very specific and robust enough to isolate, qualificate, and to quantificate the target compound (s).

In this field the best practices and improvements concerns purification and analytes extraction techniques and especially analytical instrument configuration applied to this kind of analyses.

2.1 Extraction procedures

Sample preparation steps are often sensitive to the matrix and, generally, contributed with approx. at 75% to the final error. This is particular true when multiple steps are involved into the procedure, where the final uncertainty is compounded.

The sample preparation has a straight impact on method performances, in particular trueness, precision, limits of quantitation, linearity, reproducibility, and is often the rate- limiting step for several analytical assays.

Sample processing can broadly range in complexity from simple dissolution in an appropriate solvent to a complicated extraction procedure, followed by derivatization and further extraction.

There are three general extraction methods, solvent extraction (or liquid-liquid extraction, LLE), solid phase extraction (SPE), and solid phase micro-extraction (SPME).

The use of these techniques, however, often entails the use of precipitation procedures, crude separation processes, and subsequent concentration methods.

Traditional liquid-liquid extraction procedures employ in a serial of extraction of an aqueous sample with an immiscible organic solvent resulting in a large solvent volume that must be dried and then concentrated prior to analysis, bring to an expensive, and in some cases non-reproducible procedure.

For this reason, further extraction methodologies were developed and validated. In particular, in SPE procedures, a solid sorbent material is packed into a cartridge or imbedded in a disk and performs essentially the same function as the organic solvent in liquid-liquid extraction, with a smaller organic solvent volume consumption.

SPE can be applied in several fields, from bio-analytical to environmental analyses, but it requires a sample volume adequately to extract targets analytes and the possibility to replicate the analysis because SPE is a destructive methodology.

In the early 1990s, the development of a new sampling method, non-destructive, sensitive, reproducible, relatively inexpensive and in particular solventless, allowed the trapping of volatile organic compounds (VOCs) on a silica optical fibre coated with a polymer thin layer followed by their identification by GC.

For the analyses of Limoncello the first extraction procedure used to determine the product volatile components was Liquid-Liquid Extraction (LLE).

In this methodology, extraction organic solvent generally used for this purpose is a hexane/ethyl acetate mixture (Starrantino et al., 1997, Dugo et al., 2000, Versari et al., 2003).

In particular the analyses of lemon-derived products were achieved with a preliminary dilution (1:10)-extraction step with hexane and ethyl acetate (75:25, v/v) (Starrantino et al., 1997) or pure hexane (Mondello et al., 2003) followed by direct analyses of extracts.

Solid-phase micro extraction (SPME) derives principally from SPE technologies and improves the concept of reducing solvent consumption, economic characteristic, and involved through two different extraction methods: headspace SPME (HS-SPME) and contact SPME.

The headspace SPME is based on the absorption of the analytes on a fibre coating placed in the sample's headspace volume and on the partition of the target analytes between the sampling matrix and the fibre. After exposure the fibre bearing the concentred analytes was retracted, removed from the sample vial and VOCs were thermally desorbed by insertion of the fibre into the injector port of the chromatograph.

Recently in the preparative sample scenario appeared several automated and fully independent instrumentation that allow to process high sample number (high-throughput), and especially with an improvement of efficiency and process reproducibility.

When gas-chromatographic assays were used to quantify diluted Limoncello extracts it was also necessary to decrease the concentration of ethanol that tended to immediately saturate the SPME fibres. The best election, and generally, used methodology for the analysis of volatile components was headspace SPME

(HS-SPME) on polydimethylsiloxane (PDMS) as thin layer (TL) due to non-polar characteristic of samples compounds.

This technique can be easily automated (Crupi et al., 2007) to improve analyses number (high throughput assay) and to obtain a better control on overall analysis steps, which bring to higher methodology reproducibility.

2.2 Instrumental analysis

Mono-dimensional chromatographic processes (1D) are widely applied in the analysis of food and food-derived products.

Although such methods often provide satisfying analytical results, the complexity of several matrices exceeds the resolution capacity of a single dimension separation system.

In the past years several efforts have been dedicated to the combination of independent techniques with the aim of reinforcement resolving power, until the use of multi- dimensional chromatography (MDC).

A typical comprehensive bi-dimensional chromatographic separation (2D) is achieved, generally, on two distinct columns connected in series with a special transfer system located between them. The type of interface used is connected to the specific methodology. The function of the interface is to cut and then release continuous fractions of the primary column effluent onto a fast separation column. In order to achieve comprehensive analysis and to preserve the 1D separation, the bands injected onto the secondary column must undergo elution before the following re-injection. Secondary retention times must be, at the most, equal or less than the duration of a single modulation period.

Coupled to these chromatographic techniques (both gas chromatography and high- performance liquid chromatography), several detectors are used for a clear and univocal identification and quantification of target analyte(s).

The must used are Flame Ionisation Detector (FID) coupled with GC, Mass Spectrometry Detector (MS) coupled to GC and HPLC, and Tandem Mass Spectrometry (MS/MS) interfaced with HPLC.

Due to the complex chemical composition of Limoncello extracts (comprising several classes of volatile compounds, 85-98%), the election chromatographic techniques are certainly gas- chromatographic (GC) ones.

In mono-dimensional GC the stationary phase is bonded and highly cross-linked; silphenylene polymer (polarity similar to 5% phenyl polymethylsiloxane) in programmable thermal analyses, split/splitless injector in splitless mode. This capillary column is used with both FID and MS detector, with the unique difference in helium flow rate (minor in the MS interfacing).

With this system is possible the analyses of all volatile components fraction derived from SPME extraction, while for the analyses of enantiomeric compounds chiral capillary column characterized by diethyltertbutylsilyl-β-cyclodextrins as stationary phases are generally used.

These analyses are carried out separately to obtain a complete chemical composition of volatile components. Recently Mondello and coworkers (Mondello et al., 2006) coupled these two gas-chromatographic separations in an innovative multi-dimensional GC (MDGC) system, to obtain the complete chemical composition on volatile components including enantiomeric resolution in a single analysis.

The main goal of this configuration is especially due to the possibility of evaluating the enantiomeric ratios as genuineness markers, especially in complex matrices as Limoncellos.

To detect analytes of interest flame ionisation detector (FID) and mass spectrometric detector (MS) are generally chosen.

With the first, that is a "universal" gas-chromatographic detector is possible to obtain, in a single run a complete chemical fingerprint of volatile fraction, with identification of various components by retention index (R_i). This configuration is very useful if are at disposition data bank with R_i or chemical standards. If no pure chemical standards are available, the best choice is gas-chromatography-mass spectrometry interfacing (GC-MS), because is possible to obtain R_i of several volatile components and, in addition, the mass-to-charge ratio for the correct identification of targeted compounds.

Another trend in the last years is the use of MDGC, coupled with mass spectrometry. Gas- chromatographic determination has the disadvantage that by this technique it is difficult to analyse non-volatile components of extracted samples.

High performance liquid chromatography (HPLC) is generally used for the non-volatile components. Is a well-defined technique, robust and reproducible that generally is coupled with "universal" detector as Diode Array Detector (DAD).

In the literature several paper dealing with the analysis of flavones, coumarins, and furanocoumarins composition of lemon-derived products by HPLC using different chromatographic column were reported.

In particular for the analyses of coumarins and furanocoumarins in Limoncellos samples the most widely used column are silica based (generally 300 x 4 mm, 10 µm particle size) to obtain a complete resolution in normal phase mode (Starrantino et al., 1997, Versari et al., 2003), while for the analyses of phenolic compounds Octadecylsilane (ODS) column in reversed phase mode was used (Versari et al., 2003).

Ultraviolet-visible (UV/Vis) detection was carried out at 315-330 nm for coumarins and furanocoumarins and at 280 nm for phenolic compounds.

The analyses, due to the samples matrices complexity, are carried out in gradient elution mode. In particular for the analyses of coumarins and furanocoumarins was used a mobile Phase constituted by hexane: ethyl acetate (92:8 or 88:12, v/v) and hexane: ethyl alcohol (90:10, v/v).

Only recently Crupi and coworkers (Crupi et al., 2007) developed a unique reversed phase method for simultaneous determination at 315 nm of coumarins and furanocoumarins using water and acetonitrile as mobile phase in gradient elution mode.

3. LIMONCELLO CHEMICAL COMPOSITION

It is a matter of fact that the evaluation of the organoleptic properties of limoncello is, although indirectly, connected to the analysis of the essential oil composition. The aroma of the liquor is actually one of the first consumer's perceptions that are crucial in establishing the preference among several products available in the market. For this reason, the analysis of the aromatic fraction of limoncello liquor seems to be an important item in assessing its genuineness and quality, besides "tracing" the various steps of the preparation procedure.

The current chemical composition of this Italian liqueur is mainly related to the aromatic fraction that was reported by Crupi and coworkers (Crupi et al., 2007) using previously mentioned analytical techniques.

In their comprehensive studies, after SPME extraction of volatile components and chromatographic analyses, both GC and HPLC, they detected several monoterpenes, sesquiterpenes, and oxygenated compounds, as reported in the Table 1.

Versari and coworkers reported similar analyses of several commercial Limoncello samples and chemometric elaboration with Principal Components Analysis (PCA) technique to obtain two main group: the first that showed a composition similar to lemon essential oils (high content of β–pinene, myrcene, *trans*-α-bergamotene, and β-bisabolene, and a low content in neral and geranial) (Versari et al., 2003). The composition of the second group suggested the presence of oxidative phenomena and (or) addition of flavours, in particular the presence of compounds as ethyl acetate, acetaldehyde, 2-methyl-1-propanol and glycerol indicate that fermentation process probably occurred in the sugar syrup during Limoncello dilution step after the extraction process.

The best recurrent molecules of the oil and of Limoncello (both homemade and commercial ones) belonging the volatile compounds family are α-pinene, sabinene, β-pinene, myrcene, *p*-cymene, limonene, γ-terpinene, neral, geraniol, and geranial, and as reported in the Table 1. As results of their investigation, Crupi and coworkers (Crupi et al., 2007) underline that most of analysed Limoncello samples were effectively obtained directly from fruits, but also using terpeneless oils and in some cases synthetic products of reconstituted oils.

From data reported in the Tables 1 and 2 is possible to indicate that the commercial #05 and #02 are of high quality, even if commercial #02, due to high level of *p*-cymene can indicate a long storage time.

Table 1. Amount (mg/L) of major representative volatile components in several lemon samples-derivatives in oil, homemade Limoncello, and commercial Limoncello products (Crupi et al., 2007).

Com-pounds	Oil	Home-made Limon-cello	Commercial Limoncello															
			#01	#02	#03	#04	#05	#06	#07	#08	#09	#10	#11	#12	#13	#14	#15	#16
α-Pinene	14.510	21.0	8.8	13.4	8.9	12.6	36.5	11.6	10.6	12.5	9.9	8.9	9.7	9.7	14.6	9.2	9.0	8.9
Sabinene	17.570	211.3	-	14.4	9.0	-	20.8	8.8	-	-	9.7	9.0	-	-	14.9	-	-	9.0
β-Pinene	142.090	-	9.5	66.8	11.6	37.1	247.9	42.4	12.2	44.5	23.4	11.2	18.4	9.3	78.5	13.6	11.0	11.6
Myrcene	11.900	18.8	8.9	16.1	9.0	13.0	35.7	12.1	11.0	11.0	9.2	9.1	10.5	10.2	15.0	9.5	9.3	9.1
p-Cymene	340	4.5	9.2	84.9	13.3	17.4	11.8	17.2	17.6	28.7	12.7	10.9	14.5	15.3	58.5	13.2	9.5	11.3
Limonene	626.290	837.2	26.5	634.6	43.6	353.3	1671.0	272.1	212.1	288.78	42.6	45.3	179.3	141.4	475.7	50.2	60.6	57.9
γ- Ter-pinene	104.670	143.9	9.5	11.9	10.3	47.8	281.9	31.5	30.2	34.0	12.3	10.1	13.8	21.6	25.1	8.9	10.9	15.9
Neral	13.520	390.0	30.4	10.8	-	23.6	11.7	13.5	10.2	11.0	10.1	24.8	64.0	26.5	9.9	44.6	26.3	85.0
Geraniol	270	3.2	3.3	4.9	1.9	64.7	10.9	10.9	296.1	106.2	4.6	2.5	6.2	177.5	4.6	21.8	2.8	3.6
Geranial	20.960	581.6	40.2	10.1	-	24.5	17.9	-	9.7	13.7	10.2	32.1	94.0	19.4	11.3	65.5	40.4	133.5

Table 2. Identificated and quantificated coumarins, psoralem, and phenolics composition (mg/L) of several commercial Limoncello samples (Versari et al., 2003).

Compound	Commercial Limoncello											
	#01	#02	#03	#04	#05	#06	#07	#08	#09	#10	#11	#12
Bergamottin	1.7	4.7	21.5	1.1	0.9	3.8	2.3	19.6	3.7	1.2	1.0	3.6
5-gerany-loxy-7-meth-oxycoumarine	0.7	1.6	0.8	0.5	0.5	1.3	0.8	7.0	1.2	0.4	0.2	1.2
Citroptene derivate	-	-	0.1	0.1	-	-	-	0.3	-	-	-	-

5-isopenteny-loxy-7-meth-oxycoumarine	-	-	0.4	0.2	-	-	-	1.0	-	-	-	-
Citroptene	2.0	-	1.5	1.0	1.3	0.3	1.8	4.0	0.4	1.1	0.4	1.0
Imperatorin	4.4	0.8	-	0.6	-	-	-	1.6	-	-	-	-
Eriocitrin	16.9	-	27.8	21.7	65.9	-	-	71.2	-	29.1	-	22.3
Sinapinic acid	0.4	-	0.5	-	1.0	-	-	Trace	-	1.5	-	0.5
Narirutin	5.2	14.5	1.8	0.8	2.2	-	-	2.2	-	-	-	5.9
Naringin	-	1.5	-	1.2	Trace	-	-	Trace	Trace	Trace	1.4	1.3
Hesperidin	16.7	-	25.3	12.0	32.1	-	-	42.3	-	6.7	-	7.6
Neohesperi-din	0.6	-	Trace	Trace	Trace	-	-	1.2	-	Trace	-	0.4
Diosmin	1.9	13.3	7.6	Trace	7.9	-	5.3	-	0.8	4.0	Trace	8.6

Terpenes are photosensitive compounds; in particular limonene and terpinenes are involved in this irreversible process that negatively affects the organoleptic properties of the beverage. The terpenes fraction and/or the presence of oxidized by-products can predict the origin and the quality of a Limoncello.

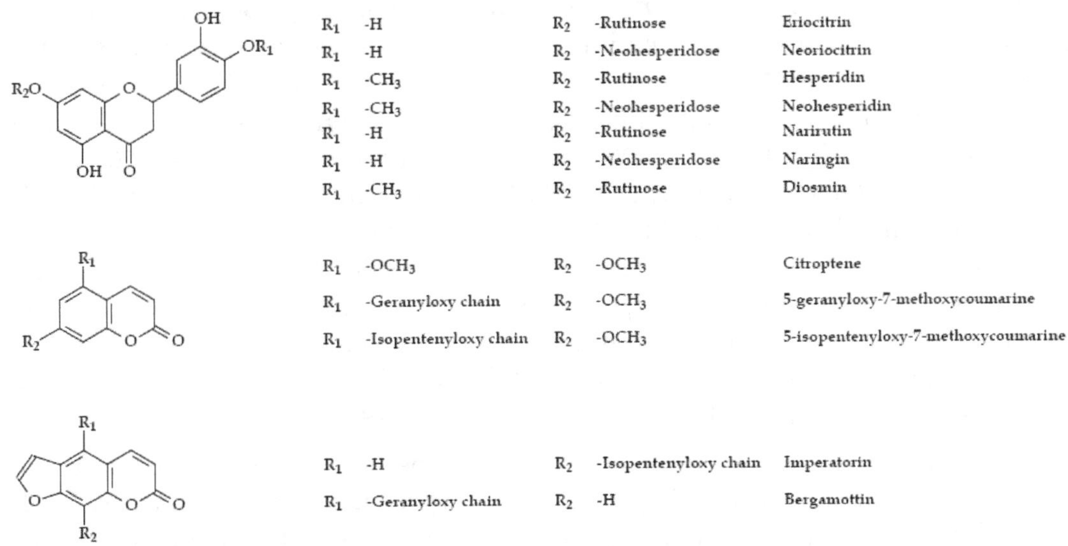

Fig. 1. Chemical structures of flavones, coumarins, and furanocoumarins identified and quantified in Limoncello by Versari and co-worker (Versari et al., 2003).

Problems connected to storage time were reported also by Da Costa and Theodore (Da Costa and Theodore, 2010) in a paper that report a complete study of over two-hundred volatile components in Limoncello and their changes that occurred with aging.

Another well-recognized question investigated by Poiana and coworkers (Poiana et al., 2006) regarded the differentiation from sample-to-sample about chemical and physical- chemical parameters of alcoholic extracts of lemons deriving from the Amalfi and Sorrento areas. These discrepancies are probably correlated with the different cultivars and growing conditions.

In particular this research underlined the evolution during the season for some classes of components. The carbonyl-to-oxygenated, alcohols-to-oxygenated and esters-to-oxygenated compounds ratios are indices of flavouring quality. These showed similar trends for both lemon types. In comparison to other lemon productions a higher amount of minor classes of components were also recorded.

4. *CITRUS* ESSENTIAL OIL HEAVY METAL COMPOSITION – FOOD QUALITY

Citrus essential oil is a very complex mixture of several classes of volatile (85-98%) and non- volatile (2-15%) compounds, as reported in the previous paragraphs. In this classification, the main present compounds are terpenes, hydrocarbons, esters, aldehydes, and ketones. The key difference between essential oils from different raw plant varieties is especially related to the composition of the volatile fraction (Steuer et al., 2001). *Citrus* essential oil is generally used as aromatising agent and additive for food and food-derived products, beverages, cosmetics and in some pharmaceuticals. In this sample extract is possible to incur into the presence of organic and inorganic contaminants, as well documented by the presence of pesticide residues (Saitta et al., 2000, Dugo et al., 1997; Verzera et al., 2004) and plasticizers (Di Bella et al., 1999, 2000, 2001; Saitta et al., 1997); however, there is a needs of available data regarding the presence of heavy metals in this products. Some reports concerning the microelements composition of *Citrus* peel extracts were published (Gorinstein et al., 2001; Simpkins et al., 2000). Metals, such as iron, copper, zinc, and cobalt, are non-toxic at modest concentrations, while cadmium, lead, mercury, and arsenic are toxic even in very low concentration level and constitute a significant health hazard (Rojas et al., 1999).

Metals levels in *citrus* essential oils mainly depend on the type of soil and treatment, but are also affected by the extraction procedures to manufacture food-derived products, such as scraping or pressing because the fruits inevitably come in contact with metallic surfaces.

In this field the best instrumental analysis concerns especially the use of electrochemical techniques, such as derivative potentiometric stripping analysis (dPSA) and Atomic Absorption Spectroscopic methods (AAS) in Graphite Furnace Atomic Absorption Spectroscopy Analysis (GFAAS) configuration to determine trace metals concentrations in a variety of matrices, such as alloys, food, biological materials, and environmental samples.

Recently La Pera and coworkers (La Pera et al., 2003) published an interesting work inherent the use of these two techniques for the simultaneous determination of cadmium, copper, lead, and zinc in *Citrus* essential oils with high recoveries values from real samples extracts.

In the Table 3 were reported the mean concentration determined by dPSA and AAS both on acid and methanol extracts.

Table 3. Heavy metal mean concentration (ng/g; n=3) and standard deviation (n=3) detected by dPSA and AAS in Lemon samples (La Pera et al., 2003).

		Cd		Cu	
Sample	Treatment	dPSA	AAS	dPSA	AAS
Lemon	Acid extracts	1.57±0.03	1.43±0.03	16.94±0.21	16.00 ± 0.20
Lemon	Methanol extracts	1.63±0.03	1.55±0.05	21.65±1.59	18.10±2.65
		Pb		Zn	
Sample		dPSA	AAS	dPSA	AAS
Lemon	Acid extracts	111.24±0.81	103.60±1.10	802.55±2.48	799.60±3.00
Lemon	Methanol extracts	113.24±0.72	103.30±7.15	809.62±2.26	788.99±9.10

These results indicates that *Citrus* lemon essential oils contains several heavy metals and that a deep control on food-derived products is required to obviate to any significant health hazard.

In particular, Verzera and coworkers (Verzera et al., 2004) showed that there are great differences between lemon biological oils and traditional ones inherent the organophosphorus pesticide content.

In particular these Authors reports that Parathion methyl, Parathion ethyl, Quinalphos, Methidathion, Clorpyphos methyl, and Azinphos ethyl are the must representative organophoshorus pesticides founded in lemon oils samples, both in biological and traditional agricultural methods.

Traditional oils were found to contain from 3.52 to 3.85 ppm of previously cited compounds, while biological ones were found to contain from 0.17 to 0.74 ppm.

Only in lemon oils obtained from fruits deriving from traditional agricultural methods were found Dicofol as organochlorine pesticides at 1.0 ppm level.

Previously reported values inherent organophoshorus pesticides were obtained by gas chromatography coupled with Flame Photometric Detector (FPD) that is similar to FID in that the sample exits the analytical column into a hydrogen diffusion flame. Where the FID measures ions produced by organic compounds during combustion, the FPD analyzes the spectrum of light emitted by the compounds as they luminesce in the flame.

Organochlorine pesticides were also obtained by gas chromatography coupled with Electron Capture Detector (ECD). The ECD measures electron-capturing compounds (frequently halogenated compounds) by creating an electrical field in which molecules exiting a GC column can be detected by the drop in current in the field.

5. CONCISE SUMMARY

A comprehensive understanding of the chemical properties of Limoncello could significantly aid in evaluating its authenticity and legitimacy. The examination of food products may focus on evaluating food quality and authenticity, monitoring technological and production processes, determining nutritional values, and identifying molecules and secondary metabolites that may confer beneficial effects on human health and safety.

Thus, a primary objective in food chemistry focuses on the ongoing enhancement, advancement, and specifically, the validation of progressively sensitive and selective analytical methodologies.

The accessibility of hyphenated analytical techniques and 2D chromatographic methods that can ascertain the origin, authenticity, and quality of Limoncello may motivate producers to create superior products, valued by consumers not due to extensive advertising, but for the attributes of their composition.

Subsequent research should focus on enhancing instrumental performance and high-throughput analyses, incorporating additional features such as cryo-trapping, and improving MS and MS/MS detection in both conventional gas chromatography and high-performance liquid chromatography, as well as in multi-dimensional gas chromatography (MDGC) applications.

6. REFERENCES

Bonomi, M.; Lubian, E.; Puleo, L.; Tateo, F. & Fasan, S. (2001). Criterio di caratterizzazione PCA di liquori "limoncello". *Industrie delle Bevande*, Vol. 30, pp. 371–374, ISSN: 0390-0541.

Cotroneo, A.; Dugo, G.; Licandro, G.; Ragonese, C. & Di Giacomo, G. (1986). On the genuineness of *Citrus* essential oils. Part XII. Characteristics of Sicilian lemon essential oil produced with the FMC extractor. *Flavour and Fragrance Journal*, Vol. 1, No. 3 (June 1986), pp. 125–134, ISSN: 1099-1026.

Cotroneo, A.; Verzera, A.; Lamonica, G.; Dugo, G. & Licandro, G. (1986). On the genuineness of *Citrus* essential oils. Part X. Research on the composition of essential oils produced from Sicilian lemons using "Pelatrice" and "Sfumatrice" extractors during the entire 1983/84 production season. *Flavour and Fragrance Journal*, Vol. 1, No. 2 (March 1986), pp. 69–86, ISSN: 1099-1026.

European Union, Council Regulation (EEC) No. 1576/89 (29 May 1989). Appellations of Origin (Spirit Drinks), 16/06/2011, Available from: < http://www.wipo.int/wipolex/en/text.jsp?file_id=126927>.

Crupi, M.L.; Costa, R.; Dugo, P.; Dugo, G. & Mondello, L. (2007). A comprehensive study on the chemical composition and aromatic characteristics of lemon liquor. *Food Chemistry*, Vol. 105, No. 2 (November 2007), pp. 771-783, ISSN: 0308-8146.

Da Costa, N.C. & Anastasiou, T.J. (2010) Analysis of volatiles in limoncello liqueur and aging study with sensory. *ACS Symposium Series, Flavor in Noncarbonated Beverages*, Vol. 1036 (March 2010), Chapter 13, pp. 177-193, ISBN: 9780841225510.

Di Bella, G.; Saitta, M.; Lo Curto, S.; Lo Turco, V.; Visco, A. & Dugo, G. (2000). Contamination of Italian *citrus* essential oils: presence of chloroparaffin. *Journal of Agricultural and Food Chemistry*, Vol. 48, No. 10 (September 2000), pp. 4460-4462, ISSN: 0021-8561.

Di Bella, G.; Saitta, M.; Lo Curto, S.; Salvo, F.; Licandro, G. & Dugo, G. (2001). Production process contamination of *citrus* essential oils by plastic materials. *Journal of Agricultural and Food Chemistry*, Vol. 49, No. 8 (July 2001), pp. 3705-3708, ISSN: 0021-8561.

Di Bella, G.; Saitta, M.; Pellegrino, M. C.; Salvo, F. & Dugo, G. (1999). Contamination of Italian essential oils: the presence of phthalate esters. *Journal of Agricultural and Food Chemistry*, Vol. 47, No. 3 (March 1999), pp. 1009-1012, ISSN: 0021-8561.

Dugo, G.; Bartle, K. D.; Bonaccorsi, I.; Catalfamo, M.; Cotroneo, A. & Dugo, P. (1999). Advanced analytical techniques for the analysis of *Citrus* essential oils. Part. 1. Volatile fraction: HRGC/MS analysis. *Essenze E Derivati Agrumari*, Vol. 9, pp. 79– 111, ISSN: 0014-0902.

Dugo, G. & Di Giacomo, A. (2002). *Citrus*. The Genus Citrus, Taylor & Francis, ISBN: 0-415- 28491-2, London and New York.

Dugo, G.; Mondello, L.; Cotroneo, A.; Bonaccorsi, I. & Lamonica, G. (2001). Enantiomeric distribution of volatile components of *Citrus* oils by MDGC. *Perfumer and Flavorist*, Vol. 26, pp. 20–35.

Dugo, G.; Saitta, M.; Di Bella, G. & Dugo, P. (1997). Organophosphorus and organochlorin pesticide residues in Italian citrus essential oils. *Perfumer and Flavorist*, Vol. 22, pp. 33-43.

Dugo, G.; Stagno d'Alcontres, I.; Donato, M. G. & Dugo, P. (1993). On the genuineness of *Citrus* essential oils. Part XXXVI. Detection of reconstituted lemon oil in genuine cold-pressed lemon essential oil by high-resolution gas chromatography with chiral capillary columns. *Journal of Essential Oil Research*, Vol. 5, No. 1 (January 1993), pp. 21–26, ISSN: 1041-2905.

Dugo, P.; Mondello, L.; Cogliandro, E.; Cavazza, A. & Dugo, G. (1998). On the genuineness of *citrus* oils. Part LIII. Determination of the composition of the oxygen heterocyclic fraction of lemon essential oils (*Citrus limon* (L.) Burm, f.) by normal-phase high performance liquid chromatography. *Flavour and Fragrance Journal*, Vol. 13, No. 5 (September/October 1998), pp. 329–334, ISSN: 1099-1026.

Dugo, P.; Russo, M.; Mondello, L.; Dugo, G.; Postorino, S. & Carnicini, A. (2000). Chemical and physicochemical parameters and composition of the aromatic fraction of limoncello. *Italian Journal of Food Science*, Vol. 12, No. 3, pp. 343–351, ISSN: 1120- 1770.

Gorinstein, S.; Martin-Belloso, O.; Park Y. S.; Haruenkit, R.; Lojek, A.; Ciz, M.; Caspi, A.; Libman, I. & Trakhtenberg, S. (2001). Comparison of some biochemical characteristics of different *citrus* fruit. *Food Chemistry*, Vol. 74, No. 3 (August 2001), pp. 309-315, ISSN: 0308-8146.

La Pera, L.; Saitta, M.; Di Bella, G. & Dugo, G. (2003). Simultaneous Determination of Cd(II), Cu(II), Pb(II), and Zn(II) in *Citrus* Essential Oils by Derivative Potentiometric Stripping Analysis. *Journal of Agricultural and Food Chemistry*, Vol. 51, No. 5 (February 2003), pp. 1125-1129, ISSN: 0021-8561.

Locatelli, M.; Melucci, D.; Carlucci, G. & Locatelli, C. (2011). Recent HPLC strategies to improve sensitivity and selectivity for the analysis of complex matrices. *Instrumentation Science and Technology*, In Press, ISSN: 1073-9149.

Mondello, L.; Casilli, A.; Tranchida, P.Q.; Cicero, L.; Dugo, P. & Dugo, G. (2003). Comparison of fast and conventional GC analysis for *Citrus* essential oils. *Journal of Agricultural and Food Chemistry*, Vol. 51, No. 19 (August 2003), pp. 5602-5606, ISSN: 0021-8561.

Mondello, L.; Casilli, A.; Tranchida, P.Q.; Furukawa, M.; Komori, K.; Miseki, K.; Dugo, P. & Dugo, G. (2006). Fast enantiomeric analysis of a complex essential oil with an innovative multidimensional gas chromatographic system. *Journal of Chromatography A*, Vol. 1 105, No. 1-2 (February 2006), pp. 11-16, ISSN: 0021-9673.

Mondello, L.; Catalfamo, M.; Cotroneo, A.; Dugo, G.; Dugo, G. & McNair, H. (1999). Multidimensional capillary GC–GC for the analysis of real complex samples. Part IV. Enantiomeric distribution of monoterpene hydrocarbons and monoterpene alcohols of lemon oils. *Journal of High Resolution Chromatography & Chromatography Communications*, Vol. 22, No. 6 (June 1999), pp. 350–356, ISSN: 0935-6304.

Moio, L.; Piombino, P.; Di Marzio, L.; Incoronato, C. & Addeo, F. (2000). L'aroma del liquore di limoni. *Industrie delle Bevande*, Vol. 29, pp. 499–506, ISSN: 0390-0541.

Naviglio, D.; Ferrara, L.; Montesano, D.; Mele, G.; Naviglio, B. & Tomaselli, M. (2001). New solid–liquid extraction technology for the production of lemon liquor. *Industrie delle Bevande*, Vol. 30, pp. 362–370, ISSN: 0390-0541.

Naviglio, D.; Pizzolongo, F.; Mazza, A.; Montuori, P. & Triassi, M. (2005a). Chemical– physical study of alcoholic extract of flavedo lemon and essential lemon oil to determine microbic charge responsible of limoncello turbidity. *Industrie delle Bevande*, Vol. 34, pp. 425–431, ISSN: 0390-0541.

Naviglio, D.; Pizzolongo, F.; Mazza, A.; Montuori, P. & Triassi, M. (2005b). Determination of microbial load responsible for limoncello turbidity. Physicochemical study of alcoholic extract of lemon flavedo and lemon essential oil. *Industrie delle Bevande*, Vol. 34, No. 199, pp. 424–430, ISSN: 0390-0541.

Naviglio, D.; Raia, C.; Bergna, M. & Saggiomo, S. (2003). Stability control and genuineness evaluation of lemon liquor: use of dynamic headspace. *Industrie delle Bevande*, Vol. 32, pp. 121–126, ISSN: 0390-0541.

Poiana, M.; Attanasio, G.; Albanese, D. & Di Matteo, M. (2006). Alcoholic extracts composition from lemon fruits of the Amalfi-Sorrento peninsula. *Journal of Essential Oil Research*, Vol. 18, No. 4 (July/August 2006), pp. 432-437, ISSN: 1041-2905.

Rojas, E.; Herrera, L.A.; Poirier, L.A. & Ostrosky-Wegman, P. (1999). Are metals dietary carcinogens? *Mutation Research/Genetic Toxicology and Environmental Mutagenesis*, Vol. 443, No. 1-2 (July 1999), pp. 157- 181, ISSN: 1383-5718.

Romano, R.; Schiavo, L.; Iavarazzo, E.; Battaglia, A. & Cassano, A. (2004). Qualitative and quantitative characteristics of lemon juice concentrate obtained by osmotic distillation. *Industrie delle Bevande*, Vol. 33, pp. 529–532, ISSN: 0390-0541.

Saitta, M.; Di Bella, G.; Bonaccorsi, I.; Dugo, G. & Della Cassa, E. (1997). Contamination of *citrus* essential oils: the presence of phosphorated plasticizers. *Journal of Essential Oil Research*, Vol. 9, No. 6 (November 1997), pp. 613-618, ISSN: 1041-2905.

Saitta, M.; Di Bella, G.; Salvo, F.; Lo Curto, S. & Dugo, G. (2000). Organochlorine pesticides residues in Italian *citrus* essential oils, 1991-1996. *Journal of Agricultural and Food Chemistry*, Vol. 48, No. 3 (February 2000), pp. 797-801, ISSN: 0021-8561.

Simpkins, W.A.; Honway, L.; Wu, M.; Harrison, M. & Goldberg, D. (2000). Trace elements in Australian orange juice and other products. *Food Chemistry*, Vol. 71, No. 4 (December 2000), pp. 423-433, ISSN: 0308-8146.

Starrantino, A.; Terranova, G.; Dugo, P.; Bonaccorsi, I. & Mondello, L. (1997). On the Genuineness of Citrus Essential Oils. Part IL. Chemical Characterization of the Essential Oil of New Hybrids of Lemon Obtained in Sicily. *Flavour and Fragrance Journal*, Vol. 12, No. 3 (May 1997), pp. 153–161, ISSN: 1099-1026.

Steuer, B.; Shulz, H. & Lager, E. (2001). Classification and analysis of *citrus* essential oils by NIR spectroscopy. *Food Chemistry*, Vol. 72, No. 1 (January 2001), pp. 113-117, ISSN: 0308-8146.

Versari, A.; Natali, N.; Russo Maria, T. & Antonelli, A. (2003). Analysis of Some Italian Lemon Liquors (Limoncello). *Journal of Agricultural and Food Chemistry*, Vol. 51, No. 17 (July 2003), pp. 4978-4983, ISSN: 0021-8561.

Verzera, A.; Trozzi, A.; Dugo, G.; Di Bella, G. & Cotroneo, A. (2004) Biological lemon and sweet orange essential oil composition. *Flavour And Fragrance Journal*, Vol. 19, No. 6 (November 2004), pp. 544–548, ISSN: 1099-1026.

5 Chapter

THE IMPACT OF NON-THERMAL TECHNOLOGIES ON PHYTOCHEMICALS

1. INTRODUCTION: AN OVERVIEW

Phytochemicals are non-nutritive compounds derived from plants that provide protective functions in the human body against disease. These phytochemicals are regarded as biologically active secondary metabolites that impart color and flavor, generally termed nutraceuticals (Kalt, 2001). Thousands of known phytochemicals, primarily generated from phenylalanine and tyrosine, provide various activities including pigmentation, antioxidation, and UV protection (Shahidi & Naczk, 2004). Evidence of the health benefits associated with the consumption of plant-derived phytochemicals has led to a heightened demand for fresh fruits and vegetables, which contain various compounds such as alkaloids (e.g., caffeine and theobromine), carotenoids (e.g., lycopene), flavonoids (e.g., flavon-3-ols), isoflavones (e.g., genistein), and phenolic acids (e.g., capsaicin, gallic acid, and tannic acid), depending on the plant species.

The health benefits of several phytochemicals are primarily ascribed to their antioxidant properties, while alternative mechanisms may possibly be involved. Fruits and vegetables are recognized for their substantial content of phytochemicals that possess both free-radical and non-radical scavenging abilities against reactive oxygen species. These harmful compounds have been recognized as toxic to cellular tissues, resulting in oxidative damage to proteins, membrane lipids, and DNA, obstructing enzymatic pathways, and provoking gene mutations. These processes are thought to underlie various chronic human diseases, including diabetes, some cancers, and certain cardiovascular and degenerative conditions (Eberhardt et al., 2000; Arab & Steck, 2000).

Alterations in phytochemical concentrations due to processing and storage can significantly undermine food quality and, ultimately, its acceptability. Although many of these molecules may be influenced by extreme temperatures, heat processing remains the predominant approach for inactivating microbes and enzymes in processed foods. However, during thermal processing, in addition to the inactivation of microorganisms, the sensory and nutritional elements of plant-based foods are adversely altered. Consumers seek products that retain their original nutritional benefits; in this regard, non-thermal technologies serve as preservation methods effective at mild temperatures (up to 40 ºC), thereby minimizing adverse thermal impacts on the nutritional content of products. These non-thermal methods guarantee the inactivation of microbes and enzymes while minimizing impacts on nutritional and quality attributes. Non-thermal technologies such as pulsed electric fields (PEF), high pressure (HP), irradiation, and ultrasound have been researched and developed over the past few decades with the ultimate goal of industrial implementation. The effects of non-thermal methods on bacteria, enzymes, and quality-related metrics have been thoroughly examined. Numerous recent endeavors have been undertaken to assess the influence of PEF, HP, irradiation, and ultrasounds on the stability of phytochemicals. This analysis examines the impact of non-thermal technologies on health-related chemicals in plant-based products..

2.NON-THERMALTECHNOLOGIESTOPRESERVEPHYTOCHEMICALS

2.1 Non-thermal processing basics

Non-thermal technologies may allow obtaining safe and shelf-stable plant-based products with minor changes or increased content in phytochemicals. Most differences between non- thermal and thermal treatments can be explained through the temperatures reached through processing. In general, temperature during processing and storage are important factors affecting the phytochemical content of the processed product. PEF processing involves the application of a high intensity electric field (20–80 kV/cm) in the form of short pulses to a food placed between two electrodes. PEF technology provides fresh-like and safe foods while reducing quality losses that can be triggered after thermal processing (Morris et al., 2007). Application of continuous PEF processing is not suitable for solid food products that do not allow pumping and is restricted to low-conductive food products without air bubbles. Liquid food products are susceptible to be treated by PEF because they are primarily composed by water and nutrients, which are electrical conductors due of the presence of large concentrations of salts and dipolar molecules. Hence, PEF technology has been suggested for the pasteurization of foods such as juices, milk, yogurt, soups, and liquid eggs (Vega-Mercado et al., 1997; Evrendilek et al., 2004; Elez-Martínez et al., 2005; Monfort et al., 2010). In general, a continuous PEF treatment system is composed of treatment chambers, a pulse generator, a fluid-handling system, and monitoring systems (Elez- Martínez et al., 2006). Temperature- and pulse-monitoring systems are used to supervise the process. The effectiveness of PEF technology not only depends on the type of equipment used but also on the treatment parameters and media to be processed (Barbosa-Cánovas et al., 1999). In addition, processing parameters such as electric field strength, total treatment time, pulse shape, pulse frequency, pulse width, polarity and temperature are involved in the degradation or generation of phytochemicals.

HP processing uses water as a medium to transmit pressures from 300 to 700 MPa to foods resulting in a reduction on microbial loads and thus extending shelf life (Patras et al., 2009). Pressure is nearly transmitted uniformly throughout the food, thus leading to very different applications in many different food products. Although pressure results are uniform, the HP technique cannot completely avoid the well-known classical limitation of heat transfer especially during pressure build up and decompression. An increase or a decrease of pressure is associated with a proportional temperature (T) change of the vessel contents, respectively, due to adiabatic heating or cooling temperature gradient. The effectiveness of HP treatment on the overall food quality and safety is not only influenced by extrinsic (process) factors such as treatment time, pressurisation/decompression rate, pressure/ temperature levels and the number of pulses, but also by intrinsic factors of the treated food product such as food composition and the physiological states of microorganisms (Knorr, 2001; Smelt et al., 2002). Effects of pressure and temperature on food constituents are governed by activation volume and activation energy. Differences in sensitivity of reactions towards pressure (activation volume) and temperature (activation energy) lead to the possibility of retaining or even diminishing some desired natural food quality attributes such as vitamins, pigments and flavour or modifying the structure of food system and food functionality, while optimizing the microbial food safety or minimizing the undesired food quality related enzymes (Barbosa- Cánovas et al., 1997; Messens et al., 1997; Hendrickx et al., 1998).

Exposure to radiation, either ionizing or non ionizing, is being regarded as one of the non thermal methods for food preservation with the best potential, especially for the decontamination of raw material for food production. The ionizing radiation source could be high-energy electrons, X-rays (machine generated), or gamma rays (from Cobalt-60 or cesium- 137), while the non-ionizing radiation is electromagnetic radiation that does not carry enough energy/quanta to ionize atoms or molecules, represented mainly by ultraviolet rays (UV-A, UV-B, and UV-C), visible light, microwaves, and infrared. Decontamination though ionizing radiation consists of the application of doses of 2-7 kGy. It has been proven that radiation can safely and effectively eliminate pathogenic nonsporeforming bacteria in foods (Alothman et al., 2009). However, radiation is not being widely used because of some misconceptions by consumers about its role in causing cancer. Radiation can influence the levels of phytochemicals and the capacity of a specific plant to produce them at different levels. Radiation treatments have been shown to either increase or decrease the antioxidant content of fresh plant produce, which is dependent on the dose delivered (usually low and medium doses have insignificant effects on antioxidants), exposure time, and the raw material used. The enhanced antioxidant capacity/activity of a plant after radiation is mainly attributed either to increased enzyme activity (e.g., phenylalanine ammonia-lyase (PAL) and peroxidase activity) or to the increased extractability from the tissues (Alothman et al., 2009).

Although ultrasound is unlike to become a commercial technology on its own, it can be applied in combination with other technologies with preservation purposes. Its lethal microbial effects have been related to cavitation, a phenomenon that generates high temperatures and pressures at a microscopic level that are responsible for the formation of highly reactive free radicals and for the mechanical damage of microorganisms (Raso et al., 1998). Ultrasound processing of juices is reported to have minimal effect on the degradation of key quality parameters such as colour and ascorbic acid in orange juice during storage at 10 ºC (Tiwari et al., 2009a). This positive effect of ultrasound is assumed to be due to the effective removal of occluded oxygen from the juice (Knorr et al., 2004).

2.2 Plant-based foods preserved by non-thermal technologies

2.2.1 Tomato

Vitamin C has received much attention when aiming at evaluating the effect of non thermal processing technologies on the phytochemicals. Vitamin C retention in PEF-treated juices depended on processing factors and thus, the lower the electric field strength, the treatment time, the pulse frequency or the pulse width, the higher the vitamin C retention in tomato juices (Odriozola-Serrano et al., 2007). According to these authors, maximal relative vitamin C content (90.2%), was attained with PEF treatments of 1 µs pulse duration applied at 250 Hz in bipolar mode at 35 kV/cm for 1000 µs. Regarding the stability of vitamin C through the storage, the concentration of vitamin C decreased over time in both heat (90ºC for 60s) and PEF-treated (35 kV/cm for 1500 µs with 4 µs bipolar pulses at 100 Hz) tomato juices following an exponential trend. In addition, these works demonstrated vitamin C is better retained in PEF treated juices than in those thermally processed after 56 days at 4 ºC. Oxidation of ascorbic acid occurs mainly during the processing of juices and depends upon many factors such as oxygen presence, heat and light. Most differences between PEF and heat treatments can be explained through the temperatures reached through processing. Ascorbic acid is a heat-sensitive bioactive compound in the presence of oxygen. Thus, high temperatures during processing can greatly affect the rates of its degradation through an aerobic pathway (Odriozola-Serrano et al., 2008a). Studies in tomato juice showed that HP processing (300 and 500 MPa) could not preserve vitamin C and the depletion of vitamin C after HP treatment is dependent mainly on temperature intensity and treating time (Hsu et al., 2008). A long exposure (up to 6 h) to extreme pressure/temperature combinations (e.g., 850 MPa combined with temperatures from 65 to 80 ºC) degraded AA to a large extent (Oey et al., 2008). In tomato puree, a 40% and 30% decrease, respectively, in the content of ascorbic acid (AA) and total AA was observed after HP treatment of 400 MPa/25 ºC/15 min (Sánchez-Moreno et al., 2006).

Processing by using non thermal technologies may be advantageous regarding the amount and stability of carotenoids. For instance, lycopene concentrations in PEF-processed tomato juice have been found to be higher than those found in untreated juices (Odriozola-Serrano et al., 2008b). Consistently, Sánchez-Moreno, et al. (2005a) observed that the content of total carotenoids in a tomato-based cold soup 'gazpacho', increased roughly a 62% after applying bipolar 4-μs pulses of 35 kV/cm for 750 μs at 800 Hz. Odriozola-Serrano et al. (2009) have observed that β-carotene in treated tomato juice undergo a significant increase (31%-38%), whereas γ-carotene content is depleted (3%-6%) after a PEF treatment (35 kV/cm for 1500 μs with 4 μs bipolar pulses at 100 Hz). Authors suggested that a plausible explanation for this fact is that γ-carotene may undergo cyclization to form six membered rings at one end of the molecule, giving β-carotene as a product. During storage, PEF-processed tomato juices better maintained the individual carotenoid content (lycopene, neurosporene and γ- carotene) than thermally-treated and untreated juices for 54 days at 4 ºC (Odriozola-Serrano et al., 2009). Individual carotenoids with antioxidant activity (β-carotene, β-cryptoxanthin, zeaxanthin and lutein) appeared to be resistant to a HP treatment of 400 MPa at 40 °C for 1 min, thus resulting into a better preservation of the antioxidant activity of a tomato-based soup with respect to the thermally pasteurized (Sánchez-Moreno et al., 2005b). It seems that high pressure influences the extraction yield of carotenoids. A significant increase in the measured carotenoid content of pressurized (400 MPa/25 ºC/15 min) compared to either thermal treated or untreated tomato purée was observed by Sánchez-Moreno et al. (2006). Due to this potential, HP technology has also been studied to extract lycopene from tomato paste waste (Jun, 2006). In contrast, other studies have not reported major effects of HP treatments on tomato products. García et al. (2001) treated a tomato homogenate for 5 min with 500 and 800 MPa and did not find any influence on the total lycopene and β-carotene concentration. Barba et al. (2010) reported that total carotenoids are particularly affected by HP, having the unprocessed vegetable beverages made with tomato, green pepper, green celery, onion, carrot, lemon and olive oil, a higher total carotenoids content than HP- processed samples. The apparently inconsistent results may be explained through the combined effect of pressure and temperature.

Tomato juices have been found to be a rich source of flavonoids, containing as the main flavonols quercetin and kaempferol, and minor phenolic acids such as ferulic, *p*-coumaric, caffeic acid, etc. PEF processing (35 kV/cm for 1500 µs with 4 µs bipolar pulses at 100 Hz) and thermal treatments (90 ºC 30 s and 90 ºC 60 s) did not affect phenolic content of tomato juices. Both PEF- and heat-treated tomato juices undergo a substantial loss of phenolic acids (chlorogenic and ferulic) and flavonols (quercetin and kaempferol) during 56 days of storage at 4ºC. Caffeic acid content was slightly enhanced over time, regardless the kind of processing, whereas PEF and heat treated tomato juices underwent a substantial depletion of *p*-coumaric acid during storage (Table 1). The increase of caffeic acid in tomato juices after 28 days of storage could be directly associated with residual hydroxylase activities, which convert coumaric acid in caffeic acid. Total phenolics in tomato based beverages and tomato purées appeared to be relatively resistant to the effect of HP (Patras et al., 2009; Barba et al., 2010).The effect of ionizing radiation on the phenolic content of tomatoes has also been studied. The gamma-radiation treatment (2, 4, and 6 kGy) markedly reduced the concentration of the phenolic compounds (*p*-hydroxybenzaldehyde, *p*-coumaric acid, ferulic acid, rutin and naringenin) in tomatoes (Schindler et al., 2005).

Table 1. Phenolic acid and flavonols retention during storage for 56 days at 4 ºC of tomato juices stabilized by heat or HIPEF treatments. Adapted from Odriozola-Serrano et al. (2009)

	Phenolic compound	Process	Phenolic retention (%) after 56 days of storage at 4 ºC
Phenolic acids	Chlorogenic	HIPEF	86
		TT	79
	Ferulic	HIPEF	67
		TT	69
	p-coumaric	HIPEF	53
		TT	53
	Caffeic	HIPEF	132
		TT	118
Flavonols	Quercetin	HIPEF	80
		TT	64
	Kaempferol	HIPEF	82
		TT	75

HIPEF: High intensity pulsed electric fields treatment at 35 kV/cm for 1000 µs; bipolar 4-µs pulses at 100 Hz; TT: thermal treatment at 90 ºC for 60 s

2.2.2 Orange

Ascorbic acid is a heat-sensitive bioactive compound in the presence of oxygen. Thus, high temperatures during processing can greatly affect the rates of its degradation through an aerobic pathway. Storage conditions such as storage temperature or oxygen concentration may have a significant influence on the rates of vitamin C degradation. Vitamin C is usually degraded by oxidative processes which are stimulated in the presence of light, oxygen, heat peroxides and enzymes (especially ascorbate oxidase and peroxidase). Many authors have reported that vitamin C in different fruit and vegetable products is not significantly affected by HP processing (Sánchez-Moreno et al., 2009). Sánchez-Moreno et al. (2006) reported 91% retention of ascorbic acid in orange juice after HP processing at 400 MPa/40 ºC/1min. In addition, HP orange juices (400 MPa/ 40 ºC/ 1 min) maintained better the vitamin C during more days of refrigerated storage than low pasteurized treated juice (70 ºC/30 s) (Plaza et al., 2006). However, differences in vitamin C pressure stability during storage could be explained by the initial oxygen content and possible endogenous pro-oxidative enzyme activity. The effects of ultrasonication on the vitamin C content of orange juice have also been studied. Degradation of vitamin C in sonicated orange juices was observed and the degradation level depended on the wave amplitude and treatment time (Tiwari et al., 2009a). Increased shelf life based on ascorbic acid retention was found for sonicated orange juice compared to thermal processed samples at 98 ºC for 21 s due to higher processing temperature (Tiwari et al., 2009a).

Some studies have demonstrated that carotenoid content is increased significantly after PEF processing compared to the untreated orange juices. Cortés et al. (2006) observed that the carotenoid concentration in orange juice rose slightly after applying intense PEF treatments of 35 and 40 kV/cm for 30-240 μs. Carotenoid concentration rose as treatment time increased when HIPEF treatments of 25 or 30 kV/cm were applied to orange-carrot juice (Torregrosa et al., 2005). It has been reported that thermal treatment may imply an increase in some individual carotenoids owing to greater stability, enzymatic degradation, and unaccounted losses of moisture, which concentrate the sample (Rodriguez-Amaya, 1997). However, carotenoids are highly unsaturated compounds with an extensive conjugated double-bonds system and they are susceptible to oxidation, isomerisation and other chemical changes during processing and storage. Cortés et al. (2006) reported a significant decrease in total carotenoids of orange juice when applying bipolar treatments of 30 kV/cm for 100 μs. This decrease in provitamin A carotenoids could be correlated with a significant decrease in vitamin A by 7.52% in high-intensity PEF-treated orange juice (30 kV/cm, 100 μs) and by 15.62% in a pasteurized orange juice (90ºC, 20 s). Moreover, PEF processing (35 kV/cm, 750 μs) or thermal treatments (70 ºC, 30 s and 90 ºC, 30 s) did not exert any effect on vitamin

A content of an orange juice (Sánchez-Moreno et al., 2005a). Research efforts have been made to obtain fruit and vegetable juices by HP processing without the quality and nutritional damage caused by heat treatments (Sánchez-Moreno et al., 2009). Individual carotenoids with antioxidant activity (β-carotene, β-cryptoxanthin, zeaxanthin and lutein) appeared to be resistant to a HP treatment of 400 MPa at 40 °C for 1 min, thus resulting into a better preservation of the antioxidant activity of orange juices with respect to those thermally pasteurized (Sánchez-Moreno et al. 2005a). Interestingly, an orange juice treated at 350 MPa/30ºC/5 min exhibited a higher carotenoid content (α-carotene, 60%; β-carotene, 50%; α-cryptoxanthin, 63%; β-cryptoxanthin, 42%) than a freshly squeezed juice (De Ancos et al., 2002), which was attributed to the desnaturation of the carotenoid-binding protein induced by pressure. Regarding the stability of carotenoids through storage of HP-pasteurized juices, non significant changes have been reported for at least 10 days of refrigerated storage in an orange juice treated at 100MPa/60ºC/5 min, whereas substantial losses were found at the end of the storage period of samples processed at 350MPa/30ºC/2.5 min or 400MPa/40ºC/1 min (20.56% and 9.16%, respectively). Plaza et al. (2010) reported that HP-treated orange juice showed a higher content in carotenoids than heat pasteurized juice during refrigerated storage at 4°C. In consequence, vitamin A values showed an increase above 40% the value of the untreated sample. The inactivation of enzymes that caused losses of carotenoids during storage and the improvement of the extraction caused by HP treatments are the reasons exposed by some authors to explain that results (De Ancos et al., 2002).

Flavonoids are the most common and widely distributed group of plant phenolics. Among them, flavones, flavonols, flavanols, flavanones, anthocyanins, and isoflavones are particularly common in fruits. In this way, Sánchez-Moreno et al. (2005a) evaluated the effect of a PEF treatment at 35 kV/cm for 750 μs with 4-μs bipolar pulses at 800 Hz on the flavanone content of orange juice. No changes in the total flavanones were observed, nor in the individual flavanone glycosides and their aglycons hesperetin and naringenin. Recent results show that HP processed orange juice (400 MPa, 40°C, 1 min) presented a significant increased on the extractability of each individual flavanone with regard untreated juice and hence on total flavanone content whereas mild pasteurization (70°C, 30 s) treatments retained similar levels to those found in untreated juices (Plaza et al., 2010). Regarding the main flavanones identified in orange juice, HP treatments (400 MPa/40°C/1 min) increased the content of naringenin by 20% and by 40% the content of hesperetin in comparison with an untreated orange juice (Sánchez-Moreno et al., 2005a). The increase in the extractability of flavanones by Plaza et al. (2010) in HP orange juice happened at beginning due to treatment. Thus, during refrigerated storage at 4 ºC, flavanone content in HP juice decreased around 50% during the first 20 days of storage at 4°C. The degradation of phenolic compounds during storage has been mainly related to the residual activity of polyphenol oxidase (PPO) and peroxidase (POD) (Odriozola-Serrano et al., 2009).

2.2.3 Berries

Different studies have proven the effectiveness of PEF in achieving higher vitamin C content in comparison with heat treatments in berries. Odriozola-Serrano et al. (2008a) reported that vitamin C retention just after treatment in heat-processed (90ºC, 60s) strawberry juice was significantly lower (94%) than that found in a juice treated at 35 kV/cm for 1700 μs in bipolar 4 μs pulses at 100 Hz (98%). Low processing temperatures reached through PEF- processing (T<40 ºC) would explain the higher retention of vitamin C in HIPEF-treated strawberry juice compared to the thermally processed samples. The concentration of vitamin C in thermally and PEF-processed juices decreased gradually with storage time. Although during 21 days of storage the concentration of vitamin C was similar among processed strawberry juices, beyond this day, juices subjected to thermal treatment at 90 ºC for 60 s exhibited lower vitamin C content compared to PEF-treated juices (Odriozola et al., 2008c). Recommended daily intake (RDI) of vitamin C is currently revised but should be never below 60 mg, as established by the U.S. Food and Drug Administration (FDA, 1999). According to this recommendation, a strawberry juice 250 mL serving size should contain 24 mg/100 mL in order to contribute to the 100% of the RDI. Vitamin C content of juice processed with either PEF or heat at 90 ºC for 30 s felt below the RDI at 35 days of storage. The concentration of vitamin C in juices treated at 90 ºC for 60 s was reduced below 24 mg/mL after 28 days of storage at 4 ºC. As compared to fruit based products, a high residual ascorbic acid concentration after HP treatment is mostly found. In berries, a high retention of AA in strawberry nectar was observed after HP treatment at 500 MPa/ room temperature/3 min (Rovere et al., 1996). Changes of vitamin C content in pressure treated food products during storage have been followed. It is suggested that further vitamin C degradation after HP processing during storage could be reduced by lowering storage temperature, for example, in pressurized (500 MPa/ room temperature/3 min) strawberry nectar (Rovere et al., 1996). A kinetic study on degradation of vitamin C in pressure treated strawberry coulis has shown that a pressure treatment neither accelerated nor slowed down the kinetic degradation of ascorbic acid during subsequent storage. Sancho et al. (1999), observed identical kinetics of vitamin C degradation in pressurized (400 MPa/20 ºC/30 min) and untreated coulis during storage at 4 ºC. In general, it can be concluded that ascorbic acid is unstable at high pressure levels combined with high temperatures (above 65 ºC) and the major degradation is caused by oxidation especially during adiabatic heating. Therefore, eliminating the oxygen content in packaging can decrease the ascorbic acid degradation during processing and subsequent storage (Oey et al., 2008). The effects of ultrasonication on the vitamin C content of juices have been also studied. Degradation of vitamin C in sonicated strawberry juices was observed and the degradation level depended on the wave amplitude and treatment time (Tiwari et al, 2009b). Ascorbic acid degradation during sonication may be due to free radical formation and production of oxidative products on the surface of bubbles (Tiwari et al., 2009a). During refrigerated storage, similar vitamin C depletion in strawberry juice was observed in both sonicated and untreated juices (Tiwari et al., 2009b).

Flavonoids are the most common and widely distributed group of plant phenolics. Among them, flavones, flavonols, flavanols, flavanones, anthocyanins, and isoflavones are particularly common in fruits. In strawberry juices, *p*-hydroxybenzoic content was enhanced slightly but significantly after PEF processing (35 kV/cm for 1700 μs in bipolar 4-μs pulses at 100 Hz) compared to the untreated juice, whereas ellagic acid was substantially reduced when the heat treatment was conducted at 90ºC for 60 s. No significant differences in flavonol (kaempferol, quercetin and myricetin) content were obtained between fresh and treated strawberry juices; thus, these phenolic compounds were not affected by processing (Odriozola-Serrano et al., 2008c). It is well known that anthocyanins are unstable pigments and can be decolorized and degraded by many factors such as temperature, pH, oxygen, enzymes, light, the presence of copigments and metallic ions, ascorbic acid, sulphur dioxide and sugars. Numerous studies have evaluated the stability of anthocyanins to different kinds of nonthermal processing. Zhang et al. (2007) observed that after processing a cyanidin-3-glucoside methanolic solution by PEF (300 pulses of 1.2-3 kV/cm, Tª ≤ 47 ºC), the anthocyanin was degraded and the formation of the colorless anthocyanin species, particularly chalcones took place. In real foods such as strawberry juice, it was suggested that the lower the treatment time and the higher the electric field strength, the greater the anthocyanin retention in strawberry juice (Odriozola-Serrrano et al., 2008b). Contrarily, Zhang et al. (2007) reported that the degradation of cyanidin-3-glucoside in blackberry increased as the electric field strength rose. Jin and Zhang (1999) indicated that the losses of anthocyanins in PEF-treated cranberry juice increased during the storage period at 4 ºC. However, the content over time was higher that of thermally-treated samples, which contained the lowest amount of anthocyanins after two weeks of storage. Pelargonidin-3- glucoside and pelargonidine-3-rutinoside content and, in turn, total anthocyanins of strawberry juices, depleted with storage time after PEF (35 kV/cm for 1,700 μs in bipolar 4- μs pulses at 100 Hz) and thermal treatments (90ºC for 60 or 30 s) (Odriozola-Serrano et al., 2008b) (Figure 1). Changes in anthocyanins content throughout storage of PEF-treated juices were associated to the presence of residual enzyme activities such as β-glucosidase (Aguiló- Aguayo et al., 2008).

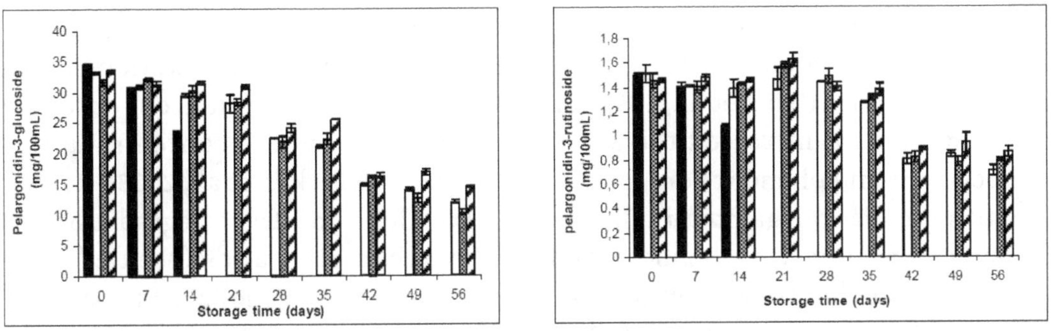

Fig. 1. Effect of HIPEF and heat processing on pelargonidin-3-glucoside (a) and cyaniding-3- glucoside of strawberry juices throughout storage at 4 ºC. Strawberry juices: (■) untreated, (□) heat treated at 90 ºC for 30 s, ( ) heat treated at 90 ºC for 60 s and ( ) HIPEF treated at 35 kV/cm for 1700 µs in bipolar 4-µs pulses and 100 Hz. Data shown are mean ± standard deviation. Adapted from Odriozola-Serrano et al. (2008c)

HP treatment at ambient temperature has been reported to have minimal effects on the anthocyanins content of various food products (Oey et al., 2008). Anthocyanins have been reported to be stable to HP treatments in different products such as strawberry juice (Zabetakis et al., 2000), blackcurrant juice (Kouniaki et al., 2004) and raspberry juice (Suthanthangjai et al., 2004). No significant changes in anthocyanin content of strawberry and blackberry purées after 15 min treatments at 500-600 MPa was also reported by Patras et al. (2009). Similarly, cyanidin-3-glucoside, a predominant anthocyanin in berries, was found to be stable in a model solution at 600 MPa for 30 min at 20 ºC. However, 30 min application of 600 MPa at 70 ºC reduced cyanidin-3-glucoside by 25%, whereas only 5% was lost after 30 min heating at the same temperature and ambient pressure (Corrales et al., 2008). Combined pressure and temperature treatment of blueberry pasteurized juice led to a slightly faster degradation of total anthocyanins during storage compared to heat treatments at ambient pressure (Buckow et al., 2010). Thus, pressure seems to accelerate anthocyanin degradation at elevated temperatures. This could be related to condensation reactions involving covalent association of anthocyanins with other flavanols present in fruit juices. During refrigerated storage, the stability of anthocyanins in HP-treated fruit juices at moderate temperatures has been related to the residual PPO and POD activities, a sufficient activity for rapid oxidation of anthocyanins and other polyphenols in the presence of oxygen. Ultrasonication may be also considered a potential technology for processing of red juices because of its minimal effect on anthocyanins. Tiwari et al. (2009b) reported a slight increase (1-2%) in the pelargonidin- 3-glucoside content of the juice at low acoustic energy density (0,33 W/mL) and treatment time (3 min) which may be due to the extraction of bound anthocyanins from the suspended pulp. Some authors have also studied the effects of non ionizing radiation in phenolic content of berries. UV-C doses at 0.25 and 1.0 kJ/m^2 increased anthocyanins concentrations in the fresh strawberries (Baka et al., 1999). Also, UV-C treatment for different durations (1, 5, and 10 min) increased the antioxidant capacity and the concentrations of anthocyanins and phenolic compounds of strawberries (Erkan et al., 2008). Related to ionizing radiation, gamma-radiation (1-10 kGy) led to the degradation of cinnamic, *p*-coumaric, gallic, and hydroxybenzoic acids (Breitfellner et al., 2002a). The hydroxylation (decomposition) of these phenolic acids has been attributed to the formation of free hydroxyl (OH·) radicals during the treatment. Catechin and kaempferol components also diminished noticeably due to gamma-radiation (1-6 kGy), whereas ellagic acid derivatives and quercetin concentrations were not affected by the treatment in strawberries (Breitfellner et al., 2002b).

2.2.4 Fruit juice-milk beverage

Fruit juice and milk beverage is a product in which the antioxidant capacity of fruit constituents can be delivered in combination with the health benefits of milk. Morales et al. (2010a) did not found significant differences in vitamin C retention (87–90%) between (35 kV/cm, 4 µs bipolar pulses at 200 Hz for 800 or 1400 µs) and thermally treated (90 ºC, 60 s) blend fruit juice-soymilk beverages. In addition, the vitamin C content of the beverages decreased gradually during storage, regardless of the treatment applied. However, throughout the first 31 days, vitamin C was better maintained in the 800 µs-PEF treated fruit juice-soymilk beverages (46.4%) than in those treated for 1400 µs (22.6%); whereas, those that where thermally treated showed the lowest retention of vitamin C (6.7%) (Morales et al., 2011). These results showed that the shorter the PEF treatment time, the higher the vitamin C retention, as previously found in other studies focused on individual fruit juices treated by PEF. The higher retention of vitamin C of PEF treated fruit juice-soymilk beverages compared to those thermally treated might be due to the lower processing temperatures achieved during PEF treatments (<32 ºC). Currently, few studies have been carried out about the effect of HP processing on phytochemicals of fruit juice–milk beverages. High ascorbic acid retention (91%) in the orange juice-milk beverage after HP (100-400 MPa/120-540 s) treatment was reported by Barba et al. (2011).

Initial degradation of ascorbic acid was less in orange juice-milk beverages treated by PEF (25 kV/cm and 280 µs) than in a heat-treated (90 ºC, 20 s) juice (Zulueta et al., 2010a). During storage, total carotenoid content of untreated and treated blend of fruit juice-soymilk beverage tended to decrease as the storage time increased (Morales et al., 2011). Moreover, thermally treated juices showed higher rate of degradation than those PEF-treated. Although the pathways of carotenoids degradation have not been well established, oxidation is the main cause of carotenoid loss, which is a spontaneous free-radical chain reaction in the presence of oxygen (Sánchez-Moreno et al., 2003). During autooxidation of carotenoids, alkylperoxyl radicals are formed and these radicals attack the double bounds resulting in formation of epoxides (Odriozola-Serrano et al., 2009). Total carotenoid content was significantly enhanced in orange juice-milk beverage treated by HP (100-400 MPa) when treatment time was 420 and 540 s in comparison with the unprocessed samples (Barba et al., 2011). According to these authors, this may be because when pressures of 100 MPa are applied, they are sufficient to cause breakage of the intracellular vacuoles and the cell walls of the plant or they also suggested that an increase in free carotenoids in juices after HP might be because there is probably an alteration in the structure of the proteins that are linked to the carotenoids. With regard to individual carotenoid concentrations, there were no significant differences for any of them (Neoxanthin + 9-cis-violaxanthin, Mutatoxanthin, Lutein, Zeaxanthin, β-Cryptoxanthin, α-Carotene, Phytoene+phytofluene, β-Carotene, ζ- Carotene, 15-cis-β-Carotene), and only in the case of the electric field strength of 35 kV/cm there was significant increase in ζ-carotene in comparison with the untreated beverage after 60 μs of treatment. Although the reductions in carotenoids with provitamin A activity are very small after pasteurization, the decreases in the concentrations of lutein (22.8%) and zeaxanthin (22.5%) after pasteurizing are considerable and must be taken into account, because these two xanthophylls play a fundamental part in sight, prevent degenerative eye diseases, and are antioxidant compounds that give quality to food products (Zulueta et al., 2010b).

Coumaric acid, narirutin and hesperidin were the most abundant phenolic compounds in a blend fruit juice-soymilk beverage. Immediately after PEF (35 kV/cm with 4 μs bipolar pulses at 200 Hz for 800 or 1400 μs) or thermal (90 ºC, 60 s) treatments, hesperidin content of the beverage showed a huge rise, resulting in a significant increase on the total phenolic concentration. In addition, total phenolic concentration seemed to be highly stable during refrigerated storage (Morales et al., 2011). According to these authors, changes observed on the phenolic content of the fruit juice-soymilk beverage after PEF or thermal treatments could be due to some of the followed reasons: (i) biochemical reactions could have occurred during the PEF or heat processing, which led to the formation of new phenolic compounds; (ii) PEF or thermal processing might have caused significant effects on cell membranes or in phenolic complexes with other compounds, releasing some free phenolic acids or flavonoids; (iii) PEF and thermal process may inactivate PPO, preventing further loss of phenolic compounds; and (iv) PEF treatment might have induced favorable conditions to increase PAL activity, resulting in an enhancement of phenolic concentration in the beverage. Levels of total phenolic compounds also increased significantly by HP in processed orange juice–milk; reaching a maximum at 100 MPa/420 s, when there was a significant increase of 22% in comparison with unprocessed samples (Barba et al., 2011). As it was mentioned before, the increase in total phenolic content may be related to an increased extractability of some of the antioxidant components following high-pressure processing.

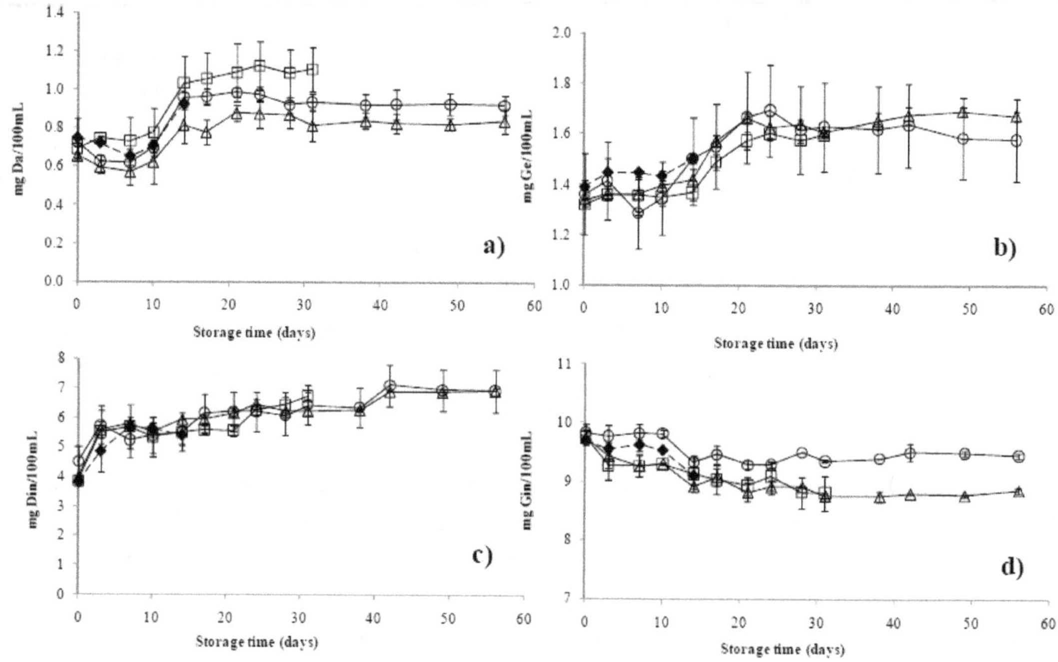

Fig. 2. Individual isoflavone profile: a) daidzein, b) genistein, c) daidzin and d) genisitin of untreated (dotted line, ♦), high intensity pulsed electric field (35 kV/cm with 4 μs bipolar pulses at 200 Hz for 800 (continue line, □) or 1400 μs (continue line, △)) and thermal (continue line, ○) (90 °C, 60 s) treated fruit juice-soymilk beverages throughout storage at 4°C (Morales et al., 2010b)

During the last years, soy beverages consumption has gradually increased due to their significant concentration of health-promoting compounds, such as isoflavones. PEF seems to be a good technology in order to obtain fruit juice-soymilk beverage with a high content of isoflavones and fresh-like characteristics. In a blend fruit juice-soymilk beverage, PEF treatment (35 kV/cm with 4 μs bipolar pulses at 200 Hz for 800 or 1400 μs) did not caused significant changes on the total isoflavone content and, during the storage period, total isoflavone content tended to increase throughout the time. Genistein, daidzein and daidzin content increased; while genistin showed a slight decrease, irrespective of the treatment applied (Figure 2) (Morales et al., 2010b). These authors suggested that the concentration of some isoflavones in the fruit juice-soymilk beverage might increase during storage from the flavonoids (mainly naringenin from orange) present in the fruits used for the elaboration of the beverage. Nevertheless, there is a need for more in-depth research to provide biochemical evidence of the observed changes.

2.2.5 Broccoli

Isothiocyanates are organosulfur compounds formed by enzyme-catalysed hydrolysis of glucosinolates, which are largely found in vegetables of the Brassicaceae family. Some isothiocyanates have shown anticarcinogenic potential (Conaway et al., 2002). Knowledge about the impact of non thermal technologies on the stability of these compounds is really scarce. Van Eylen et al. (2007) studied the pressure (600-800 MPa) and temperature (30-60ºC) stabilities of sulforaphane and phenylethyl isothiocyanate in broccoli juice. Authors concluded that isothiocyanates are relatively thermolabile and pressure stable. At the same time, mild pressure treatments where suggested as the most advantageous, because myrosinase activity is stabilized, thus leading to products with increased isothiocyanate content. In a subsequent study, Van Eylen en al. (2008) observed that the composition of glucosinolate hydrolysis products may greatly differ between different HP process conditions. Upon this base, HP treatments can be selected to optimise the health beneficial properties of plant foods.

The non ionizing radiation treatments have been shown to increase the antioxidant capacity of broccoli, which could be useful from the nutritional point of view. UV-C (4-14 kJm^{-2}) treated broccoli florets displayed lower total phenolic and flavonoid content along with higher antioxidant capacity compared to the control samples (Costa et al., 2006). On the other hand, exposure to UV-C (8 kJm^{-2}) increased total phenolic and ascorbic acid contents, as well as the antioxidant capacity of minimally processed broccoli (Lemoine et al., 2007) These authors related, an increment in the activity of PAL after treatment with UV-C to an increase in the content of phenolic compounds in treated samples since PAL is one of the key enzymes in phenolic synthesis.

2.2.6 Others

As it was mentioned before, PEF processing may allow obtaining juices with higher antioxidant potential and extended shelf-life, thus becoming a feasible alternative to heat processing. PEF processing may help to achieve fresh-like carrot juices with increased amounts of health-related phytochemicals. PEF processing (35 kV/cm for 1500 μs with 6-μs bipolar pulses at 200 Hz) resulted into a carrot juices with significantly greater vitamin C retention of 95.1% than thermal processing (90 ºC, 30 s and 90 ºC, 60 s), which exhibited a retention of 86.6–89.0% (Quitao-Teixeira et al., 2009). Watermelon juice was subjected to high-intensity pulsed electric fields (HIPEF). The effects of process parameters including electric field strength (30–35 kV/cm), pulse frequency (50–250 Hz), treatment time (50–2050 μs), pulse width (1–7 μs) and pulse polarity (monopolar/bipolar) on lycopene, vitamin C and antioxidant capacity were studied using a response surface methodology (Oms-Oliu et al., 2009). Watermelon juices treated at 25 kV/cm for 50 μs at 50 Hz using mono-

or bipolar 1-μs pulses exhibited the highest vitamin C retention (96.4–99.9%). On the other hand, vitamin C loss was higher than 50% when PEF treatment was set up at 35 kV/cm for 2050 (s at 250 Hz applying mono- or bipolar 7-μs pulses. Such severe conditions seem to greater affect vitamin C retention in watermelon juice than in other juices such as orange, orange– carrot or strawberry juices, which exhibited retention of vitamin C above 80%, because more acidic conditions are known to stabilise vitamin C (Oms-Oliu et al., 2009). During storage, vitamin C was better retained in the PEF-treated carrot juice than in the thermally processed juices for 56 days at 4 ºC. Differences in vitamin C reduction between PEF and heat treated juices throughout the storage might be due to the activity of enzymes such as ascorbate oxidase. A first-order kinetic model adequately fitted vitamin C depletion (R^2 = 0.9680– 0.838; Af = 1.039–1.068) as a function of the storage time (Quitao-Teixeira et al., 2009). Storage conditions such as temperature or the oxygen concentration may also have a significant influence on the rates of vitamin C degradation. Sonication also showed to increase vitamin C content in sonicated (ficar les condicions) guava juice compared to untreated sample, the most likely reason being the elimination of dissolved oxygen that is essential for ascorbic acid degradation during cavitation (Cheng et al., 2007). Research reporting on the impact of ionizing radiation on vitamins of plant-derived products is restricted to the effect of gamma-radiation on vitamin C. In general, most fresh-cut vegetables (iceberg, romaine, green and red leaf lettuce, spinach, tomato, cilantro, parsley, green onion, carrot, broccoli, red cabbage, and celery) can tolerate up to 1 kGy radiation without significant losses in vitamin C content (Fan and Sokorai, 2008). In the case of minimally processed irradiated cucumber and carrot, no significant differences between the vitamin C content of control and treated samples were reported through refrigerated storage (Khamat et al., 2005; Hajare et al., 2006). Minor vitamin C losses were reported for minimally processed refrigerated capsicum after gamma-radiation (1-3 kGy) during storage (Ramamurthy et al., 2004). Vitamin C content of irradiated fresh-cut celery (0.5-1.5 kGy) or lettuce (1 kGy) during refrigerated storage was higher than in non-irradiated products (Lu et al., 2005; Zhang et al., 2006). Regarding fresh-cut fruits, Fan et al. (2006), reported that vitamin C of cantaloupe was not substantially affected by treatment with non ionizing radiation. No differences have been reported between treated and untreated products regarding the stability of vitamin C during storage. On the other hand, light treatments applied as short pulses with a total fluence of 4.8 and 12 J cm^{-2} maintained amounts of vitamin C similar to those found in untreated fresh-cut mushrooms during 7 days of refrigerated storage under modified atmosphere packaging (Oms-Oliu et al., 2010).

An increase in lycopene (114%) was observed in watermelon juice treated with 7-μs bipolar pulses for 1050 μs at 35 kV/cm and frequencies ranging from 200 to 250 Hz (Oms-Oliu et al., 2009). The increase in lycopene has been related to the conversion of some carotenoids to lycopene as a result of an intense PEF treatment. β-Carotene concentration substantially increased in processed carrot juices compared to the untreated juice and thermally treated juices. PEF-treated carrot juice maintained β-carotene content better than heat treatments during 56 days of storage at 4 °C. The major cause of carotenoid losses in vegetable products is the oxidation of the highly unsaturated carotenoid structure. The severity of oxidation depends on the structure of carotenoids and the environmental conditions, and the compounds being formed may vary upon the oxidation process and the carotenoids structure (Quitao-Teixeira et al., 2009). Therefore, the higher depletion of β-carotene throughout the storage in thermally treated carrot juice compared to those PEF-processed might be due to the greater changes in carotenes structure as a consequence of high temperature. In addition, a better retention of carotenoids in HP treated carrot pureés compared to thermally processed samples was observed (Patras et al., 2009). Such effect has been well documented elsewhere and would appear to be related to an increase in extractability of antioxidant components following high pressure treatment rather than an absolute increase.

No significant differences in the amount of total phenolics were observed between untreated and PEF-treated products such as spinach puree (Yin et al., 2007) and carrot juice (Quitão- Teixeira et al., 2009) just after processing. It has been also reported that radiation treatments can generate free radicals, thus leading to an induction of stress responses in plant foods, which in turn may lead to an increase in the antioxidant synthesis. Results by Song et al. (2006) are consistent with this idea. These authors observed that total phenolic content of carrot and kale juices was substantially increased by applying a radiation treatment. However, reductions in the total phenolic content have been reported for treatments of more than 10 kGy in some irradiated products (Villavicencio et al., 2000; Ahn et al., 2005).

3. CONCISE SUMMARY

Non-thermal technologies may facilitate the production of safe, shelf-stable plant-based goods with minimal alterations or enhanced levels of health-promoting phytochemicals. The majority of distinctions between non-thermal and thermal treatments can be elucidated by the temperatures attained during processing. Temperature during processing and storage significantly influences the phytochemical composition of the treated product. The stability of these substances during storage is contingent upon the residual levels of enzymes responsible for their breakdown in each instance. Moreover, processing conditions influence the breakdown or synthesis of bioactive molecules. Comprehensive research is required to clarify the mechanisms underlying the destruction or formation of these chemicals in a food matrix processed by these innovative technologies.

Limited research evaluating the effects of non-thermal technologies on the bioavailability of bioactive compounds has indicated an elevation in plasma vitamin C levels and a reduction in oxidative stress and inflammation biomarkers in healthy individuals. Consequently, more investigation into novel applications of non-thermal processing technologies is warranted to enhance the stability, bioavailability, and biological activity of health-related phytochemicals in humans.

4. REFERENCES

Ahn, H.J.; Kim, J.H.; Kim., J.K.; Kim, D.H.; Yook, H.S.; & Byun, M.W. (2005). Combined effect of irradiation and modified atmosphere packaging on minimally processed Chinese cabbage (Brassica ropa L.). *Food Chemistry*, 89, pp. 589-597.

Aguiló-Aguayo, I.; Sobrino-López, A.; Soliva-Fortuny, R.; & Martín-Belloso, O. (2008). Influence of high-intensity pulsed electric field processing on lipoxygenase and ß- glucosidase activities in strawberry juice. *Innovative Food Science and Emerging Technologies*, 9, pp. 455-462.

Alothman, M.; Bhat, R. & Karim, A.A. (2009). Effects of radiation processing on phytochemicals and antioxidants in plant produce. *Trends Food Science and Technology*, Vol. 20, pp. 201-212.

Arab, L., Steck, S. (2000). Lycopene and cardiovascular disease. *American Journal of Clinical Nutrition*, Vol. 71, No. 6, pp. 1691S–1695S

Baka, M.; Mercier, J.; Corcuff, R.; Castaigne, F. & Arul, J. (1999). Photochemical treatment to improve storability of fresh strawberries. *Journal of Food Science*, Vol. 64, 1068-1072.

Barba, F.J.; Esteve, M.J. & Frigola, A. (2010). Ascorbic acid is the only bioactive that is better preserved by hydrostatic pressure than by thermal treatment of a vegetable beverage. *Journal of Agricultural and Food Chemistry*, Vol. 58, pp. 10070-10075.

Barba, F.J.; Cortés, C.; Esteve, M.J.; & Frígola, A. (2011). Study of Antioxidant Capacity and Quality Parameters in an Orange Juice–Milk Beverage after High-Pressure Processing Treatment. *Food Bioprocess and Technology*. DOI 10.1007/s11947-011- 0570-2.

Barbosa-Cánovas, G. V.; Pothakamury, U. R.; Palou, E., & Swanson, B. G. (1997). High pressure food processing. In: *Nonthermal preservation of foods*, G. V. Barbosa- Cánovas, U. R. Pothakamury, E. Palou, & B. G. Swanson (Eds.), Nonthermal preservation of foods, 9-52, Marcel Dekker Inc, New York.

Barbosa-Cánovas, G.V.; Góngora-Nieto, M.M.; Pothakamury, U.R.; Swanson, B.G. (1999). *Preservation of foods with pulsed electric fields*. Academic, San Diego

Breitfellner, F.; Solar, S.; & Sontag, G. (2002a). Effect of γ-irradiation on phenolic acids in strawberries. *Journal of Food Science*, Vol. 67, pp. 517-521.

Breitfellner, F.; Solar, S.; & Sontag, G. (2002b). Effect of gamma irradiation on flavonoids in strawberries. *European Food Research and Technology*, Vol. 215, pp. 28-31.

Britton, G.; & Hornero-Mendez, D. Carotenoids and colour in fruit and vegetables, in *Phytochemistry of Fruit and Vegetables*, ed. by Tomas-Barberan FA and Robins RJ. Science Publications,Oxford, pp. 11–27 (1997).

Buckow, R.; Kastell, A.; Shiferaw Terefe, N. & Versteeg, C. (2010). Pressure and Temperature Effects on Degradation Kinetics and Storage Stability of Total Anthocyanins in Blueberry Juice. *Journal of Agricultural and Food Chemistry*, Vol. 58, pp. 10076–10084.

Cheng, L.H.; Soh, C.Y.; Liew, S.C. Teh, F.F. (2007). Effects of sonication and carbonation on guava juice quality. *Food Chemistry*, Vol. 104, pp. 1396-1401.

Conaway, C.C.; Yang, Y.-M. & Chung, F.-L. (2002). Isothiocyanates as cancer chemopreventive agents: Their biological activities and metabolism in rodents and humans. *Current Drug Metabolism*, Vol. 3, pp. 233-255.

Corrales, M.; Butz, P. & Tauscher, B. (2008). Anthocyanin condensation reactions under high hydrostatic pressure. *Food Chemistry*, Vol. 110, pp. 627–635.

Cortés, C.; Torregrossa, F.; Esteve, M.J. & Frígola, A. (2006). Carotenoid profile modification during refrigerated storage in untreated and pasteurized orange juice and orange juice treated with high-intenxity pulse electric fields. *Journal of Agriculutrual and Food Chemistry*, Vol. 54, pp. 6247-6254.

Costa, L.; Vicente, A. R.; Civello, P. M.; Chaves, A. R. & Martinez, G. A. (2006). UV-C treatment delays postharvest senescence in broccoli florets. *Postharvest Biology and Technology*, Vol. 39, pp. 204-210.

De Ancos, B.; Sgroppo, S.; Plaza, L. & Cano, M.P. (2002). Possible nutricional and health- related value promotion in orange juice preserved by high pressure treatment. *Journal of the Science of Food and Agriculture*, Vol. 82, pp. 790-796.

Eberhardt, M.V., Lee, C.Y. & Liu, R.H. (2000). Antioxidant activity of fresh apples. *Nature*, Vol. 405, pp. 903-904.

Elez-Martínez, P.; Escolà-Hernández, J.; Espachs-Barroso, A.; Barbosa-Cánovas, G.V. & Martín-Belloso, O. (2005) Inactivation of *Lactobacillus brevis* in orange juice by high intensity pulsed electric fields. *Food Microbiology*, Vol. 22, No 4, pp. 311–319.

Elez-Martínez, P.; Aguiló-Aguayo, I.; Martín-Belloso, O. (2006). Inactivation of orange juice peroxidase by high-intensity pulsed electric fields as influenced by processing parameters. *Journal of the Science of Food and Agriculture*, Vol. 86, pp. 71–81.

Erkan, M.; Wang, S. Y. Wang, C. Y. (2008). Effect of UV treatment on antioxidant capacity, antioxidant enzyme activity and decay in strawberry fruit. *Postharvest Biology and Technology*, Vol. 48, pp. 163-171.

Evrendilek, G.A.; Li, S.; Dantzer, W.R.; Zhang, Q.H. (2004). Pulsed electric field processing of beer: microbial, sensory, and quality analyses. *Journal of Food Science*, Vol. 69, pp. M228–M232.

Fan, X.; Annous, B.A.; Sokorai, K.J.B.; Burke, A. & Mattheis, J.P. (2006). Combination of hot- water surface pasteurization of whole fruit and low-dose gamma irradiation of fresh-cut cantaloupe. *Journal of Food Protection*, Vol. 69, pp. 912-919.

FDA. (1999). Nutritional labeling manual: a guide for developing and using databases. Center for Food Safety and Applied Nutrition: Washington D.C.

García, A. F.; Butz, P.; Tauscher, B. (2001). Effects of high-pressure processing on carotenoid extractability, antioxidant activity, glucose diffusion, and water binding of tomato puree (lycopersicon esculentum mill.). *Journal of Food Science*, Vol. 66, No. 7, pp. 1033-1038.

Hajare, S.N.; Dokhane, V.S.; Shashidhar, R.; Sharma, A. & Bandekar, J.R. (2006a). Radiation processing of minimally processed carrot (Daucus carota) and cucumber (Cucumis sativus) to ensure safety: Effect on nutritional and sensory quailty. *Journal of Food Science*, Vol. 71, pp. S198-S203.

Hendrickx, M.; Ludikhuyze, L.; Van den Broeck, I. & Weemaes, C. (1998). Effects of high pressure on enzymes related to food quality. *Trends in Food Science and Technology*, Vol. 9, pp. 197-203.

Hsu, K.C.; Tan, F.J. & Chi, H.Y. (2008). Evaluation of microbial inactivation and physicochemical properties of pressurized tomato juice during refrigerated storage. *LWT-Food Science and Technology*, Vol. 41, pp. 367-375.

Jin, Z.T., Zhang, Q.H. (1999). Pulsed electric field inactivation of microorganisms and preservation of quality of cranberry juice. *Journal of Food Processing and Preservation*, Vol 23, pp. 481-497.

Jun, X. (2006). Application of high hydrostatic pressure processing of food to extracting lycopene from tomato paste waste. *High Pressure Research*, Vol. 26, No. 1, pp. 33-41.

Kalt, W. (2001). Health functional phytochemicals of fruit. *Horticultural Reviews*, Vol. 27, pp. 269-315.

Khamat, A.S.; Ghadge, N.; Ramamurthy, M.S. Alud, M.D. (2005). Effect of low-dose radiation on shelf life amd microbiological safety of sliced carrot. *Journal of the Science of Food and Agriculture*, Vol. 85, pp. 2213-2219.

Knorr, D. (2001). 'High pressure processing for preservation, modification and transformation of foods', *Oral presentation in XXXIX European High Pressure Research Group Meeting*, Santander (Spain), 16-19 September 2001.

Knorr, D.; Zenker, M.; Heinz, V.; & Lee, D. U. (2004). Applications and potential of ultrasonics in food processing. *Trends in Food Science & Technology*, Vol. 15, pp. 261- 266.

Kouniaki, S.; Kajda, P.; & Zabetakis, I. (2004). The effect of high hydrostatic pressure on anthocyanins and ascorbic acid in blackcurrants (*Ribes nigrum*). *Flavour and Fragance Journal*, Vol. 19, pp. 281–286.

Lemoine, M. L.; Civello, P. M.; Martnez, G. A. & Chaves, A. R. (2007). Influence of postharvest UV-C treatment on refrigerated storage of minimally processed broccoli (Brassica oleracea var. Italica). *Journal of the Science of Food and Agriculture*, Vol. 87, 6, pp. 1132-1139

Lu, Z.; Yu, Z.; Gao, X.; Lu, F. & Zhang, L. (2005). Preservation effects of gamma irradiation on fresh-cut celery. *Journal of Food Engineering*, Vol. 67, pp. 347-351.

Messens, W.; Van Camp, J. & Huygebaert, A. (1997). The use of high pressure to modify the functionality of food proteins. *Trends in Food Science and Technology*, Vol. 8, pp. 107- 112.

Monfort, S.; Gayán, E.; Saldaña, G.; Puértolas, E.; Condón, S.; Raso, J. & Álvarez, I. (2010). Inactivation of *Salmonella Typhimurium* and *Staphylococcus aureus* by pulsed electric fields in liquid whole egg. *Innovative Food Science and Emerging Technologies*, Vol. 11, pp. 306–313

Morales-de la Peña, M.; Salvia-Trujillo, L.; Rojas-Graü, M.A. & Martín-Belloso, O. (2010a). Impact of high intensity pulsed electric field on antioxidant properties and quality parameters of a fruit juice–soymilk beverage in chilled storage. *LWT - Food Science and Technology*, Vol. 43, pp. 872–881

Morales-de la Peña, M.; Salvia-Trujillo, L.; Rojas-Graü, M.A. & Martín-Belloso, O. (2010b). Isoflavone profile of a high intensity pulsed electric field or thermally treated fruit juice-soymilk beverage stored under refrigeration juice-soymilk beverage stored under refrigeration. *Innovative Food Science and Emerging Technologies*, Vol. 11, pp. 604–610

Morales-de la Peña, M.; Salvia-Trujillo, L.; Rojas-Graü, M.A. & Martín-Belloso, O. (2011). Changes on phenolic and carotenoid composition of high intensity pulsed electric field and thermally treated fruit juice–soymilk beverages during refrigerated storage. *Food Chemistry*, doi:10.1016/j.foodchem.2011.05.058

Morris, C.; Brody, A.L.; Wicker, L. (2007). Non-thermal food processing/preservation technologies: a review with packaging implications. Packaging Technology and Science, Vol 20, No 4, pp 275–286

Odriozola-Serrano, I.; Aguiló-Aguayo, I.; Soliva-Fortuny, R.; Gimeno-Añó, V. & Martín- Belloso, O. (2007). Lycopene, vitamin C, and antioxidant capacity of tomato juice as affected by high-intensity pulsed electric fields critical parameters. *Journal of Agricultural and Food Chemistry*, Vol. 55, pp. 9036-9042.

Odriozola-Serrano, I.; Soliva-Fortuny, R. & Martín-Belloso, O. (2008a). Changes of health- related compounds throughout cold storage of tomato juice stabilized by thermal or high intensity pulsed electric field treatments. *Innovative Food Science and Emerging Technologies*, Vol. 9, pp. 272-279.

Odriozola-Serrano, I.; Soliva-Fortuny, R.; Gimeno-Añó, V. & Martín-Belloso, O. (2008b). Kinetic study of anthocyanins, vitamin C, and antioxidant capacity in strawberry juices treated by high-intensity pulsed electric fields. *Journal of Agricultural and Food Chemistry*, Vol. 56, pp. 8387-8393.

Odriozola-Serrano, I.; Soliva-Fortuny, R. & Martín-Belloso, O. (2008c). Phenolic acids, flavonoids, vitamin C and antioxidant capacity of strawberry juices processed by high-intensity pulsed electric fields or heat treatments. *European Food Research & Technology*, Vol. 228, pp. 239-248.

Odriozola-Serrano, I.; Soliva-Fortuny, R.; Hernández-Jover, T. & Martín-Belloso, O. (2009). Carotenoid and phenolic profile of tomato juice processed by high intensity pulsed electric fields compared to conventional thermal treatments. *Food Chemistry*, Vol. 112, pp. 258-266.

Oey, I.; Plancken, I. V.; Loey, A. V. & Hendrickx, M. (2008). Does high pressure processing influence nutritional aspects of plant based food systems? *Trends in Food Science & Technology*, Vol. 19, No. 6, pp. 300-308.

Oms-Oliu, G.; Odriozola-Serrano, I.; Soliva-Fortuny, R.C. & Martín-Belloso, O. (2009). Effects of high-intensity pulsed electric field processing conditions on lycopene, vitamin C and antioxidant capacity of watermelon juice. *Food Chemistry*, Vol. 115, pp. 1312- 1319

Patras, A.; Brunton, N.; Pieve, S.D.; Butler, F. & Bowney. (2009). Effect of thermal and high pressure processing on antioxidant activity and instrumental colour of tomato and carrot purées. *Innovative Food Science and Emerging Technologies*, Vol. 10, pp. 16-22.

Plaza, L., Sánchez-Moreno, C.; Elez-Martínez, P., De Ancos, B., Martín-Belloso, O., Cano, M.P. 2006. Effect of refrigerated storage on vitamin C and antioxidant activity of orange juice processed by high-pressure or pulsed electric fields with regard to low pasteurization . *European Food Research and Technology*, Vol. 223, No 4, pp. 487-493

Plaza, L., Sánchez-Moreno, C.; De Ancos, B.; Elez-Martínez, P.; Martín-Belloso, O.; Cano, M.P. (2010). Carotenoid and flavanone content during refrigerated storage of orange juice processed by high-pressure, pulsed electric fields and low pasteurization. *LWT - Food Science and Technology*, doi: 10.1016/j.lwt.2010.12.013.

Quitao-Teixeira, L.J.; Odriozola-Serrano, I.; Soliva-Fortuny, R. & Martín-Belloso, O. (2009). Comparative study on antioxidant properties of carrot juice stabilised by high- intensity pulsed electric fields or heat treatments. *Journal of the Science of Food and Agriculture*, Vol. 89, pp. 2636-2642.

Ramamurthy, M.S.; Kamat, A.; Kakatkar, A.; Ghadge, N.; Bhushan, B. & Alur, M. (2004). Improvement of shelf-life and microbiological quality of minimally processed refrigerated capsicum by gamma irradiation. *International Journal of Food Sciences and Nutrition*, Vol. 55, pp. 291-299.

Raso, J.; Pagán, R.; Condón, S. & Sala, F.J. (1998). Influence of temperature and pressure on the lethality of ultrasound. *Applied Environmental Microbiology*, Vol. 64, pp. 465-471.

Rodríguez-Amaya, D.B. (1997). Carotenoids and food preservation: the retention of provitamin A carotenoids in prepared, processed and storage food. USAID, OMNI Project.

Rovere, P.; Carpi, G.; Gola, S.; Dall'Aglio, G. & Maggy, A. (1996). HPP strawberry products: an example of processing line. In: *High pressure bioscience and biotechnology*, R. Hayashi, & C. Balny (Eds.), pp. 445-450, Amsterdam: Elsevier Science B.V.

Sánchez-Moreno, C.; Plaza, L.; De Ancos, B. Cano, P. (2003). Vitamin C, provitamin A carotenoids, and other carotenoids in high-pressurized orange juice during refrigerated storage. *Journal of Agricultural and Food Chemistry*, Vol. 51, pp. 647–653.

Sánchez-Moreno, C.; Plaza, L.; Elez-Martínez, P.; De Ancos, B.; Martín-Belloso, O. & Cano, M. P. (2005a). Impact of high pressure and pulsed electric fields on bioactive compounds and antioxidant activity of orange juice in comparison with traditional thermal processing. *Journal of Agricultural and Food Chemistry*, Vol. 53, pp. 4403– 4409.

Sánchez-Moreno, C.; Cano, M. P.; De Ancos, B.; Plaza, L.; Olmedilla, B.; Granado, F.; Elez- Martínez, P.; Martín-Belloso, O. & Martín, A. (2005b). Intake of Mediterranean vegetable soup treated by pulsed electric fields affects plasma vitamin C and antioxidant biomarkers in humans. *International Journal of Food Sciences and Nutrition*, Vo. 56, pp. 115-124.

Sánchez-Moreno, C.; Plaza, L.; De Ancos, B. & Cano, M. P. (2006). Impact of high-pressure and traditional thermal processing of tomato purée on carotenoids, vitamin C and antioxidant activity. *Journal of the Science of Food and Agriculture*, Vol. 86, pp. 171- 179.

Sánchez-Moreno, C.; De Ancos, B.; Plaza, L.; Elez-Martínez, P. & Cano, M.P. (2009). Nutritional approaches and health-related properties of plant foods processed by high pressure and pulsed electric fields. *Critical Reviews in Food Science and Nutrition*, Vol. 49, pp. 552-576.

Sancho, F. ; Lambert, Y. ; Demazeau, G. ; Largeteau, A. ; Bouvier, J. M. & Narbonne, J. F. (1999). Effect of ultra-high hydrostatic pressure on hydrosoluble vitamins. *Journal of Food Engineering*, Vol. 39, pp. 247-253.

Schindler, M.; Solar, S. & Sontag, G. (2005). Phenolic compounds in tomatoes. Natural variations and effect of gamma-irradiation. *European Food Research and Technology*, Vol. 221, pp. 439-445.

Shahidi, F. & Naczk, M. (2004). Phenolics in food and nutraceuticals, In: *Sources, applications and health effects*. Boca Raton, FL: CRC Press.

Smelt, J. P.; Hellemons, J. C. & Patterson, M. (2002). Effects of high pressure on vegetative microorganisms. In: *Ultra high pressure treatments of foods*, M. Hendrickx, & D. Knorr (Eds.), pp. 55-76, New York, United States: Kluwer Academic/Plenum Publishers.

Song, H.P.; Kim, D.H.; Jo, C.; Lee, C.H.; Kim, K.S.; Byun, M.W. (2006). Effect of gamma irradiation on the microbiological quality and antioxidant activity of fresh vegetable juice. Food Microbiology, Vol 23, No 4, pp. 372-378

Suthanthangjai, W.; Kajda, P.; Zabetakis, I. (2004). The effects of high hydrostatic pressure on beta-glucosidase, peroxidase and polyphenoloxidase in red raspberry (Rubus idaeus) and strawberry (Fragaria x ananassa). *Food Chemistry*, Vol. 90, pp. 193–197.

Tiwari, B.K.; O'Donnell, C.P.; Muthukumarappan, K. & Culllen, P.J. (2009a). Ascorbic acid degradation kinetics of sonicated orange juice during storage and comparison with thermally pasteurised juice. *LWT-Food Science and Technology*, Vol. 42, pp. 700-704.

Tiwari, B.K.; O'Donnell, C.P., Patras, A., Brunton, N. & Culllen, P.J. (2009b). Stability of anthocyanins and ascorbic acid in sonicated strawberry juice during storage. *European Food Research and Technology*, Vol. 228, pp. 717-724.

Torregrosa, F.; Cortés, C.; Esteve, M.J. & Frígola, A. (2005). Effect of high-intensity pulsed electric fields processing and conventional heat treatment on orange-carrot juice carotenoids. *Journal of Agricultural and Food Chemistry*, Vol. 53, pp. 9519-9525.

Van Eylen, D.; Oey, I.; Hendrickx, M. Van Loey, A. (2007). Kinetics of the stability of broccoli (Brassica oleracea cv. Italica) myrosinase and isothiocyanates in broccoli juice during pressure/ temperature treatments. *Journal of Agricultural and Food Chemistry*, Vol. 55, pp. 2163-2170.

Vega-Mercado, H.; Martín-Belloso, O.; Qin, B.L.; Chang, F.J.; Góngora-Nieto, M.M.; Barbosa- Cánovas, G.V. & Swanson, B.G. (1997) Non-thermal food preservation: pulsed electric fields. *Trends Food Science and Technology*, Vol. 8, No 5, pp. 151–157

Villavicencio, A.L.C.H.; Mancini-Filho, J.; Delincee, H. & Greiner, R. (2000). Effect of irradiation on anti-nutrients (total phenolics, tannins and phytate) in Brazilian beans. *Radiation Physics and Chemistry*, Vol. 57, pp. 289-293.

Yin, Y.; Han, Y. & Liu, J. (2007). A novel protecting method foe visual green color in spinach puree treated by high intensity pulsed electric field. *Journal of Food Engineering*, Vol. 79, pp. 1256-1260.

Zabetakis, I.; Leclerc, D., Kajda, P. (2000). The effect of high hydrostatic pressure on the strawberry anthocyanins. *Journal of Agriculture and Food Chemistry*, Vol. 48, pp. 2749–2754.

Zhang, L. ; Lu, Z. ; Lu, F. & Bie, X. (2006). Effect of gamma irradiation on quality- maintaining of fresh-cut lettuce. *Food Control*, Vol. 17, pp. 225-228.

Zhang, Y.; Liao, X.; Ni, Y.; Wu, J.; Hu, X.; Wang, Z. & Chen, F. (2007). Kinetic analysis of the degradation and its color change of cyanidin-3-glucoside exposed to pulsed electric field. *European Food Research and Technology*, Vol. 224, pp. 597-603.

Zulueta, A.; Esteve, M.J.; & Frígola, A. (2010a). Ascorbic acid in orange juice–milk beverage treated by high intensity pulsed electric fields and its stability during storage. *Innovative Food Science & Emerging Technologies.* Vol. 11, No 1, pp. 84-90

Zulueta, A.; Barba, J.; Esteve, M.J. & Frígola, A. (2010b). Effects on the carotenoid pattern and vitamin A of a pulsed electric field-treated juice-milk beverage and behavior during storage. *European Food Research and Technology*, 231, 525-534.

6 Chapter

PHYTOCHEMICAL COMPONENTS AND FUNCTIONS OF MORINDA CITRIFOLIA L.

1. INTRODUCTION: AN OVERVIEW

Both artificial and naturally occurring antioxidants have been reported to play major roles in protecting membranes and tissues from free radical and xenobiotic-induced oxidative damage (Burton, 1989; Carini et al., 1990). Most living organisms harbour both enzymatic and non-enzymatic systems that protect them against excessive reactive oxygen species. However, various external factors (smoke, diet, alcohol and some drugs) and aging decrease the efficiency of these protective systems, thereby disrupting the redox equilibrium that is established under healthy conditions. Thus, antioxidants that scavenge reactive oxygen species may be of great value in preventing the onset and propagation of oxidative diseases (Willet, 1994). Recently, more attention has been paid to the role of natural antioxidants, mainly phenolic compounds, which may have higher antioxidant activity than do conventional antioxidants, such as vitamins C, E and β-carotene (Vinson et al., 1995). The anti-oxidative effects of natural phenolic compounds, in pure form or in extracts from various plant sources (vegetables, fruits and medicinal plants), have been studied *in vitro* using a variety of model systems (Meyer et al., 1998; Pietta et al., 1998; Yen & Hsieh, 1998). Therefore, antioxidants, which can neutralize free radicals, may be of central importance in the prevention of carcinogenicity, cardiovascular disease and neurodegenerative changes associated with aging (Halliwell, 1994; Yu, 1994). Epidemiological studies have shown that the consumption of vegetables and fruits can protect humans against oxidative damage by inhibiting and/or quenching free radicals and reactive oxygen species (Ames et al., 1993).

Oxidative stress occurs commonly in living organisms, and it is involved in the pathology of cancer, arteriosclerosis, malaria, and rheumatoid arthritis. Moreover, it may play a role in neurodegenerative diseases and ageing processes. It has been demonstrated that many vegetables, fruits, medicinal plants and other foods contain compounds with bioactivity against oxidative stress, and this activity has been attributed to vitamin C, vitamin E, α- tocopherol, β-carotene, and polyphenolic compounds (Krishnaiah et al., 2011; Moure et al., 2001). Therefore, research regarding natural antioxidants from foods and plants, particularly from folk medicinal plants, is receiving increasing attention throughout the world.

2. MORINDA CITRIFOLIA L.

The ancestors of the Polynesians are believed to have brought many plants with them, as food and medicine, when they migrated from Southeast Asia 2000 years ago. Of the 12 most common medicinal plants they brought, *Morinda citrifolia* L. was the second most popular plant used in herbal remedies to treat various common diseases and to maintain overall good health. Other names of *Morinda citrifolia* L. include M. bracteata Roxb.; M. litoralis Blanco; Indian mulberry, Bengkudu, Mengkudu (Malay). It has been reported to have a broad range of health benefits for subjects with cancer, infections, arthritis, diabetes, asthma, hypertension and pain. The Polynesians utilised the whole *Morinda citrifolia* L. plant in their medicinal remedies and as a dye for some traditional clothing. The roots, stems, bark, leaves, flowers, and fruits of the *Morinda citrifolia* L. plant are all involved in various combinations in almost 40 known and recorded herbal remedies. Additionally, the roots were used to produce a yellow or red dye for tapa cloth and fala (mats), and the fruit was eaten for health and nutrition. There are numerous Polynesian stories of heroes and heroines that used *Morinda citrifolia* L. to survive famine. *Morinda citrifolia* has a long history of use as a food in tropical regions throughout the world.

It has also been reported to have broad therapeutic effects, including anti-cancer activity, in both humans and laboratory animal models. However, the mechanisms underlying these effects remain unknown. *M. citrifolia* is unique in view of the large number of medicinal claims that have been made for its efficacy and its rapidly evolving commercial success; nevertheless, little is known about its pharmacological potential compared with other popularly used botanicals.

M. citrifolia L., a shrub originating in tropical Asia or Polynesia, has been extensively used in folk medicine and as a dye in Asian countries. In the tropics, it seems to have been greatly valued medicinally, and the plant is normally cultivated for its roots, leaves and fruits. *M. citrifolia*, which grows prevalently in tropical regions, has recently gained a great deal of interest from scientists and medical professionals due to its pharmaceutical value. Wang et al. (2002) have published a review of *M. citrifolia* L. research that summarises the therapeutic effects of various compounds in this plant.

The *M. citrifolia* L. plant is a small evergreen tree found growing in open coastal regions at sea level and in forest areas up to approximately 1300 feet above sea level. The plant is often found growing along lava flows. Polynesians are reported to have successfully used *M. citrifolia* L. to treat breast cancer and eye problems. *M. citrifolia* has been tested for a number of biological activities in animal and anti-microbial studies and found that the dried fruit has smooth muscle stimulatory activity and histaminergic effects.

The roots of these plants are reported to be good sources of anthraquinones, which are usually present as aglycones and, to lesser extent, in the form of glycosides (Thomson, 1971; Zenk, El-Shagy & Schulte, 1975). Most parts of the tree have been widely used medicinally to relieve rheumatism and other pains and for their healing effects (Perry & Metzger, 1980).

Traditionally, the roots of *M. citrifolia* L. plants were used by Polynesians to produce yellow or red dye, but more importantly, they are now known to contain medicinally active components, such as anthraquinones, which due to their anti-oxidative activity, possess various therapeutic properties. These properties include anti-bacterial, anti-viral, and anti- cancer activities, as well as analgesic effects. These factors make the compounds potentially useful in several medical applications. An increasing number of studies are focusing on finding efficient methods for producing and extracting anthraquinones from these plants. Much of the literature also involves producing the compound in root cultures of *M. citrifolia*.

Nevertheless, extraction of anthraquinones directly from plant roots is still more widely conducted and is conventionally performed by solvent extraction. Other techniques, which include super critical carbon dioxide extraction, subcritical water extraction, ultrasonic- assisted extraction (UAE), and microwave-assisted extraction (MAE) have also become of interest as alternatives to the conventional methods.

2.1 Plant description

The genus Morinda (*Rubiaceae*), which includes the species *M. Citrifolia* L., is made up of around 80 species. *M. Citrifolia* is a bush or small tree, 3-10 m tall, with abundant broad elliptical leaves (5-17 cm length, 10-40 cm width).

The small tubular white flowers are grouped together and inserted on the peduncle. The petioles leave ring-like marks on the stalks and the corolla is greenish-white (Morton, 1992; Elkins, 1998; Dixon et al., 1999; Ross, 2001; Cardon, 2003). The *M. citrifolia* L. fruit (3-10 cm length, 3-6 cm width) is ovular and fleshy with an embossed appearance (Fig. 1). It is slightly wrinkly, semi-translucent, and ranges in colour from green to yellow to almost white at the time of picking. It is covered with small reddish-brown buds containing the seeds. The ripe fruit emits a strong butyric acid-like, rancid smell (Morton, 1992; Dixon et al., 1999). The pulp is juicy and bitter, a light dull yellow or whitish colour, and gelatinous when the fruit is ripe; numerous hard triangular reddish-brown pits are found, each containing four seeds (approximately 3.5 mm) (Dittmar, 1993).

Fig. 1. Unripe fruit

Moreover, *M. citrifolia* L. leaves are well known for their strong antioxidant activity, and they have been shown to be safe in acute, subacute, and subchronic oral toxicity tests on mice (West et al., 2007). Inspired by ancient Polynesian legends, *M. citrifolia* L. leaves have been developed into therapeutic teas. The leaves are also the source for a variety of other health-promoting commercial products. Commercial *M. citrifolia* L. leaf products have been available in Japan and United States for more than seven years, used mainly for making infusions. However, some manufacturers produce capsules containing powdered *M. citrifolia* L. leaves. The major world-wide source of *M. citrifolia* L. leaves is French Polynesia because leaves from this nation having undergone a safety evaluation (West et al., 2007). Other sources include Panama, Fiji and Hawaii.

Recently, *M. citrifolia* L. fruit juice is in high demand as an alternative medicine due to its potential anti-microbial, anti-cancer, anti-inflammatory, and antioxidant effects (Wang et al., 2002). However, scientific evidence for the benefits of *M. citrifolia* L. fruit juice is still limited. In the past decade, *M. citrifolia* L. fruit juice has emerged on the worldwide market as a safe and popular health product due to its phytochemicals and nutrients. Written documentation of the consumption of this fruit as a food source precedes the twentieth century. Captain James Cook of the British Navy noted in the late 1700s that the fruit was eaten in Tahiti. An 1866 publication in London explained that *M. citrifolia* L. fruit was consumed as a food in the Fiji islands.

Later publications describe the use of this fruit as a food throughout the Pacific islands, Southeast Asia, Australia, and India. In Roratonga, "the fruit was often eaten by the natives". Australian Aborigines were reported to be "very fond" of the fruit. In Samoa, *M. citrifolia* L. fruit was common fare, and in Burma, the fruit was cooked in curries or eaten raw with salt. In 1943, Merrill described *M. citrifolia* L. as an edible plant in a technical manual of edible and poisonous plants of the Pacific islands, in which the leaves and fruits could be used as emergency food. Abbott also reported that *M. citrifolia* L. had been used as a food, drink, medicine, and colourful dye. The medicinal history and accumulated scientific studies have revealed and confirmed the Polynesians' claims regarding the health benefits of *M. citrifolia* L.

M. citrifolia L. has identifiable leaves, white tubular flowers, and a distinctive, ovoid, "grenade-like" yellow fruit. The fruit can grow in size up to 12 cm or more and has a lumpy surface covered by polygonal-shaped sections. The seeds, which are triangular and reddish brown, have an air sac contained at one end, making the seeds buoyant. This could explain, in part, the wide distribution of the plant throughout the Polynesian islands. The mature *M. citrifolia* L .fruit has a foul taste and odour. *M. citrifolia* L. is not considered to be at risk in the wild.

The fruit juice of *M. citrifolia* L. is in high demand in alternative medicine for various illnesses, such as arthritis, diabetes, high blood pressure, muscle aches and pains, menstrual difficulties, headaches, heart disease, Acquired Immune Deficiency Syndrome (AIDS), cancer, gastric ulcers, sprains, mental depression, senility, poor digestion, atherosclerosis, blood vessel problems and drug addiction (Kamiya et al., 2004; Wang et al., 2002). Therefore, one of the challenges in recent years has been to process fruit juice so as to make a more modern drug from a traditional product (Chunhieng et al., 2003). A number of *in vitro* biological activities have been reported, such as angiogenesis inhibition, antioxidant activity, inhibition of cycloxygenases-1 and -2, and tyrosine kinase inhibition. However, most of these have only been tested with crude extracts or fractions of *M. citrifolia* L., and the compound(s) responsible for these biological activities have not been fully determined, except for two compounds, neolignan and americanin A, which were identified in an n- butanol-soluble partition of the methanol extract of *M. citrifolia* L. fruits (Su et al., 2005; Zin et al., 2002).

M. citrifolia L. has recently been the object of many claims concerning its nutraceutical properties. Various publications have shown that *M. citrifolia* L. can be used to relieve multiple diseases, and its registered uses span the Pacific, Asia, and Africa. Two clinical studies have reported that relief from arthritis and diabetes are associated with *M. citrifolia* L. consumption (Elkins, 1998; Solomon, 1999). These beneficial effects may derive from certain compounds, such as scopoletin, nitric oxide, alkaloids and/or sterols, and they also may be due to the antioxidant potential of *M. citrifolia* L. As a result of this reputation, consumption of this fruit is currently high not only in the producing countries but also in the United States, Japan and Europe.

In response to this demand, some countries (such as Costa Rica and Cambodia) have increased their cultivation of *M. citrifolia* L. In these countries, the fruit is often commercialised fresh or as juice in both formal and informal markets, but it is also found as pasteurised juice, either in pure form or in combination with other juices (usually grape or blackberry juice). Commercial interest in *M. citrifolia* L. has increased tremendously in recent years, as indicated by the number of patents registered. In the United States alone, 19 patents have been registered by the US patent and Trademark Office since 1976 (USPTO, 2005). *M. citrifolia* L. juice has been accepted by the European Union as a novel food (European Commission, Scientific Committee for Food, 2002). Nevertheless, despite the real market opportunities, there has been little scientific research addressing the actual nutritional and functional properties of *M. citrifolia* L. products.

Several classes of compounds have been isolated from *M. citrifolia* L., including amino acids, anthraquinones, coumarins, fatty acids, flavonoids, iridoids, lignans and polysaccharides (Chan-Blan-co et al., 2006). Among these, scopoletin, a coumarin derivative, is one of the representative ingredients in *M. citrifolia* L. Its contribution to anti-microbial, anti- inflammatory, and antioxidative activities has been well elucidated (Deng et al., 2007). Samoylenko et al. (2006) recommended scopoletin as a constituent marker for *M. citrifolia* L. quality control. This compound has been ubiquitously found in *M. citrifolia* L. collected from Atlantic and Pacific regions and in all examined squeezed fruit juices. Determination of scopoletin in *M. citrifolia* L. might help to control the quality of *M. citrifolia* L. products. However, no reports have evaluated the antioxidative activity of scopoletin and other coumarin derivatives in *M. citrifolia* L.

Many reports on the antioxidative activity of *M. citrifolia* L. itself have been published. Assays of free-radical-scavenging activity with 1,1-diphenyl-2-picrylhydrazyl (Su et al., 2005; Yang et al., 2007), inhibition of copper-induced low-density lipoprotein oxidation (Kamiya et al., 2004), nitric oxide scavenging activity (Basu & Hazra, 2006) and quenching of H_2O_2 (Chong et al., 2004; Jeffers et al., 2007) have been performed to evaluate the antioxidative effects of *M. citrifolia* L. and its products. Polyphenols, reducing glycosides (Calzuola et al., 2006), lignin derivatives (Su et al., 2005) and anthraquinones (Chong et al., 2004) have been suggested as sources of antioxidative activity in *M. citrifolia* L. However, there is little available information with which to quantitatively evaluate the antioxidative activity of *M. citrifolia* L ingredients. Recently, a correlation between total phenol and free- radical-scavenging activity was reported (r=0.41, Yang et al., 2007).

The chemical components of *M. citrifolia* L. have not been well studied, and several anthraquinones and asperuloside are all that have been previously isolated (Levand & Larson, 1979; Srivastava & Singh, 1993). For centuries, scientists and medical professionals have been investigating the chemical constituents in all parts of *M. citrifolia* (Noni or Yor), including leaves, fruit, bark and roots. The plants contain several medicinally active components that exhibit various therapeutic effects. These include anti-bacterial, anti-viral and anti-cancer activities as well as analgesic effects. Critical reviews of the therapeutic properties of the plants are given by Chan-Blanco et al. (2006) and Wang et al. (2002). Anthraquinones have been shown to be responsible for the therapeutic properties of *M. citrifolia* L., and among this group of compounds, damnacanthal, which is present mainly in the roots, is of particular interest due to its important anticancer activity (Hiramatsu et al., 1993).

Previous phytochemical studies revealed that *M. citrifolia* L. leaves contain a variety of phytochemical constituents, including terpenoids (Ahmad & Bano, 1980; Saludes et al., 2002; Takashima et al., 2007) phytosterols, fatty acids and their glycosides (Takashima et al., 2007) iridoids and their glycosides (Sang et al., 2001 a,b,c,d; Sang et al., 2003) and flavonol glycosides (Sang et al., 2001a).

Flavonol glycosides appear to predominate in *M. citrifolia* L. leaves; rutin and other flavonol glycosides have previously been identified in raw *M. citrifolia* L. leaves (Sang et al., 2001a). However, the presence of flavonol aglycones in *M. citrifolia* L. leaves has not been previously reported. Flavonoids have been indicated to possess a variety of biological activities (Garcia- Mediavilla et al., 2006; Kampkotter et al., 2007), and they may play an important role in *M. citrifolia* L. leaves. To date, there has been no validated analytical method for determining the flavonol constituents of *M. citrifolia* L. leaves.

2.2 Yield

M. citrifolia L. is a perennial bush, and it is possible to find fruits at different stages of maturity on the same plant at the same time. The species is generally found from sea level to 400 m, although it adapts better to coastal regions (Luberck & Hannes, 2001). Under favourable conditions, the plant bears fruit approximately nine months to one year after planting. At this stage, the fruits can be harvested, but they are generally small, and the yield per tree is low. Some producers choose not to harvest in the first year, and they prune in order to let the bush grow stronger. In Hawaii, *M. citrifolia* L. fruits are harvested throughout the year, although there are seasonal patterns in flowering and fruit bearing (meteorological factors, fumigation, and irrigation) (Nelson, 2001, 2003).

In Hawaii, *M. citrifolia* L. plots are usually harvested two or three times per month, although fruit production is lower during winter. With a density of 638 plants per hectare; good soil fertility, drainage, and irrigation; appropriate pest, disease and weed control; and an appropriate fertilisation plan, it is possible to obtain yields of 7 tonnes/ha/year after the fifth year (Nelson, 2001, 2003). With a juice extraction rate of approximately 50% (w/w), one hectare can thus yield around 35 tons of juice. However, many factors may affect these yields, and most producers do not obtain such good results because of diseases and/or poor agricultural practices (growing wide plants). In Hawaii, an average annual yield of 50 tonnes/ha is generally attained (Nelson, 2001, 2003).

Table 1. Evolution of fruit skin colour and firmness in the course of ripening (adapted from Chan-Blanco et al., 2006)

Maturity stage	Colour	Firmness
1	Dark green	Very hard
2	Green-yellow	Very hard
3	Pale yellow	Very hard
4	Pale yellow	Fairly hard
5	Translucent-greyish	Soft

Depending on the post-harvest technology programme adopted, the fruits may be harvested at different stages of development and continue to mature. The evolution of the colour and firmness of fruits left to ripen naturally on the tree is reported in Table 1. Nonetheless, most processors buy *M. citrifolia* L. harvested at the "hard white" stage for juice production as the fruits become soft too quickly once this stage is reached (Nelson, 2001, 2003). The change from stage 4 to 5 occurs very quickly (within a few hours), and the pulp practically liquefies and turns from green to white, as well as develops the characteristic butyric smell.

The fruits are individually selected on the tree and harvested by hand. At the "hard white" stage, they are well able to withstand being transported in baskets or containers, and exposure of the fruits to light or high temperatures immediately after harvest does not affect their overall quality. Before processing, fruits are ripened at room temperature for one day or more, depending on the end product (such as tea, juice, pulp, dietetic products) (Nelson, 2003).

2.3 Chemical composition of M. citrifolia L.

Approximately 160 phytochemical compounds have already been identified in the *M. citrifolia* L. plant, and the major micronutrients are phenolic compounds, organic acids and alkaloids (Wang & Su, 2001). Of the reported phenolic compounds, the most important are anthraquinones (such as damnacanthal, morindone, morindin.) and also aucubin, asperuloside, and scopoletin (Wang & Su, 2001). The main organic acids are caproic and caprylic acids (Dittmar, 1993), whereas the principal reported alkaloid is xeronine (Heinicke, 1985). Chan-Blanco et al. (2006) reviewed the chemical constituents of different parts of the plant (Table 2).

Table 2. Chemical compounds of *M. citrifolia* L. (adapted from Chan-Blanco et al., 2006)

Source (plant part)	Chemical constituent	References
Flower	2-methyl-4-hydroxy-5,7-dimethoxyanthraquinone 4-O-β-D-glucopyranosyl-(1-4)-α-L-rhamnopyranoside	Sang et al. (2002)
Flower	5,8-dimethyl-apigenin 4′-O-β-D-galacatopyranoside	Sang et al. (2002), Elkins (1998)
Flower	Aracetin 7-O-β-D-glucopyranoside	
Fruit	β-D-glucopyranose pentaacetate	Sang et al. (2002), Elkins (1998)
Fruit	2,6-di-O-(β-D-glucopyranosyl-1-O-octanoyl-β-D-glucopyranose	Dittmar (1993)
Fruit	6-O-(β-D-glucopyranosyl-1-O-octanoyl-β-D-glucopyranose	Wang et al. (1999)
Fruit	Ascorbic acid	Liu et al. (2001)
Fruit	Asperulosidic acid	Morton (1992), Elkins (1998), Wang et al. (2002), McClatchey (2002)
Fruit	Asperuloside tetraacetate	Wang et al. (1999), Liu et al. (2001), Cardon (2003)
Fruit	Caproic acid	Dittmar (1993)
Fruit	Caprylic acid	Sang et al. (2002), Dittmar (1993), Elkins (1998), Wang et al. (2002), Levend and Larson (1979)
Fruit	Ethyl caprylate	Solomon (1999), Dittmar (1993), Cardon (2003), Elkins (1998), Wang et al. (2002), Levand and Larson (1979)
Fruit	Ethyl caproate	Dittmar (1993)
Fruit	Hexanoic acid	Dittmar (1993)
Fruit	Octanoic acid	Farine et al. (1996), Sang et al. (2002)
Fruit	Quercetin 3-O-α-L-rhamnopyranosyl- (1-6)-β-D-glucopyranoside	Farine et al. (1996), Sang et al. (2002), Cardon (2003), Wang & Su (2001)
Heartwood	Physcion 8-O-α-L-arabinopyranosyl-(1-3)-β-D-galactopyranosyl-(1-6)-β-D- galactopyranoside	Wang & Su (2001), Wang et al. (2002)

Leaves	Alanine	Sang et al. (2002), Srivastava & Singh (1993), Cardon (2003)
Leaves	Quercetin 3-O-α-L-rhamnopyranosyl- (1-6)-β-D-glucopyranoside	Sang et al. (2002)
Leaves	Serine	Dittmar (1993), Elkins (1998)
Leaves	Threonine	Dittmar (1993), Elkins (1998)
Leaves	Tryptophan	Dittmar (1993), Elkins (1998)
Leaves	Tyrosine	Dittmar (1993), Elkins (1998)
Leaves	Ursolic acid	Sang et al. (2002), Cardon (2003), Elkins (1998), Wang et al. (2002)
Leaves	Valine	Dittmar (1993), Elkins (1998)
Plant	2-methyl-3,5,6- trihydroxyanthraquinone	Cardon (2003), Inoue et al. (1981)
Plant	2-methyl-3,5,6- trihydroxyanthraquinone 6-O-β-D- xylopyranosyl-(1-6)-β-D-glucopyranoside	Cardon (2003), Inoue et al. (1981)
Plant	3-hydroxymorindone	Cardon (2003), Inoue et al. (1981)
Plant	3-hydroxymorindone 6-O-β-D-xylopyranosyl-(1-6)-β-D- glucopyranoside	Cardon (2003), Inoue et al. (1981)
Plant	5,6-dihydroxylucidin 3-O-β-D- xylopyranosyl-(1-6)-β-D-glucopyranoside	Cardon (2003), Inoue et al. (1981)
Plant	5,6-dihydroxylucidin	Cardon (2003), Inoue et al. (1981)
Plant	Aucubin	Elkins (1998), Wang et al. (2002)
Plant	Linoleic acid	Wang et al. (2002)
Plant	Lucidin	Cardon (2003), Inoue et al. (1981), Ross (2001)
Plant	Lucidin 3-O-β-D-xylopyranosyl-(1-6)-β-D-glucopyranoside	Cardon (2003), Inoue et al. (1981)
Plant	Scopoletin	Farine et al. (1996), Wang et al. (2002)
Leaves	Arginine	Dittmar (1993)
Leaves	Aspartic acid	Dittmar (1993)
Leaves	β-sitosterol	Sang et al. (2002), Chunhieng (2003), Elkins (1998), Wang et al. (2002)
Leaves	Citrifolinoside B	Sang et al. (2002)
Leaves	Cysteine	Dittmar (1993), Elkins (1998)
Leaves	Cystine	Dittmar (1993), Elkins (1998)

Leaves	Glutamic acid	Dittmar (1993)
Leaves	Glycine	Dittmar (1993), Elkins (1998)
Leaves	Histidine	Dittmar (1993), Elkins (1998)
Leaves	Isoleucine	Dittmar (1993), Elkins (1998)
Leaves	Kaempferol 3-O-α-L-rhamnopyranosyl- (1-6)-β-D-glucopyranoside	Sang et al. (2002)
Leaves	Kaempferol 3-O-β-D-glucopyranosyl-(1-2)- α-L-rhamnopyranosyl-(1-6)-β-D- galactopyranoside	Sang et al. (2002)
Leaves	Leucine	Dittmar (1993), Elkins (1998)
Leaves	Methionine	Dittmar (1993), Elkins (1998)
Leaves	Phenylalanine	Dittmar (1993), Elkins (1998)
Leaves	Proline	Dittmar (1993), Elkins (1998)
Leaves	Quercetin 3-O-β-D-glucopyranoside	Sang et al. (2002)
Root, heartwood, root bark	Morindone	Sang et al. (2002), Inoue et al. (1981), Dittmar (1993), Ross (2001), Cardon (2003), Wang et al. (2002)
Root, heartwood, seeds	Damnacanthal	Sang et al. (2002), Cardon (2003)
Leaves	Quercetin 3-O-β-D-glucopyranosyl-(1-2)- α-L-rhamnopyranosyl-(1-6)-β-D- galactopyranoside	Sang et al. (2002)
Root	8-hydroxy-8-methoxy-2-methyl-anthraquinone	Cardon (2003), Solomon (1999)
Root	Rubichloric acid	Elkins (1998), Morton (1992)
Root	1,3-dihydroxy-6-methyl anthraquinone	Morton (1992)
Root	Morenone 1	Solomon (1999)
Root	Morenone 2	Solomon (1999)
Root	Ruberythric acid	Cardon (2003)
Root	Rubiadin	Cardon (2003), Elkins (1998), Inoue et al. (1981), Ross (2001)
Root bark	Chlororubin	Dittmar (1993), Elkins (1998)
Root bark	Hexose	Dittmar (1993)
Root bark	Morindadiol	Dittmar (1993)
Root bark	Morindanidrine	Dittmar (1993)
Root bark	Morindine	Cardon (2003), Dittmar (1993), Elkins (1998), Morton (1992)
Root bark	Pentose	Dittmar (1993)
Root bark	Physcion	Solomon (1999)
Root bark	Rubiadin monomethyl ether	Dittmar (1993)

Root bark	Soranjidiol	Dittmar (1993), Elkins (1998), Ross (2001)
Root bark	Trioxymethylanthraquinone monoethyl ether	Dittmar (1993)
Root, root bark, fruit	Alizarin	Cardon (2003), Dittmar (1993), Elkins (1998), Ross (2001), Wang et al. (2002)
Seeds	Ricinoleic acid	Solomon (1999)

However, chemical composition differs significantly according to the part of the plant. The complete physico-chemical composition of the fruit has not yet been reported, and only partial information is available on *M. citrifolia* L. juice. The fruit contains 90% water, and the main components of the dry matter appear to be soluble solids, dietary fibre and proteins (Chunhieng, 2003). The fruit's protein content is surprisingly high, representing 11.3% of the juice dry matter, and the main amino acids are aspartic acid, glutamic acid and isoleucine (Chunhieng, 2003).

According to a book on Malaysian medicinal plants, the chemical constituents of *M. citrifolia* L. are 5,7-Acacetin-7-O-β-D(+)-glycopyranoside, ajmalicine isomers, alizarin, asperuloside, asperulosidic acid, chrysophanol (1,8-dihydroxy-3-methylanthraquinone), damnacanthol, digoxin, 5,6-dihydroxylucidin, 5,6-dihydroxylucidin-3-β-primeveroside, 5,7-dimethylapigenin- 4'-O-β-D(+)-galacto pyranoside, lucidin, lucidin-3-β-primeveroside, 2-methyl-3,5,6-trihydroxy anthraquinone, 3-hydroxymorindone, 3-hydroxymorindone-6-β-primereroside, α-methoxyalizarin, 2-methyl-3,5,6-trihydroxyanthraquinone-6-β-primeveroside, monoethoxyrubiadin, morindadiol, morindin, morindone (1,5,6-trihydroxy-2- methylanthraquinone), morindone-6-β-primeveroside, nordamnacanthal, quinoline, rubiadin, rubiadin 1-methyl ether, saronjidiol, ursolic acid, alkaloids, anthraquinones and their glycosides, caproic acid, caprylic acid, fatty acids and alcohols (C_{5-9}), flavone glycosides, flavonoids, glucose (β-D-glucopyranose), indoles, purines, and β-sitosterol.

Minerals account for 8.4% of the dry matter, and these minerals are mainly potassium, sulphur, calcium and phosphorus; traces of selenium have been reported in the juice (Chunhieng, 2003, Table 3). Vitamins have been reported in the fruit, mainly ascorbic acid (24-158 mg/100 g dry matter) (Morton, 1992; Shovie & Whistler, 2001) and provitamin A (Dixon et al., 1999). Phenolic compounds have been found to be the major group of functional micronutrients in *M. citrifolia* L. juice: damnacanthal, scopoletin, morindone, alizarin, aucubin, nordamnacanthal, rubiadin, rubiadin-1-methyl ether and other anthraquinone glycosides have been identified in *M. citrifolia* L. (Morton, 1992; Dittmar, 1993; Dixon et al., 1999; Wang & Su, 2001). Damnacanthal is an anthraquinone that has been characterised and has some important functional properties (mainly anti-carcinogenic) (Solomon, 1999). Scopoletin is a coumarin that was isolated in 1993 at the University of Hawaii and has been found to have analgesic properties as well as a significant ability to control serotonin levels in the body (Levand & Larson, 1979). Other researchers have shown that scopoletin may also have anti-microbial (Duncan et al., 1998) and anti-hypertensive effects (Solomon, 1999).

Multiple Hawaiian teams (Heinicke, 1985; Solomon, 1999) have reported the presence of a novel component, proxeronine, which is an alkaloid that is claimed to combine with human proteins and improve their functionality. The above authors attribute most of the beneficial effects of *M. citrifolia* L. to xeronine. Nonetheless, neither the chemical characterisation of this alkaloid nor the method used to assess its levels has been published to date.

Table 3. Physico-chemical composition of *M. citrifolia* L. juice (adapted from Chan-Blanco et al., 2006)

Characteristic	Chunhieng (2003)[a]	Shovic and Whistler (2001)[a]	European commission (2002)[b]
pH	3.72	-	3.4-3.6
Dry matter	9.8±0.4%	-	10-11%
Total soluble solids (°Brix)	8	-	-
Protein content	2.5%	0.4 g/100 g	0.2-0.5%
Lipids	0.15%	0.3 g/100 g	0.1-0.2%
Glucose	11.9±0.2 g/l	-	3-4 g/100 g
Fructose	8.2±0.2 g/l	-	3-4 g/100 g
Potassium	3900 mg/l	188 mg/100 g	30-150 mg/100 g
Sodium	214 mg/l	21 mg/100 g	15-40 mg/100 g
Magnesium	14 mg/l	14.5 mg/100 g	3-12 mg/100 g
Calcium	28 mg/l	41.7 mg/100 g	20-25 mg/100 g
Vitamin C	-	155 mg/100 g	3-25 mg/100 g

Approximately 51 volatile compounds have been identified in the ripe fruit (Sang et al., 2001), including organic acids (mainly octanoic and hexanoic acids), alcohols (3-methyl-3- butene-1-ol), esters (methyl octanoate, methyl decanoate), ketones (2-heptanone) and lactones ((E)-6-dodeceno-γ-lactone) (Farine et al., 1996).

Major components

A number of major components have been identified in the *M. citrifolia* L. plant, such as scopoletin, octanoic acid, potassium, vitamin C, terpenoids, alkaloids, anthraquinones (such as nordamnacanthal, morindone, rubiadin, rubiadin-1-methyl ether and anthraquinone glycoside), β-sitosterol, carotene, vitamin A, flavones glycosides, linoleic acid, alizarin, amino acids, acubin, L-asperuloside, caproic acid, caprylic acid, ursolic acid, rutin and a putative proxeronine.

A research group led by Chi-Tang Ho at Rutgers University in the United States (US) is searching for new novel compounds in the *M. citrifolia* L. plant. They have successfully identified several new flavonol glycosides, an iridoid glycoside from *M. citrifolia* L. leaves, a trisaccharide fatty acid ester, rutin, and an asperulosidic acid from the fruit. Two novel glycosides and a new unusual iridoid named citrifolinoside have been shown to have an inhibitory effect on AP-1 transactivation and cell transformation in the mouse epidermal JB6 cell line. James Duke listed 23 different phytochemicals found in *M. citrifolia* L. as well as 5 vitamins and 3 minerals in an authoritative handbook of phytochemicals.

Xeronine system

Retired biochemist Ralph Heinicke states that *M. citrifolia* L. fruit contains a natural precursor of xeronine that he named Proxeronine. Proxeronine is converted to the alkaloid xeronine in the body by an enzyme he named proxeroninase. His hypothesis is that xeronine is able to modify the molecular structure of proteins. Thus, xeronine has a wide range of biological activities. When a protein such as an enzyme, receptor, or signal transducer is not in the appropriate conformation, it will not work properly. Xeronine will interact with the protein and make it fold into its proper conformation. The result is a properly functioning protein. Whenever a problem arises in the cell due to a structural problem with a protein, xeronine's presence would be beneficial. His hypothesis may explain why Tahitian Noni @ juice (TNJ) can help in many health problems in different ways. He has obtained several patents for xeronine. He states that the active ingredient in many of the effective folklore drugs is xeronine. This alkaloid is a critical normal metabolic coregulator. The ailments that he believes are helped by *M. citrifolia* L. include high blood pressure, menstrual cramps, arthritis, gastric ulcers, sprains, injuries, mental depression, senility, poor digestion, drug addiction, and pain. "I have devoted much of my life to the study of this unique substance that I have named 'xeronine'. I am convinced of the tremendous benefits achieved by furnishing the body with a proper supply of this material" (Heinicke, 2001).

2.4 Biological activity of *M. Citrifolia* L.

2.4.1 Anti-microbial effects

The anti-microbial activity of *M. citrifolia* L. may have been its first observed property; indeed, the fruit contains relatively large amounts of sugars that do not ferment even when fruits are stored in closed containers at ambient temperature. This property is used to transport the fruit by boat from the scattered Pacific islands to processing plants without specific treatment.

It has been reported that *M. citrifolia* L. inhibits the growth of certain bacteria, such as staphylococcus aureus, Pseudomonas aeruginosa, Proteus morgaii, Bacillus subtilis, Escherichia coli, Helicobacter pylori, Salmonella and Shigella (Atkinson, 1956). The same author claims that the anti-microbial effect observed may be due to the presence of phenolic compounds such as acubin, l-asperuloside, alizarin, scopoletin and other anthraquinones. Another study showed that an acetonitrile extract of the dried fruit inhibits the growth of Pseudomonas aeruginosa, Bacillus subtilis, Escherichia coli, and Streptococcus pyrogene (Locher et al., 1995).

It has also been found that ethanol and hexane extracts of *M. citrifolia* L. have an antitubercular effect as they inhibit the growth of Mycobacterium tuberculosis by 89-95% (Saludes et al., 2002). The major components identified in the hexane extract are E-phytol, cycloartenol, stigmasterol, β-sitosterol, campesta-5,7,22-trien-3-β-ol, and the ketosteroids stigmasta-4-en-3-one and stigmasta-4-22-dien-3-one.

Other studies have reported a significant antimicrobial effect on various strains of Salmonella, Shigella, and E. coli (Bushnell et al., 1950; Dittmar, 1993). Furthermore, they showed that the anti-microbial effect is highly dependent on the stage of ripeness and on processing, being greater when the fruit is ripe and undried.

2.4.2 Anti-cancer activity

The immunomodulatory properties (the capacity to enhance the host immune system) of *M. citrifolia* L. juice have been studied by a Japanese research team (Hirazumi et al., 1996; Hirazumi & Furusawa, 1999). The ethanol precipitable fraction (ppt) of *M. citrifolia* L. juice, corresponding to a polysaccharide-rich substance composed of glucuronoic acid, galactose, arabinose and rhamnose, has been found to have immunomodulatory and anti-tumour effects against Lewis lung carcinoma (LLC). In cell models, *M. citrifolia* L.-ppt seems to stimulate the production of T-cells, thymocytes and macrophages that produce cytokines, which are important mediators of tumour cytostasis and cytotoxicity.

M. citrifolia L.-ppt also appears to stimulate murine effector cells to release several mediators such as cytokines. These mediators slow down the cell cycle in tumours, increase the response of cells to other immunised cells that fight tumour growth and have potent macrophage activator activity suspected of playing a role in the death of tumours (Hirazumi et al., 1996; Hirazumi & Furusawa, 1999).

In the same study, mice were inoculated with LLC, and those ingesting a daily dose of 15 mg of *M. citrifolia* L. juice showed a significant increase (119%) in lifespan. Nine out of 22 mice with terminal cancer survived for more than 50 days. In addition, the ingestion of *M. citrifolia* L.-ppt combined with conventional chemotherapy proved to increase the lifespan of mice with cancer, (Hirazumi et al., 1994).

Another Japanese team studied the influence of damnacanthal, an anthraquinone extracted from a chloroform extract of *M. citrifolia* L. fruits. Surprisingly, the researchers found that damnacanthal induces normal morphology in a particular type of cell found in human neoplasias (K-ras-NKR cells) that multiply uncontrollably and are highly malignant (Hiramatsu et al., 1993).

Another study showed that commercial *M. citrifolia* L. juice (Tahitian Noni juice) prevents the formation of chemical carcinogen-DNA-adducts. In the above study, rats with artificially induced cancer in specific organs were fed for one week with 10% *M. citrifolia* L. juice in their drinking water and rat food (rat chow) ad libitum. They showed reduced DNA-adduct formation depending on sex and organ. The reduction rates were: in female rats, heart 30%, liver 42%, lungs 41% and kidneys 80%; in male rats, heart 60%, liver 70%, lungs 50% and kidneys 90% (Wang & Su, 2001).

2.4.3 Anti-oxidant properties

The antioxidant properties of ethanol and ethyl acetate extracts of *M. citrifolia* L. fruit have been assessed using the ferric thiocyanate method (FTC) and thiobarbituric acid test (TBA). The authors found that ethyl acetate extract strong inhibited lipid oxidation, comparably to the same weight of pure α-tocopherol and butylated hydroxyl toluene (BHT) (Mohd et al., 2001).

Radical scavenging activity was also measured *in vitro* by the tetrazolium nitroblue (TNB) assay in commercial juice by assessing the capacity of the juice to protect cells and lipids from oxidative alteration promoted by superoxide anion radicals (SARs). The SAR scavenging activity of *M. citrifolia* L. juice was shown to be 2.8 times higher than that of vitamin C, 1.4 times that of pycnogenol (PYC) and almost the same magnitude as that of grape seed powder (Wang & Su, 2001).

2.4.4 Anti-inflammatory activity

The anti-inflammatory activity of an aqueous extract from *M. citrifolia* L.-juice was observed by inducing a locally acute inflammatory response with the help of a pro-inflammatory agent (bradykinin). It was found that the oral administration of *M. citrifolia* L. juice extract (200 mg) rapidly inhibits the formation of rat paw oedema. This effect may have resulted from interference with the B2 receptor-mediated mechanism by which bradykinin induces rat paw oedema (Mckoy et al., 2002).

Another study showed that commercial *M. citrifolia* L. juice selectively inhibits cyclo- oxygenase enzymes (COX-1 and COX-2) involved in breast, colon and lung cancer and also has anti-inflammatory activity (Su et al., 2001). The ability of noni juice to inhibit these enzymes was compared to that of traditional commercial non-steroidal inflammatory drugs, such as aspirin, Indomethacin and Celebrex. *M. citrifolia* L. juice showed selective inhibition of COX activity *in vitro* and a strong anti-inflammatory effect comparable to that of Celebrex, and presumably, this juice lacks side effects.

2.4.5 Analgesic activity

Recent research has examined the analgesic properties of commercial juice in rats. The results showed that rats fed 10% or 20% *M. citrifolia* L. juice had greater pain tolerance (162% or 212%, respectively) compared with the placebo group (Wang et al., 2002). A French research team has also studied the analgesic and sedative effects of *M. citrifolia* L. on mice through the writhing and hotplate tests. *M. citrifolia* L. root extract (1600 mg/kg) showed significant analgesic activity in the animals, similar to the effect of morphine (75% and 81% protection using *M. citrifolia* L. extract and morphine, respectively), and it also proved to be non-toxic (Younos et al., 1990).

2.4.6 Cardiovascular activity

Recent research has demonstrated the ability of *M. citrifolia* L. fruit to prevent arteriosclerosis, a disease related to the oxidation of low density lipoproteins (LDLs). Methanol and ethyl acetate extracts showed 88% and 96% inhibition, respectively, of copper- induced LDL oxidation by the thiobarbituric acid relative substance method. This beneficial effect could be due to the presence of lignans, which are phenylpropanoid dimers (Kamiya et al., 2004).

2.5 Biological activities of *M. citrifolia* L. products

2.5.1 Antibacterial activity

Acubin, l-asperuloside, alizarin in *M. citrifolia* L. fruit, and certain other anthraquinone compounds in *M. citrifolia* L. roots, are all proven antibacterial agents. These compounds have been shown to fight infectious bacteria such as Pseudomonas aeruginosa, Proteus morgaii, Staphylococcus aureus, Bacillis subtilis, Escherichia coli, Salmonella, and Shigela. These antibacterial compounds in *M. citrifolia* L. are responsible for the treatment of skin infections, colds, fevers, and other bacteria-caused health problems. Bushnell reported on the antibacterial properties of certain plants found in Hawaii, including *M. citrifolia* L. He further reported that *M. citrifolia* L. was traditionally used to treat broken bones, deep cuts, bruises, sores and wounds. Extracts from the ripe *M. citrifolia* L. fruit exhibit moderate antibacterial properties against Salmonella typhosa. Salmonella montevideo, Salmonella schottmuelleri, Shigella paradys, BH and Shigella paradys, III-Z. Leach demonstrated that acetone extracts obtained from Cycas circinalis, *M. citrifolia*, Bridelia penangiana, Tridax Procumbens, Hibiscus tiliaceus, and Hypericum papuanun show antibacterial activity. The widespread medicinal use of these plants would suggest that they do contain pharmacologically active substances, and alternative methods of extraction and screening should be utilised to find the major bioactive component in the plants for the purpose of new drug development. Locher reported that selected plants including *M. citrifolia* have a history of use in Polynesian traditional medicine for the treatment of infectious diseases. These plants have been investigated for anti-viral, anti-fungal, and anti-bacterial activity *in vitro*. Their study using *in vitro* biological assays confirmed that some of the Hawaiian medicinal plants in ethnobotanical reports have curative properties against infectious diseases.

Duncan demonstrated that scopoletin, a health promoter in *M. citrifolia* L., inhibits the activity of E. coli, which is associated with recent outbreaks resulting in hundreds of serious infections, even death. *M. citrifolia* L. also helps stomach ulcers by inhibiting H. pylori bacteria.

2.5.2 Antiviral activity

Umezawa and coworkers found that a compound isolated from *M. citrifolia* L. roots named 1-methoxy-2-formyl-3-hydroxyanthraquinone suppresses the cytopathic effect of HIV- infected MT-4 cells without inhibiting cell growth.

2.5.3 Anti-tubercular effects

At the International Chemical Congress of the Pacific Basin Societies meeting in Honolulu, Saludes and colleagues from the Philippines reported that *M. citrifolia* L. kills Mycobacterium tuberculosis. A concentration of extracts from *M. citrifolia* L. leaves killed 89% of the bacteria in a test tube, almost as effectively as the leading anti-TB drug Rifampicin, which has an inhibitory rate of 97% at the same concentration. Although there have been anecdotal reports of native use of *M. citrifolia* L. in Polynesia as a medicine against tuberculosis, this is the first report demonstrating the antimycobacterial potential of compounds obtained from *M. citrifolia* L. leaves. "I hope that pharmaceutical companies will pay attention to this research and explore the *M. citrifolia* L. plant as a potential source of drugs", said Saludes in Manila.

2.5.4 Antitumor activity

At the 83rd Annual meeting of the American Association for Cancer Research in 1992, Hirazumi, a researcher at the University of Hawaii, reported that the alcohol-precipitate of *M. citrifolia* L. fruit juice (noni-ppt) has anticancer activity on lung cancer in C57 B1/6 mice. This *M. citrifolia* L.-ppt was shown to significantly prolong (by up to 75%) the life of mice with implanted Lewis lung carcinoma. It was concluded that the *M. citrifolia* L.-ppt seems to suppress tumour growth indirectly by stimulating the immune system. Improved survival time and curative effects occurred when *M. citrifolia* L.-ppt was combined with sub-optimal doses of the standard chemotherapeutic agents, such as adriamycin (Adria), cisplatin (CDDP), 5-fluorouracil (5-FU) and vincristine (VCR), suggesting that *M. citrifolia* L.-ppt has important clinical utility as a supplemental agent in cancer treatment. These results indicate that *M. citrifolia* L.-ppt may enhance the therapeutic effect of anticancer drugs. Therefore, it may be of benefit to cancer patients by enabling them to use lower doses of anti-cancer drugs to achieve the same or even better results.

Dr. Wang and coworkers demonstrated a cytotoxic effect of TNJ on a cultured leukaemia cell line at various concentrations. TNJ showed dose-dependent cytotoxicity on cultured cancer cells by inducing cancer cell necrosis at high doses and apoptosis at lower doses. Synergistic effects of TNJ with known anticancer drugs have been found. At sub-optimal doses, both prednisolone and TNJ can induce apoptosis. When the dose of prednisolone is fixed and the dose of TNJ increases, apoptotic cells significantly increase. Therefore, TNJ is able to enhance the efficacy of anticancer drugs such as prednisolone. When a single dose of Taxol induces a lower percentage of apoptosis in leukaemia cells, TNJ enhances the rate of apoptosis to 100%. These results indicate that TNJ is able to enhance the therapeutic effects of anticancer drugs such as Taxol. These findings regarding the combination of anticancer drugs with TNJ may be significant. This approach may allow lower doses of synthetic anticancer drugs to be used, increase the tolerance of patients to the toxicity of anticancer drugs, and increase immune function, thus creating a new method for treating cancer patients.

In 1993, Hiramatsu and colleagues reported in Cancer Letters the effects of over 500 extracts from tropical plants on K-Ras-NRK cells. Damnacanthal, isolated from *M. citrifolia* L. roots, is an inhibitor of Ras function. The ras oncogene is believed to be associated with signal transduction in several human cancers such as lung, colon, and pancreatic cancer and leukaemia.

Hiwasa and coworkers demonstrated that damnacanthal, an anthraquinone compound isolated from the *M. citrifolia* L. roots, has potent inhibitory activity towards tyrosine kinases such as Lck, Src, Lyn and EGF receptors. In his research, he examined the effects of damnacanthal on ultraviolet ray-induced apoptosis in ultraviolet-resistant human UVr-1 cells. Consequently, the ultraviolet light induced a concurrent increase in both phosphorylated extracellular signal-regulated kinases and stress-activated protein kinases. After pretreatment with damnacanthal, there was a stimulatory effect on ultraviolet-induced apoptosis.

Dong reported that two glycosides extracted from *M. citrifolia* L.-ppt were effective at inhibiting cell transformation induced by TPA or EGF in the mouse epidermal JB6 cell line. This inhibition was found to be associated with the inhibitory effects of these compounds on AP-1 activity. The compounds also blocked the phosphorylation of c-jun, a substrate of JNKs, suggesting that JNKs are a critical target for the compounds in mediating Ap-1 activity and cell transformation.

2.5.5 Antihelmintic activity

An ethanol extract of tender *M. citrifolia* L. leaves was found to induce paralysis and death in the human parasitic nematode Ascaris Lumbricoides within a day. A botanist via Morton reported that *M. citrifolia* L. has been used in the Philippines and Hawaii as an effective insecticide.

2.5.6 Hypotensive activity

Dang Van Ho of Vietnam demonstrated that total extract of *M. citrifolia* L. roots has a hypotensive effect. Moorthy and coworkers found that an ethanol extract of *M. citrifolia* L. roots lowers blood pressure in anesthetised dogs. Youngken's research team determined that a hot water extract of *M. citrifolia* L. roots lowers blood pressure in anesthetised dogs. A Hawaiian physician reported that *M. citrifolia* L. fruit juice has a diuretic effect.

2.5.7 Immunological activity

Asahina found that an alcohol extract of *M. citrifolia* L. fruit at various concentrations inhibits the production of tumour necrosis factor-alpha (TNF-α), which is an endogenous tumour promoter. Therefore, the alcohol extract may inhibit the tumour promoting effect of TNF- α. Hirazumi found that *M. citrifolia* L.-ppt contains a polysaccharide-rich substance that inhibits tumour growth. It does not cause significant cytotoxicity in adopted cultures of lung cancer cells, but it can activate peritoneal exudate cells to impart profound toxicity when co-cultured with tumour cells. This suggests the possibility that *M. citrifolia* L.-ppt may suppress tumour growth through activation of the host immune system. *M. citrifolia* L.- ppt is also capable of stimulating the release of several mediators from murine effector cells, including TNF- α, interleukin-1 beta (IL-1β), IL-10, IL-12, interferon-gamma (IFN-γ) and nitric oxide (NO). Hokama separated ripe *M. citrifolia* L. fruit juice into 50% aqueous alcohol and precipitate fractions, which was found to stimulate BALB/c thymus cells in (^{3}H)thymidine analysis. It has been suggested that inhibition of Lewis lung tumours in mice, in part, may be due to the stimulation of the T-cell immune response.

Wang and coworkers at the University of Illinois college of Medicine observed that the thymus is enlarged in animals treated with TNJ. The wet weight of the thymus was 1.7 times that of control animals on the seventh day after receiving 10% TNJ in drinking water. The thymus is an important immune organ in the body that generates T cells and is involved in the aging process and cellular immune fractions. TNJ may enhance immune function by stimulating thymus growth and thereby exerting anti-aging and anticancer activities and protecting people from other degenerative diseases.

2.5.8 Mental health and improved high frequency hearing

A small human clinical trial of the effects of TNJ on auditory function and quality of life in patients with decreased bone mineral density and auditory function was conducted at the UIC College of Medicine, Rockford, IL. This study showed that TNJ improves both mental health and high frequency hearing. The results suggest that increased amounts or extended duration of TNJ intake may be required to influence this disorder.

2.6 Study of TNJ for cancer prevention

"To take medicine only when you are sick is like digging a well only when you are thirsty – is it not already too late?" (Chi Po, c 2500 B.C.). This proverb suggests that prevention is more important than treatment.

Cancer is the second leading cause of death in the US. According to the American Cancer society, 1500 people per day die from cancer in the United States. Fighting against cancer is a great task for scientists engaged in this field. The aetiology of most cases of human cancer remains unknown. Exposure to environmental carcinogens accounts for more than 90% of human cancers. Cigarette smoke is the number one high-risk environmental factor. Although some cancers are preventable, a means to prevent most cancers is not yet known. Seeking a natural way to prevent human cancer is an urgent task for cancer prevention investigators.

Studies of food, diet, and cancer have indicated that lifestyle changes such as eating more fruits and vegetables and quitting smoking will help prevent cancer. "A new plate" for America (75% vegetables, 25% meat) appeared at the 2001 annual conference of the American Institute for Cancer Research. Although TNJ possesses a broad range of therapeutic effects, its ability to prevent cancer remains unclear. Recently, new hypothesis has been investigated: whether or not TNJ can help prevent cancer during the early stages of chemical carcinogenesis.

This hypothesis was examined using two carcinogenic animal models and one human clinical study of a group of current smokers. The animal models included the following: the DMBA-induced mammary gland tumourigenesis model and an acute liver injury model induced by the liver carcinogen carbon tetrachloride (CCl_4). These are classical extrinsic carcinogenic models. DMBA-induced DNA adduct formation and histological examination by light and electron microscopy were used as sensitive biomarkers to evaluate the preventive effects of TNJ at the initiation stage of multiple-step carcinogenesis. In the mammary breast carcinogenic model, to monitor the mechanisms of carcinogenesis and DMBA DNA-adduct formation in mammary tissue, the focus was on the pathogenic changes after DMBA administration. In the acute liver injury model, the histopathological changes in liver tissue and levels of both super-oxide anion free radicals (SAR) and lipid hydroperoxide (LPO) after CCl_4 administration were the focus.

DMBA DNA-adduct formation was used as a marker to examine whether TNJ is able to prevent carcinogen-induced DNA damage. Most chemical carcinogens need to be activated by endogenous enzymes to be transformed into a form that readily binds to genetic DNA to form DNA-adducts. Carcinogen-DNA adduct formation is an important DNA damage marker that predicts the possibility of cancer development. Most scientists agree that carcinogen-induced DNA adduct formation is an early critical step in the multiple stages of carcinogenesis. Carcinogen-DNA adducts can be repaired by endogenous enzymes. Unrepaired adducts are fixed after one cell cycle. Unrepaired DNA damage is responsible for mutations and subsequent cancer development. Therefore, preventing carcinogen-DNA adduct formation is a key aspect of preventing the initial steps of carcinogenesis. If TNJ can prevent and/or block the formation of carcinogen-induced DNA adducts, it may prevent cancer at the initiation of multiple-stage carcinogenesis.

In recent years, increasing demand for higher quality and safer foods and medicines, as well as concern for environmental pollution during their commercial production, have triggered stringent regulations on toxin levels in foods and medicines as well as on the discharge of pollutants to the environment. In addition, there has been increasing consumer preference for natural substances. All these factors have provided strong motivation for the development of cost-effective new technologies, such as the eco-friendly extraction of natural substances employing green and safe solvents. In recent years, supercritical fluid extraction (SFE) has emerged as a highly promising environmentally benign technology for the production of natural extracts such as flavours, fragrances, spice oils, and oleoresins; natural anti-oxidants; natural colours; nutraceuticals and biologically active compounds. The state of a substance is called supercritical when both temperature and pressure exceed their critical point values. A supercritical fluid combines two beneficial properties, namely high density (which imparts high solvent power) and high compressibility (which permits high selectivity due to large changes in solvent power in response to small changes in temperature and pressure). In addition, SFE offers very attractive extraction characteristics owing to its favourable diffusivity, viscosity, surface tension, and other thermo-physical properties.

Since the 1980s, several potential applications of SFE have been reported. So far, the most popular SF has been carbon dioxide (CO_2), owing to its easy availability, low cost, nonflammability, nontoxicity, and its possession of a wide spectrum of solvent properties. Its critical temperature is 31.1 °C and its critical pressure is 73.8 bar. Dense or supercritical carbon dioxide could very well be the most commonly used solvent in this century due to its wide-ranging applications. Its near-ambient critical temperature makes it ideally suitable for processing thermally labile natural substances. It is generally regarded as safe (GRAS), and it yields microbial-inactivated, contaminant-free, tailor-made extracts of superior organoleptic profile and longer shelf life with highly potent active ingredients. The SFE technique ensures high consistency and reliability in the quality and safety of bioactive heat- sensitive botanical products as it does not alter the delicate balance of bioactivity of natural molecules. All of these advantages are almost impossible with conventional processes. Therefore, SFE technology using SC-CO_2 as the solvent is an ideal alternative to the conventional techniques for the extraction of bioactive ingredients from spices.

3. CONCISE SUMMARY

Morinda citrifolia L., generally referred to as noni, has a longstanding tradition of extensive consumption as a meal in tropical areas ranging from Indonesia to the Hawaiian Islands, and it is utilized as a herbal treatment for many ailments. The fruit, leaves, seeds, bark, and roots have been traditionally utilized for the prevention or amelioration of several ailments, including arthritis, infections, colds, cancer, and diabetes. Research indicates that Morinda citrifolia L. possesses antioxidant capabilities comparable to synthetic antioxidants like BHT and BHA, which are presently utilized as food additives. The antioxidants present in Morinda citrifolia L. exhibit no adverse effects, making them suitable substitutes for synthetic antioxidants in the food processing sector and holding promise for application in preventative medicine. Consequently, this fruit may serve as an antioxidant ingredient in the food manufacturing sector.

4. REFERENCES

Ahmad, V.U. & Bano, S. (1980). Isolation of β-sitosterol and ursolic acid from *Morinda citrifolia* Linn, *Journal of the Chemical Society of Pakistan*, Vol.2, pp. 71, ISSN 0253-5106

Ames, B.M.; Shigena, M.K. & Hagen, T.M. (1993). Oxidants, antioxidants and the degenerative diseases of aging, Proceedings of the National Academy of Science of the United States America, Vol.90, pp. 7915-7922

Atkinson, N. (1956). Antibacterial substances from flowering plants. 3. Antibacterial activity of dried Australian Plants by rapid direct plate test, *Australian Journal of Experimental Biology*, Vol.34, pp. 17-26, ISSN 0004-945X

Basu, S. & Hazra, B. (2006). Evaluation of nitric oxide scavenging activity, *in vitro* and *ex vivo*, of selected medicinal plants traditionally used in inflammatory diseases, *Phytotherapy Research*, Vol.20, pp. 896-900, ISSN 1099-1573

Burton, G.W. & Ingold, K.U. (1989). Mechanisms of antioxidant action: preventive and chain-breaking antioxidants, In J. Miquel, A.T. Quintanilha, & H. Weber (Eds.). Handbook of free radicals and antioxidants in biomedicine (Vol. 2, No. 29). Boca Raton, FL: CRC Press

Bushnell, O.A.; Fukuda, M. & Makinodian, T. (1950). The antibacterial properties of some plants found in Hawaii, *Pacific Science*, Vol.4, pp. 167-183, ISSN 0030-8870

Calzuola, I.; Gianfranceschi, G.L. & Marsili, V. (2006). Comparative activity of antioxidants from wheat sprouts, *Morinda citrifolia*, fermented papaya and white tea, *International Journal of Food Sciences and Nutrition*, Vol.57, pp. 168-177, ISSN 0963-7486

Cardon, D. (2003). Le Monde des Teintures Naturelles, Belin, Paris

Carini, R.; Poli, G.; Diazini, M.U.; Maddix, S.P.; Slater, T.F. & Cheesman, K.H. (1990). Comparative evaluation of the antioxidant activity of a-tocopherol, α-tocopherol poly ethylene glycol 1000 succinate and a-tocopherol succinate in isolated hepatocytes and liver microsomal suspensions, *Biochemical Pharmacology*, Vol.39, pp. 1597-1601, ISSN 0006-2952

Chan-Blanco, Y.; Vaillant, F.; Perez, A.M.; Reynes, M.; Brillouet, J.M. & Brat, P. (2006). The noni fruit (*Morinda citrifolia* L.): A review of Agricultural research, nutritional and therapeutic properties, *Journal of Food Composition Analysis*, Vol.19, pp. 645-654, ISSN 0889-1575

Chong, T.M.; Abdullah, M.A.; Fadzillah, N.M.; Lai, O.M. & Lajis, N.H. (2004). Anthraquinones production, hydrogen peroxide level and antioxidant vitamins in Marinda elliptica cell suspension cultures from intermediary and production medium strategies, *Plant Cell Reports*, Vol.22, pp. 951-958, ISSN 0721-7714

Chunhieng, M.T. (2003). Developpement de nouveaux aliments santé tropical: application a la noix du Bresil Bertholettia excels et au fruit de Cambodge *Morinda citrifolia*, Ph.D thesis. INPL, France

Deng, S.; Palu, A.K.; West, B.J.; Su, C.X.; Zhou, B.N. & Jensen, J.C. (2007). Lipoxygenase inhibitory constituents of the fruits of Noni (*Morinda citrifolia*) collected in Tahiti, *Journal of Natural products*, Vol.70, pp. 859-862, ISSN 0974-5211

Dittmar, A. (1993). *Morinda citrifolia* L.-use in indigenous Samoan medicine, *Journal of Herbs, Spices and Medicine Plants*, Vol.1, pp. 77-92, ISSN 1049-6475

Dixon, A.R.; McMillen, H.; Etkin, N.L. (1999). Ferment this: the transformation of Noni, a traditional Polynesian medicine (*Morinda citrifolia*, Rubiaceae), *Ecological Botony*, Vol.53, pp. 51-68

Duncan, S.H.; Flint, H.J.; Stewart, C.S. (1998). Inhibitory activity of gut bacteria against Escherichia coli O157 mediated by dietary plant metabolites, *FEMS Microbiology Letters*, Vol.164, pp. 258-283, ISSN 1574-6968

Elkins, R. (1998). Hawaiian Noni (*Morinda citrifolia*) Prize Herb of Hawaii and the South Pacific, Woodland Publishing. Utah

European Commission. Scientific Committee of Food (2002). Opinion of the Scientific Committee on Food of Tahitian Noni juice, SCF/CS/Dos/18 ADD 2. Belgium

Farine, J.P.; Legal, L.; Moreteau, B.; Le Quere, J.L. (1996). Volatile components of ripe fruits of *Morinda citrifolia* and their effects on Drosophila, *Phytochemistry*, Vol.41, pp. 433- 438, ISSN 0031-9422

Garcia-Mediavilla, V.; Crespo, I.; Callodo, P.S.; Esteller, A.; Sanchez-Campos, S.; Tunon, M.J. et al. (2006) The anti-inflammatory flavones quercetin and Kaempferol cause inhibition of inducible nitric oxide synthase, cyclooxygenase-2 and reactive C- protein, and down-regulation of the nuclear factor kappaB pathway in Chang Liver cells, *European Journal of Pharmacology*, Vol.557, pp. 221-229, ISSN 0014-2999

Halliwell, B. (1994). Free radicals and antioxidants: a personal view, *Nutrition Review*, Vol.52, pp. 253-265, ISSN 1753-4887

Heinicke, R.M. (1985). The pharmacologically active ingredient of Noni, Bulletin of the National Tropical Botanical Garden, Vol.15, pp. 10-14

Heinicke, R. (2001). The Xeronine system: a new cellular mechanism that explains the health promoting action of Noni and Bromelian, Direct source publishing

Hiramatsu, T.; Imoto, M.; Koyano, T.; Umezawa, K. (1993). Induction of normal phenotypes in RAS transformed cells by damnacanthal from *Morinda citrifolia*, *Cancer letters*, Vol.73, pp. 161-166, ISSN 0304-3835

Hirazumi, A.; Furusawa, E.; Chou, S.C.; Hokama, Y. (1994). Anticancer activity of *Morinda citrifolia* on intraperitoneally implanted Lewis lung carcinoma in syngenic mice, Proceedings of the Western Pharmacological Society, Vol.37, pp. 145-146

Hirazumi, A.; Furusawa, E.; Chou, S.C.; Hokama, Y. (1996). Immunomodulation contributes to the anticancer activity of *Morinda citrifolia* (Noni) fruit juice, Proceedings of the Western Pharmacological Society, Vol.39, pp. 7-9

Hirazumi, A.; Furusawa, E. (1999). An immunomodulatory polysaccharide-rich substance from the fruit juice of *Morinda citrifolia* (Noni) with antitumor activity, *Phytotherapic Research*, Vol.13, pp. 380-387, ISSN 1099-1573

Inoue K.; Nayeshiro, H.; Inouye, H. & Zenk, M. (1981). Anthraquinones in cell suspension culture of *Morinda citrifolia*, *Phytochemisry*, Vol.20, pp. 1693-1700, ISSN 0031-9422

Jeffers, P.; Kerins, S.; Baker, C.J. & Kieran, P.M. (2007). Generation of reactive oxygen and antioxidant species by hydrodynamically stressed suspensions of *Morinda citrifolia*, *Biotechnology Progress*, Vol.23, pp. 138-145, ISSN 8756-7938

Kamiya, K.; Tanaka, Y.; Endang, H.; Umar, M.; Satake, T. (2004). Chemical constituents of *Morinda citrifolia* fruits inhibit copper-induced Low-Density Lipoprotein oxidation, *Journal of Agriculture and Food Chemistry*, Vol.52, pp. 5843-5848, ISSN 0021-8561

Kampkotter, A.; Nkwonkam, C.G.; Zurawski, R.F.; Timpel, C.; Chovolou, Y.; Watjen, W. et al. (2007). Investigations of protective effects of the flavonoids quercetin and rutin on stress resistance in the model organism Caenorhabditis elegans, *Toxicology*, Vol.234, pp. 113-123, ISSN 0300-483X

Krishnaiah, D.; Sarbatly, R.; Nithyanandam, R.R., (2011). A review of the antioxidant potential of medicinal plant species. *Food and Bioproducts Processing*, Vol.89, pp. 217- 233, ISSN 0960-3085

Levand, O.; Larson, H.O. (1979). Some chemical constituents of *Morinda citrifolia*, *Planta Medica*, Vol.36, pp. 186-187, ISSN 0032-0943

Liu, G.; Bode, A.; Ma, W.Y.; Sang, S.; Ho, C.T.; Dong, Z. (2001). Two novel glycosides from the fruits of *Morinda citrifolia* (Noni) inhibit AP-1 transactivation and cell transformation in the mouse epidermal JB6 cell line, *Cancer Research*, Vol.61, pp. 5749-5756, ISSN 1538-7445

Locher, C.P.; Burch, M.T.; Mower, H.F.; Berestecky, H.; Davis, H.; Van Polel, B.; Lasure, A.; Vander Berghe, D.A.; Vlieti-Nick, A.J. (1995). Anti-microbial activity and anti- complement activity of extracts obtained from selected Hawaiian medicinal plants, *Jounal of Ethnopharmacology*, Vol.49, pp. 23-32, ISSN 0378-8741

Luberck, W.; Hannes, H. (2001). Noni. El Valioso Tesoro Curativo de Los mares del Sur. Editorial EDAF S.A., Madrid

Mc Clatchey, W. (2002). From Polynesian healers to health food stores: Changing perspectives of *Morinda citrifolia* (Rubiaceae), *Integrative Cancer Therapies*, Vol.1, pp. 110-120, ISSN 1534-7354

Mckoy, M.L.G.; Thomas, E.A.; Simon, O.R. (2002). Preliminary investigation of the anti- inflammatory properties of an aqueous extract from *Morinda citrifolia* (noni), *Pharmacological Society*, Vol.45, pp. 76-78, ISSN 1757-8175

Meyer, A.S.; Heinonen, M. & Frankel, E.N. (1998). Antioxidant interactions of catechin, cyaniding, caffeic acid quercetin and ellagic acid on human LDL oxidation, *Food Chemistry*, Vol.67, pp. 71-75, ISSN 0308-8146

Mohd. Z.; Abdul-Hamid, A.; Osman, A. (2001). Antioxidative activity extracts from Mengkudu (*Morinda citrifolia* L.) root, fruit and leaf, *Food Chemistry,* Vol.78, pp. 227- 231, ISSN 0308-8146

Morton, J.F. (1992). The ocean-going Noni, or Indian mulberry (*Morinda citrifolia, Rubiaceae*) and some of its 'colorful' relatives, *Ecological Botony,* Vol.46, pp. 241-256

Moure, A.; Cruz, J.M.; Franco, D.; Dominguez, J.M.; Sineiro, J.; Dominguez, H.; Nunez, M.J.; Parajo, J.C. (2001). Natural antioxidants from residual sources, *Food Chemistry,* Vol.72, pp. 145-171, ISSN 0308-8146

Nelson, S.C. (2001). Noni cultivation in Hawaii, *Fruit and Nuts,* Vol.4, pp. 1-4

Nelson, S.C. (2003). Noni cultivation and production in Hawaii, In: Proceedings of the 2002 Hawaii Noni conference. University of Hawaii at Nanoa. College of Tropical Agriculture and Human resources. Hawaii

Perry, L.M. & Metzger, J. (1980). Medical plant of East and South East Asia-attribute properties and uses, Cambridge: The MIT Press

Pietta, P.; Simonetti, P. & Mauri, P. (1998). Antioxidant activity of selected medicinal plants, *Journal of Agricultural and Food Chemistry,* Vol.46, pp. 4487-4490, ISSN 0021-8561 Ross, I.A. (2001). Medicinal plants of the world. Chemical Constituents, Tropical and Modern Medical Uses, Humana Press, New Jersey

Saludes, J.P.; Garson, M.J.; Franzblau, S.G. & Aguinaldo, A.M. (2002). Antitubercular constituents from the hexane fraction of *Morinda citrifolia* Linn. (*Rubiaceae*), *Phytotherapy Research,* Vol.16, pp. 683-685, ISSN 1099-1573

Samoylenko, V.; Zhao, J.; Dunbar, D.C.; Khan, I.A.; Rushing, J.W. & Muhammad, I. (2006). New constituents from Noni (*Morinda citrifolia*) fruit juice, *Journal of Agricultural and Food Chemistry,* Vol.54, pp. 6398-6402, ISSN 0021-8561

Sang, S.; Cheng, X.; Zhu, N.; Stark, R.E.; Badmaev, V.; Ghai, G.; Rosen, R.; Ho, C.T. (2001a). Flavonol glycosides and novel iridoid glycoside from the leaves of *Morinda citrifolia, Journal of Agriculture and Food Chemistry,* Vol.49, pp. 4478-4481, ISSN 0021-8561

Sang, S.; Cheng, X.; Zhu, N.; Wang, M.; Jhoo, J.W.; Stark, R.E., et al. (2001b). Iridoid glycosides from the leaves of *Morinda citrifolia, Journal of Natural products,* Vol.64, pp. 799-800, ISSN 0974-5211

Sang, S.; He, K.; Liu, G.; Zhu, N.; Cheng, X.; Wang, M., et al. (2001c). A new unusual iridoid with inhibition of activator protein-1 (AP-1) from the leaves of *Morinda citrifolia* L., *Organic Letters,* Vol.3, pp. 1307-1309, ISSN 1523-7060

Sang, S.; He, K.; Liu, G.; Zhu, N.; Wang, M.; Jhoo, J.W., et al. (2001d). Citrofolinin A, a new unusual iridoid with inhibition of activator protein-1 (AP-1) from the leaves of Noni (*Morinda citrifolia* L.), *Tetrahedron Letters,* Vol.42, pp. 1823-1825, ISSN 0040-4039

Sang, S.; Wang, M.; He, K.; Liu, G.; Dong, Z.; Badmaev, V.; Zheng, Q.Y.; Ghai, G.; Rosen, R.T.; Ho, C.T. (2002). Chemical components in Noni fruits and leaves (*Morinda citrifolia* L.) In: Ho, C.T., Zheng, Q.Y. (Eds.), Quality Management of Nutraceuticals, ASC Symposium Series 803, American Chemistry Society, Washington, DC, pp. 134-150

Sang, S.; Liu, G.; He, K.; Zhu, N.; Dong, Z.; Zheng, Q., et al. (2003).New unusual iridoids from the leaves of noni (*Morinda citrifolia* L.)show inhibitory effect on ultraviolet B- induced transcriptional activator protein-1 (AP-1) Activity, *Bioorganic and Medicinal Chemistry,* Vol.11, pp. 2499-2502, ISSN 0968-0896

Shovic, A.C.; Whistler, W.A. (2001). Food sources of provitamin A and vitamin C in the American Pacific, *Tropical Science,* Vol.41, pp. 199-202, ISSN 1556-9179

Solomon, N. (1999). The Noni Phenomenon, Direct Source Publishing, Utah

Srivastava, M.; Singh, J. (1993). A new anthraquinone glycoside form *Morinda citrifolia, Pharmaceutical Biology,* Vol.31, pp. 182-184, ISSN 1744-5116

Su, B.N.; Pawlus, A.D.; Jung, H.A.; Keller, W.J.; McLaughlin, J.L. & Kinghorn, A.D. (2005). Chemical constituents of the fruits of *Morinda citrifolia* (Noni) and their antioxidant activity, *Journal of Natural Products,* Vol.68, pp. 592-595, ISSN 0974-5211

Su, C.; Wang, M.; Nowicki, D.; Jensen, J.; Anderson, G. (2001). Selective COX-2 inhibition of *Morinda citrifolia* (Noni) *in vitro,* In: the Proceedings of the Eicosanoids and other Bioactive Lipids in Cancer. Inflammation and Related Disease. The 7th Annual Conference, 2001 October 14-17. Loews Vanderbilt Plaza, Nashville, Tennessee, USA

Takashima, J.; Ikeda, Y.; Komiyama, K.; Hayashi, M.; Kishida, A. & Ohsaki, A. (2007). New constituents from the leaves of *Morinda citrifolia,* Chemical and Pharmaceutical Bulletin, Vol.55, pp. 343-345

Thomson, R.H. (1971). Naturally occurring quinines (2nd ed.), London and New York: Academic Press (Appropriate sections)

Vinson, J.A.; Dabbag, Y.A.; Serry, M.M. & Jang, J. (1995). Plant flavonoids, especially tea flavonols, are powerful antioxidants using an *in vitro* oxidation model for heart disease, *Journal of Agricultural and Food Chemistry,* Vol.43, pp. 2800-2802, ISSN 0021-8561

Wang, M.; Kikuzaki, K.C.; Boyd, C.D.; Maunakea, A.; Fong, S.F.T.; Ghai, G.R.; Rosen, R.T.; Nakatani, N.; Ho, C.T. (1999). Novel trisaccharide fatty acid ester identified from the fruits of *Morinda citrifolia* (Noni), *Journal of Agriculture and Food Chemistry,* Vol.47, pp. 4880-4882, ISSN 0021-8561

Wang, M.Y.; Su, C. (2001). Cancer preventive effect of *Morinda citrifolia* (Noni), Annals of the New York Academy of Sciences, Vol.952, pp. 161-168, ISSN 1749-6632

Wang, M.Y.; West, B.; Jensen, C.J.; Nowicki, D.; Su, C.; Palu, A.K.; Anderson, G. (2002). *Morinda citrifolia* (Noni): a literature review and recent advances in Noni research, *Acta Pharmacologica Sinica,* Vol.23, pp. 1127-1141, ISSN 1671-4083

West, B.J.; Tani, H.; Palu, A.K.; Tolson, C.B. & Jensen, C.J. (2007). Safety tests and antinutrient analyses of noni (*Morinda citrifolia* L.) leaf, *Journal of the Science of Food and Agriculture,* Vol.87, pp. 2583-2588, ISSN 0022-5142

Willet, W.C. (1994). Diet and health-what should we eat?, *Science,* Vol.264, pp. 532-537

Yang, J.; Paulino, R.; Janke-Stedronsky, S. & Abawi, F. (2007). Free-radical-scavenging activity and total phenols of noni (*Morinda citrifolia* L.) juice and powder in processing and storage, *Food Chemistry,* Vol.102, pp. 302-308, ISSN 0308-8146

Yen, G. & Hsieh, C. (1998). Antioxidant activity of extracts from Duzhong (Eucommia ulmoides) towards various lipid peroxidation models *in vitro, Journal of Agricultural and Food Chemistry,* Vol.46, pp. 3952-3957, ISSN 0021-8561

Younos, C.; Rolland, A.; Fleurentin, J.; Lanhers, M.C.; Misslin, R.; Mortier, F. (1990). Analgesic and behavioral effects of *Morinda citrifolia, Planta Medicine,* Vol.56, pp. 430-434, ISSN 0032-0943

Yu, B.P. (1994). Cellular defences against damage from reactive oxygen species, *Physiological Reviews,* Vol.76, pp. 139-162, ISSN 0031-9333

Zenk, M.H.; El-Shagi, H. & Schulte, U. (1975). Anthraquinone production by cell suspension cultures of *Morinda citrifolia, Planta Medica. Suppl.,* pp. 79-101, ISSN 0032-0943

Zin, Z.M.; Abdul-Hamid, A. and Osman, A. (2002). Antioxidative activity of extracts from Mengkudu (*Morinda citrifolia* L.) root, fruit and leaf, *Food Chemistry,* Vol.78, pp. 227- 231, ISSN 0308-8146

7 Chapter

FLAVONOIDS IN SELECT IRANIAN ANGIOSPERMS

1. INTRODUCTION: AN OVERVIEW

Flavonoids are a category of polyphenolic chemicals found in several plant tissues, exhibiting a diverse array of biological actions [Parr and Bolwell 2000, Noori 2002, Noori et al 2009]. Their distribution in plants, synthesis, and mechanism of action have been thoroughly examined [Shirley 1996].

1.1 Structure, biosynthesis and variety

All flavonoids contain fifteen carbon atoms in their basic nucleus and these are arranged in a C_6-C_3-C_6 configuration, that is, two aromatic rings linked by a three carbon unit which may or may not form a third ring. They are divided into different groups depending on the configuration of the rings and substitutions on these rings of a variety of side-groups which characterize the individual compounds [Stace 1980] (Scheme 1).

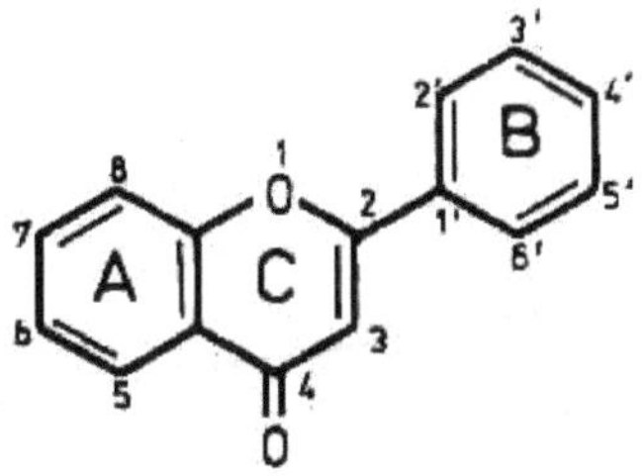

Scheme 1. The basic nucleus of flavonoids (Stace 1980)

The flavonoid variants are all related by a common biosynthetic pathway which incorporates precursors from both the "Shikimate" and "Acetate-Malonate" pathways [Hahlbrock and Grisebach 1975; Wong 1976], the first flavonoid produced immediately following confluence of the two pathways (Scheme 2).

The flavonoid initially formed in the biosynthesis is now thought to be the chalcone and all other forms are derived from this by a variety of routes [Hahlbrock 1981] (Scheme 2). More than 4000 varieties of flavonoids have been identified in different higher and lower plant species (De Groot and Rauen 1998). The main flavonoid groups are flavones (e.g. luteolin), flavanone (e.g. naringenin), flavonols (e.g. kaempferol), anthocyanidins (e.g. pelargonidin) and chalcones (e.g. butein) [Harborne et al 1975].

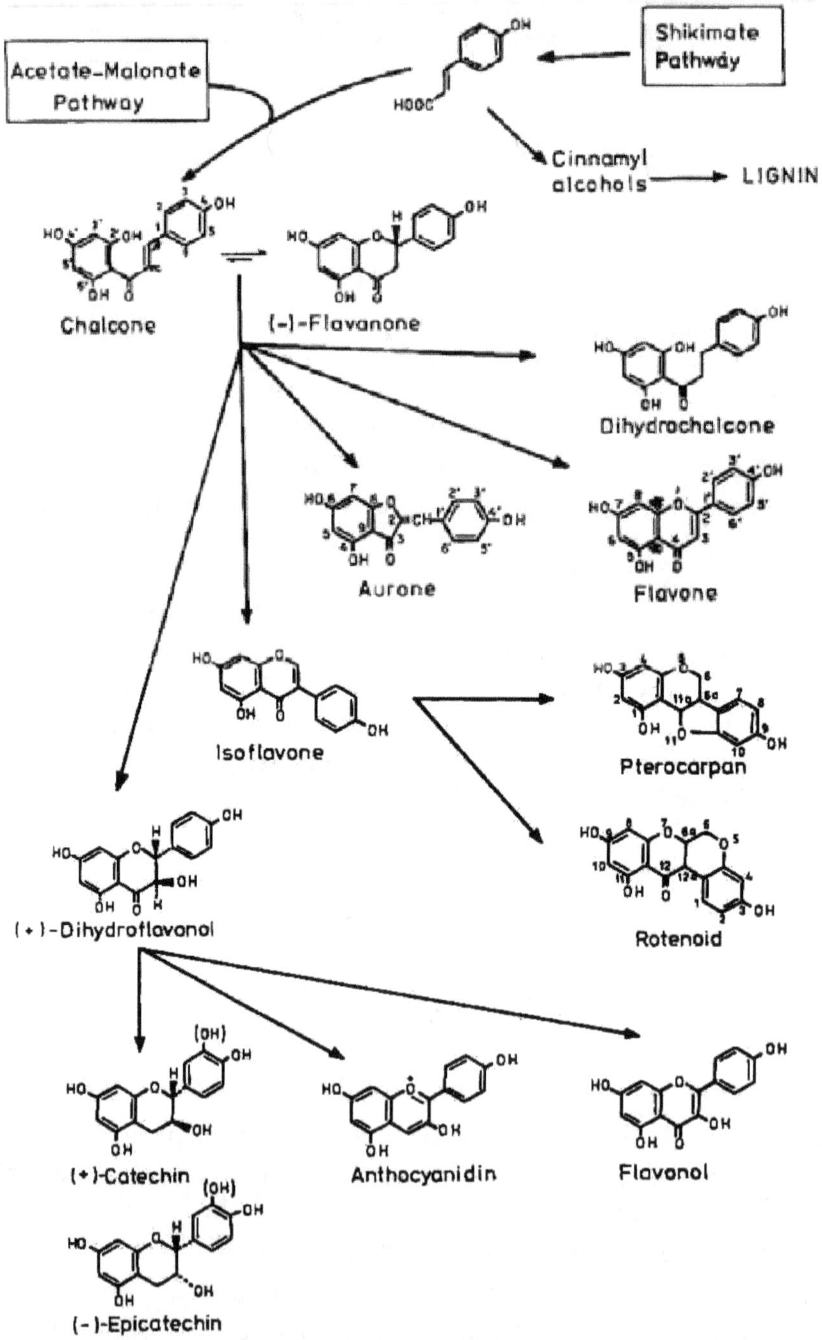

Scheme 2. Showing two biosynthetic pathways of flavonoids (Hahlbrock and Grisebach 1975; Wong 1976).

1.2 Occurance

Flavonoids are found in fruit, vegetables, grains, bark, roots, stems, leaves, flowers, tea and wine [Middleton 1998, Robles et al 2003]. The flavonoid nucleus is normally linked to a sugar moiety to form a water-soluble glycoside. Most flavonoids are stored in the plant cell vacuoles, although they also occur on the surfaces of leaves and stems (Farman 1990). In contrast to earlier studies, all these compounds are no longer judged as waste products, nor as evolutionary remnants without current function, nor as mere metabolic end products that are toxic to the plant and are therefore to be stored away in vacuoles [Parr and Bolwell 2000].

1.3 Biological activities and their usages

Flavonoids possess a wide range of biological activities, medicinal and pharmacological effects [Parr and Bolwell 2000, Noori 2002, Noori et al 2009].

1.3.1 Biological activities

A large variety of colours such as orange, scarlet, crimson, mauve, violet, blue and purple that we encounter in different part of plants, especially flowers and fruits, are caused by anthocyanins (=anthocyanidin glycosides). Chalcones and some flavones and flavonols also absorb light in the visible region and are associated with bright yellow or cream coloured flowers. Other flavones account for the whiteness in most white flowers, without which they would perhaps appear translucent. Even some of the brown and black pigments found in plants are either due to oxidative products of flavonoids or related phenolic compounds. [Farman 1990]. They are beneficial for the plant itself as physiological active compounds, as stress protecting agents, as attractants or as feeding deterrents, and, in general, by their significant role in plant resistance [Treutter 2006]. Also these compounds serve essential functions in plant reproduction by recruiting pollinators and seed disperses. They are also responsible for the beautiful display of fall color in many plant species, which has recently been suggested to protect leaf cells from photo-oxidative damage, thereby enhancing the efficiency of nutrient retrieval during senescence [Field et al 2001].

1.3.2 Medicinal and pharmacological effects

Flavonoids medicinal and pharmacological effects are their contributions to human health which has made them prominent in the past 10 years (Parr and Bolwell 2000). Many flavonoids are active principles of medicinal plants and exhibit pharmacological effects [Yilmaz and Toledo 2004].

1.4 Chemotaxonomy

Flavonoid compounds are taxonomically important. They are popular characters for chemosystematic studies because: the almost universal presence of flavonoids in vascular plants; 2. Their structural diversity; 3. The fact that each species usually contains several flavonoids; 4. The chemical stability of many flavonoids in dried plant material enabling herbarium material to be used; 5. Flavonoid profiles using different chromatographic techniques are easily obtained. 6. Flavonoids are reasonably easy to identify using published UV spectra data and available standards; 7. Flavonoids often show correlations with existing classifications at the family, genus and species level, and support revisions of existing classifications at the family, genus and species level. However, flavonoids rarely provide "key" characters (the flavonoid may be absent in one or more members of the taxon, and the same flavonoid may occur in an unrelated taxon, e.g. isoflavonoids occur in the Leguminosae and Iridaceae and biflavonyls in the Gymnospermae and some Angiospermae) [Harborne and Turner 1984].

1.5 Flavonoids in Leguminosae

The Leguminosae is economically the single most important family in the dicotyledonae, and also of major significance in nature The family is especially rich in flavonoids, producing about 28% of all known flavonoids and 95% of all isoflavonoid aglycones (Hegnauer and Grayer-Barkmeijer 1993). The importance of the phenolic constituents in the family has been stressed by Bate-Smith (Bath-Smith 1962). As Gomes et al (1981a) showed in "Advances in Legume Systematics" the Leguminosae are especially well endowed with flavonoid constituents, many of which are only known in these plants. Whithin the Leguminosae, some 850 compounds, including 362 isoflavones, are known [Dewick 1993]. There is a basic structures, Such as genistein (4′, 5, 7-trihydroxyisoflavone), 5-dexy derivatives (some 66% of structures), prenylated derivatives (some 51% of structures) and compounds with extra hydroxylation (e.g. at the 6-, 8- or 2′-positions). Isoflavonoids usually occur in the free state, and are obtained from root, wood, bark or seed rather than leaf or flower [Ingham 1981, 1983]. Flavonoids, as distinct from isoflavonoids and neoflavonoids, are widespread in the *Papilionoidaea* and there is little doubt that they occur not only in the species of the some tribes, but will eventually be found in all tribes [Gomes et al 1981b]. Harborn (1965) obtained quercetagetin from hydrolyzed petal of *Coronilla glauca* L.. He also found halogenin, 3-*O* rutinoside and limocitrin, 3- *O* rutinoside from C. glauca flower (Harborn 1981). Catechin, epigallocatechin, leucodelphinidin and 3, 3′, 4′, 5, 5′, 7- hexahydroxyflavan have been identified from *Alhaji maurorum* Medikus ground parts [Islambeko et al 1982]. Malvidin from hydrolyzed flower, myricitrin from flower and laef of *Cercis siliquastrum* L. have been isolated [Torck et al 1969, Sagareishvili and Ananiya 1990].

1.6 Polygonaceae flavonoids

Based on Isobe and Noda (1987) flavonoids and flavonol glycosides are of wide-spread occurrence in the genus *Polygonum*. Among them, glycosylation at C-3 of the quercetin nucleus has been found to be the most common trend, and present in all species of this genus [Park, 1987]. While rhamnose, gloucose, arabinose and rhamnosyl-rhamnose are the most common sugars found as aglycones of the flavonol glycosides [Mun and Park 1995], galactosylation is rather uncommon in the genus *Polygonum* or in the family polygonaceae [Collins et al 1975]. Kawasaki et al (1986) isolated thirty-three kinds of flavonoids from Polygonaceae species leaves. Quercetin glycosides were commonly found in the family. In the quercetin glycosides, 3-*O*-rhamnoside was most frequently found, 3-*O*-glucuronide is also distributed widely. Myricetin glycosides were rare. Methylated flavonols were found in some species of the section *Echinocaulon* and *Persicaria* [Kawasaki et al 1986]. The aerial exudate of *Polygonum senegalense* has been reported to contain 12 flavonoids of the chalcone and flavanone types, and they are distinctly different from internal tissue aglycones [Midiwo et al 2007]. Also Hsu (2006) studies revealed that *Polygonum aviculare* L. extract has high phenolics and flavonoid contents.

Trichopoulou et al (2000) showed that some wild edible species of *Rumex* such as *R. acetosa* L. and *R. japonicas* Houttuyn have a very high flavonol content. Hasan et al (1995) studies showed besides rutin, quercetin 3-rhamnoside and kaempferol 3-rhamnosyl (1 -> 6) galactoside, a new flavonol glycoside, quercetin 3-glucosyl (1 -> 4) galactoside, and 1, 6, 8- trihydroxy-3-methyl anthraquinone (emodin) have been characterized from leaves of *R. chalepensis.*

1.7 Euphorbiaceae flavonoids

Several studies indicated that flavonoids occurred in various species of *Euphorbia.* Nagase reported the isolation of 5, 7, 4'-trihydroxy-flavone-7-glucoside from the leaves and stems of *E. thymifolia* [Nagase et al 1942]. Ten years later, quercetin was isolated from ethyl acetate extract of aqueous solution of hydroalcoholic extract of *E. pilulifera* [Hallett and Parks 1951]. Sotnikova and his coworkers identified 3', 4'-pentahydroxyflavone 3β-D-galactopyranoside, steppogenin stepposide, isomyricitrin and nineteen other flavonoids in *E. stepposa* [Sotnikova and Litvinenko 1968, Sotnikova et al. 1968]. Muller and Pohl (1970) isolated six new flavonoids all being glycosides of rhamnetin from *E. amygdaloides.* The qualitative composition of flavonoids in alcoholic extract of *E. helioscopia* indicated fifteen substances with flavonoidal nature, by two-dimentional paper chromatography. Acid hydrolysis of *E. helioscopia* alcoholic extract by Volobueva (1970) yielded quercetin and kaempferol. Quercetin-3-xylosidoglucoside has been identified as one of the two flavonoids found in methanolic extract of *E. chamaesyce* and quercetin-3β-D-galactopyranoside gallate has been reported in *E. verrucosa* and *E. platiphyllos* [Singla and Pathak 1990]. Chromatography on cellulose of methanolic extract of *E. lucida* yielded quercetin and its derivatives, viz., isoquercetin, avicularoside, hyperoside and rutoside [Burzanska 1975]. The hypotensive principales of *E. maddeni* were found to be kaempferol-4'-*O*-glucose and hyperin (Sahai et al 1981). In an effort to identify the constituents responsible for the antiviral activity of *E. grantii*. Van Hoof et al. (1984) isolated derivatives of 3-methylquercetin. Polyphenolic components from the aerial parts of *E. soongarica* and *E. alatavica* have been identified as the esters of gallic acid, luteolin-3-rhamnoside and luteoline-3-galactoside [Omurkhamzinova and Erzhanova 1985]. Gautam and Mukhraya (1981) have isolated quercetin-3-O-β-D- glucopyranosyl (1-4)-O-α-L-rhamnopyranoside from the leaves of *E. dracuncoloides*. Kaempferol 3-O-glucoside and quercetin 3-O-glucoside were obtained from *E. larica*, *E. virgata*, *E. chamaesyce* and *E. magalanta*. Also all these taxa, except *E. chamaesyce* contained kaempferol 3-rutinoside and rutin. *E. larica* also yielded 6-methoxyapigenin while *E. virgata* and *E. magalanta* yielded kaempferol. There is an unknown acytylated kaempferol derivative in *E. chamaesyce* (Ulubelen et al 1983). Murillo and Jakupovic (1998) identified myricetin-3- rhamnoside and one flavonoid glycosides in *E. aucherii* which was collected in Iran. Aimova et al (1999) found quercetin 3-(2'''-Galloylglucosyl) (1→2)-α-L-arabinofuranoside in *E. pachyrrhiza*. Studies of Halaweish et al (2003) are the first report of quercetin-3-O-β glucuronic acid in *E. esula*. They separated and identified kaempferol-3-O-β glucuronic acid and quercetin-3-O-β glucuronic acid from the species. Papp et al (2005) studies showed arial parts of *Euphorbia cyparissias* had 2 main flavonoids: kamfpherol-3-glucuronide and quercetin-3-glucuronide. Adedapo et al (2005) showed arial branch extract of *E. hirta* contains kaempferol, quercitol and quercitrin. The phytochemical studies of Falodun et al (2006) on *E. heterophylla* extract revealed the presence of flavonoids in the extract.

Abdel-Sattar (1985) reported existing flavonoids in *Chrozophora*. Then Hashim et al (1990) found kampferol, acacetin, luteolin and apigenin glycosides in *Chrozophora* species. Isorhamnetin and quercetin glycosides were separated from *C. oblique* [Mohamed 2001]. Talischi et al (2005) reported apigenin and quercetin glycosides from metanolic extract of *C. tinctoria* aerial parts. Also Delazar et al (2006) separated 5-flavonoid glycosid from aerial parts metanolic extract of *C. tinctoria*. Shi et al (2006) has reported 7-flavonoid glycoside from *C. sabulosa* species. Vassallo et al (2006) were separated three new flavon glycosids from *C. senegalensis* leaf metabolic extract. Apigenin and loteulin glycosids were reported from leaf aqueous-etanolic extract of *C. brocchiana* species [Hawas 2007].

1.8 Resedaceae flavonoids

Several studies indicated that flavonoids occurred in various species of *Reseda*. Eight flavone, 15 flavonols and one isoflavone have been reported from the *Reseda*. *Reseda luteola* contains 40% flavonoids, primarily luteolin, but also luteolin-7-O glucoside and apigenin [Woelfle et al 2009]. Moiteiro et al (2008) found luteolin 4́-O-glucoside in *Reseda luteola* for first time. Berrahal et al (2006) reported five flavonoid glycosides, quercetin-7-O-α-L-rhamnosyl-3-O-β-D- glucoside, isorhamnetin-3 O-β-

D-glycosyl-7-O-α-L-rhamnoside, kaempferol-7-O-α-L- rhamnoside, kaempferol-7-O-α-L-rhamnosyl-3-O-β-D-glucoside and kaempferol-3, 7-O-α-L- dirhamnoside from aerial parts of *R. villosa* for first time. El-Sayad et al (2001) isolated aglycone flavonols, kaempferol and quercetin from the Mediterranean *Reseda* species. Also Yuldashev et al (1996) reported flavonol diglycosides of kaempferol, quercetin and isorhamnetin from four other *Reseda* species. Rzadkowska (1969) isolated four 3-O-glycosides from *R. lutea*.

1.9 Cyperaceae flavonoids

Clifford and Harborne (1969) studies showed identification of the flavonoid pigment aureusidin from *Scirpus nodosus*. Quercetin, kaempferol, apigenin and luteolin were reported from *S. wichurai* [Ahmed et al 1984]. Naser et al (2000) identified lupeol betulin, betulinalaldehyde and apigenin from *Scirpus tuberosus*. Also β-sitosterol, quercetin 3- β- glucoside, quercetin 3, 7- β-diglucoside and isorhamnetin 3, 7- β-glucoside were identified from *Scirpus litoralis* using spectroscopic analyses [Naser et al 2000]. Yang et al (2010) used a developed capillary electrophoresis with amperometric detection method for the determination of some phenolic compounds in the rhizome of *Scirpus yagara* Ohwi. Their work determined existing four phenolic compounds: transresveratrol, scirpusin A, scirpusin B, and p-hydroxycinnamic acid in the species rhizome.

1.10 Aim

The aim of this study was to compare the leaf flavonoids profiles of some Iranian Angiosperm species from Leguminosae, Polygonaceae, Euphorbiaceae, Resedaceae and Cyperaceae.

2. MATERIALES AND METHODES

2.1 Collection of plant material and praperation

Mature fresh leaves of eight Legumes, seven *Polygonum*, seven *Rumex*, seventeen *Euphorbia*, two *Chrozophora*, four *Reseda* and five *Scirpus* species from different parts of Iran were collected during 2006-2010 as described in Table 1. Plants identified using available references [Rechinger 1964, Mobayen 1979, 1980, Ghahreman 1979-2006]. Specimens of each sample were prepared for reference as herbarium vouchers that were deposited at the Arak University herbarium. Samples were air dried for detection and identification of flavonoids.

2.2 Extraction of the plant material

For a comparative analysis of the flavonoids, small extracts of all the accessions were prepared by boiling 200 mg of powdered air dried leaf material for 2 min in 5 ml of 70% EtOH. The mixture was cooled and left to extract for 24 h. The extract was then filtered, evaporated to dryness by rotary evaporation at 40°, and taken up in 2 ml of 80% MeOH for analysis by 2-Dimensional Paper Chromatography (2-D PC).

2.3 Flavonoid analysis by 2-Dimensional Paper Chromatography (2-D PC)

For the detection of flavonoids, ca 20 µl of each of the small extracts was applied to the corner of a quarter sheet of Whatman No 1 chromatography paper as a concentrated spot (10 applications of 2µl). The chromatogram for each sample was developed in BAW (n-BuOH- HOAc-H_2O=4:1:5; V/V; upper layer), 1st direction, and HOAc (=15% aqueous acetic acid), 2nd direction, with rutin (= quercetin 3-O-rutinoside) as a standard. After development, the chromatograms were viewed in longwave UV light (366 nm) and any dark absorbing and fluorescent spots were marked. R_f-values in BAW and 15% HOAc were calculated.

2.4 Methods of identification of the flavonoids

When sufficient amounts of purified flavonoids had been obtained, as in the cases of the flavonoids from studied samples, they were identified by means of UV spectroscopy using shift reagents to investigate the substitution patterns of the flavonoids [Mabry et al. 1970, Markham 1982] and by acid hydrolysis to identify the aglycone and sugar moieties. Cochromatography with standards was also performed where possible. Flavonoid standards available for comparison during the study obtained commercially from Merck, Sigma and Fluka.

2.5 Acid hydrolysis and identification of flavonoid aglycones

A small amount of each purified flavonoid (ca 0.5 mg) was dissolved in 0.5 ml of 80% MeOH in a test tube. To this sample 2 ml of 2M HCl were added and the mixture was heated in a water bath at 100°C for 0.5 h. The solution was cooled, 2 ml of EtOAc were added and thoroughly mixed with the aqueous layer using a whirley mixer. The upper EtOAc layer was removed with a pipette, evaporated to dryness, dissolved in 0.5 ml of MeOH and applied as spots on thin layer chromatograms (cellulose). The TLC plates were run in three solvents alongside standards to identify the aglycone moiety [Harborne 1998].

Table 1. The sampling and also two-dimensional paper and thin layer chromatographically data of 48 studied plant samples from Markazi Province, Iran.

Voucher data	Taxon	Number of total flavonoids	Number of flavonoid sulphates	Number of flavone C-and C-/O-glucosides	Number of aglycones
	Chrozophora				
*CAM2	C. tinctoria	3	3	2	8

*CAM22	C. hierosolymitana	3	2	3	8
	Euphorbia				
*CMK 23	*E. bungei* Boiss.	5	2	3	0
CMK 65	*E. chamaesyce* L.	8	4	4	0
CMK 57	*E. cheiradenia* Boiss. et Hohen.	7	5	2	0
CMK 63	*E. cordifolia* Ell.	9	3	6	0
CMK 60	*E. esula* L.	7	3	4	0
CMK 59	*E. falcata* L.	7	3	4	0
CMK 32	*E. helioscopia* L.	5	3	2	0
CMK 26	*E. heteradena* Jaub,et Spach.	9	3	6	0
CMK 54	*E. macroclada* Boiss.	6	1	5	0
CMK 70	*E. microsciadeae* Boiss.	9	5	4	0
CMK 69	*E. ozyridiforma* Parsa.	7	4	3	0
CMK 62	*E. peplus* L.	8	4	4	0
CMK 74	*E. petiolata* Banks et Soland	9	6	3	0
CMK 16	*E. seguieriana* Necker.	8	5	3	0
CMK 10	E. splendida Mobayen.	7	3	4	0
CMK 48	*E. szovitsii* Fisch.& Mey.	8	5	3	0
CMK 34	E. tehranica Boiss.	8	6	2	0
	Papilionoideae				
*CMJ148	*Alhagi camelorum* Fisch.	2	1	1	0
CMJ149	*Cercis siliquastrum* L.	2	1	1	0
CMF1	*Coronilla varia* L.	5	4	1	0
CMJ150	*Glycyrrhiza glabra* L.	2	1	1	0
CMJ151	*Medicago sativa* L.	0	0	0	0
CMJ152	Robina peseudo-acacia L.	4	3	1	0
CMJ153	Sophora alopecuroides ssp. alopecuroides	0	0	0	0
CMN1	Sophora alopecuroides ssp. tomentosa	0	0	0	0
	Polygonum				
*CEM1	*P. aviculare*	4	0	4	0
CEM2	*P. convolvolus*	1	0	1	0
CEM3	*P. hyrcanicum*	4	0	4	0
CEM4	*P. patulum*	3	0	3	0
CEM5	*P. alpestre*	3	0	3	0
CAM6	*P. arenastrum*	4	0	4	0
CAM7	*P. persicaria*	3	1	2	0
	Reseda				
*CMG27	Reseda aucheri	8	5	3	0

CMG21	*R. buhseana* Mull-Arg.	10	5	4	1
CMG22	*R. bungei Boiss.*	8	7	1	0
CMG11	*R. lutea L.*	7	6	1	0
	Rumex				
*CMR2	*R. chalpensis*	10	4	6	0
CMR4	*R. crispus*	6	2	4	0
CMR6	*R. obtusifolius*	10	4	6	0
CMR7	*R. tuberosus*	10	5	5	0
CMR9	*R .pulcher*	8	2	6	0
CMR12	*R. acetosella*	8	2	6	0
CMR14	R.*conglomeratus*	9	2	6	1
	Scirpus				
*CNM4	*S. holoschenus* L.	8	4	1	3
CMN23	*S. lacustris* L.	6	5	0	1
CMN8	*S. littoralis* Kuntze	10	8	1	1
CMN6	*S. maritimus* L.	9	5	3	1
CMN18	S. multicaule	12	5	2	5

Voucher data	Api-genin	Chry-cin	Kaemp-ferol	Luteo-lin	Identifica-tionMyricetin	Naringenin	Querce-tin	Rhamne-tin	Rutin	Vitex-in
Chrozophra										
*CAM2	++++	-	-	-	-	-	+++	-	++	-
CAM22	++++	-	-	-	-	-	++++	-	-	-
Euphorbia										
*CMK 23	-	-	++	-	-	-	+++	-	++	-
CMK 65	-	-	-	-	-	-	+++	-	+++	-
CMK 57	-	-	-	-	-	-	+++	-	+++	-
CMK 63	-	-	-	-	-	-	-	-	++	-
CMK 60	-	-	+++	-	-	-	+++	-	+	-
CMK 59	-	-	+++	-	-	-	++	-	+++	-
CMK 32	-	-	-	-	-	-	+++	-	+	-
CMK 26	-	-	+++	-	-	-	++	-	++	-
CMK 54	-	-	-	-	+++	-	++	-	-	-
CMK 70	-	-	-	-	-	-	+++	-	+++	-
CMK 69	-	-	+++	-	-	-	++	-	+	-
CMK 62	-	-	++	-	-	-	+	-	++	-
CMK 74	-	-	-	-	+++	-	+++	-	-	-
CMK 16	-	-	+++	-	-	-	+++	-	+++	-
CMK 10	-	-	-	-	-	-	+	-	+	-
CMK 48	-	-	++	-	-	-	+++	-	++	-
CMK 34	-	-	-	-	-	-	-		+	-
Papilionoideae										

*CMJ148	-	-	++	-	-	-	--	-	++	-
CMJ149	-	-	++		+++		+++		++	-
CMF1	-	-	-	-	±	-	+	+	±	-
CMJ150	-	-	--	-	-	-	++	-	++	-
CMJ151	-	-	-	-	-	-	-	-	-	-
CMJ152	++		++		-	-	-	-	++-	-
CMJ153	±	-	-	-	-	-	-	-	-	-
CMN1	±	-	-	-	-	-	-	-	-	-
Polygonum										
*CEM1	-	-	+	-	+	-	+	-	+	-
CEM2	-	-	-	-	-	-	+	-	-	-
CEM3	-	-	+	-	+	-	+	-	-	-
CEM4	-	-	+	-	-	-	+	-	-	-
CEM5	-	-	+	-	-	-	+	-	+	-
CAM6	-	-	+	-	+	-	+	-	+	-
CAM7	-	-	-	-	+	-	+	-	-	-
Reseda										
*CMG27	-	-	+	-	+	-	+++	++	++	-
CMG21	-	-	+++	+	++	-	-	-	-	-
CMG22	-	-	+++	+	++	-	-	-	-	-
CMG11	-	-	+++	++	-	-	+++	±	++	
Rumex					+					
*CMR2	+	-	+	+	-	-	+	+		
CMR4	-	-	-	±	-	++	++	+	-	-
CMR6	-	-	±	++	-	±	±	+	+	-
CMR7	++	-	±	-	-	±	±	++	-	-
CMR9	+	-	+	+	-	-	++	+	+	-
CMR12	-	-	+	-	-	±	+++	-	-	-
CMR14	+	-	-	±	±	±	++	+	-	-
Scirpus										
*CNM4	++	+	-	+	-	+++	+	+	+	+
CMN23	-	±	-	+	±	±	+++	+	++	+++
CMN8	++	+	-	+	-	+++	-	+++	+	+
CMN6	-	+	-	+	-	++	-	++	+	+
CMN18	++	+	-	++	-	+++	++	++	+	++

*C=Collection number -(non flavonoid), ± (non or a few flavonoid), + (few flavonoid), ++ (middle concentration of flavonoid), +++ (high concentration of flavonoid)

3. RESULTS

All studied plant species exceptional two subspecies *Sophora alopecuroides* and *Medicago sativa* contained flavonoid compounds in their leaves. Their flavonoid profiles show a wide variety between the species. Data in Table 1 shows the sampling and also two-dimensional paper and thin layer chromatographical data of 48 studied *plant samples* from Markazi Province, Iran.

4. DISCUSSION

Studies of leaf flavonoids showed some phytochemical characters such as total number of flavonoids, flavonoid group such as aglycone, flavones *C*- and *C-/O* glycoside and flavonoid sulphate and kind of flavonoids such as kaempherol, quercetin, myricetin are valuable for chemotaxonomy and their usage.

Chemical study of two *Chrozophora* species using two dimentional paper chromatography (2-DPC) and thin layer chromatography (TLC) showed both *Chrozophora* species contain flavonoid sulphates, flavone *C* and *C-/O*-glycosides and aglycon. Also all of studied species have apigenin and quercetin while rutin was just found in *C. tinctoria* species that is recorded first time for Markazi Province. All of studied species have flavonoid compounds that have variation in their flavonoid type and number (Table 1).

Phytochemical studies of the Euphorbiaceae have been extremely useful in clarifying systematic relationships within the family (Simpson and Levin 1994). Flavonoids occur widely in plants and are a biologically major and chemically diverse group of secondary metabolites that are popular compounds for chemotaxonomic surveys of plant genera and families [Harborne 1994]. There are some studies in this connection. Mues and Zinsmeister (1988) have discussed about variation of occurrence phenolic compounds in mosses and liverworts. Also they showed there is a clear flavonoid distinction between the subclasses Marchantiidae and Jungermanniidae. Another important chemotaxonomic programme has concerned the ferns and fern allies [Harborne 1986]. The phenolic patterns appear to be more useful for studying relationships within relatively narrow taxonomic limits, e. g. at the species and genus level. Turning to the angiosperms, a chemotaxonomic survey of 255 species of the family Iridaceae has been carried out by Williams et al (1986), who found that flavone *C*-glycosides were present in 66% of the samples [Harborne 1986]. Another family survey has been carried out in the *Polygonaceae*, in which 28 species were analysed for their flavonoid pattern [Harborne 1986]. Studying flavonoid pattern can be used for chemosystematic and lower taxonomic levels. 25 *Avena* species (Poaceae) were investigated for the flavonoid content of leaf tissue [Saleh et al 1988]. Diploid *triticum* species could be divided into two groups depending on the presence or absence of two major di-*C*- glycosylflavones (Harborne et al 1986). Flavonoid data of the genus *Vitis* indicate three chemical groups [Moore and Giannasi 1994]. Several studies indicated that flavonoids occurred in various species of *Euphorbia*. They may be useful taxonomic markers within the genus. Also *Euphorbia* flavonoids are very important for their toxicity and some different potential clinical applications such as their antiatherosclerotic, antiinflammatory, antitumor, antithrombogenic, antiosteoporotic and antiviral effects [Nijveldt et al 2001]. Papp et al (2005) showed populations of *Euphorbia cyparissias* can be separated clearly from each other according to their morphology and flavonoid pattern.Our results showed all studied *Euphorbia* species contained flavonoid compounds in their leaves that their flavonoid profiles show a wide variety between the taxa. There are flavonoid sulphate and flavone *C* and *C-/O*-glycosides in all species, but *E. bungei*, *E. heteradena* and *E. microsciadea* in addition these two flavonoid types have dihydroflavonol 3-O-monoglycosides. *E. cordifoila*, *E. heteradena*, *E. microsciadeae* and *E. petiolata* have the highest number of total flavonoid compounds (9) and *E. bungei* and *E. helioscopia* have the lowest number of flavonoid compounds (2) in their leaves (Table 1). Identification of flavonoids by standards showed all of studied *Euphorbia* species contain rutin with the exception of *E. macroclada* and *E. petiolata*. Also all taxa studied, except 2 species (*E. cordifolia* and *E. tehranica*) have quercetin. Harborne and Baxter (1999) reported that quercetin is widely distributed in various plant families. Kaempferol found in 8 species and myricetin was found just in *E. macroclada* and *E. petiolata* (Table 1). As Volobueva (1970) showed two-dimentional paper chromatography and acid hydrolysis of *E. helioscopia* alcoholic extract yielded quercetin and kaempferol. Also Gautam and Mukhraya (1981) isolated quercetin 3-O glucoside and kaempferol 3-O glucoside from *E. larica*, *E. virgata*, *E. chamaesyce* and *E. magalanta* and rutin obtained with the exception *E. chamaesyce*. Both quercetin and kaempferol are flavonols. The flavonols may be among the most important flavonoids, they are the most ancient and widespread of the flavonoids, synthesized even in mosses and ferns, and have a wide range of potent physiological activities [Stafford 1991]. Chemical study of 17 *Euphorbia* species using two dimentional paper chromatography (2-DPC) and thin layer chromatography (TLC) showed rutin, quercetin and kaempferol are the most representative compounds for the genus and the presence of myricetin is a taxonomic character for separation of some *Euphorbia* species. It is believed that *Euphorbia* species can be separated from each other according to their flavonoid pattern.

Application of plant flavonoides data can revealed similarity and relationship between plants and inferring phylogeny and used in their taxonomy. 2-dimentional paper chromatography (2-D PC), on leaves of *Alhagi camelorum* Fisch, *Cersis silliquastrum* L., *Coronilla varia* L., *Glycirhiza glarba* L. and *Robnia peseudoacia* from Markazi Province showed all of five named species contain aglycones. *A. camelorum* and *G. glarba* had the most flavonoid variation and concentration having flavon glycoside and two subspecies of *Sophora alopecuroides* and *Medicago sativa* had not or had the least. The most flavonoid compounds similarity was between *C. silliquastrum* and *R. peseudoacia* (Table 1).

Phytochemical examination of the studied *polygonum* species showed all of *Polygonum* species contain flavon *C*- and *C-/O*-glycosides. *P. hyrcanicum* had the most flavonoid variation and concentration and *P. convolvolus* species with having just one flavonoid had the least. Flavonoid sulphates was found just in *P. persicaria* species (Table 1).

Chemical study of four *Reseda* species using two dimentional paper chromatography (2-DPC) and thin layer chromatography (TLC) showed all studied *Reseda* species contained flavonoid compounds in their leaves and kempferol is the most representative compound for the genus (Table 1). They may be useful taxonomic markers within the genus. Also *Reseda* flavonoids are very important for their toxicity and some different potential clinical applications such as their antiatherosclerotic, antiinflammatory, antitumor, antithrombogenic, antiosteoporotic and antiviral effects [Nijveldt et al 2001]. The presence of quercetin and absence of myrestin in *R. lutea* are taxonomic characters for separation of the species from two other species (*R. buhseana* and *R. bungei*). Among the many functions of flavonoids at the interface between plant and environment, their activity as signals was intensively studied. Flavonoids are also beneficial for the plant itself as physiological active compounds, as stress protecting agents, as attractants or as feeding deterrents, and, in general, by their significant role in plant resistance [Treutter 2006].

Chemical studies of seven *Rumex* species using two dimentional paper chromatography (2- DPC) and thin layer chromatography (TLC) showed all of studied *Rumex* species contain flavonoid compounds with wide variation. *R. chalpensis*, *R. obtosifolius* and *R. tuberusos* species had the most flavonoid number and *R. crispus* species had the least. Identified flavonoid compounds in all of studied species with the exception *R. crispus* (lack flavonoid sulphate) are flavones *C* and *C-/O* glucoside. *R. acetosella* and *R. conglomerates* had aglycon. Rutin and luteolin found in all of studied species exceptional *R. chalpensis*, *R. obtosifolius* and *R. pulcher*. All of studied species showed wide variation in existing and concentration of myricetin, apigenin, narengenin, rhamnetin, quercetin and kaempferol. All of studied species with the exception *R. chalpensis* and *R. tuberusos* had quercetin and also kaempferol found in 3 species (*R. chalpensis*, *R. pulcher* and *R. acetosella*) (Table 1).

Phytochemical studies on five species of *Scirpus* (*S. holoschenus* L., *S. lacustris* L., *S. littoralis* Kuntze, *S. maritimus* L. and *S. multicaule*) from different parts of Markazi Province, Iran area using two-dimentional paper chromatography (2-DPC) and thin layer chromatography (TLC) showed all of studied taxa contain vitexin, luteolin, rutin and rhamnetin. There were chrysin and naringenin in all of populations with the exception of *S. lucustris* and apigenin was found in 3 species weheras others lack. Quercetin was not found in *S. maritimus* and *S. littoralis* where as three other species had (Table 1).

5. CONCISE SUMMARY

Our research indicated that the majority of the gathered plant species are weeds that thrive in impoverished soils and degraded pastures. Advancements persist in elucidating the functions of flavonoids in stress mitigation, alongside delineating the processes that regulate the quantity and diversity of flavonoids synthesized by plants in response to various environmental stimuli [Chalker-Scott 1999]. Ultimately, additional research is required employing high-performance liquid chromatography with diode array detection and atmospheric pressure chemical ionization liquid chromatography-mass spectrometry to assess the flavonoid profiles in the examined and other species.

6. REFERENCES

Abdel-Sattar EA. (1985). A Pharmacognostical Study of Chrozophora plicata (Vahl.) Growing in Egypt. MSc thesis. Faculty of Pharmacy. Cairo University. Cairo Egypt.

Adedapo, A. A., Abatan, M. O., Idowu, S. O. & Olorunsogo, O. O. (2005). Effects of chromatographic fractions of *E. hirta* on the rat serum biochemistry, *African Journal of Biomedical Research*, 8: 185-189.

Ahmed, J., Shamsuddin, K. M. & Zamman, A. (1984). *J. Indian Chem. Soc.*, 61, 92.

Aimova, M. Zh., Rakhmadieva, S. B., Erzhanova, M. S. & Abilov. Zh. A. (1999). *Akad. Nauk. Resp. Kaz. Ser. Khim*, 26 (see Williams, A. and Grayer, J. R. 2004, Anthocyanins and other flavonoids, *Nat. Prod. Rep.*, 21: 539-573.

Bath-Smith, E. C. (1962) *J. Linn. Soc.* (Bot.), 58, 95.

Berrahal, D., Kabouche, A., Kabouche, Z & Bruneau, C. (2006). Flavonoid glycosides from *Reseda villosa* (Resedaceae), Biochemical Systematics and Ecology, 34: 777-779.

Burzanska, Z. (1975). Quercetin derivatives of *Euphorbia* indica W. K., *Acta Pol. Pharm.*, 32 (6): 703-708; *Chem. Abstr.*, 85, 189224v (1976).

Chalker-Scott, L. 1990. Environmental significance of anthocyanins in plant stress responses, *Photochem. Photobiol.*, 70: 1–9.

Clifford, H. T. & Harborne , J. B. (1969). *Phytochemistry*, 8, 123.

Collins, F. W., Bohm, B. A. & Wilkine, C. K. (1975). *Phytochemistry*, 14, 1099.

De Groot, H. & Rauen, U. (1998): Tissue injury by reactive oxygen species and the protective effects of flavonoids. *Fundam. Clin. Pharmacol.*, 12, 249-255.

Delazar A Talischi B Nazemiyeh H Rezazadeh H Nahar L. & Sarker SD. (2006). Chrozophorin: a new acylated flavone glucoside from Chrozophora tinctoria (Euphorbiaceae). Revista Brasileira de Farmacognosia. *Brazilian J of Pharmacog,* 16 (3): 286- 290.

Dewick, P. M. (1993). In J. B. Harborne (ed.) *The Flavonoids: Advances in Research since 1986*, pp. 117-238. Chapman & Hall, London.

El-Sayad, N. H., Omara, N. M., Yosef, A. K., Farag, A. M. & Mabry, T. J. 2001. Kaempferol triosides from Reseda muricata, Phytochemistry, 57 (4): 575-578.

Falodun, A., Okunrobo, L. O. & Uzoamaka, N. (2006). Phytochemical screening and anti- inflammatory evaluation of methanolic and aqueous extracts of *Euphorbia heterophylla* L. (Euphorbiaceae), *African Journal of Biotechnology*, 5 (6): 529-531.

Farman, M. (1990). Isolation and characterization of flavonoids from *Indigofera hebepetala*, Quaid-i- A Azam University, Islamabad.

Field, T. S., Lee, D. W. & Holbrook, N. M. (2001). Why leaves turn red in autumn, The role of anthocyanins in senescing leaves of Red-Osier Dogwood, *Plant Physiol.*, 127: 566–574.

Gautam, R. K. and Mukhraya, D. K. 1981. *Natl. Acad. Sci. Lett.*, 10, 95 (see Singla and Pathak 1990).

Ghahreman, A. 1979-2006. Flore de l'Iran, A join project by the Research Institute of Forests and rangelands (Iran) and Tehran University, Published by RIFR, Ministry of Reconstruction Jahad, Volumes 1-24.

Gomes, C. M. R., Gottlieb, O. R., Bettolo, G. B. M., Monache, F. D. & Polhill, R. M. (1981b). Systematic significance of flavonoids in *Derris* and *Lonchocarpus*, *Biochemical Systematics and Ecology*, 9 (2-3): 129-147.

Gomes, C. M. R., Gottlieb, O. R., Gottlieb, R. G. & Solatino, A. (1981a) Advances in Legume Systematics, 2, 465.

Hahlbrock H. & Grisebach H (1975) Biosynthesis of flavonoids. *In* JB Harborne, TJ Mabry, H Mabry, eds, The Flavonoids. Academic Press, San Diego, pp 866–915.

Hahlbrock,K. (1981) in Stumpf,P.K. and Conn,E.E. (eds.), The Biochemistry of Plants, Vol. 7, Academic Press, NY, pp. 425456.

Halaweish, F. Kronberg, S. & Rice, J. A. (2003). Rodnet and ruminant ingestive response to flavonoids in *Euphorbia esula*, *Journal of Chemical Ecology*, 29 (5): 1073-1082.

Hallett, F. P. & Parks, L. M. (1951). A note on isolation of quercetin from *E. pilulifera*, *J. Am Pharm Assoc.*, 40: 56-57.

Harborne, J. B (1981). *Phytochemistry*, 20, 1117.

Harborne J. B. (1965) Plant polyphenols-XIV: Characterization of flavonoid glycosides by acidic and enzymic hydrolyses. *Phytochem.*, 4: 107-120.

Harborne, J. B. (1986). *The Flavonoids. Advances in Research*, Chapman & Hall/CRC, Boca Ratón (1986).

Harborne, J. B. (1994). *The Flavonoids: Advance in Research Since 1986*, Chapman and Hall, New York (1994).

Harborne, J. B. and Baxter, H. 1999. *The handbook of natural flavonoids*, Vol. 1, Wiley, Chichester.

Harborne, J. B. & Turner, B. L. (1984). *Plant chemosystematics*, Academic Press, London. Harborne, J. B., (1998). *Phytochemistry Methods*, 3rd ed. Chapman and Hall, London.

Harborne, J. B., Heywood, V. H. & Chen, X. Y. (1986). Separation of ostericum from *Angelica* on the basis of leaf and mericarp flavonoids, *Biochem. Syst. Ecol.*, 14, 81-83.

Harborne, J.B., Mabry, T.J. & Mabry, H. (Eds.), 1975. The Flavonoids. Chapman & Hall, London.

Hasan A, Ahmed I, Jay M. & Voirin B. (1995). Flavonoid glycosides and an anthraquinone from *Rumex chalepensis*. *Phytochemistry* 39: 1211–1213.

Hashim OK Abouzaid MM Abdelgalil FM & Saleh NAM. (1990). The flavonoids of Egyptian Chrozophora species. *Biochem. Syst. Ecol.*, 152 -151 :18.

Hawas UW. (2007). Antioxidant activity of brocchlin carboxylic acid and its methyl ester from Chrozophora brocchiana. *Natural Product Research.*; 640 –632 :(7) 21.

Hegnauer, R. & Grayer- and Barkmeijer, R. J. (1993). *Phytochemistry*, 34, 30.

Hsu, C. Y., (2006). Antioxidant activity of extract from *Polygonum*, Biological Research, 39 (2): 281-288.

Ingham, J. L. (1981). In R. M. Polhill and P. H. Raven (eds.) *Advances in Legume Systematics*, pp. 599-626. HMSO, London.

Ingham, J. L. (1983). *Fortschritte d. Chem. org. Naturst.*, 43, 1.

Islambeko, Sh. Yu. Et al. (1982). *Khim.Prir. Soedin.*: 653; *CA* 98: 50389c.

Isobe, T., Noda, Y. (1987). Yakugaku Zasshi, 107, 1001.

Kawasaki, M., Kanomata, T. & Yoshitama, K., 1986. Flavonoids in the leaves of twenty-eight polygonaceous plants. *Bot. Mag.* (Tokyo) 99, 63}74.

Mabry, T. J., Markham, K. R. & Thomas, M. B., 1970. *The Systematic Identification ofFlavonoids*, Springer Verlag, Berlin.

Markham, K. R., 1982. *Techniques of Flavonoid Identification*, Academic Press, London.

Middleton, E. J., (1998). Effect of plant flavonoids on immune and inflammatory cell function, *Adv. Exp. Med. Biol.*, 439: 175-182.

Midiwo, J. O., F. M. Omoto, A. Yenesew, H. M. Akala, J. Wangui, P. Liyala, C. Wasuna, N. C. Waters (2007). The first 9-hydroxy isoflavanone and anti-plasmodial chalcones from the aerial exudates of *Polygonum senegalense.* Arkivoc (ix) 21-27.

Mobayen, S. 1979. Iran Vegetation (Vascular Plant Flora), Tehran University Publications, no. 1500/2, Vol. 2: 85-152.

Mobayen, S. 1980. Iran Vegetation (Vascular Plant Flora), Tehran University Publications, no. 1500/2253, Vol. 1: 209-245.

Mohamed KS. (2001). Phenylpropanoid glucosides from Chrozophora oblique. *Phytochem.*; 58: 615-618.

Moiteiro, C. Gaspar, H., Rodrigues, Al., Lopes, J. F. & Carnide, V. (2008). HPLC quantification of dye flavonoids in *Reseda luteola* L. from Portugal, J. Sep. Sci., 31 (21): 3683-3687.

Moore, M. O. & Giannasi, D. E. (1994). , *Plant systematics and evolution*, 193 (1-4): 21-36.

Mues, R. & Zinsmeister, H. D. (1988). The chemotaxonomy of phenolic compounds in Bryophytes. *Journal of Hattori Botanical Laboratory*, 64: 109–141.

Muller, R. & Pohl, R. (1970). Flavonol glycosides of *Euphorbia amygdaloides* and their quantitative determination, *Planta Med.*, 18 (2), 114-129.

Mun, J. H. & Park, C. W. (1995). Flavonoid chemistry of *Polygonum* sect. Tovara (Polygonaceae): A systematic survey, *Plant Systematic And Evolution*, 196 (3-4): 153- 159.

Murillo, R. and Jakupovic, J. 1998. Glycosides from *Euphorbia aucherii*, *Ing. Cienc. Quim.*, 18: 57-60.

Nagase, M. (1942). *J. Agr. Chem. Soc. Japan*, 17, 183; *Chem. Abstr.*, 36, 3525 (see Singla and Pathak).

Naser, M. I, Abu-Mustafa, E. A., Abdel-Razik & Dawidar, A. M. (2000). . Lipid and flavonoids from some Cyperaceae plants and their anti-microbial activity, *Bul. Nat. Re. Cen. (Cario)*, 25 (2): 105-113.

Nijveldt, R. J., Van Nood, E., Van Hoof, EC., Boelens, P. G., Van Norren, K. & Van Leeuwen, P. AM. (2001). Flavonoids: a review of probable mechanisms of action and potential applications, *Am. J. Clin. Nutr.*, 74: 418-425.

Noori, M. (2002). *Characterization of the Iranian species of Sophora and Ammodendron (Leguminosae; Sophoreae)*, PhD Thesis, University of London and Royal Botanic Gardens, Kew, UK.

Noori, M., Chehreghani, A. & Kaveh, M. (2009). Flavonoids of 17 species of *Euphorbia*, (Euphorbiaceae) in Iran, *Toxicological and Environmental Chemistry*, 91 (3): 409-418.

Omurkhamzinova, V. B. & Erzhanova, M. S., E. E. C. S. (1985). *Int. Conf. Chem., Biotechnol. Biol. Act. Nat. Prod. (Proc.)*, 3rd, *Chem. Abstr.*, 110, 4671g (1989), (see Singla & Pathak 1990).

Papp, N., Vasas, A., Hohmann, J. & Szabo, L. G. (2005). Morphological and flavonoid pattern variations within some *Euphorbia cyparissias* L. populations, *Acta Biologica Szegediensis*, 49 (1-2): 171-172.

Park, C. (1987), Flavinoid chemistry of *Polygonum* Sect Echinocaulon: a systematic survey. *Syst. Bot*, 12: 461-462.

Parr, A. J. & Bolwell, G. P. (2000). Phenols in the plant and in man. The potential for possible nutritional enhancment of the diet by modifying the phenols content or profile, *Journal of the Science of Food and Agriculture*, 80: 985–1012.

Rechinger, K. H. 1964. *Flora Iranica*, Akodemische Druck–U. Verlagsanstalt Graz–Austria.

Robles, C., Greff, S., Pasqualini, V., Garzino, S., Bousquet-Melou, A., Fernandez, C., Korboulewsky, N. & Bonin, G. (2003). Phenols and flavonoids in Aleppo pine needles as bioindicators of air pollution, *J. Environ. Qual.*, 32: 2265-2271.

Rzadkowska-Bodalska, H. (1969). *Pharm. Pharmacol*, 21:169.

Sagareishvili, T. G. & Ananiya, M. D. (1990). *Izv. Akad. Nauk Gruz. SSR, Ser. Khim.*. 16: 155- 156; *CA* 113: 188036v.

Sahai, R., Dube, M. P. and Rastogi, R. P. 1981. Chemical and pharmacological study of *Euphorbia maddeni*, *Indian J. Pharm. Sci.*, 43, 216.

Saleh A. M, Nozzolillo C. & Altosaar, I. (1988). Flavonoid variations in *Avena* species, *Biochem. Sys. Ecol.*, 16: 597–599.

Shi XH Liu YQ & Kong LY. (2006). Studies on the flavones in of Chrozophora sabulosa. *Zhongguo Zhong Yao Za Zhi.*, 31 (5): 395- 397.

Shirley, B.W. (1996) Flavonoid biosynthesis: «new» functions for an «old» pathway. Trends in Plant Science, 1, 377-382.

Simpson, M. G. & Levin, G. A. (1994). Pollen ultrastructure of the biovulate Euphorbiaceae, *International Journal of Plant Sciences*, 155: 313-341.

Singla, A. K. & Pathak, K. (1990). Phytochemistry of *Euphorbia* species, *Fitoterapia*, LXI (6): 483-516.

Sotnikova, O. M. & Litvinenko, V. I. (1968). Isomyricitrin from *Euphorbia stepposa*, *Chemistry of Natural Compounds*, 4 (1): 42-43.

Sotnikova, O. M., Chagovets, R. K. & Litvinenko, V. I. (1968). New flavanone compounds from *Euphorbia stepposa*, *Chemistry of Natural Compounds*, 4 (2): 71-74.

Stace C (1980). Plant taxonomy and biosystematics. Edward Arnold publisher Ltd, London.

Talischi B Modarresi M Bamdad Moghadam S Asnaashari S Nazemiyeh.H Rezazadeh H & Delazar A. (2005). Antioxidant Activity of Methanol Extract of Chrozophora tinctoria and Identification of Two O-Glycoside Flavons Isolated from it. In: First Seminar of Medicinal and Natural Products Chemistry Shiraz. Iran.

Torck, M. et al (1969) *Ann.Pharm. Fr.* 27: 419-426; *CA* 72: 63577a.

Treutter, D. (2006) Significance of flavonoids in plant resistance: a review, *Environmental Chemistry Letters*, 4 (3): 147-157.

Trichopoulou, A., Vasilopoulou, E., Hollman, P., Chamalides, C., Foufa, E., Kaloudis, T., Kromhout, D., Miskaki, P., Petrochilou, I., Poulima, E., Stafilakis, K., & D. Theophilou. (2000). Nutritional composition and flavonoid content of edible wild greens and green pies: a potential rich source of antioxidant nutrients in the Mediterranean diet. *Food Chemistry*, 70, 319-323.

Ulubelen, A., Öksuz, S., Halfon, B., Aynehchi, Y. & Mabry, T. J. (1983). Flavonoids From *Euphorbia larica, E. virgata, E. chamaesyce* and *E. magalanta, Journal of Natural Products*, 49: 598.

Van Hoof, L., Vanden, Berqhe, D. A., Hatifield, G. M. & Vlietinck, A. J. (1984). Plant antiviral agents, 3-methoxyflavones as potent inhibitors of viral-induced block of cell synthesis, *Planta Med.*, 50: 513-517.

Vassallo A Cioffi G De Simone F Braca A Sanogo R Vanella A Russo A & De Tommasi N. (2006). New flavonoid glycosides from Chrozophora senegalensis and their antioxidant activity. *Nat product commun,* 1 (12): 1089- 1095.

Volobueva, M. A. (1970). Phytochemical study of *Euphorbia helioscopia*, *Trudy Alma Atinskogo Meditsinsko Instituta*, 26: 451 -455, *Chemical Abstracts*, 77, 7254 (1972).

Williams, C. A., Harborne, J. B. and Goldblatt, P. 1986. Correlation between phenolic patterns and tribal classification in the family Iridaceae, *Phytochemistry*, 25: 2135-2154.

Woelfle, U., Simon-Haarhaus, B., Merfort, I & Schempp, C. M. 2009. *Reseda luteola* L. extract displays antiproliferative and pro-apoptptic activities tht are related to its major flavonoids, Phytotherapy Research, 10.1002/ptr.3069.

Wong, E. 1976. Biosynthesis of flavonoids. In: Goodwin TW (ed). Chemistry and biochemistry of plant pigments, Academic Press, NY, London, 1: 464-526.

Yang. G., Zhang, L. & Chen, G. (2010). Determination of four phenolic compounds in *Scirpus yagars* Ohwi by CE with amperometric detection, *Chromatographia*, 71 (1-2): 143-147.

Yilmaz, Y. & Toledo, R. T. 2004. Health aspects of functional grape seed constituents, *Trends Food Sci. Technol.*, 15422-15433.

Yuldashev, M. P., Batirov, E. K., Malikov, V. M. & Yuldashev, N. P. (1996). Khim. Prir. Soedia, 6: 949.

8 Chapter

PHYTOCHEMICALS DERIVED FROM BEILSCHMIEDIA ANACARDIOIDES AND THEIR BIOLOGICAL IMPORTANCE

1. INTRODUCTION: AN OVERVIEW

Medicinal plants offer a wide variety of basic resources for primary healthcare in Africa and globally. The World Health Organization (W.H.O) estimates that over 80% of Africans residing on the continent utilize traditional medical practitioners and traditional medicine for the treatment of various maladies. This practice holds significant importance within Africa's economic and cultural context.

It is believed that fewer than 10% of the world's genetic resources have been rigorously examined as potential sources of pharmaceuticals. From this minuscule portion, humanity has garnered substantial advantages.

The exploration of bioactive natural products from higher plants is intensifying, as they possess the potential to yield novel lead compounds or serve direct applications. The urgency of this search is escalating due to the depletion of natural resources. It is estimated that 25-30 million hectares of the world's rainforests are lost annually. The critical issue articulated by numerous experts is the effective and speedy identification of bioactive compounds from the extensive array of unexamined plants. A component of the solution is to refine the search criteria.

There are six methodologies for selecting plant materials for research: local random selection, taxonomic classification, ethnobotanical analysis, phytochemical evaluation, information-based criteria, and serendipitous discovery. The ethnomedical technique seems to be the preferred way for natural product chemists in Africa and other poor nations. This method exclusively involves the collection of plants utilized in traditional medicine.

Minimal attention has been accorded to Beilshmiedia species. Prior research focuses on the trees and herbs of Beilshmiedia species, aiming to cultivate herbs that contain the same endiandric acid compounds as the trees. Subsequent efforts resulted in a patent for the intriguing synthesis of endiandric acid derivatives.

In our pursuit of potential pharmacological products derived from ethnobotanical data, we have been examining traditional medicines with scientifically validated therapeutic efficacy in relation to biological assessments of patients undergoing treatment in a specialist clinic. We have chosen a traditional medicine derived from the plant Beilshmiedia anacardioides (Lauraceae) due to its demonstrated efficacy against genital infections and rheumatisms, as evidenced by clinical study. To our knowledge, there are no phytochemical investigations of Beilshmiedia anacardioides available in the literature. We recommend that phytochemists seeking novel bioactive natural compounds examine medicinal plants with clinically validated therapeutic efficacy in African medicine.

The genus Beilschmiedia includes approximately 200 species that are extensively spread in the intertropical zone (Fouilloy, 1974). The stem bark of B. anacardioides is utilized in the Western Province of Cameroon for the treatment of uterine tumors (Tchouala, 2001). Various species of the genus Beilschmiedia are utilized in traditional African medicine for the treatment of multiple diseases (Tchouala, 2001; Iwu, 1993). Previous phytochemical studies of Beilschmiedia species have identified bioactive lignans (Chen et al., 2006; Chen et al., 2007), flavonoids (Harbone et al., 1969), triterpenoids (Chen et al., 2006), tetracyclic endiandric acid (Bandaranayake et al., 1981; Banfield et al., 1994), and alkaloids (Clezy et al., 1966; Kitagawa et al., 1993; Chouna et al., 2011).

We have commenced a systematic phytochemical analysis of the extracts of Beilshmiedia anacardioides and evaluated the antibacterial efficacy of eight newly isolated compounds against five microbial strains: Bacillus subtilis, Micrococcus luteus, Streptococcus faecalis, Pseudomonas palida, and Escherichia coli.

The techniques employed for the isolation of the compounds were primarily column chromatography and preparative thin-layer chromatography. The structures of all compounds were determined using contemporary spectroscopic techniques, including 1D-NMR (1H-NMR, 13C-NMR with DEPT studies), 2D-NMR (1H-1H-COSY, HMQC, HMBC, NOESY), mass spectrometry, infrared spectroscopy, and X-ray crystallography.

The antibacterial efficacy of the novel compounds was assessed by the dilution method concerning the zone of inhibition (ZI) and minimum inhibitory concentration (MIC).

We provide the data we have achieved and published in three esteemed scientific journals. (Chouna et al.,2009; 2010; 2011).

2. STUDY OF THE ETHNOMEDICAL PREPARATION

The ethnomedical preparation is a decoction.The decoction is prepared as follows: Boil about 80 g dry stem bark powder in 3 litres of water for 15 minutes. Filter when lukewarm. Drink a glass twice daily for ten days. A treatment for fibromes could last about two to three months, depending on the patient's age.

3. STUDY SETTING

Cameroon is a bridge between Central Africa and West Africa, humid Africa and dry sahelian Africa, French speaking and English speaking Africa (French and English are official languages). The country is open to the Gulf of Guinea in his south-west border. Lake Chad is at the extreme –North border. A country of 475.442 square kilometers, Cameroon is borded in the west by Nigeria, on the east by Chad and the Central African Republic, and on the south by Congo, Gabon, and Equatorial Guinea.

The Bamoun are a Bantu people living in the west Region of Cameroon. They number more than half a million. They have a rich cultural Heritage, including famous traditional Healers. Sultan Njoya wrote a book on Bamoun traditional medicine. Important Bamoun towns are Foumban, Foumbot, Koutaba, Massangam, Magba, Malantouen. Among important villages are Mahoua, Manki 1 and Manki 2, where the plant *Beilshmiedia anacardioides* was collected.

3.1 Generalities on *Beilschmiedia* anacardioides

B. anacardioides is found in Central Africa, especially in Cameroon, Tchad and Gabon. In Cameroon, this species is found in the Adamaoua and the West Region (Eyog et al., 2006; Fouilloy, 1974). It is synonymous with *B. ngriki* and *B. Jacques-felixii* and it is commonly named *ntseum* (in Bamoun language) in the Noun subdivision of the West region of Cameroon (Eyog et al., 2006; Fouilloy, 1974; Tchouala, 2001).

3.2 Uses of *Beilschmiedia* species in traditional medicine

B. anacardioides stem bark is used in the Noun sub-division of the West Region of Cameroon to treat uterine tumours, rubella, rheumatisms, bacterial and fungal infections (Tchoula, 2001). Seeds are used as spices (Eyog et al., 2006). *B. lancilimba* is used in the same region to cure skin bacterial infections (Tchouala, 2001). *B. manii* is used to treat dysentery and headache. It is also used as an appetite stimulant (Iwu, 1993).

4.PHYTOCHEMISTRYOFPLANTCONSTITUENTSOFBEILSHMIEDIA SPECIES

A review of the literature revealed that no phytochemial studies have been carried out on *Beilshmiedia anacardioides* prior to the initiation of our study. The various phytochemical and pharmacological studies performed and reported in the literature on the beilshmiedia genus are discussed below.

4.1 Alkaloids isolated from the Beilschmiedia genus

Very few alkaloids have been isolated from the *Beilschmiedia* genus.

Table 1. Structure of some alkaloids isolated from the *Beilschmiedia* genus

	Structure and name	Source and references
9: Dehatrine	Structure and name 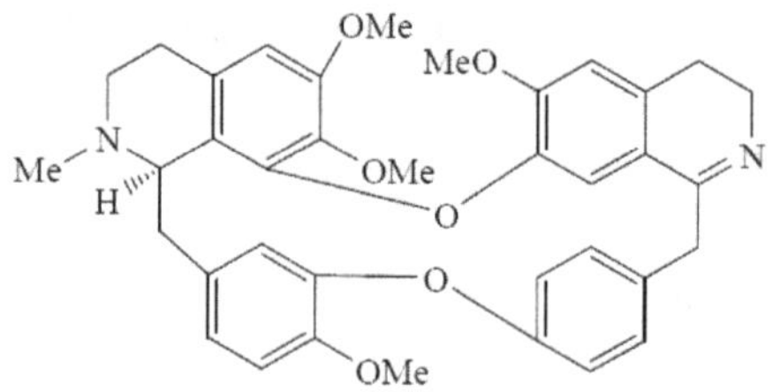	Source and references Wood of *B. madang* (Kitagawa et al., 1993)
10: Laurelliptine	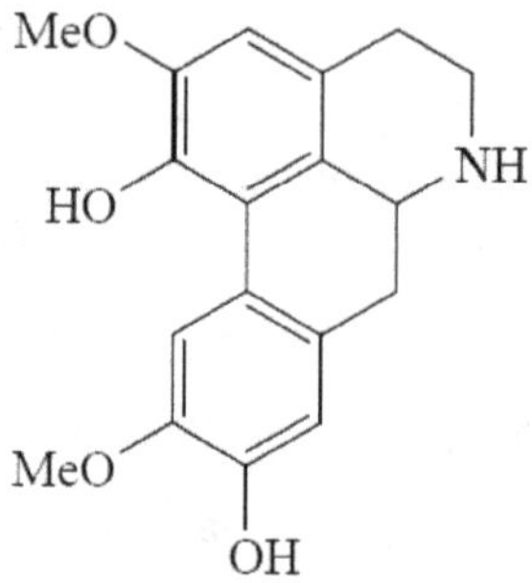	Stem bark of *B. elliptica* (Clezy et al., 1966)
11: (6,7-Diméthoxy-4-methylisoquinolinyl)-(4'-methoxyphenyl)-methanone	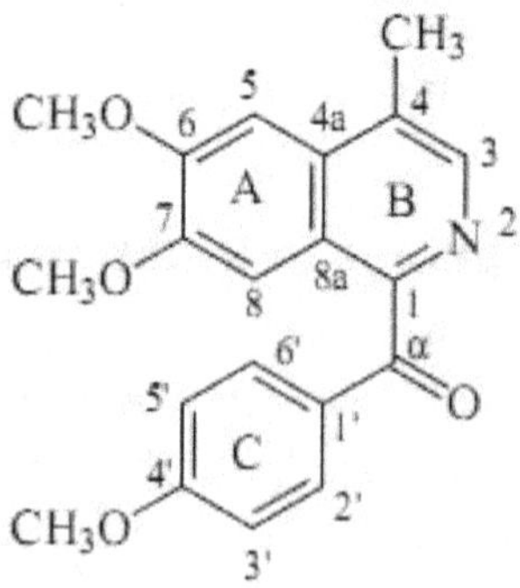	Leaves of *B. Brevipes* (Pudjiastuti et al.,2010)
12: Obscurine	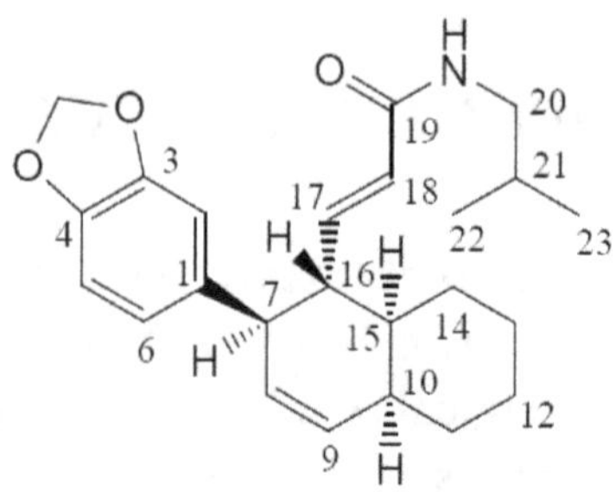	Stem bark of *B. Obscura* (Lenta et al.,2011)

Pharmacological importance of alkaloids isolated from the Beilschmiedia genus

A bisbenzylisoquinoline alkaloid dehatrine (**9**) isolated from the wood of *B. madang*, exhibited potent inhibitory activity (IC_{50} value of 0.017 μM) against the proliferation of the malaria pathogen *P. falciparum* (Kitagawa et al., 1993). Paulo and coworkers (1992) demonstrated the antimicrobial properties of laurelliptine (**10**).

4.2 Phenolic and phenolic derived compounds

4.2.1 Lignans and neolignans

Lignans and neolignans and flavonoids are the main phenolic compounds encountered in the *Beilschmiedia* genus.

Table 2. Structure of some lignans and neolignans isolated from *Beilschmiedia* genus

	Structure and name	Source and references
	Structure and name	Source and references
13: Beilschmin A	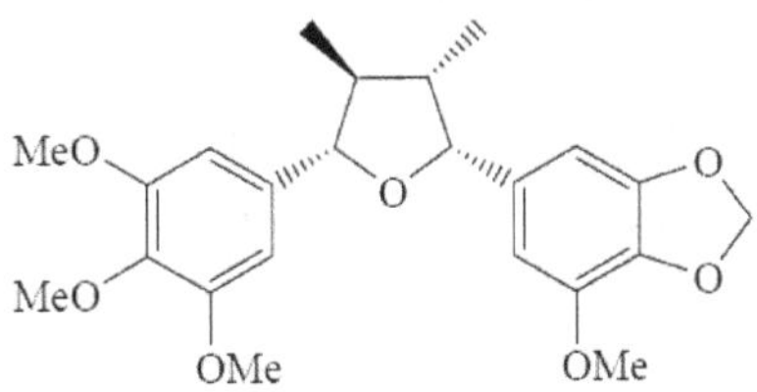	Stem of *B. tsangii* (Chen et al., 2006)
14: Beilschmin B	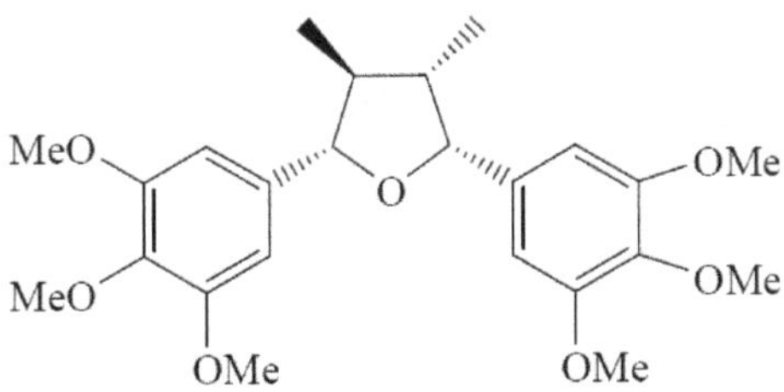	Stem of *B. tsangii* (Chen et al., 2006)
15: 4α,5α,-Epoxybeilschmin A	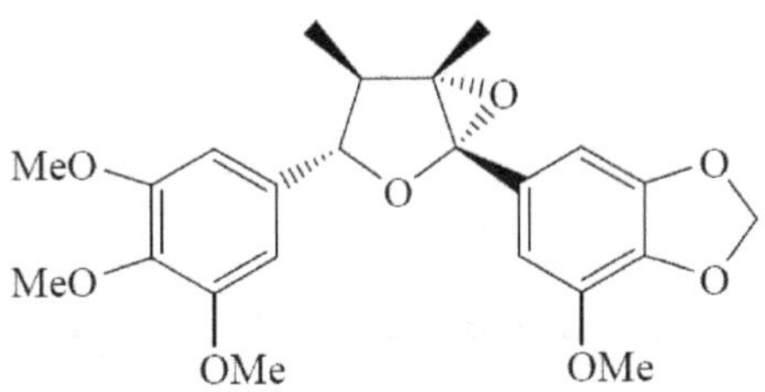	Leaves of *B. tsangii* (Chen et al., 2007)

16: 4α,5α,-Epoxybeilschmin B	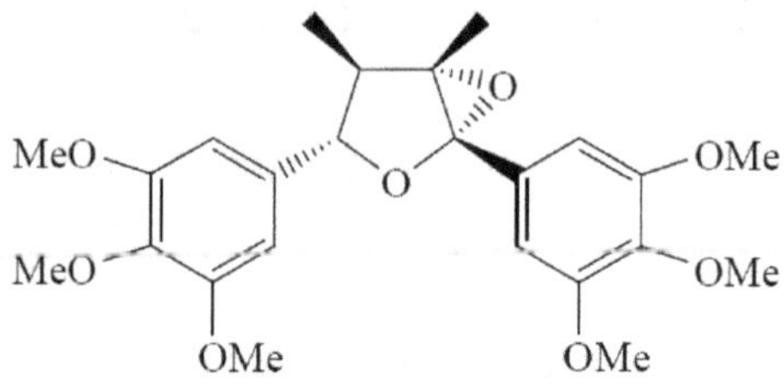	Stem of *B. tsangii* (Chen et al., 2006)
17: Beilschmin C	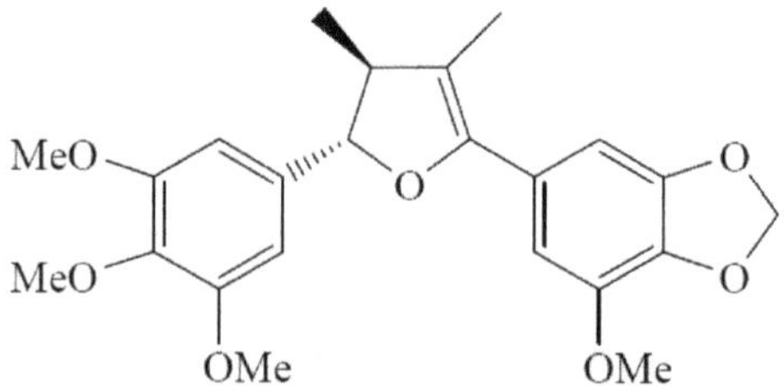	Stem of *B. tsangii* (Chen et al., 2006)
18: Tsangin A	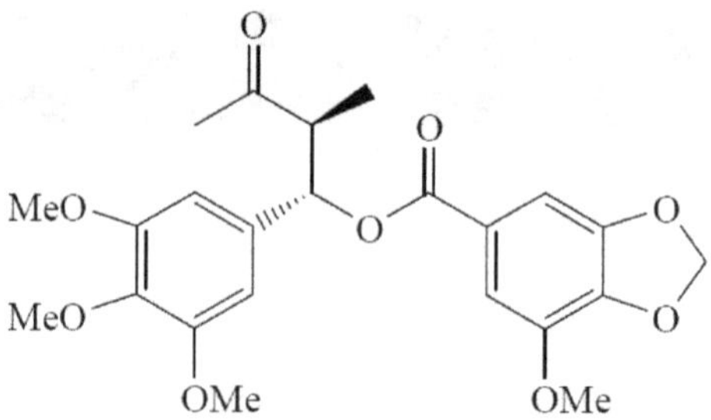	Stem of *B. tsangii* (Chen et al., 2006)
19: Tsangin B	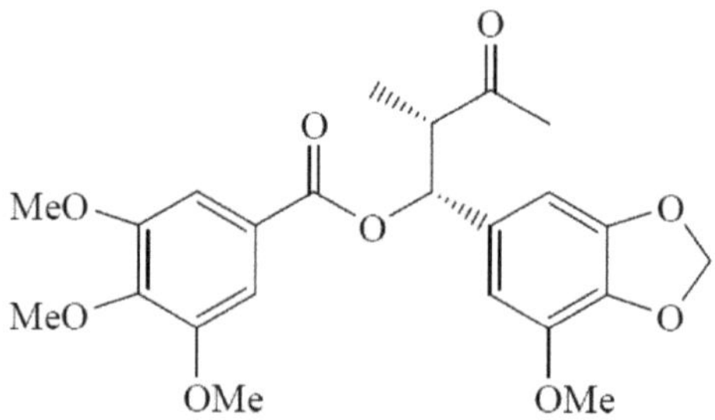	Stem of *B. tsangii* (Chen et al., 2006)
20: Magnolol	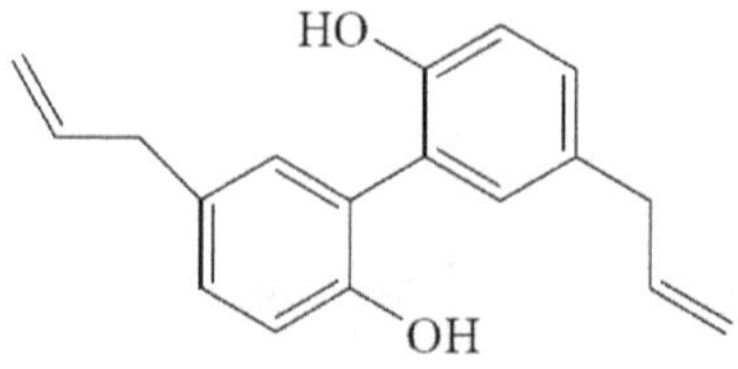	Leaf of *B. volckii* (Banfield et al., 1994)

Pharmacological importance of lignans and neolignans isolated from the Beilschmiedia genus

Tetrahydrofuran-type lignans beilschmin A (**13**) and B (**14**), dihydrofuran-type lignan beilschmin C (**17**) together with tsangin A (**18**) and B (**19**) were found cytotoxic

(IC_{50} value below 4 µg/mL) in P-388 and/or HT-29 cell lines *in vitro* (Chen et al., 2006). In addition, beilschmin A (**13**) and B (**14**) exhibited potent antitubercular activity (MIC values of 2.5 and 7.5 µg/mL, respectively) against *Mycobacterium tuberculosis* 90-221387 in *vitro* (Chen et al., 2007). A neolignan, magnolol (**20**) displayed wide biological properties, mainly cytotoxic (Li et al., 2007), antidepressant (Li et al., 2007), antimicrobial (Park et al., 2004) and anti- inflammatory (Lee et al., 2005).

4.2.2 Some flavonoids isolated from Beilschmiedia genus

Pharmacological importance of flavonoids isolated from the Beilschmiedia genus

Lenta and coworkers (2009), evaluated the antibacterial activities of the extract and flavonoids isolated from the stem of *B. zenkeri*, in vitro against three strains of microbes, *pseudomonas agarici, Bacillus subtilis,* and *streptococus minor.* Their activities were moderate compare to reference drugs ampicillin and gentamicin. (2S,4R)-5,6,7-trimethoxyflavan-4-ol (**22a**) exhibited the best potency against *S. minor* (IC_{50} of 197.5 µM) (Lenta et al., 2009).

Table 3. Structure of flavonoids isolated from the *Beilschmiedia* genus

Source and references	Structure and name	Source and references
21: 5-Hydroxy-7,8-dimethoxyflavanone	Structure and name 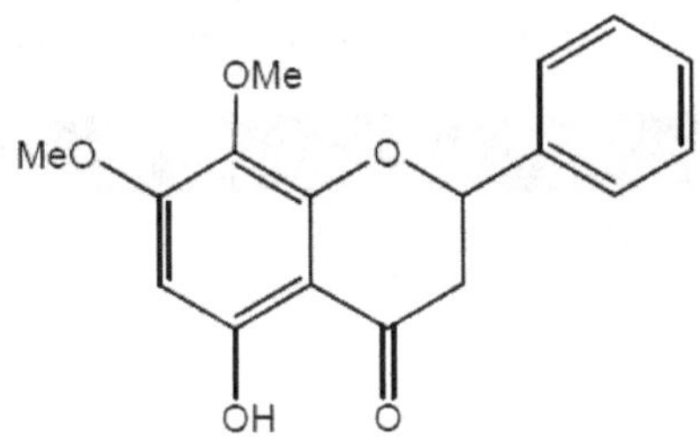	Source and references Stem of *B. zenkeri* (Lenta et al., 2009)
22a: R = H: (2S,4R)-5,6,7-trimethoxyflavan-4-ol **22b:** R = CH3: (2S,4R)-4,5,6,7-trimethoxyflavan	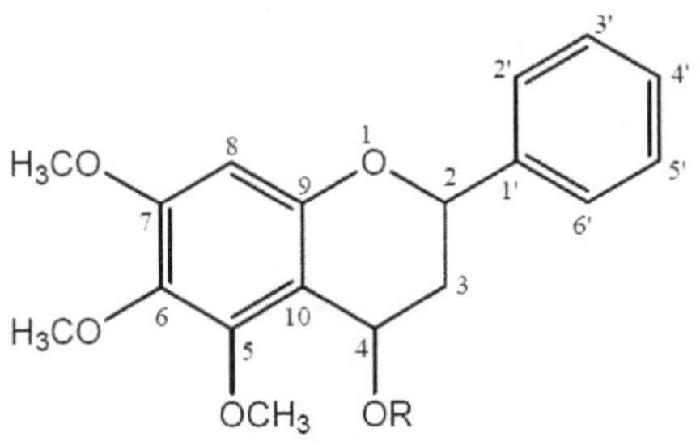	Stem of *B. zenkeri* (Lenta et al., 2009)

23a: R = CH3: Beilschmieflavonoid A **23b**: R = H: Beilschmieflavonoid B	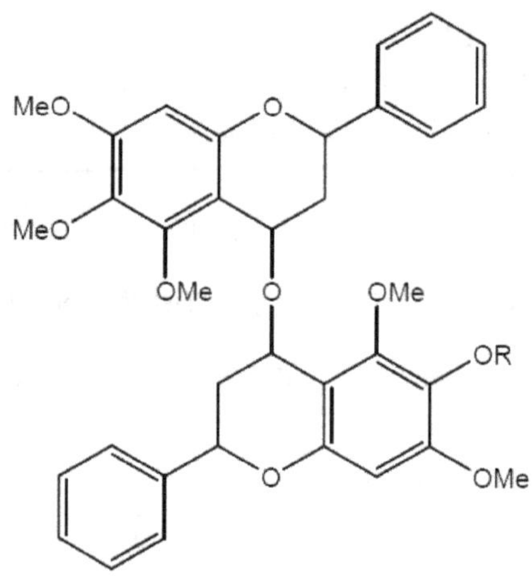	Stem of *B. zenkeri* (Lenta et al., 2009)

4.2.3 Other phenolic and phenolic derived compounds from the Beilschmiedia genus and their pharmacological importance

Vanillin (**21a**) and 4-hydroxybenzaldehyde (**24b**) were isolated from *Beilschmiedia tsangii* (Chen et al., 2006). Both compounds were reported to exhibit analgesic, anti-inflammatory and antifungal activities (Lee et al., 2005; Lee et al., 2006; Fitzgerald et al., 2005).

Table 4. Structure of other phenolic and phenolic derived compounds isolated from the *Beilschmiedia* genus

	Structure and name	Source and references
24a: Vanillin	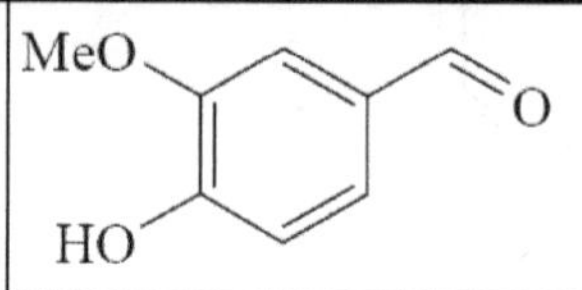	Stem of *B. tsangii* (Chen et al., 2006)
24b: 4-hydroxybenzaldehyde	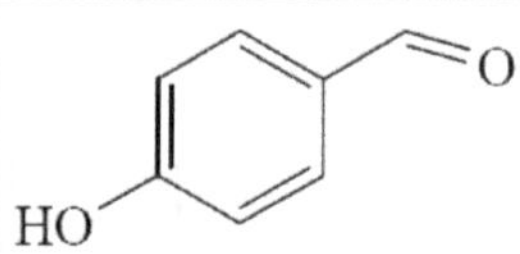	Stem of *B. tsangii* (Chen et al., 2006)
25a: Oligandrol	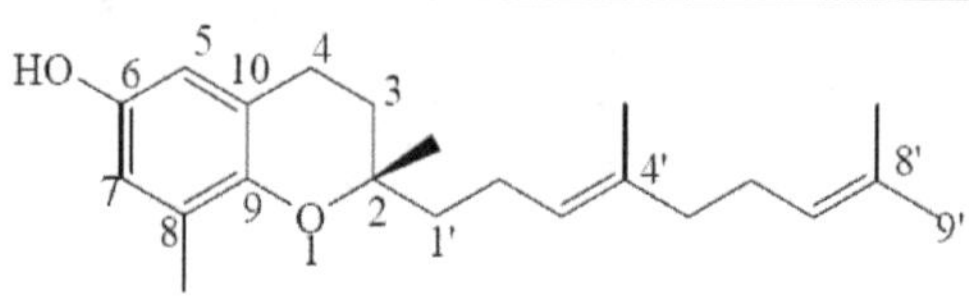	Bark of *B. oligandra* (Banfield et al., 1994)
25b: Oligandrol methyl ether	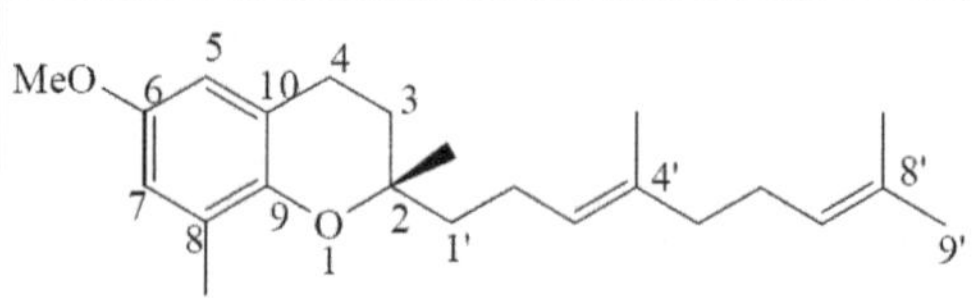	Root of *B. erythrophloia* (Yang et al., 2008)

25c: 3,4-Dehydrooligandrol methyl ether	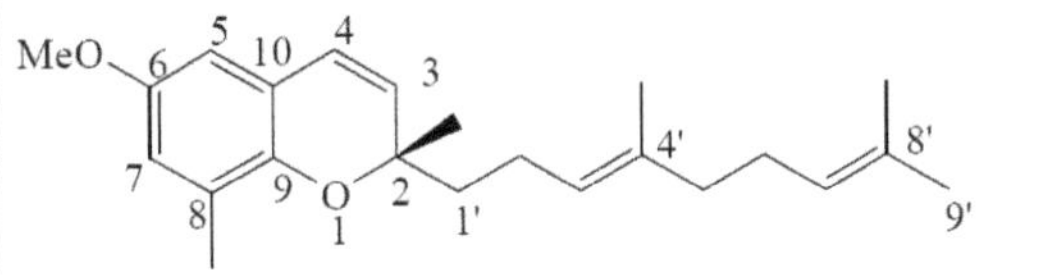	Root of *B. erytrhophloia* (Yang et al., 2008)
26: Farnesylol	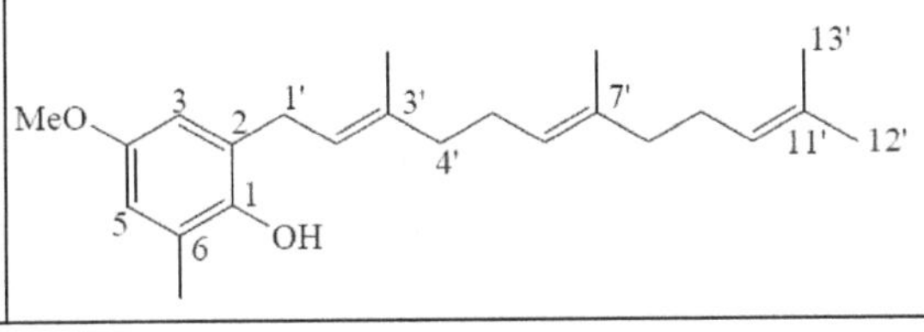	Root of *B. erytrhophloia* (Yang et al., 2008)

4.3 Endiandric acids

Endiandric acids are a rare class of secondary tetracyclic metabolites generally encountered in *Beilschmiedia* and *Endiandra* species of the Lauraceae family. Endiandric acids are products of electrocyclic cyclization of naturally occurring polyketides (Bandanarayake et al., 1980).

4.3.1 Some endiandric acids previously isolated

Pharmacological importance of the endiandric acids

Very few pharmacological studies have been done in this class of metabolites. Endiandric acid H (**41**) is used for the manufacture of medication, in particular for the treatment of asthmatic disorders or concomitant inflammatory symptoms of asthma (Eder et al., 2004).

Erytrophloin C (**34**) exhibited antitubercular activity (MIC value of 50 µg/mL) against *Mycobacterium tuberculosis* H37Rv in *vitro* (Yang et al., 2009).

Table 5. Structure of some endiandric acids previously isolated

Structure and name		Source and references
27: Endiandric acid A	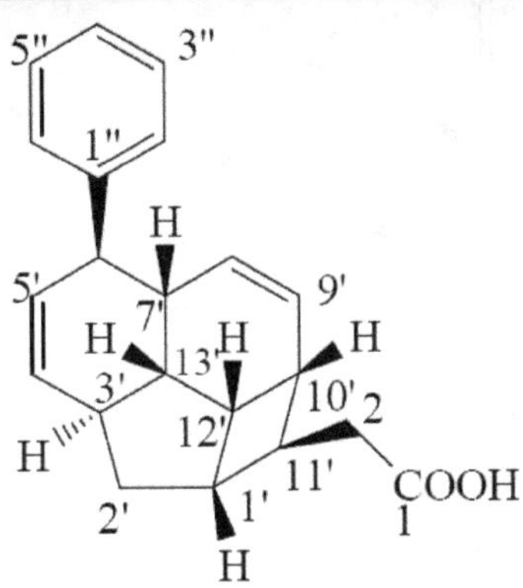	Leaves of *Endiandra entrorsa* and *Endiandra oligandra* (Bandaranayake et al., 1981) Bark of *B. oligandra* (Banfield et al., 1994)

28: Methylenedio-xyendiandric acid A	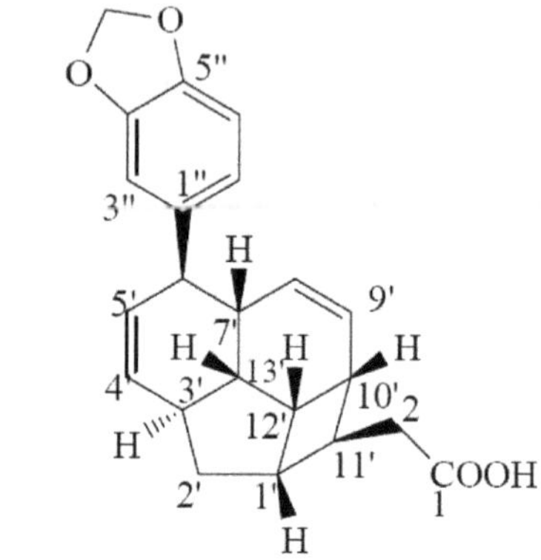	Leaves of *Endiandra entrorsa* (Banfield et al., 1994) and Stem bark of *B. manii* (Mpetga, 2005)
29: Endiandric acid B	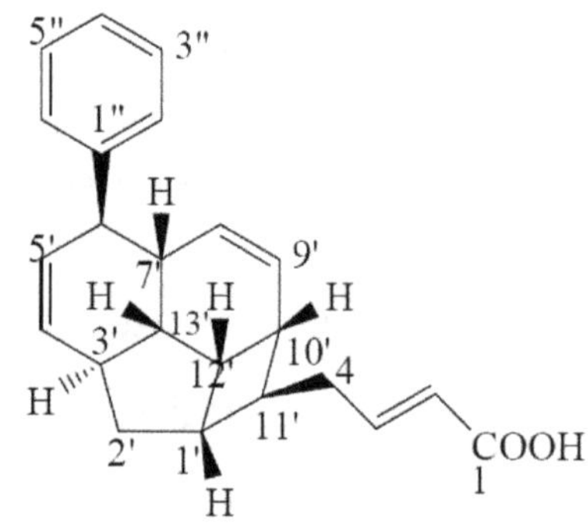	Leaves of *Endiandra entrorsa* (Bandaranayake et al., 1982) Barks and leaves of *Endiandra jonesii* (Banfield et al., 1994)
30: Endiandric acid C	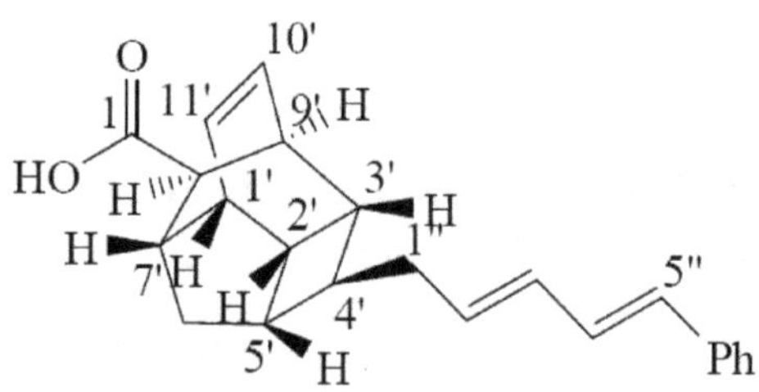	Leaves of *Endiandra entrorsa* (Bandaranayake et al., 1982) Bark and leaves of *Endiandra jonesii*(Banfield et al., 1994)
31: Endiandric acid D	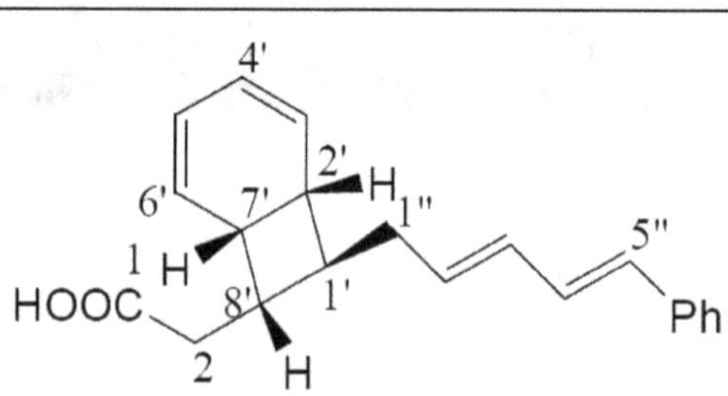	Leaves of *Endiandra entrorsa* (Banfield et al., 1983)
32: Erytrophloin A	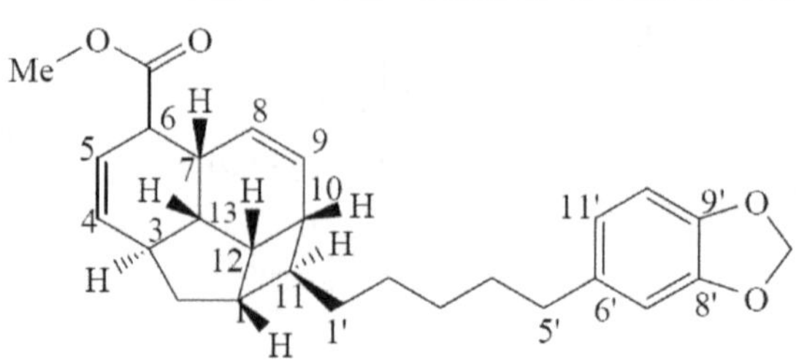	Root of *B. erytrhophloia* (Yang et al., 2009)

33: Erytrophloin F	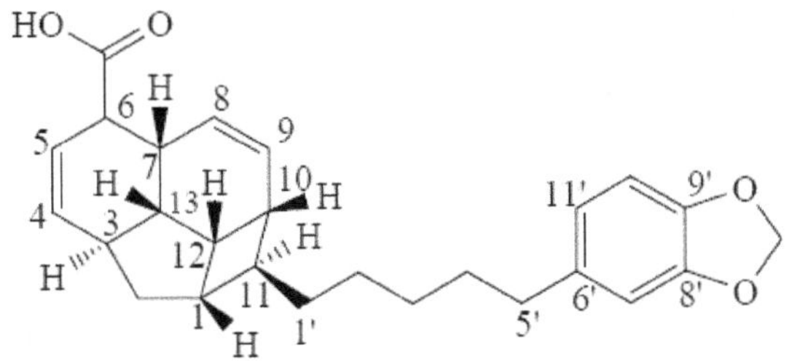	Root of *B. erytrhophloia* (Yang et al., 2009)
34: Erytrophloin C	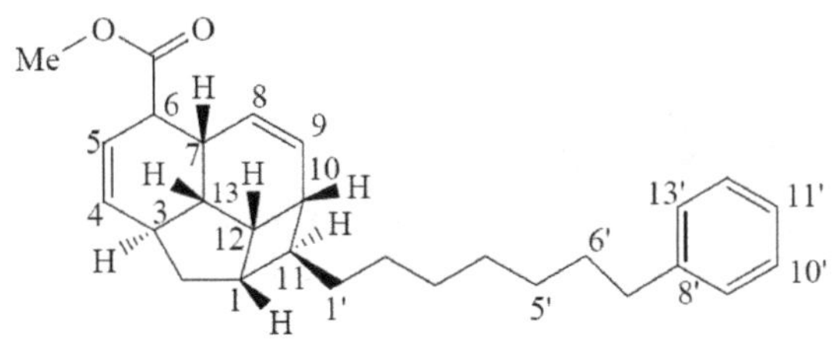	Root of *B. erytrhophloia* (Yang et al., 2009)
35: Erytrophloin E	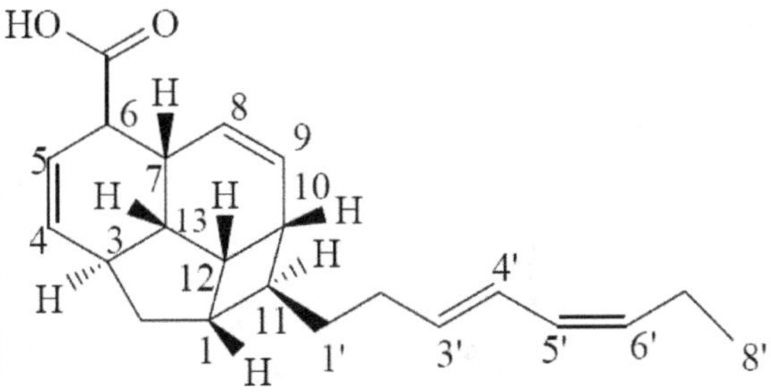	Root of *B. erytrhophloia* (Yang et al., 2009)
36: Erytrophloin B	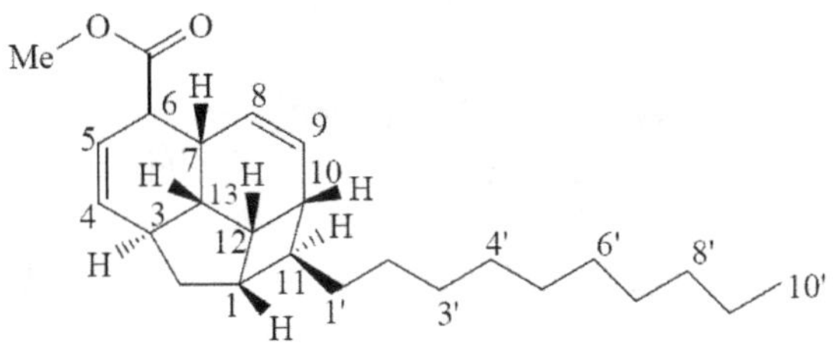	Root of *B. erytrhophloia* (Yang et al., 2009)
37: Erytrophloin D	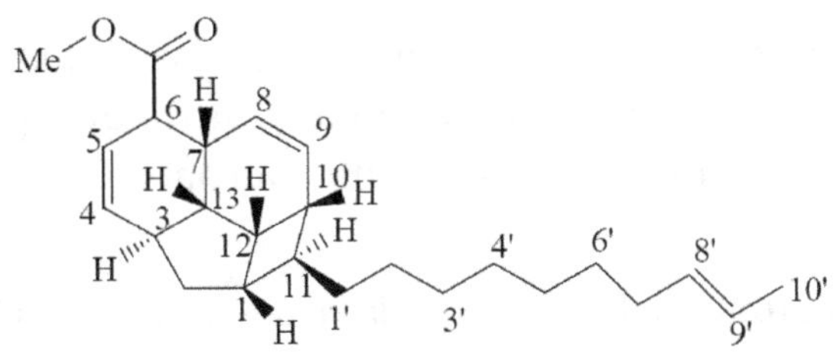	Root of *B. erytrhophloia* (Yang et al., 2009)
38: Beilcyclone	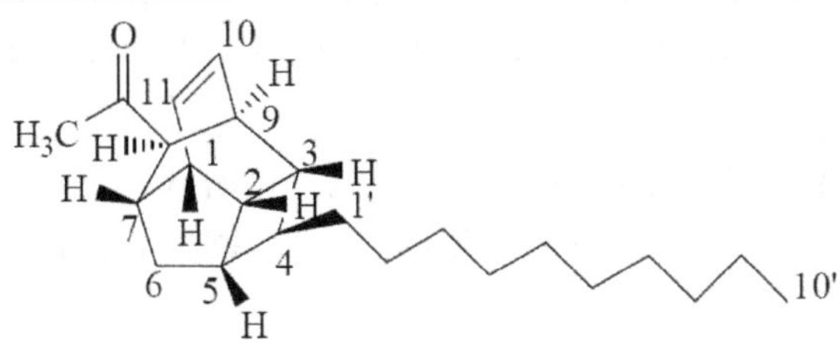	Root of *B. erytrhophloia* (Yang et al., 2009)

39: Endiandric acid J	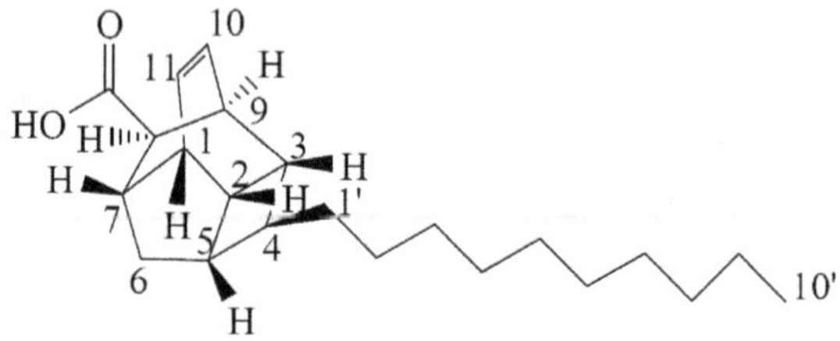	Root of *B. erytrhophloia* (Yang et al., 2008)
40: Endiandric acid I	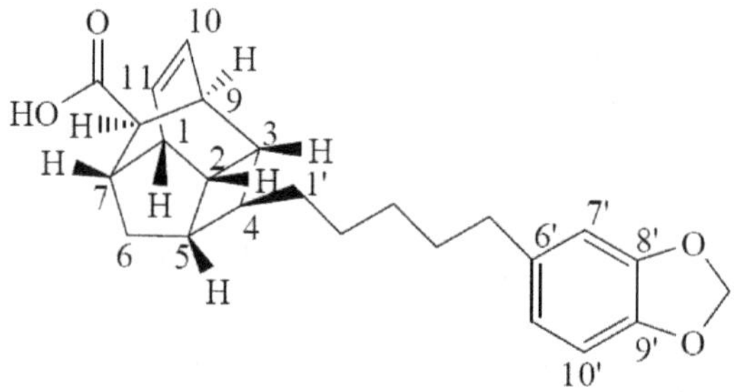	Root of *B. erytrhophloia* (Yang et al., 2008)
41: Endiandric acid H	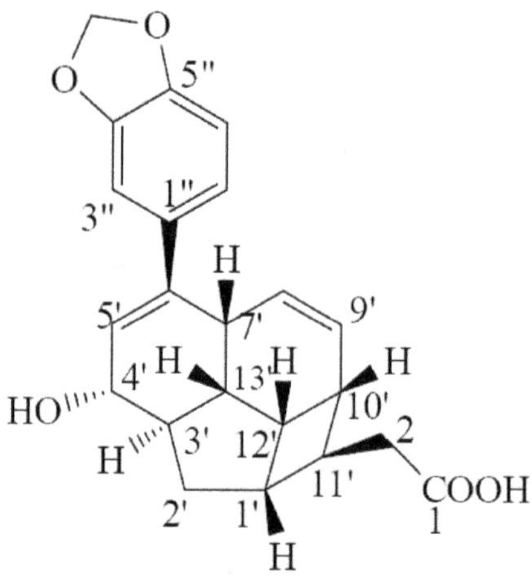	*Stem of B. fulva* (Eder et al., 2004)

5. RESULTS

We have initiated a systematic phytochemical investigation of the extracts of *Beilshmiedia anacardioides* and have so far obtained the following results which have led to three publications in renowned scientific journals (Chouna et al.,2009; 2010; 2011).

Air-dried and ground stem bark of *B. anacardioides* was extracted successively at room temperature with MeOH. The methanol extract was re-extracted in turn with CH_2Cl_2 and EtOAc. These extracts were concentrated to dryness under reduced pressure.

The CH_2Cl_2 extract was submitted to repeated column chromatography on silica gel, yielding beilschmiedic acids A (**1**), B (**2**) and C (**3**) and the known β-sitosterol (Chouna et al.,2009) .

Further Successive purifications by column chromatography over silica gel and preparative TLC afforded three new endiandric acid derivatives: beilschmiedic acids D(**4**) and E(**5**), and Beilshmiedin (**8**) (Chouna et al.,2010), together with the known compounds bisabolene (Mossa et al 1992; Barrero et al. 1990), and tricosanoic acid (Erdemoglu et al., 2008).

The ethyl acetate soluble part of the MeOH extract of the stem bark of *B. anacardioides* was fractionated by column chromatography over silica gel. Successive purifications by column chromatography and preparative TLC afforded two new endiandric acid derivatives: beilschmiedic acids F(**6**) and G(**7**) Chouna et al.,2011, along with the known constituents beilschmiedic acid A(**1**), beilschmiedic acid C(**3**) [6] and sitosterol-3-*O*-*β*-*D*-glucopyranoside (Chouna et al.,2011).

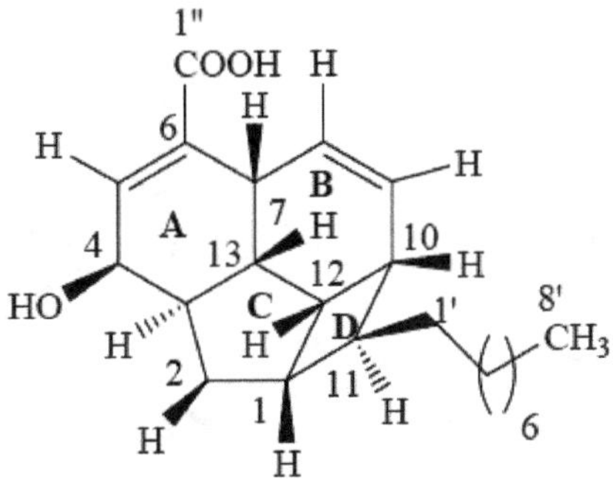
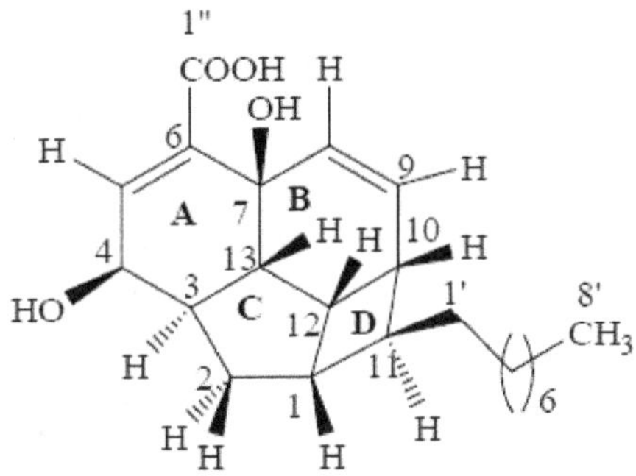

Beilshmiedic acid A (**1**) Beilshmiedic acid B (**2**)

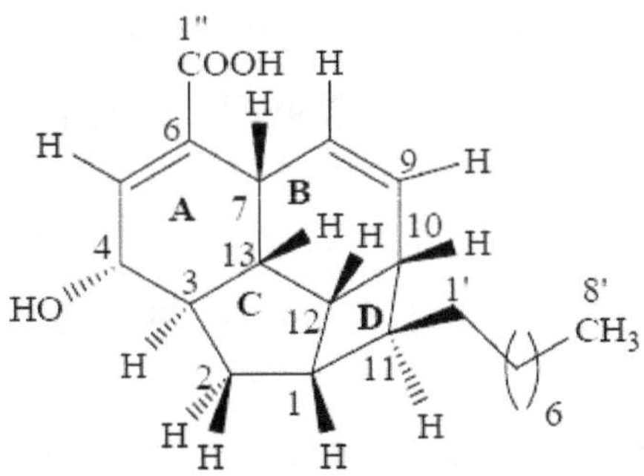
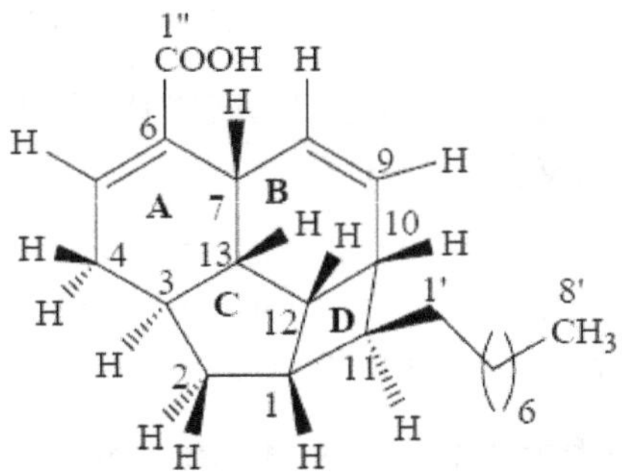

Beilshmiedic acid C (**3**) Beilshmiedic acid D (**4**)

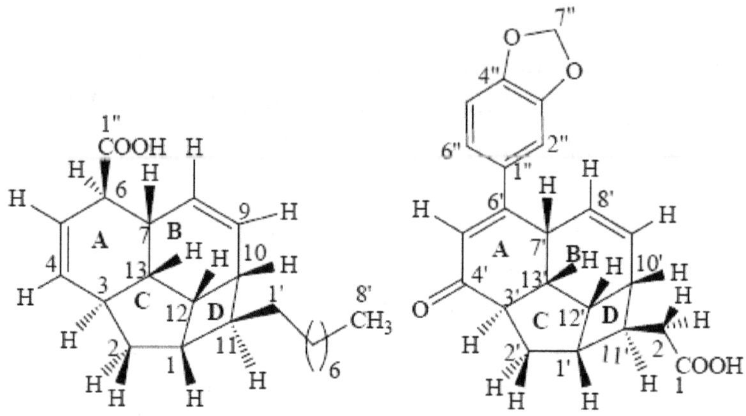

Beilshmiedic acid E (5) Beilshmiedic acid F (6)

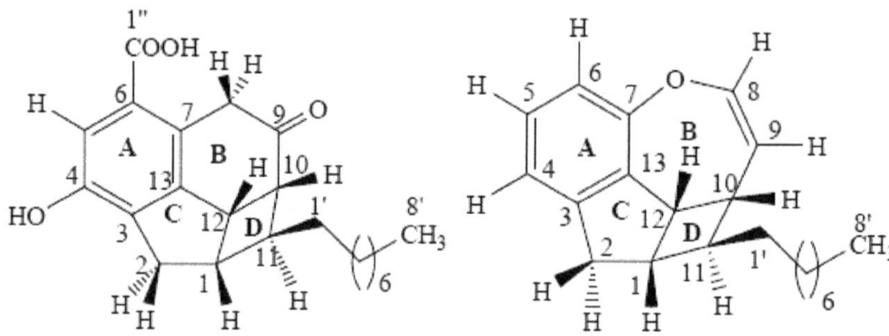

Beilshmiedic acid G (7) Beilshmiedin (8)

Scheme 1. endiandric acid derivatives from *Beilshmiedia anacardioides*.

6. BIOLOGICAL ACTIVITY AND THE SIGNIFICANCE OF SOME COMPOUNDS

Our preliminary antibacterial studies on the new endiandric acid derivatives have yielded chemical entities that have been shown to possess significant activities (Chouna et al., 2009) .

Antibacterial assay on some compounds isolated from *B.anacardioides*

Compounds **1-8** were tested in vitro for their antibacterial activity against *Bacillus subtilis, Streptococcus ferus, Streptococcus minor, Micrococcus luteus, Escherichia coli,* and *Pseudomonas agarici*, using the dilution technique.

The ZI (Table 1) and MIC (Table 2) obtained for these compounds indicated that they possessed strong to weak antibacterial activity against *gram* positive bacteria

Beilshmiedic acid C **(3)** demonstrated the best potency against *B. subtilis* and *M. luteus*, compared to the reference drug ampicillin. The MIC values (Table 2) of Beilshmiedic acids B(**2)**, C **(3)** and G(**7)**, against *B. subtilis* and Beilshmiedic acid C **(3)** against *M. luteus* were found to be greater than that of standard drug ampicillin, indicating that this series of compounds might be possible candidates as antibacterial drugs.

None of the tested compounds was active against *Gram* negative *P. palida* and *E. Coli.* Therefore, they might be well tolerated as antibiotics even for long term treatments.

Table 6. Antibacterial activity (Zone of inhibition of compounds in mm) of compounds 1-8 (500μg/mL) against *B. Subtilis, M. luteus, S. faecalis, S. minor, S. ferus, P. palida* **and** *E. Coli.*

Compound tested	*B. subtilis*	*M. luteus*	*S. faecalis*	*S. minor*	*S. ferus*	*P. palida*	*E. coli*
1	15	12	14	n.t.	n.t.	-	-
2	16	15	15	n.t.	n.t.	-	-
3	13	30	18	n.t.	n.t.	-	-
4	10	n.t.	n.t.	10	-	-	-
5	12	n.t.	n.t.	-	12	-	-
6	10	n.t.	n.t.	-	-	-	-
7	20	15	16	-	-	-	-
8	10	n.t.	n.t.	-	-	-	-
Ampicillin	29	26	25	22	23	-	-

(-) inactive, n.t. (not tested)

Table 7. Antibacterial activity (MIC in µM) of compounds 1-8 against *B. Subtilis, M. luteus, S. faecalis, S. minor, S. ferus, P. palida* **and** *E. coli.*

Compound tested	*B. subtilis*	*M. luteus*	*S. faecalis*	*S. minor*	*S. ferus*
1	181.60	173.60	363.30	n.t.	n.t.
2	11.30	347.20	45.30	n.t.	n.t.
3	5.60	< 0.70	22.70	n.t.	n.t.
4	381.00	n.t.	n.t.	190.50	-
5	381.00	n.t.	n.t.	190.50	-
6	343.40	n.t.	n.t.	-	-
7	87.78	10.95	87.78	-	-
8	422.20	n.t.	n.t.	-	-
Ampicillin	89.5	1.95	3.9	1.05	5.25

(-) inactive, n.t. (not tested)

Beilshmiedic acid C (**3**) was more active than Beilshmiedic acid D (**4**) . The enhanced activity may be due to the additional hydroxyl group at C-4 position in Beilshmiedic acid C (**3**) . Beilshmiedic acid B(**2**) which possesses one hydroxyl group more than Beilshmiedic acid A

and Beilshmiedic acid C (**3**) was less active. Beilshmiedic acid A (**1**) was more active than Beilshmiedic acid C (**3**) . They are epimers at C-4 position; the modification of the configuration at this position influences significantly the activity.

Based on the skeletal features, it is difficult at this stage to define the contribution of the different functional groups with respect to the activity. The mechanism of action of this class of metabolites on these strains is not yet known. Further investigations will help to establish the mode of action of this particular skeleton. These interesting results highlight the potency of this rare class of metabolites that might be investigated for the search of new antibacterial drugs.

7. CONCISE SUMMARY

We hypothesize that phytochemists seeking novel bioactive natural compounds should explore medicinal plants with clinically validated therapeutic efficacy in African medicine.

We proposed that natural compounds derived from Beilshmiedia anacardioides may be effective in addressing vaginal infections caused by B. subtilis and M. luteus, as well as rheumatism attributed to Streptococcus ferus and S. minor. The biological actions of several compounds discovered in our investigations, particularly Beilshmiedic acid C, strongly corroborate this idea.

As the availability of data from phytochemical and biological analyses of clinically researched therapeutic plants increases, the role of these plants in illness therapy will become increasingly delineated. Consequently, African Traditional Medicine will attain global recognition.

8. REFERENCES

Banfield, J.E.; Black, D.S.C; Fallon, G.D.; Gatehouse B.M.: Constituents of *Endiandra* species.V. 2-[3',5'-Dioxo-4'-phenyl-10'-{(E,E)-5''-phenyl-penta-2'',4''-dien-1''-yl}-2',4',6'-triazatetracyclo[5,4,2,$0^{2,6}$,$0^{8,11}$]tridec12'-en-9'-yl]-acetic acid derived from *Endiandra introrsa* (Lauraceae). *Aust. J. Chem.*1983, 36, 627-632.

Banfield J.E.; Black D.C.; Collins D.J.; Hyland B.P.M.; Lee J.J., Pranowo R.S. Constituents of some species of *Beilschmiedia* and *Endiandra* (Lauraceae): New Endiandric acid and benzopyran derivatives isolated from *B. oligandra*. *Aust. J. Chem.* 1994, *47*, 587-607.

Bandaranayake, W.M.; Banfield, J.E.; Black, D.S.C.; Fallon G.D.; Gatehouse, B.M.: Constituents of *Endiandra* species. III. 4-[(E,E)-5'-Phenylpenta-2',4'-dien-1'-yl]tetra- cyclo[5,4,0,$0^{2,5}$,$0^{3,9}$]undec-10-ene-8-carboxylic acid from *Endiandra introrsa* (Lauraceae). *Aust. J. Chem.* 1982. 35, 567 – 579

Bandaranayake, W.M.; Banfield, JE; Black D.C.; Fallon, G.D.; Gatehouse B.M.: Constituents of Endiandra species. Endiandric acid, a novel carboxylic acid from *Endiandra introrsa* (Lauraceae) and a derived lactone. *Aust. J. Chem.* 1981, *34*, 1655-1667.

Bandaranayake, W. M., Banfield, J. E., Black, D. C. Postulated electric reactions leading to endiandric acid and related natural products. *J. Chem. Soc., Chem. Com.* 1980, 13, 902-903.

Barrero A. F.; Alverez-Manzaneda, E.J.R.; Alverez-Manzaneda, R.R.: Bisabolene derivatives and other constituents from *Achillea odorata. Phytochemistry* 29, 3213-3216.

Chen, J.J.; Chou, E.T.; Duh, C.Y.; Yang, S.Z.; Chen, I.S.: New cytotoxic tatrahydrofuran- and dihydrofuran-type lignans from the stem of *Beilschmiedia tsangii. Planta Medica.* 2006. *72*, 351-357.

Chen, J.J.; Chou, E.T.; Peng, C.F.; Chen, I.S.; Yang, S.Z.; Huang, H.Y.: Novel epoxyfuranoid lignans and antitubercular constituents from the leaves of *Beilschmiedia tsangii. Planta Medica.* 2007, *73*, 567-571.

Clezy, P.S.; Gellert, E.; Lau, D.Y.; Nichol, A.W.: The alkaloids of *Beilschmiedia elliptica. Aust. J. Chem.*1966, *19*, 135-142.

Chouna, J.R.; Nkeng-Efouet, P.A.; Lenta, B.N.; Devkota, K.P.; Neumann, B.; Stammler, H.G.; Kimbu, S.F.; Sewald, N. Antibacterial endiandric acid derivatives from *Beilschmiedia anacardioides. Phytochemistry.* 2009,70: 684-688.

Chouna, J. R.; Nkeng-Efouet, P. A.; Lenta, B. N.; Wansi, J. D.; Fon, K. S.; Sewald, N.. Endiandric acid derivatives from the stem bark of *Beilschmiedia anacardioides. Phytochemistry Letters* 3.2010, Pages 13-16.

Chouna,J. R.; Nkeng-Efouet, P. A. Lenta,B. N.; Wansi, J. D.; Stammlerf, B. N. H.; Fon, K. S.; Sewald, N.: Beilschmiedic Acids F and G, Further Endiandric Acid Derivatives from *Beilschmiedia anacardioides. Helvetica Chimica Acta.* 2011, 94, 1-7.

Nurgün, E. E. A.; Akgöç M.; Simay, Ç.; Gökhan, B.: Comparison of the Seed Oils of *Ferulago trachycarpa Boiss.* From different localities with respect to Fatty Acids. *Rec. Nat. Prod.* 2008. 2:1, 13-18.

Eder, C.; Kogler, H.; Haag-Richter, S. 2004. Endiandric acid H and derivatives, procedure for its production and use of the same in the treatment of asthma. *German patent 10235624.*

Eyog Matig, O.; Ndoye, O.; Kengue, J.; Awano, A. *Les fruitiers forestiers comestibles du Cameroun.* 2006. Cotonou: IPGRI/SAFORGEN/IRAD/CIFOR.

Fouilloy, R. Flore du Cameroun (Lauracées, Myristicaées et Momimiacées). Paris, Muséum National d'Histoire Naturelle 18. 1974, p 8.

Fitzgerald, D.J.; Stratford, M.; Gasson, M.J.; Narbad, A. Structure function analysis of the vanillin molecule and its antifungal properties*J. Agric. Food Chem.* 2005, 53, 1769- 1775.

Harborne,J.B.; Mendez, J. Flavonoids of Beilsmiedia miersil. *Phytochemisty.* 1969, 8, 763-764. Iwu, M.M. *Handbook of African Medicinal Plants.* CRC Press Inc, Boca Raton, Ann Arbour, London, Tokyo. 1993, p. 435.

Kitawaga, I.; Minagawa,K.; Zhang, R.S.; Hori, K.; Doi, M.; Ishida, T.; Kimura, M.; Uji, T.; Shibuya, H.; Dehatrine, H.: An antibacterial bisbenzylisoquinoline alkaloid from the Indonesian medicinal plant *Beilsmiedia madang*, isolated as a mixture of two rotational isomers. *Chem. Pharm. Bull.*1993, 41, 997- 999.

Lee, J.; Jung, E.; Park, J.; Yung, K.; Lee, S.; Hong, S.; Park, J.; Park, E.; Kim, J.; Park, S.; Park, D.: Anti-inflammatory effects of magnolol and honokiol are mediated through inhibition of the downstream pathway of MEJKK-1 in NF-kappaB activation signalling . *Planta Medica.*2005, 71, 338-343.

Lee, J. Y.; Jang, Y. W.; Moon, H.; Sim, S.; Kim, S Kang, H. S.: Anti-inflammatory action of phenolic compounds from *Gastrodia elata* root., *Arch. Pharm. Res.* 2006, 10, 849-859

Lenta, N.B.; Tantangmo, F.; Devkota, P.K.; Wansi, J.D.; Chouna, J.R.; Soh, F.R.C.; Neumann, B.; Stammler, H.G.; Tsamo, E.; Sewald, N.: Bioactives Constituents of the Stem Bark of *Beilschmiedia zenkeri*. *Journal of Natural Products*. 2009, 72, 2130-2134.

Lenta, B. N.; Chouna, J. R.; Nkeng-Efouet, P. A.; Fon, K. S.; Tsamo, E.; Sewald, N.: Obscurine, a New Cyclostachine Acid Derivative from *Beilschmiedia obscura*, 2011, *Natural Product Communication*. 2011,6:11,1591-1592.

Li, H. B.; Yi, X.; Gao, J. M.; Ying, X. X.; Guan, H. Q.; Li, J. C.: Magnoliol-induced H-460 cells death via autophagy but not apoptosis. *Arch. Pharm. Res*. 2007, 30, 1566-1574.

Moghaddam, M.F.; Farimani, M.M.; Salahvarzi, S.; Amin, F.: Chemical constituents of dichloromethane extract of cultivated *Satureja khuzistanica*. *eCam* . 2007, 4, 95-98.

Mpetga, S. D.J. Contribution à l'étude phytochimique d'une plante médicinale du Cameroun: *Beilsmiedia mannii* (Lauracée). Thèse de Doctorat en Chimie organique, Université de Dschang. 2005, p. 53.

Mossa, S.J.; Muhammad, I.; El-Feraly, S.F.; Hufford, D.C.; Mcphall, R.D.; Mc phall, T.A. Bisabolene and gualene sesquiterpenes from *Policarta glutinosa*, *Phytochemistry*. 1992, 31, 575-5780.

Pudjiastuti, P.; Mukhtar, M.R.; Hadi, A.H.A; Saidi, N.; Morita, H.; Litaudon, M.; Awang, K.: (6,7-Dimethoxy-4-methylisoquinolinyl)-(4' methoxyphenyl)-methanone, a New Benzylisoquinoline Alkaloid from *Beilschmiedia brevipes*. *Molecules*. 2010, 1-8.

Paulo, M. Q.; Barbosa-Filho, J. M.; Lima, E.O.; Maria R. F.; Barbosa, R. C.; Kaplan, M.A. Antimicrobial activity of benzylisoquinoline alkaloids from *Annona salzmanii*. *J. of Ethnopharmacol*. 1992, 36, 39-41.

Park, J.; Lee, J.; Jung, E.; Park, Y.; Kim, K.; Park, B.; Yung, K.; Park, E.; Kim, J.; Park, D.: In vitro antibacterial and anti-inflammatory effects of honokiol and magnolol against *Propionibacterium sp*. *Eur. J. Pharmacol*. 2004, 12, 189-195.

Tchouala, J.M.: Inventaire ethnobotanique des plantes médicinales du Noun (Ouest- Cameroun) utilisées pour soigner les infections fongiques et bactériennes. Mémoire de maitrise de Biologie Végétale, Université de Dschang. 2001, p. 33.

Yang, P.S.; Cheng, M.J.; Chen, I.J.; Chen, I.S.: Two new endiandric analogs, a new benzopyran and a new benzenoid from the root of *Beilshmiedia erytroploia*. *Helv.Chim. Act*. 2008, 91, 2130-2138.

9 Chapter

PHENOLIC COMPOUNDS AND ANTIOXIDANT CHARACTERISTICS OF SELECTED THAI FLORA

1. INTRODUCTION: AN OVERVIEW

Thai flora has been utilized as medicinal agents for ages due to the presence of active phytochemicals, especially phenolic compounds. These components serve as antibiotics, render cell walls impermeable to gas and water, operate as structural materials to confer stability to plants, and offer protection against ultraviolet (UV) radiation. Consequently, flora in the tropical region, such as Thailand, possess a significant concentration of phenolic compounds that are produced as secondary metabolites in plants (Shahidi & Naczk, 2003).

Various components of edible plants from tropical and subtropical regions are recognized for their high content of phenolic compounds, which are garnering growing interest from consumers for multiple reasons (Leong & Shui, 2002). Epidemiological studies indicate a correlation between the intake of phenolic-rich foods or beverages and the prevention of illnesses, including cancer, stroke, and coronary heart disease. This correlation is partially elucidated by the ability of phenolic compounds to inhibit the oxidative alteration of low-density lipoproteins (LDL), which is involved in the onset of arteriosclerosis. Recently, other mechanisms have been suggested for the involvement of antioxidants in decreasing the incidence of cardiovascular disease, beyond merely protecting LDL from damage generated by reactive oxygen species (ROS). Various phenolic antioxidants substantially influence cellular reactions to diverse stimuli, such as cytokines and growth factors. Despite numerous studies documenting the composition and antioxidant activity of phenolic compounds in tropical edible plants (Auddy et al., 2003; Nũnez sellés et al., 2002; Habsah et al., 2000), knowledge regarding the phenolics and antioxidant potential of Thai edible plant species remains scarce in relation to the extensive biodiversity of cultivated edible plants. Furthermore, numerous studies have been published in the Thai language, rendering them inaccessible to English-speaking researchers.

This chapter aims to critically analyze the composition and antioxidant capabilities of phenolic compounds found in several Thai plants. Moreover, factors influencing the extraction of these components from Thai flora are documented.

2. PHENOLIC COMPOUNDS OF SOME THAI PLANTS

Plant materials contain many phytochemicals including compounds with antioxidant activity, which are mostly phenolic in structure(Johnson, 2001). Compounds with antioxidant activity are mainly phenolic acids, flavonoids and polyphenols (Dillard & German, 2000). Phenolic acids such as caffeic acid and gallic acid are widely distributed in the plant kingdom. The most widespread and diverse phenolics are the flavonoids which have the same C15 (C6-C3-C6) skeleton and retard oxidationof a variety of easily oxidizable compounds (Zheng & Wang, 2001). Flavonoids include catechins, proanthocyanins, anthocyanidins, flavones, flavonols and their glycosides (Ho, 1992). Flavonoids are ubiquitous in plants, since almost all plant tissues are able to synthesize flavonoids. Among the most widely distributed are the flavonols quercetin and rutin.

Investigations have increased considerably in recent years in order to find natural plant antioxidants to replace synthetic compounds whose use is being restricted due to possible side effects such as carcinogenecity (Zeng & Wang, 2001). Many papers have reported that phenolic plant constituents provide protection against oxidation (Amarowicz et al., 2003; Pokorny, 2001). Phenolic substances inhibit propagation of the oxidation chain reactions due to their resonance stabilized free-radical forms (Lindsay, 1996).

Phenolic compounds possess one or more aromatic rings bearing two or more hydroxyl groups (Ho, 1992). They are closely associated with the sensory and nutritional quality of fresh and processed foods. In general, the leaves, flowers, fruits and other living tissues of the plant contain glycosides, woody tissues contain aglycones, and seeds may contain either (Huang & Ferraro, 1992).

Agricultural based manufacturers in Thailand produce and export many fruit and vegetable products. These Thai plants are widely distributed throughout the tropics particularly in Southeast Asia. Many researchers have shown that several parts of tropical and subtropical plants contain large amounts of natural phenolic phytochemicals, such as flavonoids (Leong & Shui, 2002; Kähkonen et al., 1999; Demo et al., 1998). Hence, there is a potential for Thai plants to be used as sources of phenolic antioxidants and commercial extracts could be prepared from numerous available raw plant materials. Cost, simplicity and safety should be considered in the development of an acceptable extraction procedure (Pokorny & Korczak, 2001).

Due to the diversity and complexity of natural mixtures of phenolic compounds in plant extracts, it is rather difficult to characterize every phenolic compound. Each plant generally contains different mixtures of phenolic compounds. The Folin-Ciocalteu method is a rapid, widely used assay to determine the total concentration of phenolic compounds. It is known that different phenolic compounds vary in their responses in the Folin-Ciocalteu method. Many researchers have reported the total phenolic content of Thai plants as shown in Table 1.

The data clearly indicate that some of these plants are rich in natural phenolic compounds. Many plants with a high phenolic content have an astringent taste (for leaves) or strong colors (for flowers and fruits) due to flavonoid components. Proanthocyanidins contribute astringency to plants and other flavonoids contribute red or violet colors (anthocyanins) or yellow colors (flavonols). Higher total phenolic and flavonoid contents have been found in seeds compared to other tissues. Typically, leaf photosynthesis products including essential nutrients such as sucrose are translocated from leaves to fruits and seeds which are the food storage organs of the plants (Salisbury & Ross, 1992). This leads to a concentration of phenolic compounds in seeds. Thai plant samples (Table 1) may be classified into two groups with high and low contents of polyphenolic phytochemicals, Samples having relatively high concentrations of phenolic phytochemicals (more than 100 mg GAE/g dry weight of material) included seeds of *Antidesma velutinum* Tulas, *Cleistocalyx operculatus* var. paniala (Roxb.), *Eugenia siamensis* Craib., bark of *Acacia catechu* (L.F) Wild., kernel of *Areca catechu* Linn. and stem of *Cassia fistula* Linn. Chanwitheesuk et al. (2005) reported the total phenolic content of Thai plants using Folin Denis reagent and pyrocatechol as reference, with data calculated as mg% of pyrocatechol equivalents. For comparison, the total phenolic content of Leucaena was reported as 405 mg% (Chanwitheesuk et al., 2005) and 51.20 mg GAE/g db plant (Maisuthisakul et al., 2008). The data showed that the different reference compound and the units used required a conversion factor of around eight to convert the values (Table 1).

Quantitative determination of individual flavonoid glycosides is difficult because most standards are not commercially available. Hence, the total flavonoid content determined using a colorimetric method (Bonvehí et al., 2001) is commonly used for flavonoid evaluation. Generally, the calculation of the total flavonoid content is quoted in units of mg rutin equivalent of flavonoid compounds in one gram of plant extract based on the dry weight of the original plant sample. It is well known that flavonoids possess antioxidant properties both in vitro and in vivo. The flavonoids contain a number of phenolic hydroxyl groups attached to aromatic ring structures, which confer the antioxidant activity. Flavonoids were found in leaves, seeds and fruits and were good antioxidants (Bonvehí et al., 2001). The data clearly indicate that some plants in Thailand are rich in natural antioxidant flavonoids. Some data in Table 1 were used to find a relationship between total phenolic and flavonoid content. The flavonoid content correlates moderately with the phenolic content as shown in Fig. 1.

Table 1. Ranges of some phenolic constituents in some Thai plants

Scientific name	Common name	Plant part	Moisture Content (%)	Total phenolics (mgGAE/g db plant)¥	Total flavonoids (mgRE/gdb plant)	Total carotenes Total (mg%)	xanthophyll (mg%)	Tannin (mg% of tannic acid equivalent)
	Herb and	vegetable						
Acacia pennata	Acacia leaf	Young leaves		121.00,		1.27	1.59	11.1
Acanthopanax trifoliatum *Allium ascalonicum* Linn.¥	 Onion	Leaves Flower	 94.70	275.00° 55.70	 20.20	2.54	3.17	57.30
Artemisia dubia Wall. ex DC.		Stem and leaves		14.24				
(Syn. A. *vulgaris* L. var. *indica*								
Maxim.)								
Aspidistra sutepensis K. Larsen		Flower		5.06				
Azadirachta indeca A. Juss Var.		Flower	77.20	40.30	29.80			
siamensis valeton¥								
Basel/a alba Linn.¥	Ceylon spinash	Leaves	93.5	15.50	6.2			
Ridens bipinnata, L.		Stem and leaves		34.18				
Ridens pilosa Linn.		Stem and leaves		24.62				
Buddieia asiatica Lour	Rachawadi pa	Stem and leaves		19.17				
***Cassia siamea* Britt.** ¥,	Thai copper pod	Flower	74.8	51.50	24.8			
		Leaves		384.00•		1.92	1.59	110.00
Careya sphaerica Roxb.¥	Tummy wood	Young leaves and	75.3	54.50	20.5			

		Leaves						
Centella asiatica Linn.¥,	Pennywort	Leaves	86.6	12.40	10.6	12.80	10.60	24.30
Cratoxylum fonnosum Dyer.¥		Young leaves and	85.5	63.40	25.5			
		Leaves						
Coccinia grandis	Ivy gourd	Leaves		74.70•		1.94	2.65	17.70
Coleus amboinicus	Country borage	Leaves		54.80		2.54	4.24	24.30
Commelina diffusa Burm.£.		Stem and leaves		19.73				
Conyza sumatrensis (Retz.) Walker		Stem and leaves		15.66				
Coriandrum sativum	Coriander	Leaves		33.0•		2.52	1.05	24.30
Cucurbita moschata	Pumpkin	Leaves		87.8'		1.92	1.59	4.48
Cuscuta australis R. Br.		Stem		33.21				
Diplazium esculentum (Retz.) Sw.		Stem and leaves		19.48				
Dolichandrone serrulata (DC.) Seem.		Flower		13.25				
Dregea volubilis		Leaves		100.00,		6.14	1.07	17.70
Embelia ribes Burm.£.		Leaves		57.89				
Embelia sessilijlora Kurtz		Leaves		65.08				
EnJthrina crista Galli.¥	Coral tree	Leaves	81.7	67.50	20.2			
Gymnema inodorum		Leaves		188.00a		1.31	1.07	11.1
Hydrocharis dubia (Bl.) Back.*	Frogs bit	Bud	95.0	20.40	8.9			

Scientific name	Common nan1e	Plant part	Moisture Content (%)	Total phenolics (mgGAE/g db plant)¥	Total flavonoids (mgRE/gdb plant)	Total carotenes Total (mg%)	xanthophyll (mg%)	Tannin (mg% of tannic acid equivalent)
Lasia spinosa (Linn.) Thw. ¥		Leaves	94.8	6.40	4.4			
Leucaena glauca Benth¥	Lead tree	Young leaves and	79.9	51.20	22.3			
		Leaves						
Lirnnocharis Jlava Buch.¥		Leaves	94.7	5.40	3.7			
Macropanax dispermus		Leaves		651.00•		3.89	1.06	37.4
Marsdenia glabra		Leaves		51.50°		8.92	7.42	4.47
Melicope pteleifolia (Champ.ex		Hartley Leaves		40.31				
Benth.)								
Mentha aroensis	Japanese mint	Leaves		70.0•		4.48	26.5	21.0
Mentha cordifolia *Micromelum minutum Wight & Arn*	Kitchen mint	Leaves Stem and leaves		280.00a 61.15		2.58	4.24	73.7
Momordica charantia **Linn.*,**	Balsum pear	Bud and Leaves	87.0	50.90	21.6	1.31	0.54	4.48
-Musa spiantum **Linn.¥**	**Banana**	Flower	92.8	45.30	20.3			
Neptunia oleracea	Water cress	Leaves		104.00•		3.18	1.06	21.00
Ocimum americanum	Hairy basil	Leaves		43.6•		5.12	9.52	11.10
Ocimum basilicurn Linn.¥,	Sweet basil	Leaves	89.8	50.50	15.3	10.80	13.30	30.90
Ocimum basilicum **cultivar Purple**		Leaves		3.40 (by fresh				
Petra' ***Ocimum sanctum*** **Linn.¥,**	Holy basil	Young leaves and	87.6	weight) 41.90	12.6	5.13	3.18	40.80

		Leaves						
Oenanthe stolonifera	Chenese celery	Leaves		329.00•		3.83	14.8	34.2
Orthosiplwn grandiflorus	Cat's whisker	Leaves		145.00•		3.20	25.40	30.90
Piper retrofracturn	Long pepper	Flower		57.5•		1.28	5.31	7.78
Polycia fruticosa		Leaves		46.30•		2.52	2.13	24.30
Sauropus androgynus Linn. ¥		Young leaves and	89.9	11.50	10.4			
		Leaves						
Schima wallichii (DC.) Korth.		Leaves		206.10				
Sechiurn edule	Chayote	Leaves		66.1°		3.83	2.13	4.48
Sesbania grandiflora Desv. ¥	Cork wood	Flower	91.1	58.60	13.1			
Spondias pinnata Kurz.¥	Hog plum	Young leaves and	76.4	42.60	14.8			
		Leaves						
Suaeda maritima		Red mature Leaves		38.60				
		Green young Leaves		66.90				
		Green flower		59.30				
Syzygium gratum (Wight) S.N.Mitra		Young leaves and	79.6	57.30	23.6			
var. gratum¥ Tarnarindus indica	Tamarind	Leaves Young leaves		121.00,		0.64	1.05	77.00

ne	Common name	Plant part	Moisture Content (%)	Total phenolics (mg GAE/g db plant)¥	Total flavonoids (mg RE/g db plant)	Total carotenes (mg%)	Total xanthophyll (mg%)	(mg acid
r	Tonkin jasmine	Flowers	-	98.40[a]	-	1.29	4.24	
rengelii (G.Don) Sleum.		Leaves	-	95.42	-	-	-	
	Berries and fruits							
rvanh	Siam cardamon	Fruit	-	46.30[a]	-	1.29	1.07	
tescus Linn. ¥	Chilli pepper	Fruit	85.4	40.30	13.3	-	-	
ensis Craib. ¥	Jambolan plum	Fruit	85.1	82.40	44.3	-	-	
ccenses Linn. ¥	Malay apple	Fruit	93.7	69.20	28.7	-	-	
harantia Linn. ¥	Balsam pear	Fruit	75.9	50.90	21.6	-	-	
mblica¥	Indian gooseberry	Fruit	82.6	69.10	23.4	-	-	
nata Kurz. ¥	Hog plum	Fruit	77.3	47.20	12.6	-	-	
	Berry and fruit seeds							
lutinum Tulas*		Seed	38.4	123.30	50.3	-	-	
perculatus var. paniala		Seed	55.1	173.60	44.2	-	-	
ensis Craib.*	Jambolan Plum	Seed	50.3	180.50	50.4	-	-	
ica Benth. *	Leadtree	Seed	76.5	20.40	5.3	-	-	
ppaceum Linn. ¥	Rambutan	Seed	36.3	43.50	13.3	-	-	
a Hassk. ¥		Seed	70.7	51.90	20.3	-	-	
Linn. ¥	Pepper	Seed	72.5	53.10	22.8	-	-	
ndica Linn¥	Tamarind	Seed	49.5	40.70	23.2	-	-	
	Chewing plants							
u (L.F) Wild.*	Black catechu	Bark	16.3	177.70	41.8	-	-	
i Linn.*	Betel nut	Whole fruit	90.2	52.50	12.6	-	-	
		Kernel	91.2	137.30	42.8	-	-	
Linn.*	Golden shower	Stem core	11.4	103.60	25.4	-	-	
nn*	Betel leaf	Leaf	82.6	57.50	14.9	-	-	

Source: Adapted from ¥Maisuthisakul et al. (2008); *Maisuthisakul et al. (2007a); ϵLee & Scagel (2009); ✪ Chuenarom et al. (2010); ⵋPhomkaivon & Areekul (2009); ⋕Chanwitheesuk, Teerawutgulrag & Rakariyatham (2005)

- means not reported or not determined.

[a] The data were calculated as mg% of pyrocatechol equivalent.

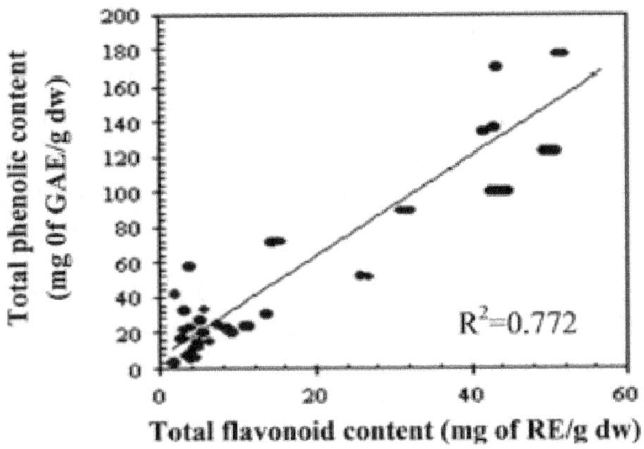

Fig. 1. Relationship between total phenolic and total flavonoid content in Thai plant extracts.

Chanwitheesuk et al. (2005) reported the total carotene, total xanthophyll and tannin contents of edible plants grown in Thailand as shown in Table 1. Total carotenes and total xanthophylls were determined according to Helrich (1990). Tannin contents were determined colorimetrically using the Folin-Dennis reagent according to Helrich (1990).

Identification of the phenolic components in Thai plants has not commonly been reported because it is expensive and standards required for identification are often not available. Hence, this chapter will report the phenolic profile of only four of the Thai plants investigated; Piper betel Linn, *Careya sphaerica*, *Cratoxylum formosum* Dyer., and *Leucaena leucocephala* de Wit.

Piper betel Linn.; which is commonly known as Betel leaf, is chewed with Betel nut and lime by some people in Asia. Its Thai name is different in different areas, for instance, Plu- Cheen, Se-ke, Bul-plao-yuan, she-ke (South), Pu (North-West). The leaves are chewed alone or with other plant materials including the areca nut (Areca catechu Linn.) and lime. Many researchers have focused on the red lime betel quid in the past few years. A little information about the Betel leaf was found. The Betel leaf itself has a spicy taste and yields an essential oil widely used as a medicine. In Thailand, it is used to treat bruises, heal urticaria, cure ringworm and joint pain as well as relieving toothache. It is also used in cough and mucus remedies and infusions to cure indigestion, as a topical cure for constipation, as a decongestant and as an aid to lactation. The characteristics and chemical composition of 100 grams of Betel leaves are 44 kcal for energy, 85-90 g of water, 3-3.5 g of protein, 2.3 g of fiber, 0.63-0.89 mg of nicotinic acid, 0.005-0.01 g of vitamin C, 1.9-2.9 mg of Vitamin A, 10-70 μg of thiamine, 1.9-30 μg of riboflavin, 0.1-1.3 g of tannin, 0.05-0.6 g of phosphorus, 1.1-4.6 g of potassium, 0.2-0.5 g of calcium, 0.005-0.007 g of iron, 3.4 μg of iodine (Guha, 2006).

Other biological activities described for the essential oil include antifungal, antiseptic and anthelmintic effects (Evans et al., 1984). It was reported that Betel leaf was rich in carotenes (80 IU/g fresh wt.) and phenolics. Data on the phenolic compounds of this plant have been reported for chavicol (Amonkar, et al., 1986), chavibetol, chavibetol acetate (Rimando, et al., 1986) and eugenol (Nagabhushan, et al., 1989). The major bioactive phenolic compoundsin Thai Betel leaf extracted with ethyl acetate were found to be relatively low in polarity. They are chavicol and two-unknown compounds which show higher polarity than chavicol (Maisuthisakul, 2008). The chemical structures of phenolic compounds found in betel leaf are shown in Fig. 2.

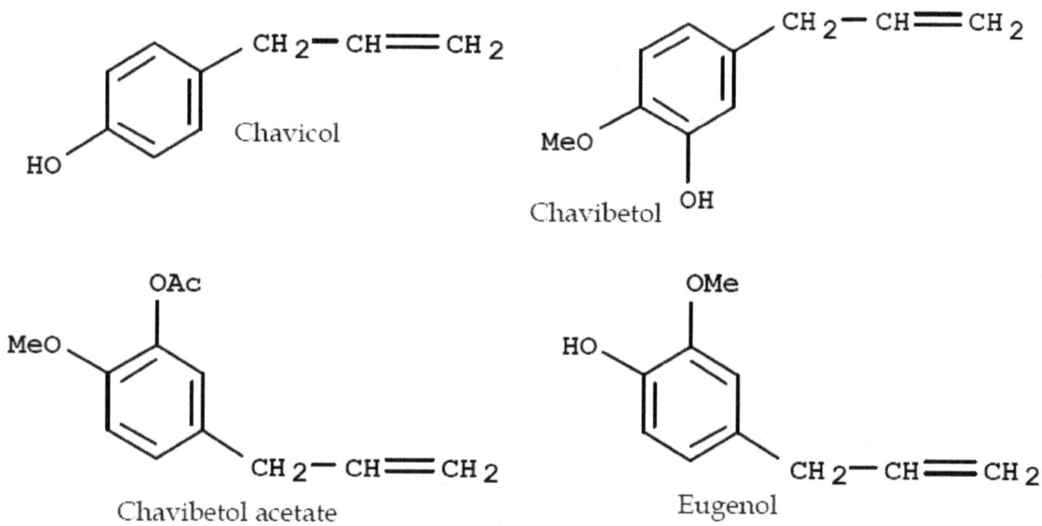

Fig. 2. Some phenolic compounds found in *Piper betel* Linn.

Careya sphaerica Roxb.; This plant is also known as *Careya arborea* Roxb. This plant is normally consumed fresh and mostly found in the North – East of Thailand. Its name is different in different areas, for instance, Kradon, Phak–Kradon, Kradonbok, Kradonkhon (North – East), Khui (Khanchanaburee), Phuk-Pui (North), Puikradon (South), Pui - khao (Chiangmai). Thai people traditionally eat shoots, young leaves and young flowers of this plant. It tastes a little astringent due to the phenolic compounds present. The harvesting season of Kradonbok is during March to May of each year. Kradonbok trees are planted commercially in Sakolnakhon, Kalasin, Yasothorn, Mahasarakham, and Bureerum which are provinces in the North – East of Thailand. Kradonbok has some health benefits such as the use of Kradonbok leaf for healing a wound and flowers for remedying a cough. The characteristics and chemical composition of 100 grams of Kradonbok leaves are 83 Kcal for energy, 1.9 g of fiber, 13 mg of calcium, 18 mg of phosphorus, 17 mg of iron, 3958 IU of riboflavin, 1.8 mg of niacin and 126 mg of vitamin C (Nutrition division, 1992).

There have been few reports about the phenolic compounds of *Careya sphaerica* Roxb. during the last 30 years. Gupta et al. (1975) reported that the phenolic consituents present in the leaf extracts when extracted with petroleum ether at room temperature were lupeol, hexacosanol ($C_{26}H_{54}O$), α-spinosterol ($C_{29}H_{48}O$), teraxerol ($C_{30}H_{50}O$), β-sitosterol ($C_{29}H_{50}O$), quercetin ($C_{15}H_{10}O_7$), taraceryl acetate ($C_{32}H_{52}O_2$) and ellagic acid ($C_{14}H_6O_8$). Careaborin, β- amyrin, careyagenolide, maslinic acid and α-hydroxyursolic acid were also found in the leaves (Das & Mahato, 1982 Das & Mahato, 1982; Talapatra et al., 1981). The chemical structures of some componentsextracted from *Careya sphaerica* Roxb.leaf are shown in Fig. 3.

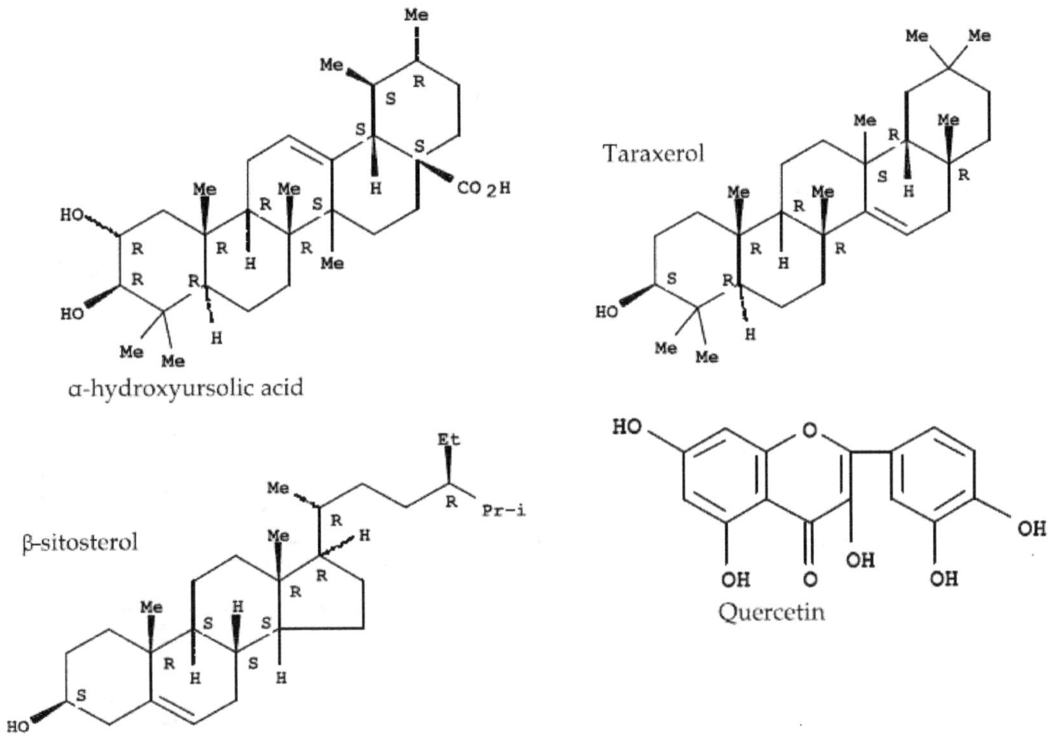

Fig. 3. Some phytochemicals extracted from *Careya sphaerica* **Roxb.**

The HPLC chromatogram of the Kradonbok extract is shown in Fig. 4. The ethanol extract was dissolved in methanol and passed through a Sep-Pak C18 cartridge (Waters, Milford, MA.). The C-18 cartridge was first conditioned by suction with 1 column volume of methanol followed by 2 column volumes of a 3% HCl solution (v/v) in HPLC grade water. The cartridge was not allowed to dry out during conditioning. The aqueous extract was then transferred to the cartridge. The cartridge bed was then rinsed with HCl (3%, 5 mL) and air- dried under vacuum for ~10 min. Phytochemicals were eluted with HPLC grade methanol. Samples were filtered through a 0.20 mm Millipore filter (type HA) into a 2 mL autosampler vial for subsequent analysis by HPLC. The solution (10 mL) was injected into the

HPLC and analyzed according to the following conditions: column, Synergi Hydro RP column (150 × 4.6 mm id., 4mm, Phenomenex), fitted with a Allsphere ODS-2 guard column (10 × 4.6 mm id., Alltech). The HPLC system was equipped with diode array detector (Dionex PDA 100 photodiode array, USA) controlled by Chromeleon software version 6.60 Build 1428 (Dionex Corporation, Sunnyvale, USA). Chromatograms were recorded at 260, 280 and 320 nm.

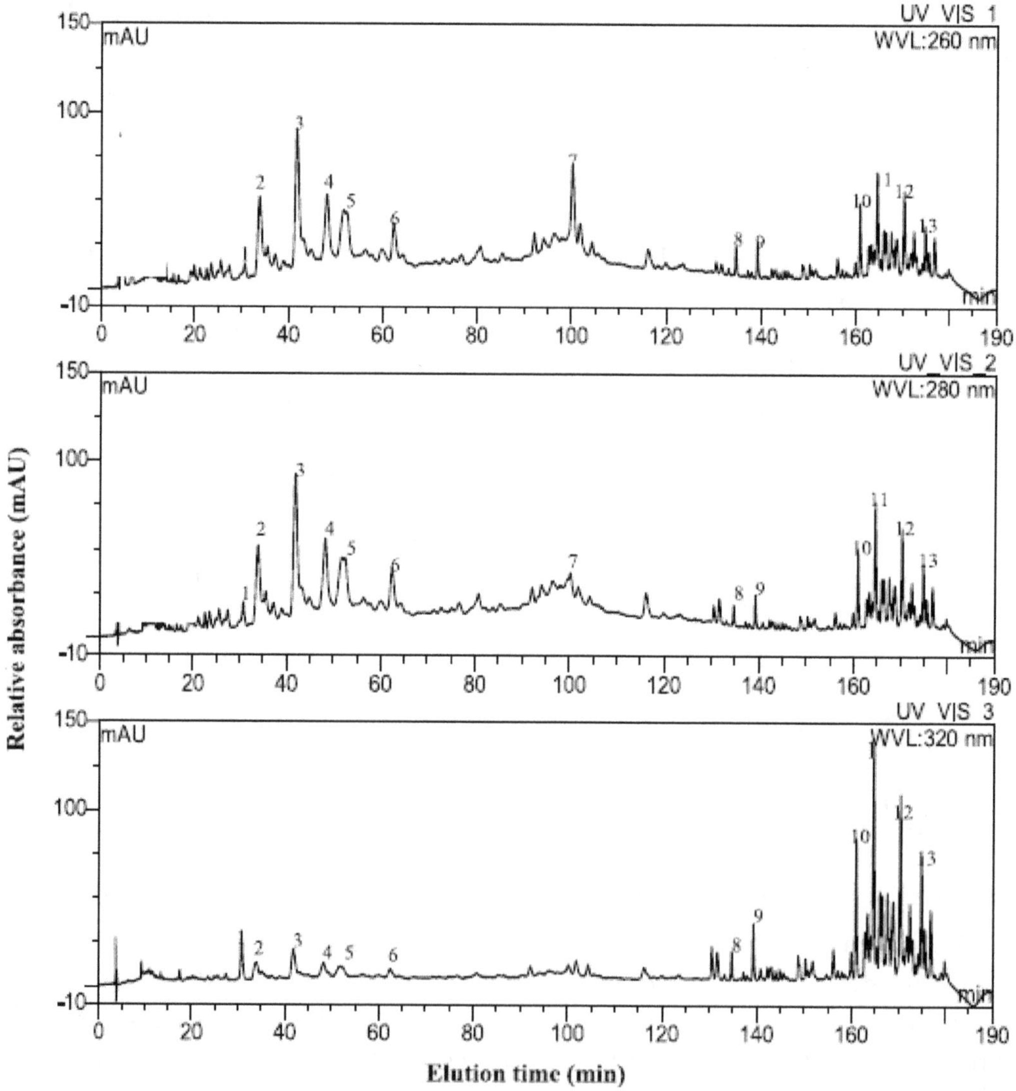

Source: Maisuthisakul (2007)

Fig. 4. HPLC chromatogram for the extract from *Careya sphaerica* Roxb. (Kradonbok) detected at 260, 280 and 320 nm.

Thirteen main peaks were detected in Thai Tummy wood leaf extracted with ethanol and the components corresponding to peaks 1, 2, 3, 4, 5, and 6 eluted at 11.06, 33.91, 41.79, 48.35, 51.91 and 62.36 min in the more hydrophilic region (short retention time). The other main components eluted at 100.19 min (peak 7), 134.71 min (peak 8), 139.27 min (peak 9), 160.93 min (peak 10), 164.63 min (peak 11), 170.44 min (peak 12) and 174.90 min (peak 13) in the more hydrophobic region (long retention time).

The identity and purity of peak 1 in the HPLC chromatogram of the extract of *Careya sphaerica* Roxb. was determined by comparison of the retention time and UV spectrum (Fig. 5) with that of pure gallic acid. The identification of peak 1 as gallic acid was confirmed by co-injection of gallic acid with the plant extract. HPLC-ESI-MS confirmed the identification of this compound (MS $[M + H]^+$ at m/z 171, MS $[M + Na]^+$ at m/z 193 and MS $[M+4Na]^+$ at m/z 262), which is consistent with the molecular weight of 170.12 for gallic acid. Compound 7 was identified by comparison of the retention time and UV spectrum (Fig. 6) with that of pure ellagic acid. The identification of peak 7 as ellagic acid was confirmed by co-injection of ellagic acid with the plant extract. HPLC-ESI-MS confirmed the identification of this compound (MS $[M + H]^+$ at m/z 303, MS $[M + Na]^+$ at m/z 325), confirming the molecular weight of 302.19 for ellagic acid. Talapatra et al. (1981) reported that ellagic acid was isolated from the leaves of *Careya arborea*, which is the synonym of *Careya sphaerica* Roxb.

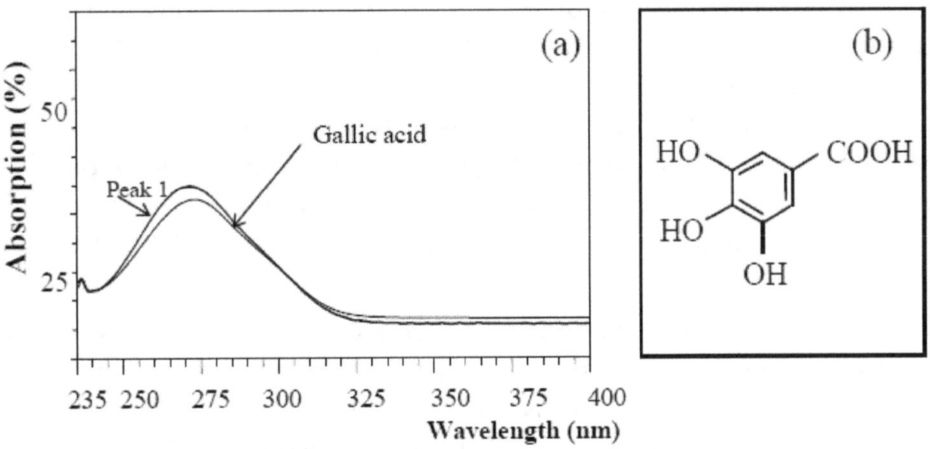

Source: Maisuthisakul (2007)

Fig. 5. UV spectrum of (a) peaks 1 of *Careya sphaerica* Roxb. (Kradonbok) extract, gallic acid and (b) structure of gallic acid.

The known antioxidant components 1 (gallic acid) and 7 (ellagic acid) were present at 0.93 % and 2.37 % of the extract. These concentrations were calculated according to the peak areas from the HPLC chromatogram. The thirteen phenolic compounds represented about 56.16% of the total phenolic compounds of Kradonbok extract.

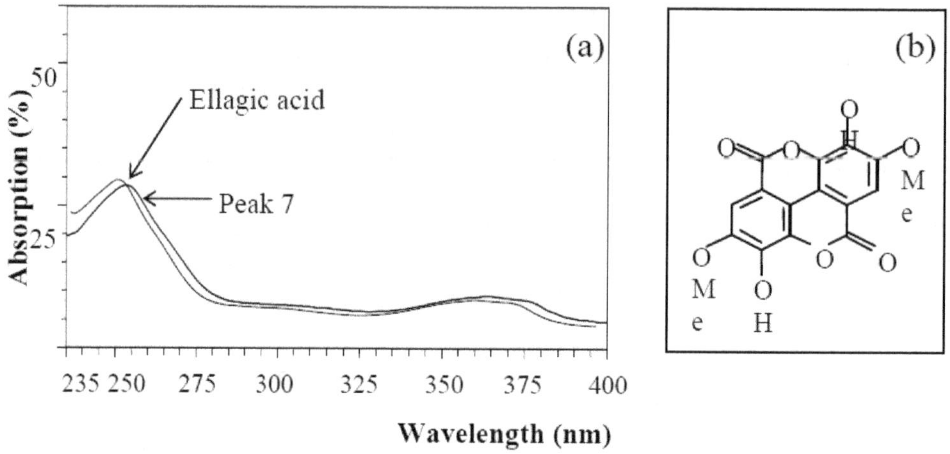

Source: Maisuthisakul (2007)

Fig. 6. UV spectrum of (a) peak 7 of *Careya sphaerica* Roxb. (Kradonbok) extract, ellagic acid and (b) structure of ellagic acid.

Cratoxylum formosum **Dyer.**; The name of this plant differs in different local areas, for instance, Teawkon (Central), Teawdang (North) and Tao (South). Thai people traditionally eat shoots and young leaves of this plant. It tastes sour and a little astringent due to the phenolic phytochemicals present. The harvesting season of *Cratoxylum formosum* Dyer. is during March to May of each year. Trees of this plant are planted commercially in Sakolnakhon, Kalasin, Yasothorn, Mahasarakham, Bureerum, which are the provinces in North - East Thailand. The chemical composition per 100 grams of Teaw leaves which provides 58 Kcal for energy includes 1.5 g of fiber, 67 mg of calcium, 19 mg of phosphorus, 205 mg of iron, 4500 μg of β-carotene, 750 μg vitamin A as retinol, 10.04 mg of thiamin, 0.67 mg of riboflavin, 3.1 mg of niasin and 58 mg of vitamin C (Nutrition division, 1992).

Relatively little work has been done on the phytochemicals in *Cratoxylum* sp. The only studies concerning the phytochemistry of *Cratoxylum* sp. was published by Kitanov & Assenov (1988), and Kumar et al. (2004) who reported that the phenolic compounds in *Cratoxylum pruniflorum* Kurz were quercetin ($C_{15}H_{10}O_{12}$), hyperoside ($C_{21}H_{20}O_{12}$), 1,3,6,7- tetrahydroxyxanthone, mangiferin ($C_{19}H_{18}O_{11}$) and isomangiferin ($C_{19}H_{18}O_{12}$), (Kitanov & Assenov, 1988). The phenolic constituents in *Cratoxylum neriifolium* Kurz were biflavonol GB-2, pentahydroxyflavanone chromone and stigmasterol (Kumar et al., 2004). The chemical structures of some phytochemicals found in *Cratoxylum* sp. leaf are shown in Fig. 7.

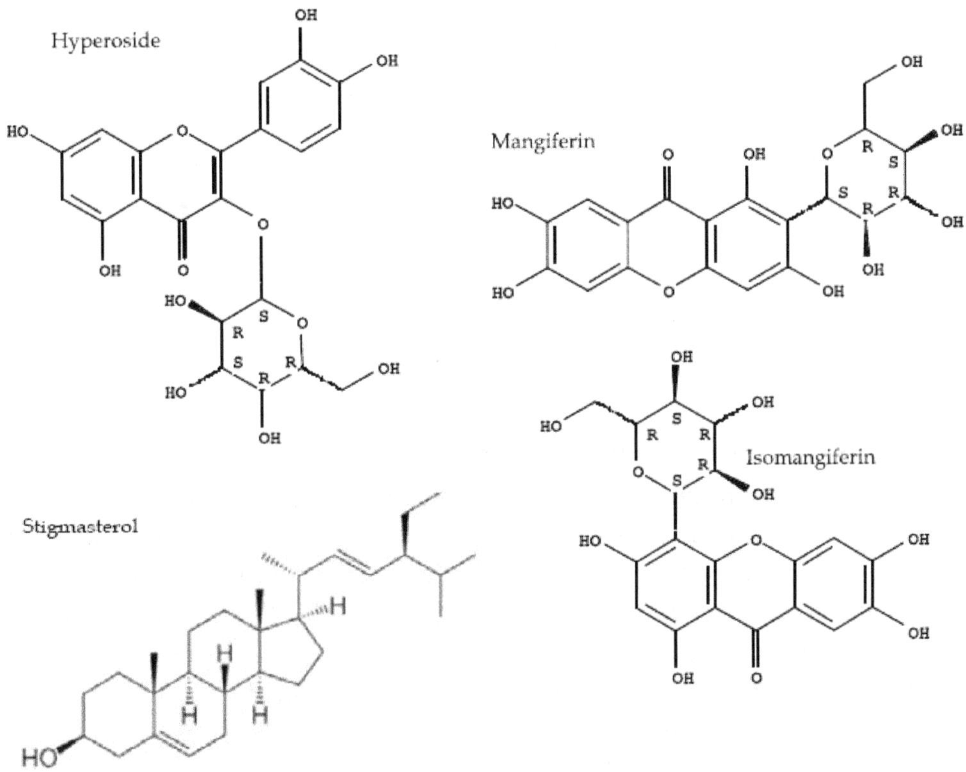

Fig. 7. Some phytochemicals found in *Cratoxylum* sp.

With regards to Thai plants, three active phenolic ingredients were chlorogenic acid, dicaffeoylquinic acid and ferulic acid hexose derivative (Maisuthisakul et al., 2007b). Chlorogenic acid was present at 60 % of the extract. Two minor components (dicaffeoylquinic acid and ferulic acid hexose derivative) were present at 7 % and 2 %, and other components that were present at lower concentrations were also detected. Some chemical structures of phenolic compounds found in Thai *Cratoxylum formosum* leaf are shown in Fig. 8.

Leucaena glauca **Benth**; the plant is found throughout Thailand in the settled areas at low and medium altitudes. It occurs widely and is abundant. Its name is different in different areas, for instance, Kratin-Thai (Central), Satorban (South), Katong and Kratin. Thai people traditionally eat young leaves and the young pod of this plant. The young leaf is found in all seasons, however, it is most abundant during March to May of each year. Kratin trees are planted commercially in Roi-ed, Amnat Charoen, Pichit, Nakhonsawan, Songkla, Krabi, Pattani and Trang. Kratin leaves contain leucine which can absorb selenium. The charcateristics and chemical composition of 100 grams of Kratin leaves are 62 Kcal for energy, 8.4 g of protein, 3.8 g of crude fiber, 137 mg of calcium, 11 mg of phosphorus, 9.2 g of iron, 7883 IU of total vitamin A, 0.33 mg of thiamin, 0.09 mg riboflavin, 1.7 mg of niacin and 8 mg of vitamin C (Nutrition division, 1992).

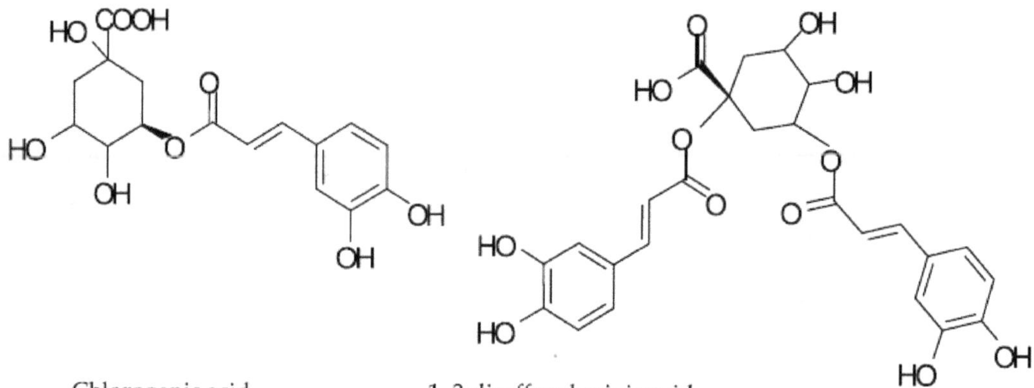

Chlorogenic acid 1, 3 dicaffeoylquinic acid

Fig. 8. Some phenolic compounds found in *Cratoxylum formosum* in Thailand.

Relatively little information on the phenolic constituents of *Leucaena glauca* Benth has been published during the last 30 years. Chen (1979) found foeniculin ($C_{14}H_{18}O$) and kaempferol- 3-xyloside ($C_{20}H_{18}O_{10}$) in the leaves. Guaijaverin ($C_{20}H_{18}O_{11}$), juglania ($C_{20}H_{18}O_{10}$), kaempferol-3-O-β-xyloside and quercitrin ($C_{21}H_{20}O_{11}$) were also found in the leaves (Morita et al., 1977). The chemical structures of phenolic compounds found in Thai *Leucaena glauca* Benth leaf are shown in Fig. 9.

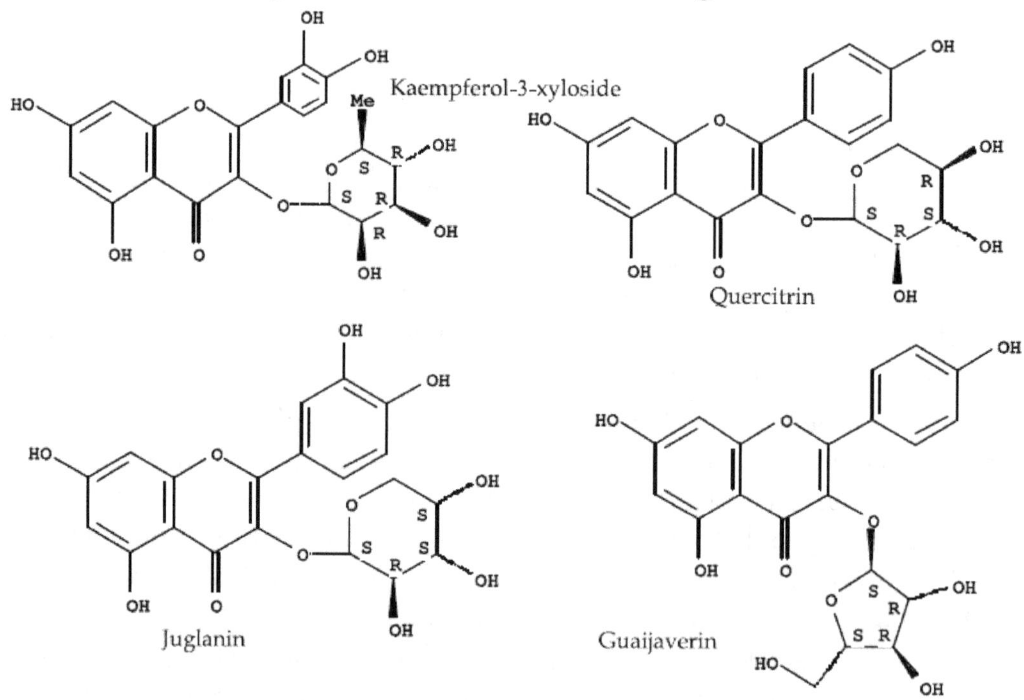

Kaempferol-3-xyloside Quercitrin Juglanin Guaijaverin

Fig. 9. Some phenolic compounds found in *Leucaena glauca* Benth

The phenolic components of Leucaena glauca Benth in the HPLC chromatogram appeared as 4 peaks which were present at 15.28%, 6.50%, 9.71% and 36.91%, respectively, and other components that were present at lower concentrations were also observed. These concentrations were calculated from each area by dividing by the total peak area from the HPLC chromatogram. The four phenolic compounds were about 68.4% of the total phenolic content of Leucaena glauca Benth extract. The compounds corresponding to peaks 1, 2, 3, and 4 had a similar UV spectrum to that of flavonoids such as quercetin, and kaempferol which are flavonols and luteolin which is a flavone (Fig. 10). The identification of each peak was confirmed by using comparisons of their UV spectra and LC-MS in both positive and negative mode in order to obtain more information on the structural features of the conjugated forms of the phenolic compounds.

3. ANTIOXIDANT PROPERTIES OF SOME THAI PLANTS

In general, the antioxidant activity of plant extracts is associated with specific compounds or classes of compounds, such as flavones, flavonols and proanthocyanidins in plant materials native to the Mediterranean area (Skerget et al., 2005), carotenoids (Stahl & Sies, 2003) and melatonin (Chen et al., 2003). Most of the antioxidant substances in plants are phenolic compounds. Phenolic compounds serve as oxidation terminators by scavenging radicals to form resonance stabilized radicals (Rice-Evans et al., 1997).

Although the antioxidant capacities are influenced by many factors, which cannot be fully described with a single method, the DPPH radical scavenging activity is the most commonly used method for assessment of antioxidant properties of natural products. The DPPH• assay overcomes the limitations of monitoring the activity of the numerous samples over a specified period of time. It is reproducible and strongly correlated with phenolic compounds (Maisuthisakul et al., 2007; Katalinic et al., 2006; Miliauskas et al., 2004; Matsuda et al., 2001). In addition the radical scavenging method had many benefits compared to the lipid oxidation method (Nuutila et al., 2002). Gallic acid, which has been used as a standard, has been reported to be the most abundant phenolic compound in plants (Witzell et al., 2003., Nuutila et al., 2002). The EC_{50} is defined as the amount of antioxidant required to cause a 50% reduction in the absorbance of DPPH. These values were changed to antiradical activity defined as $1/EC_{50}$, since this parameter increases with antioxidant activity. The antiradical activity of α-tocopherol was 0.67 (Maisuthisakul et al., 2008). Many researchers have reported the total phenolic content of Thai plants as shown in Table 2. The plants which had strong antiradical activity (Table 2) had high total phenolic and flavonoid contents (Table 1).

In Thailand, there are many plants which show antiradical activity higher than α-tocopherol such as *Careya sphaerica* Roxb., *Cratoxylum formosum* Dyer., *Erythrina crista* Galli., *Leucaena glauca* Benth, *Momordica charantia* Linn., *Ocimum basilicum* Linn., *Ocimum sanctum* Linn., *Syzygium gratum* (Wight) S.N.Mitra var. gratum,

Allium ascalonicum Linn., *Azadirachta indeca* A. Juss Var. siamensis valeton, *Cassia siamea* Britt., *Musa spiantum* Linn., *Sesbania grandiflora* Desv. The antiradical activity of Thai fruits and fruit seeds were higher than that of α-tocopherol except for seeds of *Leucaena glauca* Benth (Table 2).

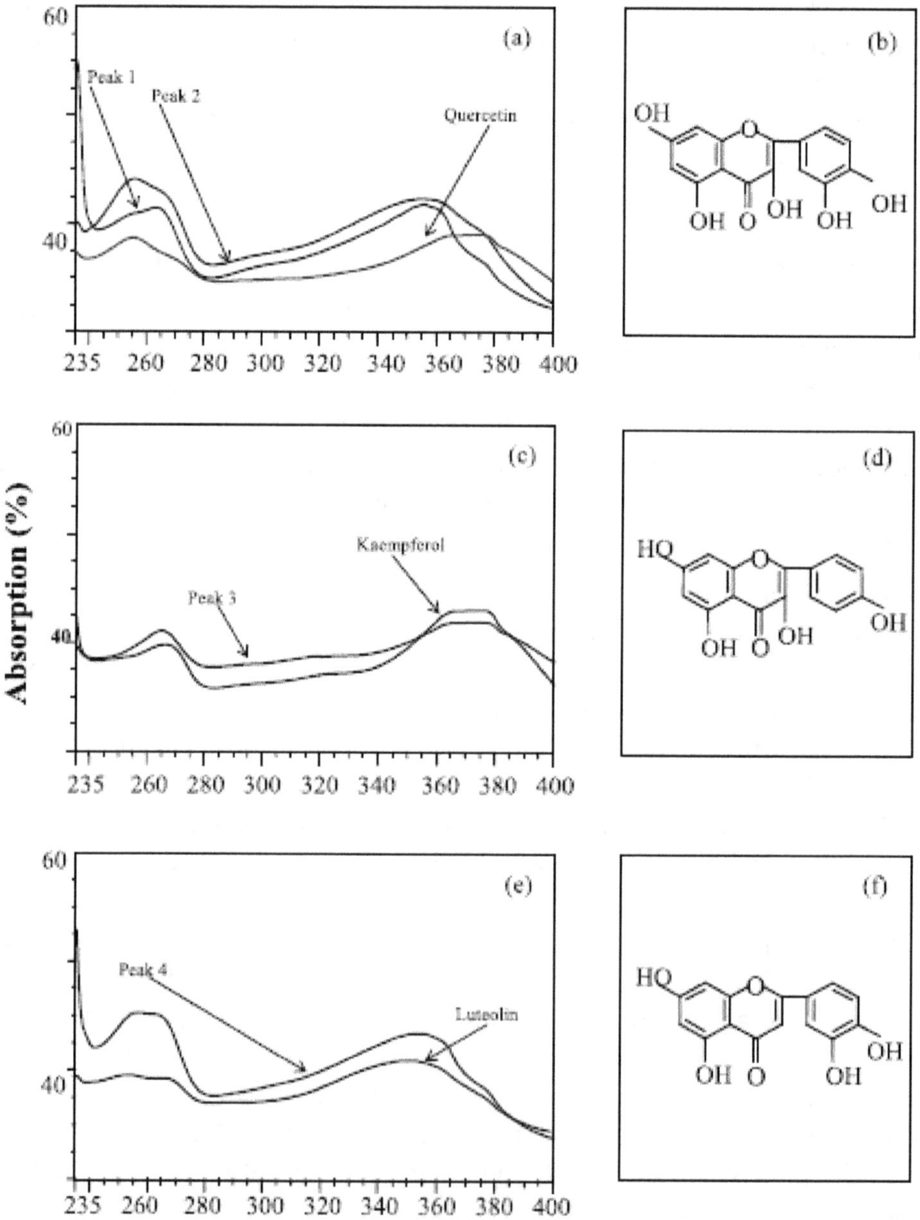

Fig. 10. UV spectrum of (a) peaks 1, 2 of *Leucaena glauca* Benth (Kratin) extract, quercetin, (b) structure of quercetin, (c) UV spectrum of peaks 3 of Kratin extract, kaempferol, (d) structure of kaempferol, (e) UV spectrum of peaks 4 of Kratin extract, luteolin and (f) structure of luteolin.

Table 2. Ranges of antiradical activity in some Thai plants

Scientific name	Local name	Plant part	Antiradical activity (1/ EC50)
	Herb and vegetable		
Allium ascalonicum Linn. ¥	Hom	Flower	2.6
Artemisia dubia Wall. ex DC. (Syn. *A. vulgaris* L. var. *indica* Maxim.)⁂	Hia	tem and leaves	0.19
Aspidistra sutepensis K. Larsen ⁂	Nang-laeo	Flower	0.06
Azadirachta indeca A. Juss Var. . siamensis valeton¥	Sa-dao	Flower	1.2
Basella alba Linn. ¥	Plang	Leaf	0.7
Bidens bipinnata , L. ⁂	Ya-Puen-Laem Nok-Sai	Stem and leaves	1.04
Bidens pilosa Linn. ⁂	Peen-nok-sai	Stem and leaves	1.22
Buddieia asiatica Lour ⁂	Ra-cha-wa-di-pa	Stem and leaves	0.26
Careya sphaerica Roxb. ¥	Kra-don	Young leaf and leaf	2.3
Cassia siamea Britt. ¥	Kee-lek	Flower	2.4
Centella asiatica Linn. ¥	Bua-bok	Leaf	0.7
Cratoxylum formosum Dyer. ¥	Tew	Young leaf and leaf	4.4
Commelina diffusa Burm.f. ⁂	Plap	Stem and leaves	0.64
Conyza sumatrensis (Retz.) Walker ⁂	Ya-khamai	Stem and leaves	0.46
Cuscuta australis R. Br. ⁂	Khruea-kham	Stem	1.08
Diplazium esculentum (Retz.) Sw. ⁂	Kut-khao	Stem and leaves	0.63
Dolichandrone serrulata (DC.) Seem. ⁂	Khae-pa	Flower	0.06
Embelia ribes Burm.f. ⁂	Som-jee	Leaves	2.86

Embelia sessiliflora Kurtz ж	Som-kui	Leaves	5.0
Erythrina crista Galli. ¥	Tong-lang	Leaf	3.1
Hydrocharis dubia (Bl.) Back.*	Tub-tao	Bud	0.82
Lasia spinosa (Linn.) Thw. ¥	Nham	Leaf	0.1
Leucaena glauca Benth¥	Kra-tin	Young leaf and leaf	1.5
Limnocharis flava Buch. ¥	Pai	Leaf	0.1
Melicope pteleifolia (Champ.ex Benth.) ж	Sa- Riam -Dong	Hartley Leaves	0.18
Micromelum minutum Wight & Arn ж	Sa-mui	Stem and leaves	0.83
Momordica charantia Linn.*	Ma-ra-khee-nok	Bud and leaf	1.70
Musa spiantum Linn. ¥	Hua-plee	Flower	1.8
Ocimum basilicum Linn. ¥	Ho-ra-pa	Young leaf and leaf	1.80
Ocimum sanctum Linn. ¥	Ka-prow	Young leaf and leaf	1.8
Sauropus androgynus Linn. ¥	Whan-ban	Young leaf and leaf	0.7
Schima wallichii (DC.) Korth. ж	Talo	Leaves	12.5
Sesbania grandiflora Desv. ¥	Kae	Flower	1.7
Spondias pinnata Kurz. ¥	Ma-kok	Young leaf and leaf	0.7
Stachytarpheta jamaicensis (L.) Vahl (*S.indica* Pun-ngu-keaw Vahl)		Leaf	0.016
		Stem	0.018
		Root	0.018
		Inflorescence	0.020

Syzygium gratum (Wight) S.N.Mitra var. gratum¥	Mek	Young leaf and leaf	1.8
Tiliacora triandra (Colebr.) Diels ж	Ya-nang	Leaves	0.15
Vaccinium sprengelii (G.Don) Sleum. ж	Som-pi	Leaves	1.89
	Berries and fruits		
Capsicum frutescus Linn. ¥	Prik	Fruit	1.8
Scientific name	Local name	Plant part	Antiradical activity (1/ EC50)
Eugenia siamensis Craib. ¥	Chom-pu-nam	Fruit	5.0
Euginia malaccenses Linn.	Chom-pu-ma- meaw	Fruit	2.2
Momordica charantia Linn. ¥	Mara-khee-nok	Fruit	1.7
Phyllanthus emblica¥	Ma-kham-pom	Fruit	2.0
Spondias pinnata Kurz. ¥	Ma-kok	Fruit	1.6
	Seeds		
Antidesma velutinum Tulas	Ma-mao	Seed	14.28
Cleistocalyx operculatus var. paniala (Roxb.)*c	Ma-kieng	Seed	11.11
Eugenia siamensis Craib	Chom-pu-nam	Seed	6.67
Leucaena glauca Benth	Kra-tin	Seed	0.14
Nephelium lappaceum Linn.	Ngo	Seed	2.2
Parkia speciosa Hassk. ¥	Sa-tor	Seed	1.5
Piper nigrum Linn. ¥	Prik-Thai-dum	Seed	3.0
Tamarindus indica Linn¥	Ma-kham	Seed	2.0

Chewing plants			
Acacia catechu (L.F) Wild.*	See-sead	Bark	20.0
Areca catechu Linn.*	Mhak	Whole fruit	2.13
		2.13	5.56
Cassia fistula Linn.*	Kaen-khun	Stem core	6.25
Piper betel Linn*	Bai-plu	Leaf	3.13

Source: Adapted from ¥Maisuthisakul et al. (2008); *Maisuthisakul et al., 2007a; €Lee & Scagel, 2009;✪ Kerdchoechuen & Laohakunjit, 2010; ✡Ongard & Dara, 2010; ✂Phomkaivon & Areekul (2009)- means not reported or not determined.

4. FACTORS AFFECTING EXTRACTION OF PLANT PHENOLICS

Natural antioxidants are available from raw materials of variable composition. Both the content of active substances and the content of various other compounds may vary. The quality of natural extracts and their antioxidative activity depends not only on the quality of the original plant, date and storage, but also the extraction conditions which affect the plant phenolic compounds extracted (Moure et al., 2001).

4.1 Sample preparation for storage

Sample preparation is required to keep samples for a certain period of time before analysis since even the total phenolic content was shown to decrease during storage compared with the fresh leaves. In addition, the availability of raw materials is usually limited to a harvesting season. The preliminary processing of plants is necessary for storage.

Temperature is the major factor influencing changes in antioxidant activity during storage (Moure et al., 2001). From numerous publications, the preservative processes before storage vary such as drying citrus peel and seed in an oven at 40 ºC and keeping at room temperature (Bocco et al., 1998) or drying Indian Laburnum by air at 25ºC and keeping at room temperature (Siddhuraju et al., 2002) or keeping berries in a still air freezer at -25 to-30ºC (Amakura et al., 2000).

Sample preparation methods gave significant effects on total phenolic content and antioxidant activity but had no marked effects on yield of the extract of *Careya sphaerica* or Kradonbok (Maisuthisakul & Pongsawatmanit, 2005). Freezing, especially fast freezing, for keeping the leaves until extraction gave a higher total phenolic content compared with those obtained from other drying methods and slow freezing (Fig. 11). The drying methods studied were (1) hot air drying by tray dryer at 40°C for 18 h with air velocity about 0.5 m/s; (2) vacuum drying by vacuum dryer at 40°C, 100 mmHg (EYELA, model VOS-300SD, Japan) for 10 h; (3) air drying at room temperature 25°C for 12 h with air velocity about 3.2 m/s. The rate of freezing affected the total phenolic content obtained because larger ice crystals grew during slower freezing which would damage plant cells and cause a loss of antioxidant activity. Enzymes from plant cells such as lipoxygenase can oxidize polyphenols (Akoh & Min, 1997). The total phenolic content obtained from air drying was higher than those obtained from vacuum drying and hot air drying because some phenolic components may be degraded by higher temperature (Moure et al., 2001). This effect also found in fresh Mulberry leaves, where the amount of flavonoids was higher in air-dried samples than that in oven-dried samples, probably due to decomposition after storage (Zhishen et al., 1999).

Normally, temperature affects the compounds' stability due to chemical and enzymatic degradation. These mechanisms were reported as mainly responsible for a reduction in phenolic content (Larrauri et al., 1997). Maisuthisakul & Pongsawatmanit also reported that the reduction in antioxidant activity was higher than that expected from the reduction in phenolic contents, probably due to the synergistic effect of natural phenolics (Moure et al., 2001). In addition, phenolics can react with other plant components, and prolonged exposure at moderate temperatures can also cause phenolic degradation. Therefore, sample preparation conditions including temperature and time before storage should be controlled.

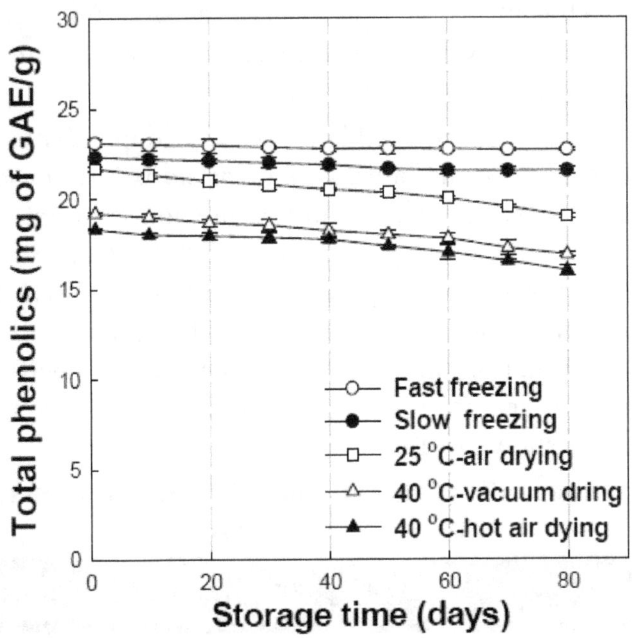

Source: Adapted from Maisuthisakul & Pongsawatmanit (2005)

Fig. 11. Total phenolic content of the dried extracts from Careya sphaerica leaves obtained by various sample preparation methods.

4.2 Extraction conditions

Sovent effect; solvent extraction is more frequently used for isolation of antioxidants and both extraction yield and antioxidant activity of extracts are strongly dependent on the solvent, due to the different antioxidant potential of compounds with different polarity (Marinova & Yanishlieva, 1997). Apolar solvents are among the most common solvents for removing polyphenols from water. Ethyl acetate and diethyl ether have been used for extraction of low molecular weight phenols from oak wood (Fernández de Simón et al., 1996). Ethanol and water are the most widely employed solvents for reasons of lack of toxicity and abundance, respectively.

With regards to Thai plants, Maisuthisakul (2007) reported that the solvent used had significant effects on antioxidant activity, total phenolic content, yield and partition coefficient of Thai betel leaf extract ($p < 0.05$). The antioxidant activity assessed by DPPH and ABTS radicals was stronger with less polar solvents (Table 3). The results showed that the DPPH activity of the extract obtained with ethyl acetate was significantly higher than that obtained with the other solvents. Total phenolic content confirmed this finding. The extract with a solvent which has higher polarity was found to contain rather small amounts of phenolic compounds. The EC50 and TEAC value of α- tocopherol was also measured and gave values of 14.95±0.23 μg.mL-1 and 2.30±0.03 mmol of Trolox/g sample, respectively. Betel leaf extracted with ethyl acetate (Table 3) had a weak antioxidant activity compared to α- tocopherol. The antioxidant activity and total phenolic contents were significantly different from various solvent extractions. The antioxidant activity values were consistent with those of Chen et al. (2001).

The effectiveness of phenolic antioxidants is often dependent on their polarity. Decker (1998) used the term "antioxidant paradox" to describe how polar antioxidants are most effective in bulk lipids while nonpolar antioxidants are most effective in dispersed lipids. The polarity of the Betel leaf extract was assessed by determination of the oil-water partition coefficient by HPLC. The oil-water partition coefficient was calculated by summing the areas of the three phenolic peaks in the HPLC chromatogram. The oil-water partition coefficient of the Betel leaf extract from ethyl acetate was significantly different from those extracted with other solvents as shown in Table 4.

Table 3. Antioxidant activity and total phenolic content of Betel leaf extracted by different solvents¥

Solvent used	DPPH activity (EC_{50}, μg.mL-1)	TEAC (mmol Trolox/g sample)	Total phenolic content (mg GAE/g sample)
Methanol	36.65 ± 2.60^{a}	2.01 ± 0.06^{c}	49.89 ± 0.21^{d}
Ethanol	33.85 ± 2.81^{ab}	2.12 ± 0.03^{bc}	50.38 ± 0.08^{c}
Acetone	30.09 ± 1.21^{b}	2.21 ± 0.04^{ab}	53.28 ± 0.19^{b}
Ethyl acetate	17.04 ± 0.51^{c}	2.34 ± 0.06^{a}	55.35 ± 0.14^{a}

Note: ¥ Data followed by different letters within each column are significantly different according to Duncan's multiple range tests at $P < 0.05$. Data were represented as means from three replicate measurements.

Source: Maisuthisakul (2007)

Table 4. Partition coefficient of solvent used and Betel leaf extract extracted by different solvents¥

Solvent used	Partition coefficient of solvent#	Partition coefficient of extract
Methanol	-0.77	2.02± 0.01[a]
Ethanol	-0.32	2.09± 0.02[ab]
Acetone	-0.24	2.15± 0.01[b]
Ethyl acetate	0.66	2.31± 0.03[c]

Note: ¥ Data followed by different letters within each column are significantly different according to Duncan's multiple range test at $P < 0.05$. Data were represented as means from three replicate measurements.

Data obtained from literature review.

Source: Maisuthisakul (2007)

The phenolic compounds in Betel leaf from the literature include low polarity compounds such as chavicol (log P = 2.50±0.21), chavibetol (log P = 2.30±0.23), chavibetol acetate (log P = 2.39±0.24), and eugenol (log P = 2.20±0.23). Normally, a frequently used descriptor for the estimation of the lipophilicity of phenolic compounds is the partition coefficient. The partition coefficients of Betel leaf extracts were those of low polarity compounds since the value is higher than 0 (Munishwar, et al., 1997) (Table 4). Less polar solvents showed higher extraction efficacy due to the low polarity of the phenolic compounds of Betel leaf extract. Compounds, which have a polarity similar to the solvent, are able to dissolve more than compounds with different polarity. It can be noted here that Betel leaf extract is rich in less polar phenolic compounds. The solvent used for extraction also affected the total phenolic content in the extracts.

The ratios of solvents used in mixed solvents affected the phenolic content extracted from some Thai plants. *Careya sphaerica* Roxb. (Kradonbok), *Cratoxylum formosum* Dyer. (Teaw) and *Sauropus andrugynus* Merr. (Phak whan ban) leaves were used to study the effect of ethanol concentration used in the extraction. The total phenolic content of plant extracts readily increased with increasing concentration of ethanol from 80% to 95%, but there was no significant difference between the extracts using ethanol concentrations of 95% and 99% for each plant. The effect of ethanol concentration on antioxidant activity and total phenolic content was similar (Fig. 12).

The inflorescence, leaves, stem and root of *Stachytarpheta indica* Vahl were extracted with three methanol and water solvent mixtures, namely water, 75% methanol and 50% methanol. The results showed that the leaf extract from 75% methanol had the highest antioxidant activity in both fresh and dry samples (Ongard & Dara, 2010).

Extraction temperature effect; the temperature of extraction affects the compounds' stability due to the decomposition of phenolic compounds. The effect of temperature has been studied in the extraction of anthocyanins. They were shown to be degraded since the visible spectrum showed both a reduction in the peak at 400-500 nm and reduction in the red color. The temperature during extraction can affect extractable compounds to different extents; boiling and resting increases the total phenol content extracted from *Quercus suber* cork (Conde et al., 1998). Milder extraction temperatures are desirable in those cases where some compounds can be degraded, e.g. carnosic acid (Ibañez et al., 1999).

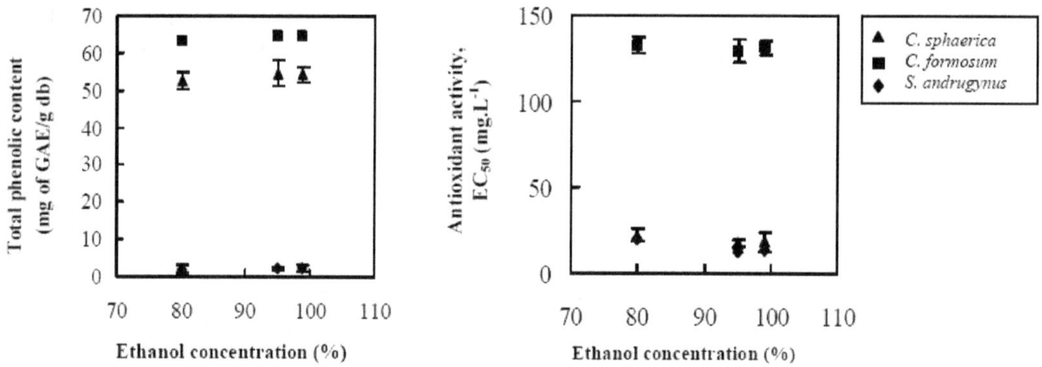

Source: Adapted from Maisuthisakul (2007)

Fig. 12. Total phenolic contents and antioxidant activity of the extracts from *Careya sphaerica* **Roxb.,** *Cratoxylum formosum* **Dyer. and** *Sauropus andrugynus* **Merr. leaves obtained with various ethanol concentrations.**

Other factors; such as extraction time, pH, the particle size of materials, and extraction methods were reported to affect the antioxidant activity and concentrations of phenolic compounds extracted. Sheabar & Neeman (1988) reported the maximum solubility of phenolic compounds from olive rape at pH 4 in the organic phase. The yield of extracted phenolics was correlated with plant cell wall breakdown. Particle size reduction significantly increased the antioxidant activity as a result of both increased extractability and enhanced enzymatic degradation of polysaccharides (Weinberg et al., 1999). Various process conditions (refluxing, shaking and ultrasonic extraction) also affected the concentrations of antioxidants in extracts from balm leaves (Herodež et al., 2003).

5. CONCISE SUMMARY

Numerous plants in Thailand exhibit potential as sources of extracts abundant in phenolic compounds and natural antioxidants. Phenolic molecules serve as the primary antioxidants in plants. Additionally, practical factors pertinent to the utilization of this class of chemicals must be evaluated, including extraction efficiency, availability of adequate raw materials, and toxicity or safety concerns. Further characterisation of the phenolic content is required to effectively exploit these substantial sources of natural antioxidants.

6. REFERENCES

Amakura, Y.; Umino, Y.; Tsuji, S. & Tonogai, Y. (2000). Influence of jam processing on the radical scavenging activity and phenolic content in berries. *Journal of Agricultural and Food Chemistry,* Vol. 48, pp 6292-6297.

Amarowicz, R.; Pegg, R. B.; Rahimi-Moghaddam, P.; Barl, B. & Weil, J. A. (2004). Free- radical scavenging capacity and antioxidant activity of selected plant species from the Canadian prairies. *Food Chemistry* Vol.84, pp 551-562.

Amonkar, A. J. ; Nagabhushan, M. ; D'Souza, A. V. & Bhide, S. V. (1986). Hydroxychavicol: a new phenolic antimutagen from Betel leaf. *Food and Chemical Toxicology,* Vol. 24, Number 12, pp 1321-1324.

Auddy, B.; Ferreira, M.; Blasina, F.; Lafon, L.; Arredondo, F.; Dajas, F.; Tripathi, P.C.; Seal, T. & Mukherjee, B. (2003). Screening of antioxidant activity of three Indian medicinal plants, traditionally used for the management of neurodegenerative diseases. *Journal of ethnopharmacology*, Vol. 84, Number (2-3), pp 131-8.

Bocco, A.; Cuvelier, M. E.; Richard, H. & Berset, C. (1998). Antioxidant activity and phenolic composition of citrus peel and seed extracts. *Journal of Agricultural and Food Chemistry*, Vol. 46, pp 2123-2129.

Bonvehí, J. S.; Torrent, M. S. & Lorente, E. C. (2001). Evaluation of polyphenolic and flavonoid compounds in honey-bee collected pollen produced in spain. *Journal of Agricultural and Food Chemistry, Vol.* 49, pp 1848-1853.

Chanwitheesuk, A.; Teerawutgulrag, A. & Rakariyatham, N. (2005). Screening of antioxidants activity and antioxidant compounds of some edible plants of Thailand. *Food Chemistry*, Vol. 92, Number 3, pp 491-7.

Chen, G.; Huo, Y.; Tan, D.X.; Liang, Z.; Zhang, W. & Zhang, Y. (2001). Melatonin in Chinese medicinal herbs. *Life Science*, Vol. 73, Number 1, pp 19-26.

Conde, E.; Cadahía, E.; García-Vallejo, M.C. & Fernández de Simón, B. (1998). Polyphenolic composition of *Quercus suber* cork from different Spanish provenances, *Journal of Agricultural and Food Chemistry,*Vol. 46, Number 8, pp 3166–3171.

Das, M. C. & Mahato, S. B. (1982). Triterpenoid sapogenols from the leaves of *Careya arborea*: structure of careyagenolide. *Phytochemistry*, Vol. 21, Number 8, pp 2069-2073.

Demo, A.; petrakis, C.; Kefalas, P. & Boskou, D. (1998). Nutrient antioxidants in some herbs and Mediteranean plant leaves. *Food Research International*, Vol. 31, pp 351-354.

Dillard, C.J. & German, J.B. (2000). Review phytochemicals : nutraceuticals and human health. *Journal of Science, Food and Agriculture*, Vol. 80, pp 1744-1756.

Evans, P. H.; Bowers, W. S. & Funk, E. J. (1984). Identification of fungicidal and nematocidal components in the leaves of Piper Betel (*Piperaceae*). *Journal of Agricultural and Food Chemistry*, Vol. 32, pp1254-1256.

Fernández de Simón, B.; Cadahía, E., Conde, E. & García-Vallejo, M.C. (1996). Low molecular weight phenolic compounds in spanish oak woods. *Journal of Agricultural and Food Chemistry*, Vol. 44, pp 1507–1511.

Guha P. (2006). Betel leaf: the neglected green gold of India. *Journal of Hum Ecology*, Vol.19, pp 87–93.

Gupta, R. K.; Chakraboty, N. K. & Dutta, T. R. (1975). Crystalline constituents from *Careya arborea* Roxb. leaves. *Indian Journal of Pharmacy*, Vol. 37, Number 6, pp 161-162.

Habsah, M.; Amran, M.; Mackeen, M.M.; Lajis,N.H.; Kikuzaki, H.; Nakatani, N.; Rahman, A.A.; Gahfar, A. & Ali, A.M. (2000). Screening of *Zingiberaceae* extracts for antimicrobial and antioxidant activities. *Journal of ethnopharmacology*, Vol. 72, pp 403-410.

Helrich, K. (1990). *Official methods of analysis of the association of official analytical chemist*. 15th ed. Virginia: Association of Official Analytical Chemists. pp. 703, 1048–1049.

Herodež, S.S.; Hadolin, M.; Škerget, M. & Knez, Z. (2003). Solvent extraction study of antioxidants from Balm (*Melissa officinalis* L.) leaves. Food Chemistry, Vol. 80, pp 275–282.

Ho, C.T. (1992). Phenolic compounds in food. In *Phenolic Compounds in Food and their Effects on Health I*, ed. C. T. Ho, C. Y. Lee & M-T. Huang. ACS Symposium series 506, Washington, DC. pp. 2–7.

Huang, H & Ferraro, T. (1992). Phenolic compounds in food and cancer prevention. In: M.J. Huang, C. Ho & C.Y. Lee, Editors, *Phenolic Compounds in Food and Their Effects on Health. II. Antioxidants and Cancer Prevention*. American Chemical Society, Washington, DC. pp. 8–34.

Ibáñez E.; Oca, A.; Murga, G.; López-Sebastián, A.; Tabera, J. & Reglero, G. (1999). Supercritical fluid extraction and fractionation of different preprocessed rosemary plants. *Journal of Agricultural and Food Chemistry*, Vol. 47, pp 1400–1404.

Kähkonen M.P.; Hopia, A.I.; Vuorela, H.J.; Rauha, J.P.; Pihlaja, K.; Kujala, T.S. & Heinonen, M. (1999). Antioxidant activity of plant extract containing phenolic compounds. *Journal of Agricultural and Food Chemistry*, Vol. 47, pp 3954-3962.

Katalinic, V.; Milos, M.; Kulisic, T. & Jukic, M. (2006). Screening of 70 medicinal plant extracts for antioxidant capacity and total phenols. *Food Chemistry*, Vol. 94, pp 550- 557.

Kerdchoechuen, O. & Laohakunjit, N. (2010). Antioxidant Capacity and Total Phenolics of Plant Extract from Annual Seablite (*Suaeda maritima*). *Agricultural Science Journal*, Vol 41, Number (3/1)(Suppl.), pp 621-624.

Kitanov, G. & Assenov. I. (1988). Flavonols and xanthones from *Cratoxylum pruniflorum* Kurz. (Guttiferae). *Pharmazie*, Vol. 43, Number 12, pp 879-880. (Abstract).

Kumar, V.; Brecht, V. & Frahm, A. W. (2004). Conformational analysis of the biflavonoids GB2 and a polyhydroxylated flavanone-chromone of *Cratoxylum nerifolium*. *Planta Medica*, Vol. 70, Number 7, pp 646-651. (Abstract).

Larrauri, J.A.; Ruperez, P. & Saura-Calixto, F. (1997). Effect of drying temperature on the stability of polyphenols and antioxidant activity of red grape pomace peels. *Journal of Agricultural and Food Chemistry*, Vol. 45, pp 1390–1393.

Lee, J. & Scagel, C.F. (2009). Chicoric acid found in basil (*Ocimum basilicum* L.) leaves. *Food Chemistry*, Vol. 115, pp 650-656.

Leong, L.P. & Shui, G. (2002). An investigation of antioxidant capacity of fruits in Singapore markets. *Food Chemistry*, Vol. 76, pp 69-75.

Lindsay, R.C. (1996). *Food additives.* Marcel Dekker Inc., New York. pp. 778–780. Maisuthisakul, P. (2007). *Evaluation of Antioxidant Potential and Phenolic con stituents in Selected Thai Indigenous Plant Extracts.* Dissertation. Kasetsart University. 241 p.

Maisuthisakul, P. & Pongsawatmanit, R. (2005). Effect of sample preparation methods andextraction methods and extraction time on yield and antioxidant activity from Kradonbok (*Careya sphaerica* Roxb.) leaves. *Kasetsart Journal: Natural Science,* Vol. 38, Number 5, pp 8-14.

Maisuthisakul, P.; Pongsawatmanit, R. & Gordon, M.H. (2007a). Assessment of phenolic content and free radical-scavenging capacity of some Thai indigenous plants. *Food Chemistry,* Vol. 100, pp 1409-1418.

Maisuthisakul, P.; Pongsawatmanit, R. & Gordon, M.H. (2007). Characterization of the phytochemicals and antioxidant properties of extracts from Teaw (*Cratoxylum formosum* Dyer). *Food Chemistry,* Vol. 100, pp 1620-1629.

Maisuthisakul, P.; Pasuk, S. & Ritthiruangdej, P. (2008a). Relationship of antioxidant properties and chemical composition of some Thai plants. *Journal of Food Composition and Analysis,* Vol. 21, pp 229-240.

Maisuthisakul, P. (2008). Phenolic antioxidants from Betel leaf (*Piper betel* Linn.) extract obtained with different solvents and extraction time. *UTCC journal,* Vol. 28, Number 4, pp 52-64.

Marinova, E. M. & Yanishlieva, N. V. (1997). Antioxidative activity of extracts from selected species of the family Lamiacea in sunflower oil. *Food Chemistry,* Vol. 58, pp 245–248.

Matsuda, H.; Morikawa, T.; Toguchida, I.; Park, J.Y.; Harima, S. & Yoshikawa, M. (2001). Antioxidant constituents from rhubarb: structural requirements of stilbenes for the activity and structures of two new anthraquinone glucosides. *Bioorganic and Medicinal Chemistry.* Vol. 9, pp 41–50.

Miliauskas, G.; Venskutonis, P.R. & van Beek, T.A. (2004). Screening of radical scavenging activity of some medicinal and aromatic plant extracts. *Food Chemistry,* Vol. 85, pp 231-237.

Morita, N.; Arisawa, M.; Nagaes, M.; Hsu, H. Y. & Chen, Y. (1977). Studies on the constituents of formorsan leguminosae. III. Flavonoids from *Leucaena glauca, Cassia fistula* and 8 other species. *Shoyakura Zasshi,* Vol. 31, pp 172-174. (Abstract).

Moure, A.; Cruz, J. M.; Franco, D.; Dominguez, J. M.; Sineiro, J.; Dominguez, H. & Núñez,

M. J. (2001). Natural antioxidants from residual sources. *Food Chemistry,* Vol. 72, pp 145-171.

Munishwar, N. G. (1997). Polarity Index: the Guiding Solvent Parameter for Enzyme Stability in Aqueous-Organic Cosolvent Mixtures. *Biotechology Progress,* Vol. 13, pp 284-288.

Nagabhushan, M.; Amonkar, A. J.; Nair, U. J.; D'Souza, A. V. & Bhide, S. V. (1989). Hydroxy-Chavicol: a New Anti-Nitrosating Phenolic Compound from Betel Leaf. *Mutagenesis,* Vol. 4, Number 3, pp 200-204.

Núñez Sellés, A.; Vélez Castro, H. T. & Agüero-Agüero, J. (2002). Isolation and Quantitative Analysis of Phenolic Antioxidants, Free Sugars, and Polyols from Mango (*Mangifera indica* L.) Stem Bark Aqueous Decoction Used in Cuba as a Nutritional Supplement. *Journal of Agricultural and Food Chemistry,* Vol. 50, pp 762–766.

Nutrition Division. Department of Health Ministry of Public Health. (1992). *Nutritive Values of Thai Foods.* Thai veterans oganisation. Bankok.

Nuutilla, A.M.; kammiovirta, K. & Oksman-Caldentey, K.M. (2002). Comparison of methods for the hydrolysis of flavonoids and phenolic acid from onion and spinash for HPLC analysis. Food Chemistry, Vol. 76, pp 519-525.

Phomkaivon, N. & Areekul, V. (2009). Screening for antioxidant activity in selected Thai wild plants. *Asian Journal of Food Ago-Industry*, Vol. 2, Number 4, pp 433-440

Pokorny, J. & Korczak, J. (2001). Preparation of natural antioxidants. In J. Pokony, N. Yanishlieva, M. Gordon, eds. *Antioxidants in food, Practical applications.* Woodhead Publishing Limited, Cambride. pp. 311-330.

Pokorny, J. 2001. Introduction, In J. Pokorny, N. Yanishlieva & M. Gordon, eds. *Antioxidants in Food, Practical applications.* Woodhead Pulbishing Limited, Cambridge. pp 1-3.

Povichit, N.; Phrutivoraponkul, A.; Suttajit, M.; Chaiyasut, C. & Leelapornpisid, P. (2010). Phenolic content and in vitro inhibitory effects on oxidation and protein glycation of some Thai medicinal plants. *Pakistan Journal of Pharmaceutical Science*, Vol. 23, Number 4, pp 403-408.

Rice-Evans, C.A.; Miller N. J. & Paganga, G. (1997). Structure-antioxidant activity relationships of flavonoids and phenolic acids. *Free Radical Biology and Medicine,* Vol. 20, pp 933-956.

Rimando, A. M.; Han, B. H.; Park, J. H. & Cantoria, M. C. (1986). Studies on the constituents of Philippine Piper Betel leaves. *Archiv of Physical Medicine and Rehabilitation*, Vol. 9, Number 2, pp 93-97.

Salisbury, F.B. & Ross, C.W. (1992). *Plant Physiology*. Wadsworth Publishing Co., Belmont, CA. 682 pp.

Shahidi, F. & Naczk, M. (2003). *Phenolics in food and nutraceuticals.* Boca Raton, FL, USA: CRC Press.

Sheabar, F.Z. & Neeman, I. (1988). Separation and concentration of natural antioxidants from the rape of olives. *Journal of American Oil Chemist's Society,* Vol. 65, pp 990– 993.

Siddhuraju, P.; Mohan, P. S. & Becker, K. (2002). Studies on the antioxidant activity of Indian Laburnum (*Cassia fistula*. L): a preliminary assessment of crude from stem, bark, leaves and fruit pulp. *Food Chemistry*, Vol. 79, pp 61-67.

Skerget, M.; Kotnik, P.; Hadolin, M.; Hra, A.R.; Simoni, M. & Knez, A. (2005). Phenols, proanthocyanidins, flavones and flavonols in some plant materials and their antioxidant activities. *Food Chemistry*, 89, pp 191–198.

Stahl, W. & Sies, H. (2003). Antioxidant activity of carotenoids. *Molecular Aspects of Medicine*, Vol. 24, pp 345–351.

Talapatra, B.; Basak, A. & Talapatra, S. K. (1981). Terpenoids and related compounds; Careaborin, a new titerpene ester from the leaves of *Careya arborea. Journal of the Indian Chemical Society,* Vol. 58, Number 8, pp 814-815. (abstract).

Weinberg, Z.G.; Akiri, B.; Potoyevski, E. & Kanner, J. (1999). Enhancement of polyphenol recovery from rosemary (*Rosmarinus officinalis*) and sage (*Salvia officinalis*) by enzyme-assisted ensiling (ENLAC). *Journal of Agricultural and Food Chemistry,* Vol. 47, pp 2959-2962.

Witzell, J.; Gref, R. & Nüsholm, T. (2003). Plant part specific and temporal variation in phenolic compounds of boreal bilberry (*Vaccinium myrtillus*). *Biochemical Systematic and Ecology*, Vol. 31, pp 115-127.

Zheng, W. & Wang, S.Y. (2001). Antioxidant activity and phenolic compounds in selected herbs. *Journal of Agricultural and Food Chemistry,* Vol. 49, pp 5165-5170.

Zhishen, J., Mengcheng, T. & Jianming, W. (1999). The determination of flavonoid contents in mulberry and their scavenging effects on superoxide radicals. *Food Chemistry*, Vol. 64, pp 555-559.

10
Chapter

LIGNANS: CHEMICAL AND BIOLOGICAL CHARACTERISTICS

1. INTRODUCTION: AN OVERVIEW

The plant world has served as the foundation of traditional medicine for millennia and is a significant source for the discovery of novel physiologically active chemicals (Fabricant & Farnsworth, 2001; Gurib-Fakim, 2006; Newman, 2008). The investigation, advancement, and application of natural goods as medicinal agents, particularly those sourced from higher plants, have escalated in recent years (Gurib-Fakim, 2006). Numerous primary metabolites, including vincristine, vinblastine, taxol, and morphine, have been extracted from plants, with many subsequently changed to produce enhanced counterparts exhibiting improved action, reduced toxicity, or increased solubility. Nonetheless, despite the efficacy of this drug discovery approach, a mere fraction of plants have undergone phytochemical investigation and evaluation for their medicinal potential (Ambrosio et al., 2006; Hostettmann et al., 1997; Soejarto, 1996).

The initial phase in the quest for novel plant-derived pharmaceuticals or lead chemicals involves the extraction of secondary metabolites. Historically, researchers of natural products focused primarily on elucidating the structures and stereochemistry of these molecules; however, contemporary studies have increasingly emphasized their biological activity (Ambrosio et al., 2006). The integrative approach was bolstered by significant advancements in the creation of innovative bioassay techniques. Consequently, numerous chemicals extracted from plants in the past have been "rediscovered" (Ambrosio et al., 2008; Ambrosio et al., 2006; Houghton, 2000; Porto et al., 2009a; Porto et al., 2009b; Tirapelli et al., 2008).

Plants manufacture various kinds of secondary metabolites, among which lignans are acknowledged as a category of natural compounds exhibiting a broad range of significant biological functions. Table 1 encapsulates the primary biological features documented in the literature for lignans.

The word "Lignan" was initially coined by Haworth in 1948 to denote a class of dimeric phenylpropanoids characterized by the linkage of two C6-C3 units via a central carbon (C8), as illustrated in Figure 1. Gotlieb (1978) has suggested that micromolecules including two phenylpropanoid units linked in other configurations, such as C5-C5', should be designated as "neolignans" (Umezawa, 2003). Gordaliza et al. (2004) state that lignans are present in over 60 families of vascular plants and have been extracted from various plant parts, exudates, and resins..

Table 1. Main biological activities of lignans

Biological activity	Reference
Antiviral	(Charlton, 1998; Cos et al., 2008; McRae & Towers, 1984; Yousefzadi et al., 2010)
Anticancer	(McRae & Towers, 1984; Pan et al., 2009; Saleem et al., 2005; Yousefzadi et al., 2010)
Cancer prevention	(Huang et al., 2010; Webb & McCullough, 2005)
Anti-inflammatory	(Saleem et al., 2005)
antimicrobial	(Saleem et al., 2005)
antioxidant	(Fauré et al., 1990; Pan et al., 2009; Saleem et al., 2005)
immunosuppressive	(Saleem et al., 2005)
Hepatoprotective	(Negi et al., 2008)
Osteoporosis prevention	(Habauzit & Horcajada, 2008)

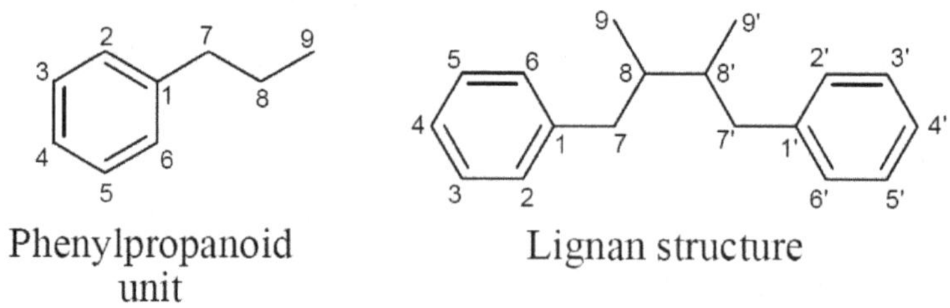

Fig. 1. Phenylpropanoid unit and lignan structure

Most of the known natural lignans are oxidized at C9 and C9′ and, based upon the way in which oxygen is incorporated into the skeleton and on the cyclization patterns, a wide range of lignans of very different structural types can be formed. Due to this fact, lignans are classified in eight subgroups and (Chang et al., 2005; Suzuki & Umezawa, 2007), among these subgroups, the furan, dibenzylbutane and dibenzocyclooctadiene lignans can be further classified in "lignans with C9 (9′)-oxygen" and "lignans without C9 (9′)-oxygen". **Figure 2** displays the main classes of lignans, as well as their subgroups. It is noteworthy that, despite its structural variation, lignans also display a substantial variation on its enantiomeric composition (Umezawa et al., 1997). In this sense, these metabolites can be found as pure enantiomers and as enantiomeric compositions, including racemates (Macias et al., 2004).

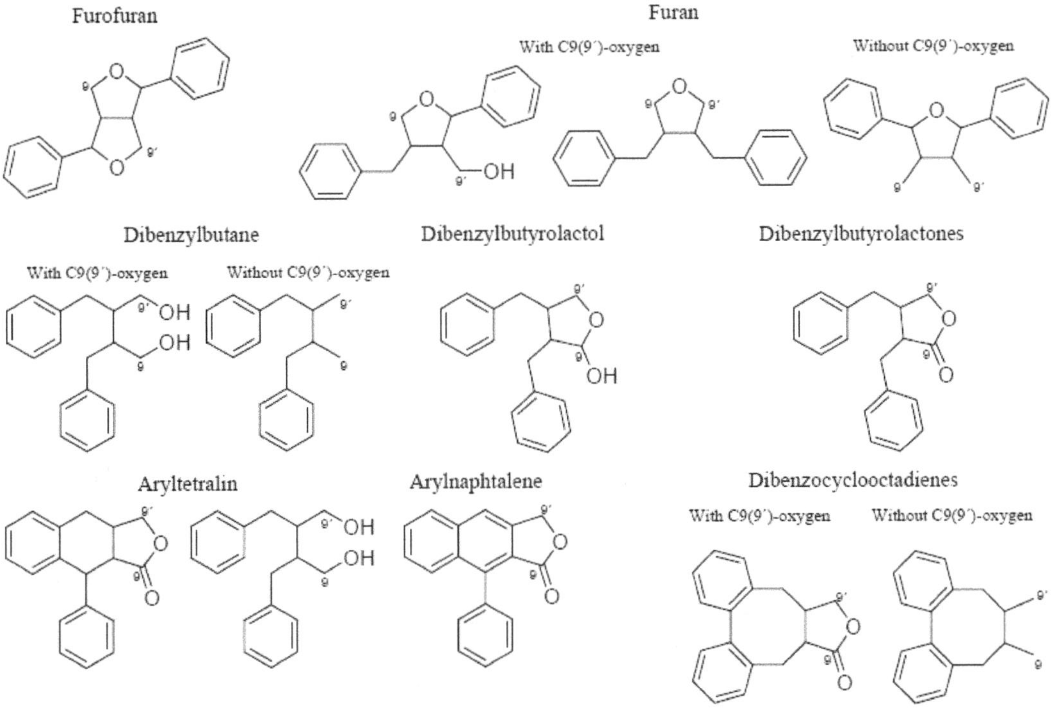

Fig. 2. Main subclasses of lignans. Adapted from Suzuki & Umezawa (2007).

2. CHEMICAL ASPECTS OF LIGNANS

As mentioned before, lignins and lignans are both originated from C_6-C_3 units, thus indicating that these metabolites are biosynthesized through the same pathway in the earlier steps. As seen in **Figure 3**, aromatic aminoacids *L*-phenylalanine and *L*-tyrosine are produced from shikimic acid pathway, and then converted in a series of cinnamic acid derivatives. The reduction of these acids via coenzyme A of related esters and aldehydes forms three alcohols (*p*-coumaryl alcohol, coniferyl alcohol and sinalpyl alcohol) that are the main precursors of all lignins and lignans.

The peroxidase induces one-electron oxidation of the phenol group allowing the delocalization of the unpaired electron through resonance forms. In these hydroxycinamyl alcohols, conjugation allows the unpaired electron to be delocalized also into the side chain. After this point, radical pairing of these resonance structures originates reactive dimeric systems susceptible to nucleophilic attack from hydroxyl groups, leading to a wide range of lignans, as shown in **Figure 2**.

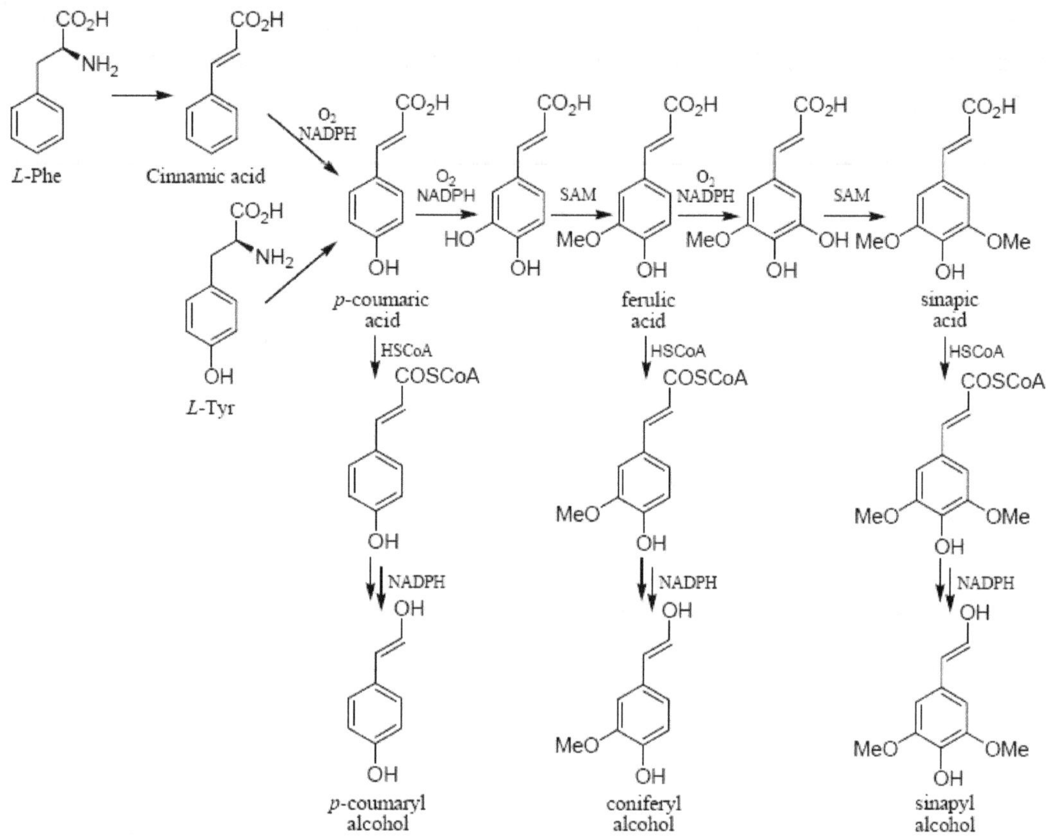

Fig. 3. Biosynthesis of hydroxycinamyl alcohol monomers, the precursors of lignans according to Dewick (2002).

Among these subgroups, the biosynthesis of C9 (9′)-oxygen lignans is the most well known. This type of lignan is formed through the enantioseletive dimerization of two coniferyl alcohol monomeric units (D resonance form of coniferyl alcohol radical, **Figure 4**) into pinoresinol via intermolecular 8,8′ oxidative coupling with the aid of dirigent protein (Dewick, 2002; Suzuki & Umezawa, 2007).

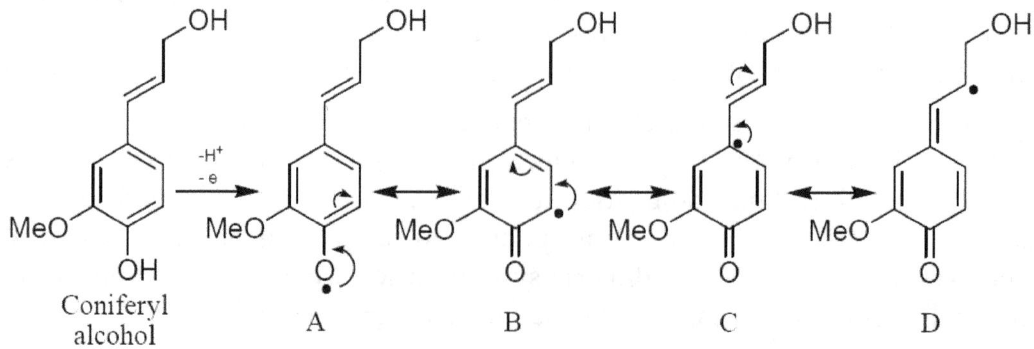

Fig. 4. Resonance forms of coniferyl alcohol radical

The following steps involve sequencial stereoselective enzymatic reduction of pinosresinol by pinoresinol/lariciresinol reductase to generate lariciresinol and then secoisolariciresinol by secoisolariciresinol dehydrogenase. The main steps of this biosynthetic proposal are depicted in **Figure 5**. Secoisolariciresinol gives the presumably common precursor of all dibenzylbutyrolactol lignans and, through the formation of matairesinol and yatein, also forms the aryltetralin lignans. These subclasses of lignans includes some important bioactive compounds such as cubebin (**1**) and podophyllotoxin (Canel et al., 2000; de Souza et al., 2005; Gordaliza et al., 2004; Saraiva et al., 2007; Silva et al., 2007; Silva et al., 2009; Srivastava et al., 2005; You, 2005; Yousefzadi et al., 2010).

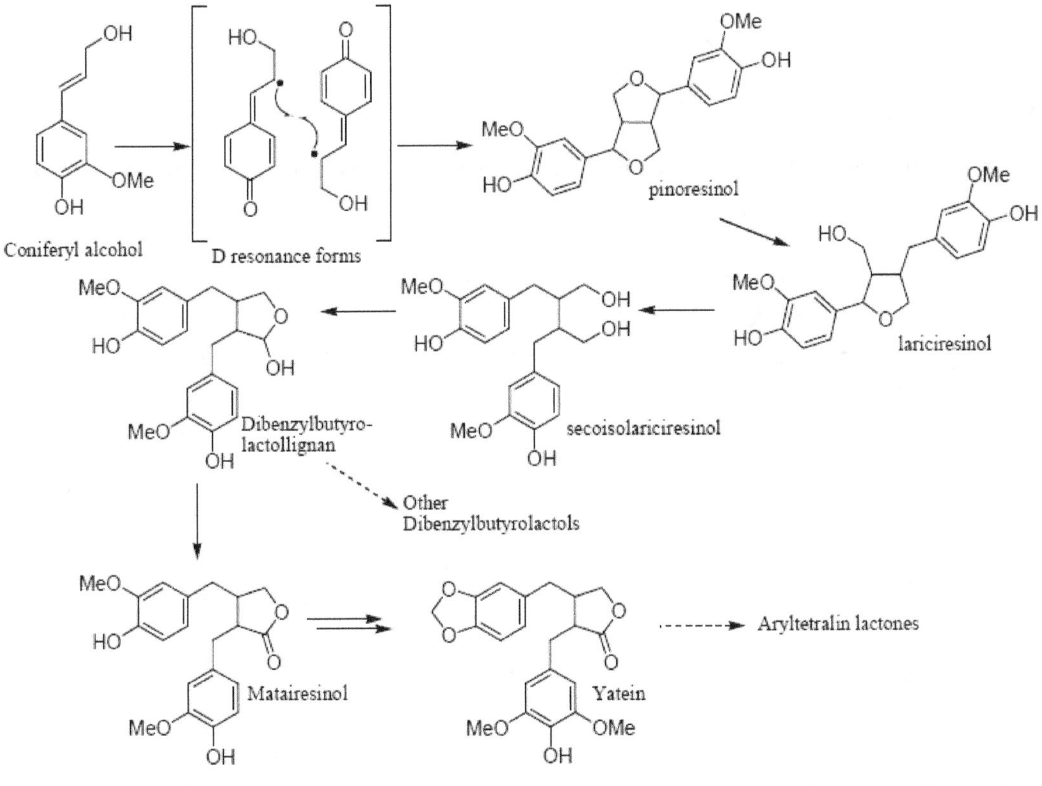

Fig. 5. Biosynthesis of dibenzylbutyrolactols and aryltetralin lactones (Canel et al., 2000).

3.PODOPHYLLOTOXIN:CHEMICALANDBIOLOGICALAPPROACHES

Podophyllotoxin (**Figure 6**), a naturally occurring aryltetralin lignin, is one of the most important compound due to its high toxicity and current use as a local antiviral agent (Yousefzadi et al., 2010). Moreover, this metabolite has been used to obtain structural analogues which are employed as anticancer drugs (Ayres & Loike, 1990; Yousefzadi et al., 2010) and several semi-synthetic podophyllotoxin-related derivatives showed to be topoisomerase II inhibitor, acting as an antimitotic compound (You, 2005; Yousefzadi et al., 2010). **Figure 6** also shows the clinically valuable anticancer agents, etoposide, teniposide and etoposide phosphate, obtained from podophyllotoxin.

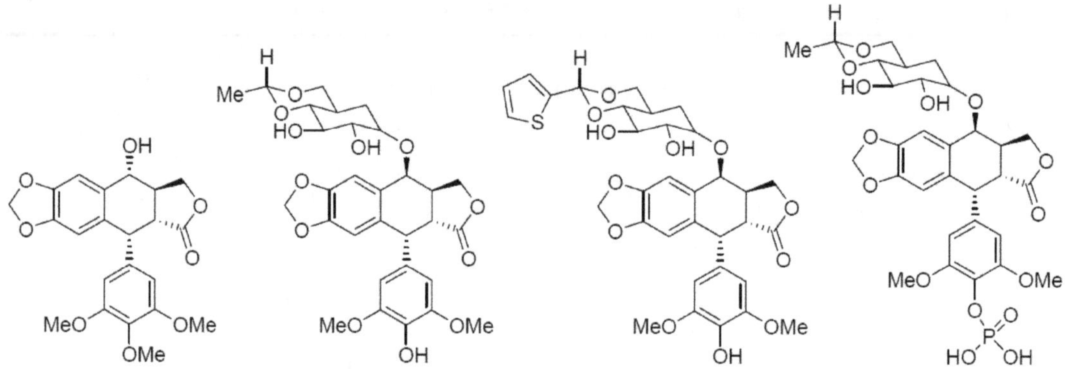

Podophyllotoxin Etoposide Teniposide Etoposide phosphate

Fig. 6. Chemical structures of clinically available podophyllotoxin derivatives.

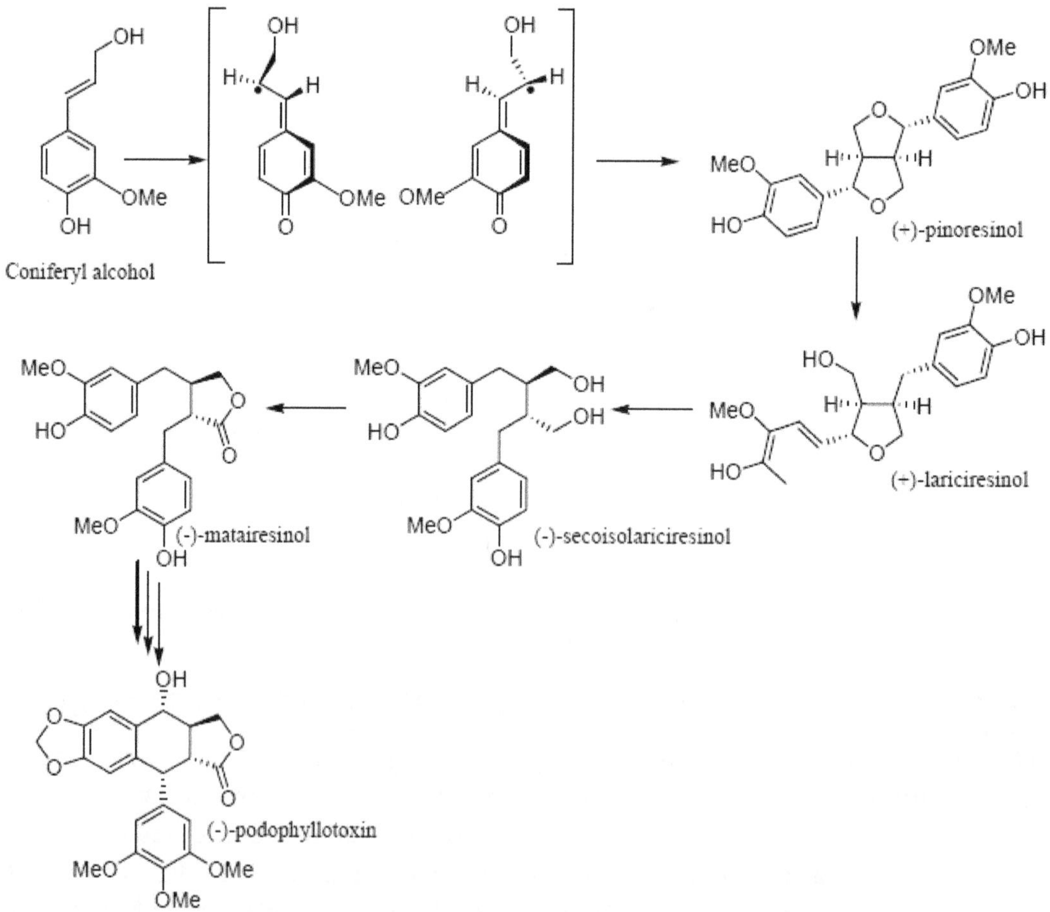

Fig. 7. Biosynthesis proposal of podophyllotoxin according to Canel et al., 2000.

According to You (2005), podophyllotoxin still can be considered a hot prototype for discovery and development of novel anticancer agents, even in the 21st century. This leading compound has been isolated from the roots of *Podophyllum* species and more recently from other genus, such as *Linum* (Yousefzadi et al., 2010). Due to its importance in anticancer therapy, several biotechnological approaches including the use of cell cultures, biotransformation processes and metabolic engineering techniques to manipulate the biosynthetic pathway **(Figure 7)**, have been currently developed and are alternatives for the production of podophyllotoxin.

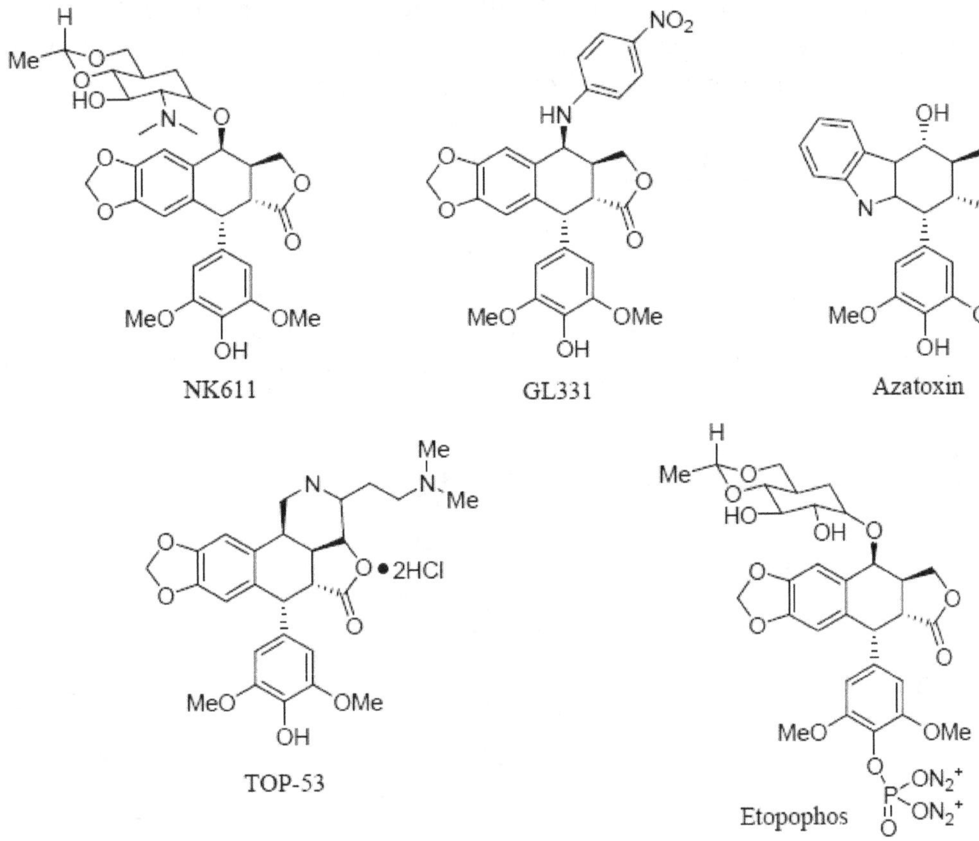

Fig. 8. Examples of antineoplasic candidates developed from podophyllotoxin chemical skeleton.

Despite the fact that etoposide, teniposide and etoposide phosphate are clinically valuable anticancer agents, several adverse effects and drug resistance have been associated with the use of these drugs (You, 2005). In this sense, several studies focusing to prepare novel derivatives and to understand the structure-activity relationship (SAR) of podophyllotoxins have been published (You, 2005). Based on these data a great number of potential drug candidates were synthesized (You, 2005; Yousefzadi et al., 2010). **Figure 8** shows the structures of new antineoplasic candidates developed from podophyllotoxin chemical skeleton.

In order to better explore the biological potential of this class of metabolites, our research group has concentrated efforts to investigate the biological activity of some dibenzylbutyrolactone lignans, mainly cubebin (**Figure 9, 1**) and its semi-synthetic derivatives. In this sense, the most significant achievements in our investigations are described in the following sections.

4. TRYPANOCIDAL ACTIVITY OF CUBEBIN AND ITS DERIVATIVES

The Chagas' disease, or American trypanosomiasis, is endemic in Central and South America and it is estimated that 16–18 million people are currently infected with the protozoan flagellate *Trypanosoma cruzi* (Molfetta et al., 2005) and more than 100 million are exposed to the risk of infection (Takeara et al., 2003).

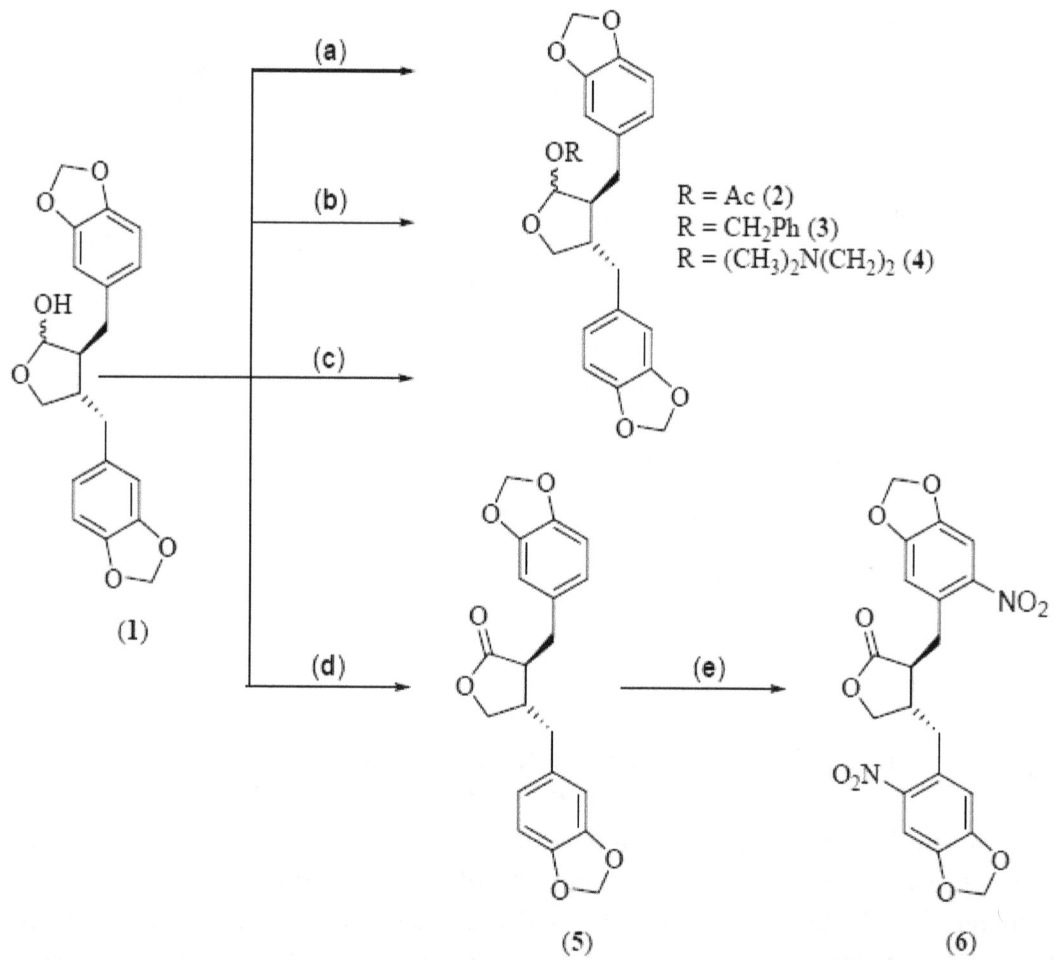

Fig. 9. Reagents and conditions: (a) Ac_2O, Py, room temperature, 24 h; (b) NaH, BnBr, THF, room temperature, 24h; (c) EtONa, $(CH_3)_2CH_2Cl$, EtOH, reflux, 6h; (d) PCC, CH_2Cl_2, room temperature, 12h; (e) HNO_3, 2h, -10° C.

Since it was discovery in 1909, Chagas′ disease infection has been difficult to control due to its multiple characteristics (de Souza et al., 2005). One of the main causes of these difficulties are to find an efficient compound to combat the aetiologic agent (*T. cruzi*) is directly linked to the morphologic characteristics of its strains, mainly due to the occurrence of various sub- populations of the parasite, leading to a different host tissue′s tropism (de Souza et al., 2005).

Clinical treatment of infected patients is relied on two nitroheterocyclic drugs, the nifrofuran nifurtimox, Lampit®, which production has now been discontinued, and the 2- nitroimidazole benznidazole, Rochagan® (Paulino et al., 2005). Both drugs, if administered during the acute phase of the disease, could cure 50-70% of the patients. However, these drugs display limited efficacy in the treatment of the chronic phase of the disease and are quite toxic for the patients (de Souza et al., 2005). Therefore, there is an urgent demand for the discovery and development of novel therapeutic compounds to treat Chagas′ disease De Souza et al. (2005) (de Souza et al., 2005) have reported the trypanocidal activity of cubebin (**1**) and its semi-synthetic derivatives against free amastigote forms of *T. cruzi*. **Figure 9** also shows the compounds obtained by partial synthesis from cubebin (**1**), as well as the reagents and conditions used in these reactions.

The natural cubebin (**1**), used as the starting compound to obtain the evaluated dibenzylbutyrolactone derivatives, did not display activity against trypomastigote forms of *T. cruzi* (Bastos et al., 1999). Hence, the biological evaluation against amastigote forms was undertaken only for lignans **2**, **3**, **4**, **5** and **6**. Cubebin was selected as starting compound because of its availability, being easily isolated in large amounts from the seeds of *Piper cubeba* (de Souza et al., 2005). **Table 2** shows the results of the trypanocidal activity evaluation of compounds **2**, **3**, **4**, **5**, **6** and benznidazole, against amastigote forms of Y strain of *T. cruzi*.

Table 2. Results of the trypanocidal activity evaluation of compounds 2, 3, 4, 5, 6 and benznidazole, against amastigote forms of Y strain of *T. cruzi* (de Souza et al., 2005).

Compounds	Concentration (μM) X % of lyse (± SD)				IC_{50} (μM)
	0.5	2.0	8.0	32.0	
2	14.5 ± 1.9	26.0 ± 2.5	29.0 ± 5.2	26.4 ± 1.2	1.5 X 10^4
3	37.0 ± 1.4	38.0 ± 7.0	46.8 ± 6.4	68.8 ± 2.9	5.7
4	32.6 ± 2.6	55.4 ± 4.3	50.8 ± 1.0	57.8 ± 8.0	4.7
5	47.6 ± 9.5	57.0 ± 1.1	57.6 ± 8.9	63.6 ± 5.2	0.7
6	34.6 ± 7.9	48.7 ± 1.4	38.9 ± 2.0	48.5 ± 6.1	95.3
Benznidazole	38.4 ± 3.0	67.0 ± 7.2	69.0 ± 4.0	68.6 ± 1.7	0.8

The production of compound **2** by substitution of the lactol hydrogen of cubebin by an acetyl group led to a strong reduction of its trypanocidal ctivity, in comparison with all other evaluated compounds belonging to the same group (IC_{50} = 1.5 X 10^4 µM; **Table 2**). Furthermore, the comparison of compounds **2** and **3** indicate that the biological activity against the amastigote forms of *T. cruzi* was significantly affected by the nature of the substituting group at position C-9, which played an important role in the reduction of the calculated IC_{50} value for compound **3** (IC_{50} = 5.7 µM; **Table 2**). Likewise, cubebin derivative **4**, bearing an amino group at the lactol ring, displayed an activity quite similar to compound **3** Analysis of the obtained results, displayed in **Table 2**, indicate that compound 5 was the most active, with an IC_{50} value of 0.7 µM similar to that displayed by benznidazole (IC_{50} =0.8 µM), a standard drug used as the positive control. On the other hand, most of the other evaluated compounds displayed much lower activity, with the exception of compounds **3** (IC_{50} = 5.7 µM) and **4** (IC_{50} = 4.7 µM), which showed significant activity.

In this study, De Souza et al. (2005) also pointed out that hinokinin (**HK, 5**) is a promising compound to continue examining, since at 0.5 µM it displayed higher activity than benznidazole and at the other assayed concentrations (2.0, 8.0 and 32.0 µM) it showed similar activity.

Table 3. Results of the trypanocidal activity evaluation of HK, benznidazole and nifurtimox against epimastigote forms of CL strain of *T. cruzi* (Saraiva et al., 2007).

Compounds	Concentration (µM) X % of lyse (± SD)					IC_{50} (µM)
	0.5	2.0	8.0	32.0	128.0	
HK	21.79±1.82	99.03±0.15	100.0±0.35	100.0±0.35	100.0±0.64	0.67
Nifurtimox	11.78±13.92	49.38±6.71	65.46±5.36	81.54±2.71	97.54±1.80	3.08
benznidazole	0	1.23±5.87	26.18±10.71	51.13±5.23	76.09±2.74	30.89

Table 4. Results of the trypanocidal activity evaluation of HK, benznidazole and nifurtimox against intracellular amastigote forms of CL strain of *T. cruzi* (Saraiva et al., 2007).

Compounds	Concentration (µM) X % of lyse (± SD)				IC_{50} (µM)
	2.0	8.0	32.0	128.0	
HK	25.84±1.09	31.92±9.29	61.72±9.17	100.0±0.40	18.36
Nifurtimox	44.6±0.99	63.78±1.25	83.31±0.79	90.63±1.13	3.54
benznidazole	14.33±2.65	35.81±0.65	57.28±1.99	78.75±0.67	20.00

In view of higher trypanocidal activity displayed by **HK** (**5**) against free amastigote forms of *T. cruzi* (**Table 2**; (de Souza et al., 2005), this lignan was selected to be assayed against epimastigote and intracellular amastigote forms of *T. cruzi*, both in vitro and *in vivo* assays (Saraiva et al., 2007). The results of the trypanocidal activity against epimastigote and intracellular amastigote forms of *T. cruzi* are shown in **Tables 3 and 4**, respectively.

As it can be observed in **Table 3**, **HK** showed a very significant activity against epimastigote forms of *T. cruzi*, displaying IC_{50} value (0.67 μM) much lower than benznidazole and nifurtimox, used as positive controls (Saraiva et al., 2007). **HK**, also showed to be very active against intracellular amastigote forms, displaying IC_{50} value of 18.36 μM, which was similar to benznidazole (IC_{50} = 20.0 μM, **Table 4**).

The *in vivo* assays (Saraiva et al., 2007) were performed using five groups of five BALB/c males, weighing approximately 20 g each. The groups were as follows: group 1, animals without infection; group 2, control infected animals; group 3, animals treated with solvent; group 4, animals treated with benznidazole 40 mg kg^{-1} day^{-1}; group 5, animals treated with **HK** 40 mg kg^{-1} day^{-1}. The animals were inoculated with 2 X 10^4 trypomastigote forms of *T. cruzi* (Y strain). The treatment was initiated 48 h after infection and maintained for 20 days. The animals were treated twice a day with 20 mg kg^{-1} benznidazole and **HK** orally. The results obtained showed that the treatment with **HK** promoted 70.8% of parasitaemia reduction in the parasitaemic peak, while benznidazole displayed approximately 29.0% of parasite reduction (Saraiva et al., 2007). In addition, **HK** was able to reduce the number of parasites more than benznidazole not only in the parasitaemic peak, but also in all curse of infection (Saraiva et al., 2007).

Moreover, it was observed that the groups treated with **HK** displayed better survival rates than the group treated with benznidazole, with survival until the 22nd and 16th day after the beginning of the infection, respectively (Saraiva et al., 2007). Despite the obtained significant results for the *in vivo* assays, the treatment with **HK** or benznidazole did not cause parasitological cure. Overall, considering the promising results displayed by **HK** against both the epimastigote and amastigote forms of the parasite in the *in vitro* assay, as well as the good result displayed in the *in vivo* assay, this lignan has been considered as a lead compound for the development of new drugs for the treatment of Chagas´disease (Saraiva et al., 2007).

In order to obtain better efficacy of **HK** towards the intracellular forms of the parasite, our research group prepared and investigated the effect of **HK** load poly($_{D,L}$-lactide-co- glycolide) microparticules.

5.HINOKININ-LOADPOLY(D,L-LACTIDE-CO-GLYCOLIDE)MICROPARTICLES FOR CHAGAS´ DISEASE

The drug delivery system were developed for the purposes of bringing, uptaking, retaining, releasing, activating, localizing, and targeting the drugs at the right timing, period, dose, and place (Ueda & Tabata, 2003). The use of biodegradable polymers, as poly($_{D,L}$-lactic-co- glycolic acid; PLGA), for the controlled release of therapeutic agents is now well established.

These systems have been extensively utilized for oral and parenteral administration (Saraiva et al., 2010). The physical properties and the Food and Drug Administration approval of poly(lactide-co-glycosides) make them the most extensively studied commercially available biodegradable polymers (Birnbaum et al., 2000).

The microparticles can be able to sustain the release of the drug for a considerable period of time, to reduce the required frequency of administration increasing patient compliance, to avoid plasmatic fluctuations, to decrease side effects, and to facilitate dosage administration (Hans & Lowman, 2002). In this sense, our research group prepared **HK**-loaded PLGA microparticles to protect HK of biological interactions and promote its sustained release for treatment of Chagas´ disease. Moreover, the trypanocidal effect of microparticles containing **HK** was evaluated *in vivo*.

The **HK**-loaded PLGA microparticles were prepared with success (Saraiva et al., 2010) and presented narrow distribution size and a mean diameter of 0.862 μm, with PDI of 0.072 mm. Scanning electron micrographs of PLGA microparticles obtained showed that **HK** loaded microparticles presented, smooth and spherical surface. Due to their small diameter, the **HK** microparticles obtained are better suited for parenteral delivery (Cegnar et al., 2005).

The trypanocidal *in vivo* experiments were performed using Female Swiss mice (weigh, 20- 22 g) which were infected intraperitoneally with 2 X 10^4 trypomastigotes forms of *T. cruzi*. The treatment (20 days) was performed through subcutaneous route and initiated 48 h after infection, according to Saraiva et al. (2010).

The treatment of infected mice with 40 mg kg^{-1} of **HK**-loaded microparticles each 2 days was able to provoke significant decrease in parasitemia levels compared with those recorded in untreated controls ($P<0.05$ at days 12, 14, 16, 19 and 21 post-infection with *T. cruzi*). The treatment with an equivalent amount of empty microparticles (without **HK**) had no effect on the parasitemia compared to untreated controls (Saraiva et al., 2010). Moreover, administration of **HK**-loaded microparticles was able to reduce the number of parasites more than the treatment with 20 mg kg^{-1} day^{-1} of **HK** not only in the parasitemic peak, but also in the course of infection ($P<0.05$ at days 14, 16, 19 and 21 post-infection with *T. cruzi*) (Saraiva et al., 2010). The use of PLGA micropartiles as vehicle for HK delivery can improve HK trypanocidal activity. It may be attributed to the fact that it can protect HK of biological interactions and promote its sustained release, with maintenance of its plasmatic concentration in therapeutic levels (Saraiva et al., 2010).

The **HK**-loaded microparticles developed by our research group can be considerable a promising system for sustained release of **HK** for therapeutic use and could be used in future clinical studies (Saraiva et al., 2010). Also, it is very important to point out that other *in vivo* assays have been developed by our research group in order to evaluate the parasitological cure of infection and the activity of this delivery system coating **HK** against other strains of *T. cruzi*.

6. INFLUENCES OF STEREOCHEMISTRY ON TRYPANOCIDAL ACTIVITY OF DIBENZYLBUTYROLACTONE LIGNANS

Wehavebeenreportingthesignificanttrypanocidalactivityofdibenzylbutyrolactone lignans, mainly the semi-synthetic **HK**. Such results aroused the interest within our group to study the effect of stereochemistry in this biological property. For this purpose, methylpluviatolide, one of the most powerful compounds regarding trypanocidal activity (Bastos et al., 1999) was synthesized in its *trans* and *cis* racemic forms. Thus, allowing us to evaluate the trypanocidal activity not only of a mixture of these two stereoisomers, but also of the pure enantiomers, which were separated by chiral HPLC (da Silva et al., 2008).

Trans (tM) and *cis* (cM)racemic forms of methylpluviatolide were prepared by a procedure described by Landais et al. (1991) and Charlton and Chee (1997) (**Figure 10**).

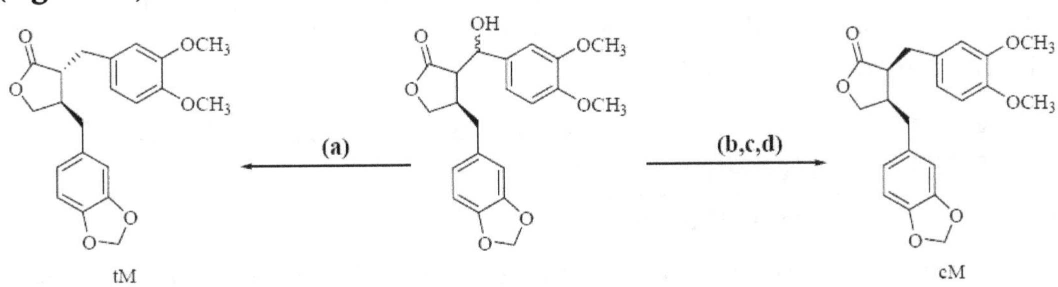

Fig. 10. Reagents and conditions: (a) H_2, 4 atm, Pd/C, ETOH, $HClO_4$, 60h, room temperature; (b) THF, AC_2O, Et_3N, DMAP, 2h, room temperature; (c) DBU, CH_2Cl_2, 5h, room temperature; (d) H_2, 4 atm, Pd/C, ETOH, 60h, room temperature.

The results obtained for the racemic mixture of *trans* and *cis* methylpluviatolide against *T. cruzi* showed that racemic *cis*-stereoisomer (**2**) is inactive, while the racemic *trans*- stereoisomer (**1**) display significant trypanocidal activity, with an IC_{50} of 89.3 μM (da Silva et al., 2008).

On the basis of these results, a separation of the *trans*-stereoisomer from the racemic mixture was undertaken by chiral HPLC using an analytical Chiracel OJ (4.6 x 250 mm) column, aiming to evaluate the trypanocidal activity of each enantiomer separately. The chromatogram gave a well resolved peak separation, allowing the isolation of both enantiomers (da Silva et al., 2008), which were evaluated against trypomastigote forms of the Y strain of *T. cruzi*. **Table 5** shows the results of the trypanocidal activity evaluation of (+)-*trans*- methylpluviatolide (**+tM**) and (-)-*trans*-methylpluviatolide (**-tM**) against trypomastigote forms of the Y strain of *T. cruzi*.

Table 5. Results of the trypanocidal activity evaluation of +tM or -tM against trypomastigotes forms of the Y strain of *T. cruzi*. (da Silva et al., 2008).

Compounds	Concentration (μM) X % of lyse (± SD)			IC_{50} (μM)
	8.0	32.0	128.0	
+tM	5.3±2.5	7.6±3.4	9.9±2.5	1.3 X 10^6
-tM	40.6±3.6	52.3±4.4	79.7±0.0	18.7

The results show that **+tM** is completely inative, whereas the **–tM** displayed good activity, with an IC_{50} of 18.7 μM (da Silva et al., 2008). These results indicate that despite being completely inactive, the **+tM** blocks the action of the **–tM** when they are present in a racemic mixture. It should be taken into consideration that the **+tM** might bind to the active sites as a competitive antagonist, which may be confirmed by comparison of the IC_{50} value of the racemic mixture with that of the **–tM** itself (da Silva et al., 2008).

In conclusion, this study pointed the importance of the stereochemistry on trypanocidal activity of dibenzylbutyrolactone lignans and brings new perspective in the importance to understand the trypanocidal structure-activity relationship for this class of natural compounds.

7. ANTIMICROBIAL POTENTIAL OF SOME NATURAL AND SEMI-SYNTHETIC LIGNANS AGAINST

Mycobacteria and oral pathogens

The lignans possess a wide spectrum of biological activities, including antimicrobial (Saleem et al., 2005). Considering this fact, our research group also decided to investigate the potential of some natural and semi-synthetic lignans against mycobateria and oral pathogens (Silva et al., 2007; Silva et al., 2009).

Tuberculosis is a severe infectious disease caused by mycobacteria belonging to the *Mycobacterium tuberculosis* complex. According to WHO, tubercolosis affects nearly 30% of the world´s population and is responsible for 3 million deaths worldwide each year, mainly in developing countries (Raviglione, 2003). The current chemotherapy of this pathology has been based on the use of combined drug therapy with rifampicin, isonizid, and pyrazinamide. However, the incorrect use and long drug administration, as well as the high cost and countless side-effects have led people to abandon the treatment before being completely cured, leading to resistant bacilli (Timmins & Deretic, 2006). In addition, the existence of drug-resistant tuberculosis reinforces the need to develop new safe and effective antimycobacterial drugs. In this sense, our research group evaluated the antimycobacterial activity of several lignans obtained from cubebin (Silva et al., 2009).

As shown in **Figure 11**, (-)-cubebin (**1**) was isolated from powdered seeds of *Piper cubeba* and then submitted to various semi-synthetic procedures to obtain hinokinin (**HK, 5**), (-)-O- acetyl-cubebin (**2**), (-)-O-methyl-cubebin (**7**), (-)-O-(N,N-dimethylamine-ethyl)-cubebin (**4**) and (-)-6,6′-dinitrohinokinin (**6**). All these compounds were assayed *in vitro* by the microdilution technique on a Resazurin microtiter assay (REMA) plate, using a procedure adapted from (Palomino et al., 2002).

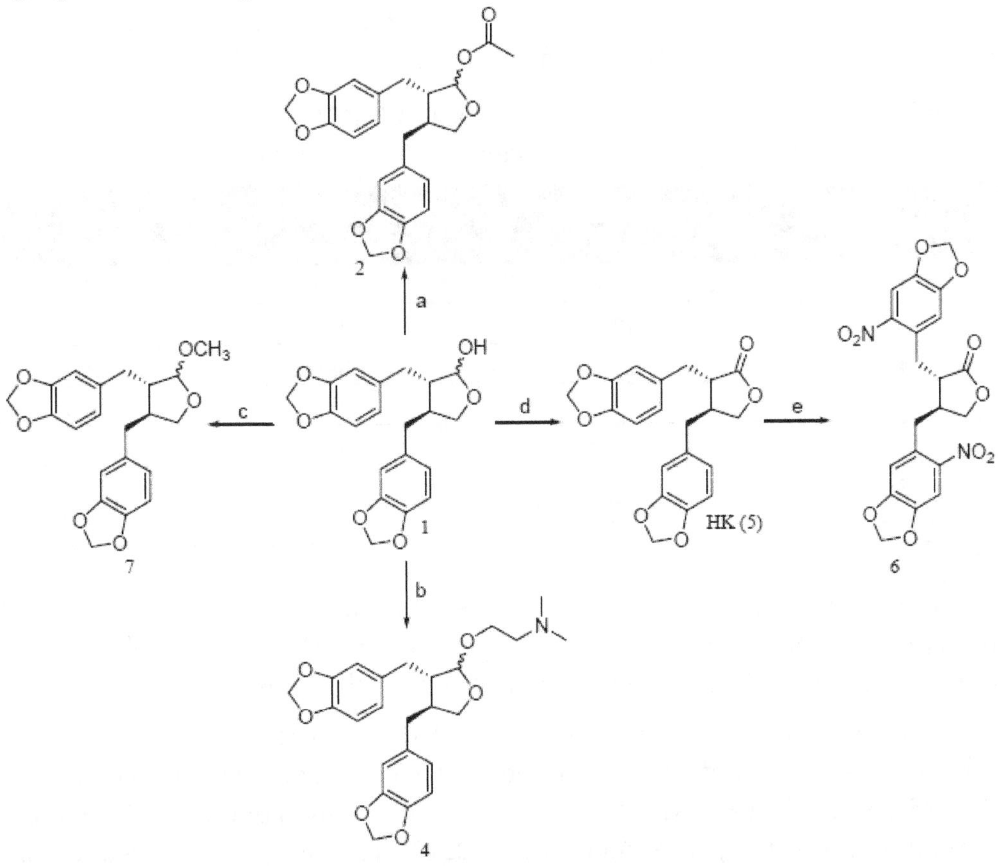

Fig. 11. Chemical structures and conditions of the reactions (a) Acetic anhydride, room temperature, 24h. (b) Dimethyllethylammonium chloride, EtONa, dry THF, room temperature, N_2 atmosphere, 6h. (c) Methyl iodide, NaH, dry THF, room temperature, 6h. (d) PCC (pyridinium chlorochromate) in dry methylene chloride, 24h, in an ice bath with continuous stirring. (e) HNO_3, chloroform, -6°C, 2h.

Cubebin (1) did not display any activity against the investigated strains (Table 6). HK (5) was moderately active against *Mycobacterium tuberculosis*, with a MIC value equal to 62.5 μg mL^{-1}. Compound (2), whose lactol group is acetylated, displayed activity against *M. tuberculosis* (MIC = 125 μg mL^{-1}) and *M. avium* (MIC = 62.5 μg mL^{-1}). The best result was achieved with compound 7, whose lactol group is methylated, leading to a MIC value equal to 31.25 μg mL^{-1} against *M. avium*. The other compounds were not active against any of the studied mycobacteria. In the case of the lactol-containing compounds evaluated here, it seems to be essential that the lactol group is absent, and the substituent of this group should be small, as in the case of 2 and 7. *M. kansaii* (ATCC 12478) was the most resistant mycobacterium concerning the evaluated compounds, with MIC values varying between 1000 and 2000 μg mL^{-1}.

To sum up, cubebin and hinokinin semi-synthetic derivatives were prepared and evaluated for their antimycobacterial activity. Some derivatives were active against *M. tuberculosis* and *M. avium*, suggesting that this class of compounds may lead to a new generation of antituberculosis agents (Silva et al, 2009).

Table 6. Minimal inhibitory concentration (MIC) of cubebin (1) and its derivatives against *M. tuberculosis*, *M. kansasii*, **and** *M. avium*.

Compounds	MIC [μg mL^{-1}]		
	M. tuberculosis	*M. kansaii*	*M. avium*
1	500	2000	1000
2	125	1000	62.5
4	250	2000	250
5	62.5	2000	500
6	1000	2000	1000
7	250	2000	31.25
Rifampicin[a]	0.031	0.015	0.062

a Standard antibiotic

Recently, our research group also investigated the antimicrobial activity of cubebin and related derivatives against oral pathogens, mainly those responsible for caries disease, which are intimately related with the dental plaque formation.

Dental plaque is defined as a biofilm consisting of cariogenic bacteria adhered on the tooth surface and plays an important role in the development of dental caries (Chung et al., 2006; Xie et al., 2008), one of the main oral diseases that affect humankind (More et al., 2008; Souza et al., 2010). This destructive infection of the dental hard tissues can progress and if untreated, lead to the death of vital pulp tissue and tooth loss (Allaker & Douglas, 2009). Bacteria from the genus *Streptococci* are commonly isolated from the oral cavity (Hirasawa & Takada, 2002) and have been responsible for this infectious disease. Among them, *Streptococcus mutans* is considered one of the main cariogenic microorganisms, due to its ability to synthesize extracellular polysaccharides from sucrose, mainly water- insoluble glucan, and initiate plaque formation (Koo et al., 2000). Other aerobic bacteria such as *Enterococcus faecalis*, *Lactobacillus casei*, *Streptococcus mitis*, *S. sanguinis*, *S. sobrinus* and *S. salivarius* are also important in the latter formation of the dental biofilm (Chung et al., 2006).

The mechanical removal of the dental plaque is the most efficient procedure to prevent caries, but the majority of the population does not perform this removal efficiently (Ambrosio et al., 2008). Moreover, dental treatment is often very expensive and not readily accessible, especially in developing countries (More et al., 2008). In this sense, the use of chemicals as a complementary measure is necessary and has demonstrated to be of great value in the prevention of the formation and in the decreasing of the tooth surface biofilm (Furiga et al., 2008).

Extensive efforts have been made toward the search for anticariogenic compounds that can be incorporated into dental products, aiming at complementing the mechanical removal. Several antibiotics, such as ampicillin, chlorhexidine, sanguinarine, metronidazole, phenolic-antiseptics and quaternary ammonium-antiseptics have been used to prevent dental caries. Among these compounds, chlorhexidine is considered a gold standard anticariogenic and has received the approval of the American Dental Association Council on Dental Therapeutics (Ambrosio et al., 2008). However, the regular use of oral care products containing this chemical are often associated with tooth and restoration staining, changes in the taste of food, and a burning sensation at the tip of the tongue (Greenberg et al., 2008; More et al., 2008; Porto et al., 2009b). In addition, chlorhexidine is much less effective in reducing the levels of *Lactobacillus* species, which are strongly related to caries evolution (Ambrosio et al., 2008). All these problems, therefore, denote that finding new, safe and effective anticariogenic coumpounds is still needed.

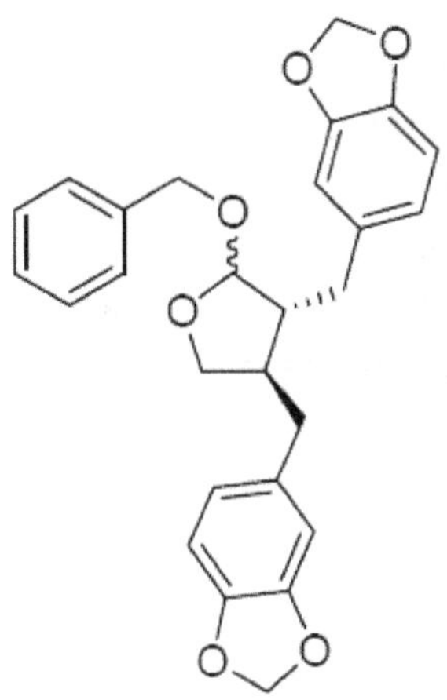

Fig. 12. Structure of O-benzyl cubebin (compound 8)

Thus, our research group tested compounds **1, 4, 5,** and **6** (**Figure 11**) and another semi- synthetic derivative (O-benzyl cubebin, **8, Figure 12**)using the broth microdilution method (Andrews, 2001) against the following microorganisms: *Enterecoccus faecalis* (ATCC 4082), *Streptococcus salivarius* (ATCC 25975), *Streptococcus mitis* (ATCC 49456), *Streptococcus mutans* (ATCC 25275), *Streptococcus sobrinus* (ATCC 33478), *Streptococcus sanguinis* (ATCC 10556) and *Candida albicans* (ATCC 28366) (Silva et al. 2007). **Table 7** displays the minimum inhibitory concentration values obtained for these compounds

Table 7. Values of minimum inhibitory concentrations (in milimolar) of cubebin and its semi-synthetic derivatives against oral pathogens

Compound	Microorganism						
	E. faecalis	S. salivarius	S. sanguinis	S. mitis	S. mutans	S. sobrinus	*C. albicans*
1	0.35	0.25	0.22	0.20	0.32	0.27	0.28
4	0.31	0.21	0.21	0.19	0.28	0.23	0.23
5	0.38	0.25	0.25	0.25	0.32	0.28	0.28
6	0.30	0.20	0.21	0.18	0.27	0.23	0.23
8	0.31	0.20	0.23	0.18	0.29	0.23	0.28
CHD[a]	5.9×10^{-3}	1.7×10^{-3}	3.9×10^{-3}	5.9×10^{-3}	5.9×10^{-3}	1.5×10^{-3}	7.9×10^{-3}

a Chlorhexidine

The semi-synthetic derivative **6** was the most active one against all the evaluated microorganisms (**Table 7**). Compounds **5** and **6** are lignan-lactones and differ from cubebin by the presence of a carbonyl group at C9 (**Figure 11**). Analysis of the obtained results suggested that the presence of the carbonyl group at C9 with introduction of polar groups in the aromatic rings is beneficial for the antimicrobial activity.

The obtained results for antimicrobial activity are in accordance to those obtained for anti- inflammatory and analgesic activities. Compounds possessing a lactone ring bearing two methylendioxyaryl groups display significant anti-inflammatory and analgesic activities, and the introduction of polar groups in the aromatic rings is advantageous for these activities . However, with regard to trypanocidal activity, the introduction of nitro groups at the aromatic rings is harmful for this activity. Besides, the lignan-lactone **HK** (**5**) was the most active compound against *T. cruzi* (de Souza et al., 2005).

8. CONCISE SUMMARY

Despite the extensive range of biological activities associated with lignans, the literature has primarily highlighted their antioxidant qualities and their role in cancer treatment and prevention. Fauré et al. (1990); McRae & Towers (1984); Pan et al. (2009); Saleem et al. (2005); Yousefzadi et al. (2010). In recent years, our research group has emphasized the significance of metabolites, particularly cubebin and its semi-synthetic derivatives, as possible antichagasic drugs (da Silva et al., 2008; de Souza et al., 2005; Saraiva et al., 2010; Saraiva et al., 2007). The encouraging findings achieved against T. cruzi indicate that additional research on these lignans for other parasite infections is warranted. Consequently, our lab is currently concentrating on the assessment of these compounds against Schistosoma mansoni and Fasciola hepatica, in addition to the synthesis of novel cubebin-related semi-synthetic derivatives.

Furthermore, our findings on the antibacterial properties of these metabolites underscored their potential as novel antimicrobial agents (Silva et al., 2007; Silva et al., 2009). The literature additionally presents experiments aimed at exploring other aspects of antimicrobial activity, including time-kill curve experiments as per D'Arrigo et al. (2010) and inquiries into a potential synergistic effect between the most effective lignans tested and currently employed antimicrobial agents. (White et al., 1996).

9. REFERENCES

Allaker R. P.; C. W. I. Douglas. (2009). Novel anti-microbial therapies for dental plaque- related diseases. *International Journal of Antimicrobial Agents*, 33, 1, (2009), pp. 8-13, 0924-8579

Ambrosio S. R.; C. R. Tirapelli; F. B. da Costa; A. M. de Oliveira. (2006). Kaurane and pimarane-type diterpenes from the *Viguiera* species inhibit vascular smooth muscle contractility. *Life Sciences*, 79, 10, (2006), pp. 925-933, 0024-3205

Ambrosio S. R.; N. A. J. C. Furtado; D. C. R. De Oliveira; F. B. Da Costa; C. H. G. Martins; T. C. De Carvalho; T. S. Porto; R. C. S. Veneziani. (2008). Antimicrobial activity of kaurane diterpenes against oral pathogens. *Zeitschrift Fur Naturforschung Section C-a Journal of Biosciences*, 63c, 5-6, (2008), pp. 326-330, 0939-5075

Andrews J. M. (2001). Determination of minimum inhibitory concentrations. *Journal of Antimicrobial Chemotherapy*, 48 Suppl 1, (2001), pp. 5-16, 0305-7453

Ayres D. C.; J. D. Loike. (1990). *Lignans: Chemical, Biological and Clinical Properties* (1), Cambridge University Press, Cambridge

Bastos J. K.; S. Albuquerque; M. L. Silva. (1999). Evaluation of the trypanocidal activity of lignans isolated from the leaves of *Zanthoxylum naranjillo*. *Planta Medica*, 65, 6, (1999), pp. 541-544, 0032-0943

Birnbaum D. T.; J. D. Kosmala; D. B. Henthorn; L. Brannon-Peppas. (2000). Controlled release of beta-estradiol from PLAGA microparticles: the effect of organic phase solvent on encapsulation and release. *Journal of Controlled Release*, 65, 3, (2000), pp. 375-387, 0168-3659

Canel C.; R. M. Moraes; F. E. Dayan; D. Ferreira. (2000). Podophyllotoxin. *Phytochemistry*, 54, (2000), pp. 115-120, 0031-9422

Cegnar M.; J. Kristl; J. Kos. (2005). Nanoscale polymer carriers to deliver chemotherapeutic agents to tumours. *Expert Opinion on Biological Therapy*, 5, 12, (2005), pp. 1557-1569, 1744-7682

Chang J.; J. Reiner; J. Xie. (2005). Progress on the chemistry of dibenzocyclooctadiene lignans. *Chemical Reviews*, 105, 12, (2005), pp. 4581-4609, 0009-2665

Charlton J. L.; G.-L. Chee. (1997). Asymmetric synthesis of lignans using oxazolidinones as chiral auxiliaries. *Canadian Journal of Chemistry*, 75, 8, (1997), pp. 1076-1083, 0008-4042

Charlton J. L. (1998). Antiviral activity of lignans. *Journal of Natural Products*, 61, 11, (1998), pp. 1447-1451, 0163-3864

Chung J. Y.; J. H. Choo; M. H. Lee; J. Hwang. (2006). Anticariogenic activity of macelignan isolated from *Myristica fragrans* (nutmeg) against *Streptococcus mutans*. *Phytomedicine*, 13, 4, (2006), pp. 261-266, 0944-7113

Cos P.; L. Maes; A. Vlietinck; L. Pieters. (2008). Plant-derived leading compounds for chemotherapy of human immunodeficiency virus (HIV) infection - an update (1998- 2007). *Planta Medica*, 74, 11, (2008), pp. 1323-1337, 0032-0943

D'Arrigo M.; G. Ginestra; G. Mandalari; P. M. Furneri; G. Bisignano. (2010). Synergism and postantibiotic effect of tobramycin and *Melaleuca alternifolia* (tea tree) oil against *Staphylococcus aureus* and *Escherichia coli*. *Phytomedicine*, 17, 5, (2010), pp. 317-322, 1618-095X

da Silva R.; J. Saraiva; S. de Albuquerque; C. Curti; P. M. Donate; T. N. C. Bianco; J. K. Bastos; M. L. A. Silva. (2008). Trypanocidal structure-activity relationship for cis- and trans-methylpluviatolide. *Phytochemistry*, 69, 9, (2008), pp. 1890-1894, 0031-9422

de Souza V. A.; R. da Silva; A. C. Pereira; A. Royo Vde; J. Saraiva; M. Montanheiro; G. H. de Souza; A. A. da Silva Filho; M. D. Grando; P. M. Donate; J. K. Bastos; S. Albuquerque; M. L. e Silva. (2005). Trypanocidal activity of (-)-cubebin derivatives against free amastigote forms of *Trypanosoma cruzi*. *Bioorganic & Medicinal Chemistry Letters*, 15, 2, (2005), pp. 303-307, 0960-894X

Dewick P. M. (2002). *Medicinal Natural Products: a Biosynthetic approach* Willey, Chichester, 978-0-470-74168-9

Fabricant D. S.; N. R. Farnsworth. (2001). The value of plants used in traditional medicine for drug discovery. *Environmental Health Perspectives*, 109 Suppl 1, (2001), pp. 69-75, 0091-6765

Fauré M.; E. Lissi; R. Torres; L. A. Videla. (1990). Antioxidant activities of lignans and flavonoids. *Phytochemistry*, 29, 12, (1990), pp. 3773-3775, 0031-9422

Furiga A.; A. Lonvaud-Funel; G. Dorignac; C. Badet. (2008). In vitro anti-bacterial and anti-adherence effects of natural polyphenolic compounds on oral bacteria. *Journal of Applied Microbiology*, 105, 5, (2008), pp. 1470-1476, 1365-2672

Gordaliza M.; P. A. Garcia; J. M. del Corral; M. A. Castro; M. A. Gomez-Zurita. (2004). Podophyllotoxin: distribution, sources, applications and new cytotoxic derivatives. *Toxicon*, 44, 4, (2004), pp. 441-459, 0041-0101

Greenberg M.; M. Dodds; M. Tian. (2008). Naturally occurring phenolic antibacterial compounds show effectiveness against oral bacteria by a quantitative structure- activity relationship study. *Journal of Agricultural and Food Chemistry*, 56, 23, (2008), pp. 11151-11156, 1520-5118

Gurib-Fakim A. (2006). Medicinal plants: traditions of yesterday and drugs of tomorrow. *Molecular aspects of medicine*, 27, 1, (2006), pp. 1-93, 0098-2997

Habauzit V.; M. N. Horcajada. (2008). Phenolic phytochemicals and bone. *Phytochemistry Reviews*, 7, (2008), pp. 313-344, 1568-7767

Hans M. L.; A. M. Lowman. (2002). Biodegradable nanoparticles for drug delivery and targeting. *Current Opinion in Solid State & Materials Science*, 6, (2002), pp. 319-327, 1359-0286

Hirasawa M.; K. Takada. (2002). Susceptibility of *Streptococcus mutans* and *Streptococcus sobrinus* to cell wall inhibitors and development of a novel selective medium for *S- sobrinus*. *Caries Research*, 36, 3, (2002), pp. 155-160, 0008-6568

Hostettmann K.; J. L. Wolfender; S. Rodriguez. (1997). Rapid detection and subsequent isolation of bioactive constituents of crude plant extracts. *Planta Medica*, 63, 1, (1997), pp. 2-10, 0032-0943

Houghton P. J. (2000). Use of small scale bioassays in the discovery of novel drugs from natural sources. *Phytotherapy Research*, 14, 6, (2000), pp. 419-423, 0951-418X

Huang W. Y.; Y. Z. Cai; Y. Zhang. (2010). Natural phenolic compounds from medicinal herbs and dietary plants: potential use for cancer prevention. *Nutrition and Cancer*, 62, 1, (2010), pp. 1-20, 1532-7914

Koo H.; A. M. Vacca Smith; W. H. Bowen; P. L. Rosalen; J. A. Cury; Y. K. Park. (2000). Effects of *Apis mellifera* propolis on the activities of streptococcal glucosyltransferases in solution and adsorbed onto saliva-coated hydroxyapatite. *Caries Research*, 34, 5, (2000), pp. 418-426, 0008-6568

Landais Y.; J. P. Robin; A. Lebrun. (1991). Ruthenium dioxide in fluoro acid medium: I. A new agent in the biaryl oxidative coupling. Application to the synthesis of non phenolic bisbenzocyclooctadiene lignan lactones. *Tetrahedron*, 47, 23, (1991), pp. 3787-3804, 0040-4020

Macias F. A.; A. Lopez; R. M. Varela; A. Torres; J. M. G. Molinillo. (2004). Bioactive lignans from a cultivar of *Helianthus annuus*. *Journal of Agricultural and Food Chemistry*, 52, 21, (2004), pp. 6443-6447, 0021-8561

McRae D. W.; N. G. H. Towers. (1984). Biological activities of lignans. *Phytochemistry*, 23, 6, (1984), pp. 1207-1220, 0031-9422

Molfetta F. A.; A. T. Bruni; K. M. Honorio; A. B. da Silva. (2005). A structure-activity relationship study of quinone compounds with trypanocidal activity. *European Journal of Medicinal Chemistry*, 40, 4, (2005), pp. 329-338, 0223-5234

More G.; T. E. Tshikalange; N. Lall; F. Botha; J. J. M. Meyer. (2008). Antimicrobial activity of medicinal plants against oral microorganisms. *Journal of Ethnopharmacology*, 119, 3, (2008), pp. 473-477, 0378-8741

Negi A. S.; J. K. Kumar; S. Luqman; K. Shanker; M. M. Gupta; S. P. Khanuja. (2008). Recent advances in plant hepatoprotectives: a chemical and biological profile of some important leads. *Medicinal Research Reviews*, 28, 5, (2008), pp. 746-772, 1098-1128

Newman D. J. (2008). Natural products as leads to potential drugs: An old process or the new hope for drug discovery? *Journal of Medicinal Chemistry*, 51, 9, (2008), pp. 2589- 2599, 0022-2623

Palomino J. C.; A. Martin; M. Camacho; H. Guerra; J. Swings; F. Portaels. (2002). Resazurin microtiter assay plate: Simple and inexpensive method for detection of drug resistance in *Mycobacterium tuberculosis*. *Antimicrobial Agents and Chemotherapy*, 46, 8, (2002), pp. 2720-2722, 0066-4804

Pan J. Y.; S. L. Chen; M. H. Yang; J. Wu; J. Sinkkonen; K. Zou. (2009). An update on lignans: natural products and synthesis. *Natural Products Reports*, 26, 10, (2009), pp. 1251-1292, 1460-4752

Paulino M.; F. Iribarne; M. Dubin; S. Aguilera-Morales; O. Tapia; A. O. Stoppani. (2005). The chemotherapy of chagas' disease: an overview. *Mini-Reviews in Medicinal Chemistry*, 5, 5, (2005), pp. 499-519, 1389-5575

Porto T. S.; N. A. J. C. Furtado; V. C. G. Heleno; C. H. G. Martins; F. B. da Costa; M. E. Severiano; A. N. Silva; R. C. S. Veneziani; S. R. Ambrosio. (2009a). Antimicrobial *ent*-pimarane diterpenes from

Viguiera arenaria against Gram-positive bacteria. *Fitoterapia*, 80, (2009a), pp. 432-436, 0367-326X

Porto T. S.; R. Rangel; N. Furtado; T. C. de Carvalho; C. H. G. Martins; R. C. S. Veneziani; F. B. Da Costa; A. H. C. Vinholis; W. R. Cunha; V. C. G. Heleno; S. R. Ambrosio. (2009b). Pimarane-type Diterpenes: Antimicrobial Activity against Oral Pathogens. *Molecules*, 14, 1, (2009b), pp. 191-199, 1420-3049

Raviglione M. C. (2003). The TB epidemic from 1992 to 2002. *Tuberculosis (Edinb)*, 83, 1-3, (2003), pp. 4-14, 1472-9792

Saleem M.; H. J. Kim; M. S. Ali; Y. S. Lee. (2005). An update on bioactive plant lignans. *Natural Product Reports*, 22, 6, (2005), pp. 696-716, 0265-0568

Saraiva J.; C. Vega; M. Rolon; R. da Silva; E. S. ML; P. M. Donate; J. K. Bastos; A. Gomez- Barrio; S. de Albuquerque. (2007). In vitro and in vivo activity of lignan lactones derivatives against *Trypanosoma cruzi*. *Parasitology Research*, 100, 4, (2007), pp. 791-795, 0932-0113

Saraiva J.; A. A. Lira; V. R. Esperandim; D. da Silva Ferreira; A. S. Ferraudo; J. K. Bastos; E. S. ML; C. M. de Gaitani; S. de Albuquerque; J. M. Marchetti. (2010). (-)-Hinokinin- loaded poly(D,-lactide-co-glycolide) microparticles for Chagas disease. *Parasitology Research*, 106, 3, (2010), pp. 703-708, 1432-1955

Silva M. L.; H. S. Coimbra; A. C. Pereira; V. A. Almeida; T. C. Lima; E. S. Costa; A. H. Vinholis; V. A. Royo; R. Silva; A. A. Filho; W. R. Cunha; N. A. Furtado; C. H. Martins; T. C. Carvalho; J. K. Bastos. (2007). Evaluation of *Piper cubeba* extract, (-)- cubebin and its semi-synthetic derivatives against oral pathogens. *Phytotherapy Research*, 21, 5, (2007), pp. 420-422, 0951-418X

Silva M. L.; C. H. Martins; R. Lucarini; D. N. Sato; F. R. Pavanb; N. H. Freitas; L. N. Andrade; A. C. Pereira; T. N. Bianco; A. H. Vinholis; W. R. Cunha; J. K. Bastos; R. Silva; A. A. Da Silva Filho. (2009). Antimycobacterial activity of natural and semi-synthetic lignans. *Zeitschrift Fur Naturforschung Section C-a Journal of Biosciences*, 64, 11-12, (2009), pp. 779-784, 0939-5075

Soejarto D. D. (1996). Biodiversity prospecting and benefit-sharing: perspectives from the field. *Journal of Ethnopharmacology*, 51, 1-3, (1996), pp. 1-15, 0378-8741

Souza A. B.; C. H. G. Martins; M. G. M. Souza; N. A. J. C. Furtado; V. C. G. Heleno; J. P. B. d. Sousa; E. M. P. Rocha; J. K. Bastos; W. R. Cunha; R. C. S. Veneziani; S. R. Ambrósio. (2010). Antimicrobial activity of terpenoids from *Copaifera langsdorffii* Desf. against cariogenic bacteria. *Phytotherapy Research*, 25, n/a, (2010), pp. 215- 220, 1099-1573

Srivastava V.; A. S. Negi; J. K. Kumar; M. M. Gupta; S. P. Khanuja. (2005). Plant-based anticancer molecules: a chemical and biological profile of some important leads. *Bioorganic & Medicinal Chemistry*, 13, 21, (2005), pp. 5892-5908, 0968-0896

Suzuki S.; T. Umezawa. (2007). Biosynthesis of lignans and norlignans. *Journal of Wood Science*, 53, (2007), pp. 273-284, 1435-0211

Takeara R.; S. Albuquerque; N. P. Lopes; J. L. Lopes. (2003). Trypanocidal activity of *Lychnophora staavioides* Mart. (Vernonieae, Asteraceae). *Phytomedicine*, 10, 6-7, (2003), pp. 490-493, 0944-7113

Timmins G. S.; V. Deretic. (2006). Mechanisms of action of isoniazid. *Molecular Microbiology*, 62, 5, (2006), pp. 1220-1227, 0950-382X

Tirapelli C. R.; S. R. Ambrosio; F. B. da Costa; A. M. de Oliveira. (2008). Diterpenes: a therapeutic promise for cardiovascular diseases. *Recent Patents on Cardiovascular Drug Discovery*, 3, 1, (2008), pp. 1-8, 1574-8901

Ueda H.; Y. Tabata. (2003). Polyhydroxyalkanonate derivatives in current clinical applications and trials. *Advanced Drug Delivery Reviews*, 55, 4, (2003), pp. 501-518, 0169-409X

Umezawa T.; T. Okunishi; M. Shimada. (1997). Stereochemical diversity in lignan biosynthesis. *Wood Research: bulletin of the Wood Research Institute Kyoto University*, 85, (1997), pp. 96-125, 0049-7916

Umezawa T. (2003). Diversity in lignan biosynthesis. *Phytochemistry Reviews*, 2, 3, (2003), pp. 371-390,

Webb A. L.; M. L. McCullough. (2005). Dietary lignans: potential role in cancer prevention. *Nutrition and Cancer*, 51, 2, (2005), pp. 117-131, 0163-5581

White R. L.; D. S. Burgess; M. Manduru; J. A. Bosso. (1996). Comparison of three different *in vitro* methods of detecting synergy: time-kill, checkerboard, and E test. *Antimicrobial Agents and Chemotherapy*, 40, 8, (1996), pp. 1914-1918, 0066-4804

Xie Q.; J. Li; X. Zhou. (2008). Anticaries effect of compounds extracted from *Galla chinensis* in a multispecies biofilm model. *Oral Microbiology and Immunology*, 23, 6, (2008), pp. 459-465, 1399-302X

You Y. (2005). Podophyllotoxin derivatives: current synthetic approaches for new anticancer agents. *Current Pharmaceutical Design*, 11, 13, (2005), pp. 1695-1717, 1381-6128

Yousefzadi M.; M. Sharifi; M. Behmanesh; E. Moyano; M. Bonfill; R. M. Cusido; J. Palazon. (2010). Podophyllotoxin: Current approaches to its biotechnological production and future challenges. *Engineering in Life Sciences*, 10, 4, (2010), pp. 281-292, 1618-2863

11 Chapter

THE GENUS GALANTHUS: A SOURCE OF BIOACTIVE CONSTITUENTS

1. INTRODUCTION: AN OVERVIEW

The Amaryllidaceae family is among the 20 most significant plant families that contain alkaloids (Zhong, 2005). It consists of approximately 1100 perennial bulbous species categorized into 85 genera, found in the tropics and mild temperate zones globally (Willis, 1988). The particular alkaloids generated by amaryllidaceous plants have garnered significant interest owing to their notable pharmacological properties. Galanthamine is a long-acting, selective, reversible, and competitive inhibitor of the acetylcholinesterase enzyme (Thomsen et al., 1998), marketed as a hydrobromide salt under the names Razadyne® (formerly Reminyl®) and Nivalin® for the treatment of Alzheimer's disease, poliomyelitis, and other neurological disorders (Heinrich and Teoh, 2004). Following its identification in Galanthus woronowii by Proskurina and colleagues in 1955 (Proskurina et al., 1955), the pharmacological attributes of galanthamine rapidly garnered interest from the pharmaceutical sector. Initially synthesized by Sopharma (Bulgaria) as Nivalin® from G. nivalis in the early 1960s, this species was subsequently supplanted by alternative plant sources due to its diminutive size and inconsistent galanthamine concentration (Berkov et al., 2009b).

The genus Galanthus (Snowdrop; Greek gála "milk", ánthos "flower") has around 19 species (World Checklist of Selected Plant Families), of which 11 have been studied for their alkaloid content. Despite the genus being just partially researched, phytochemical investigations have uncovered a remarkable variety of alkaloid structures, several of which are described for the first time and possess unknown bioactivity. This article offers a concise summary of the phytochemical research conducted on the genus Galanthus.

2. GEOGRAPHICAL DISTRIBUTION, TAXONOMICAL ASPECTS AND ECOLOGY OF *GALANTHUS*

The genus *Galanthus* L. is distributed around Europe, Asia Minor and the Caucasus region. The limits of its area of distribution are the Pyrenees in the west, the Caucasus and Iran in the east, and Sicily, the Peloponnese and Lebanon in the south. The northern distribution limit cannot be assessed due to human introduction and cultivation (Davis, 1999). Some species are widespread, while others are restricted to small areas. *G. nivalis,* for example, is native to a large area of Europe, stretching from the Pyrenees to Italy, Northern Greece, Ukraine, and European Turkey, while *G. trojanus* is a rare plant in the wild, found in a single location (an area less than 10 km^2) in western Turkey (Davis and Ozhatay, 2001). Turkey is the country where most species (14) are geographically concentrated (Ünver, 2007).

All species of *Galanthus* are perennial, herbaceous plants that grow from bulbs. They have two or three linear leaves and an erect, leafless scape. The scape bears a pair of bract-like spathe valves at the top, from which emerges a solitary, bell-shaped white flower, held on a slender pedicel. The flower of *Galanthus* consists of six tepals, the outer three being larger and more convex than the inner series. The inner flower segments are marked with a green, or greenish-yellow, bridge-shaped mark at the tip of each tepal. The ovary is three-celled, ripening into a three-celled capsule. Each whitish seed has a small, fleshy tail (elaiosome) containing substances attractive to ants, which distribute the seeds (Davis, 1999). The genus *Galanthus* is closely related to the genus *Leucojum* L. but its plants can be easily distinguished because *Leucojum* has flowers with six equal tepals, from 2 to 6-7 flowers per scape and several leaves (Meerow and Snijman, 1998).

Species of the genus *Galanthus* L. (Amaryllidaceae) are difficult to distinguish and classify because of a lack of clearly definable morphological characteristics and a high level of variability. The search for other useful systematic information has produced little consensus in the enumeration of the species, divisions within the genus and relationships among their various components (Davis and Barnet, 1997). Besides morphological features, cariological (Kamari, 1981), anatomical (Davis and Barnet, 1997) and DNA (Zonneveld *et al.,* 2003) methods have been used to clarify the taxonomy of the genus.

It is generally accepted that the genus *Galanthus* comprises 19 species, 6 varieties and 2 natural interspecies hybrids (World Cheklist of Selected Plant Families):

1. *Galanthus alpinus* Sosn., Vestn. Tiflissk. Bot. Sada 19: 26 (1911).

   Galanthus alpinus var. alpinus.

   Galanthus alpinus var. bortkewitschianus (Koss) A.P.Davis, Kew Bull. 51: 750 (1996).

2. *Galanthus angustifolius* Koss, Bot. Mater. Gerb. Bot. Inst. Komarova Akad. Nauk S.S.S.R. 14: 134 (1951).

3. *Galanthus cilicicus* Baker, Gard. Chron. 1897(1): 214 (1897).

4. *Galanthus elwesii* Hook.f., Bot. Mag. 101: t. 6166 (1875), nom. cons.

   Galanthus elwesii var. elwesii

   Galanthus elwesii var. monostictus P.D.Sell in P.D. Sell & G.Murrell, Fl. Great Britain Ireland 5: 363 (1996).

5. *Galanthus fosteri* Baker, Gard. Chron., III, 5: 458 (1889).
6. *Galanthus gracilis* Celak.,Sitzungsber. Königl. Böhm. Ges. Wiss., Math.-Naturwiss. Cl. 1891(1): 195 (1891).
7. *Galanthus ikariae* Baker, Gard. Chron. 1893(1): 506 (1893).
8. *Galanthus koenenianus* Lobin, C.D.Brickell & A.P.Davis, Kew Bull. 48: 161 (1993).
9. *Galanthus krasnovii* Khokhr., Byull. Moskovsk. Obshch. Isp. Prir., Otd. Biol., n.s., 68(4): 140 (1963).
10. *Galanthus lagodechianus* Kem.-Nath., Zametki Sist. Geogr. Rast. 13: 6 (1947).
11. *Galanthus nivalis* L., Sp. Pl.: 288 (1753).
12. *Galanthus peshmenii* A.P.Davis & C.D.Brickell, New Plantsman 1: 17 (1994).
13. *Galanthus platyphyllus* Traub & Moldenke, Herbertia 14: 110 (1948).
14. *Galanthus plicatus* M.Bieb., Fl. Taur.-Caucas., Suppl.: 225 (1819).
15. *Galanthus reginae-olgae* Orph., Atti Congr. Int. Bot. Firenze 1874: 214 (1876).

    Galanthus reginae-olgae subsp. reginae-olgae.

    Galanthus reginae-olgae subsp. vernalis Kamari, Bot. Jahrb. Syst. 103: 116 (1982).

16. *Galanthus rizehensis* Stern, Snowdrops & Snowflakes: 37 (1956).
17. *Galanthus transcaucasicus* Fomin, Opred. Rast. Kavk. Kryma 1: 281 (1909).
18. *Galanthus trojanus* A.P.Davis & Özhatay, Bot. J. Linn. Soc. 137: 409 (2001).
19. *Galanthus woronowii* Losinsk. in V.L.Komarov (ed.), Fl. URSS 4: 749 (1935).
20. *Galanthus × allenii* Baker, (*G. alpinus × G. woronowii*) Gard. Chron., III, 9: 298 (1891).
21. *Galanthus × valentinei* Beck, (*G. plicatus × G. nivalis*) Wiener Ill. Gart.-Zeitung 19: 57 (1894).

The habitats of *Galanthus* species are varied, ranging from undisturbed broad-leaved or coniferous woodlands of, for example oak (*Quercus* spp.), beech (*Fagus orientalis*), maple (*Acer* spp.), pines (*Pinus* spp.), Cilician fir (*Abies cilicia*), and cedar of Lebanon (*Cedrus libani*), woodland edges, river banks, scrub, grassland, amongst large rocks, and pockets of soil on rocks and cliff faces. *G. peshmenii* can sometimes be found only 10 m from the sea-shore on Kastellorhizo, a typical hot and dry Aegean island. In contrast, *G. platyphyllus* is a plant of the subalpine to alpine zone, and occurs mainly at altitudes of 2,000 - 2,700 m in alpine grasslands and meadows above the tree-line and at the edges of high-altitude woodlands (Davis, 1999). Typically, the *Galanthus* species are winter-to-spring flowering plants, but some species, like *G. cilicicus*, *G. peshmenii* and *G. reginae-olgae*, flower in autumn.

G. nivalis and *G. elwesii* are two of the best known and most frequently cultivated bulbous plants. Their popularity is due to their beauty, longevity and because they flower when little else is in season. A vast number of cultivars and clones are available (Davis, 1999). Huge numbers of wild-collected bulbs are exported annually from Turkey. In the early 1980s onwards this trade increased, with many millions of *G. elwesii* bulbs being exported via the Netherlands. The large numbers of *Galanthus* bulbs coming into commerce caused great concern because it was uncertain whether the collection of bulbs in such high numbers was sustainable. For this reason, *Galanthus* was placed on Appendix II of CITES in 1990. The wild harvesting of *G. elwesii* bulbs is now carefully controlled and monitored, and export quotas are set each year. Some snowdrop species are threatened in their wild habitats, and in most countries it is now illegal to collect bulbs from the wild. Under CITES regulations, international trade in any quantity of *Galanthus*, whether bulbs or plants, live or dead, is illegal without a CITES permit. This applies to hybrids and named cultivars as well as species. CITES does, however, allow a limited trade in wild-collected bulbs of just three species (*G. nivalis*, *G, elwesii* and *G. woronowii*) from Turkey.

3. BIOSYNTHESIS AND STRUCTURAL TYPES OF AMARYLLIDACEAE ALKALOIDS

A particular characteristic of the Amaryllidaceae plant family is a consistent presence of an exclusive group of isoquinoline alkaloids, which have been isolated from plants of all the genera of this family. As a result of extensive phytochemical studies, over 500 alkaloids have been isolated from the amaryllidaceous plants (Zhong, 2005). The Amaryllidaceae type alkaloids have been structurally classified into nine main subgroups, namely lycorine, crinine, haemanthamine, narciclasine, galanthamine, tazettine, homolycorine, montanine and norbelladine (Bastida *et al.*, 2006). In the genus *Galanthus*, however, two new structural subgroups, graciline and plicamine type alkaloids, have been found (Ünver, 2007). The following new subgroups have also been reported: specific augustamine-type structures in *Crinum kirkii* (Machocho *et al.*, 2004), a carboline alkaloid in *Hippeastrum vittatum* (Youssef, 2001), mesembrane (*Sceletium*)-type compounds in *Narcissus pallidulus* and *N. triandrus* (Bastida *et al.*, 2006), and phtalideisoquinoline-, benzyltetrahydroisoquinoline- and aporphine-type alkaloids in *G. trojanus* (Kaya *et al.*, 2004*b*, 2011). Mesembrane-type compounds are typical of the genus *Sceletium* of the Aizoaceae, while phtalideisoquinoline-, benzyltetrahydroisoquinoline- and aporphine-type alkaloids are found in the Papaveraceae, both families being dicotyledonous. Tyramine-type protoalkaloids, which are biosynthesized in Poaceae, Cactaceae, some algae and fungi, have also been found in *Leucojum* and *Galanthus* species (Berkov *et al.*, 2009*a*, 2011).

Amaryllidaceae alkaloids are formed biogenetically by intramolecular oxidative couplingofnorbelladinesderivedfromtheaminoacidsL-phenylalanineandL-tyrosine (Bastida *et al.*, 2006). The key intermediate metabolite is O-methylnorbelladine. *Ortho-para′* phenol oxidative coupling of O-methylnorbelladine results in the formation of a lycorine-type skeleton, from which homolycorine-type compounds proceed. The galanthamine-type skeleton originates from *para-ortho′* phenol oxidative coupling. *Para-para′* phenol oxidative coupling leads to the formation of crinine, haemanthamine, tazettine, narciclasine and montanine structures (Bastida *et al.*, 2006). In the present article, for the structures reported by different authors we have adopted the numbering system according to Bastida *et al.*, (2006, Fig. 1).

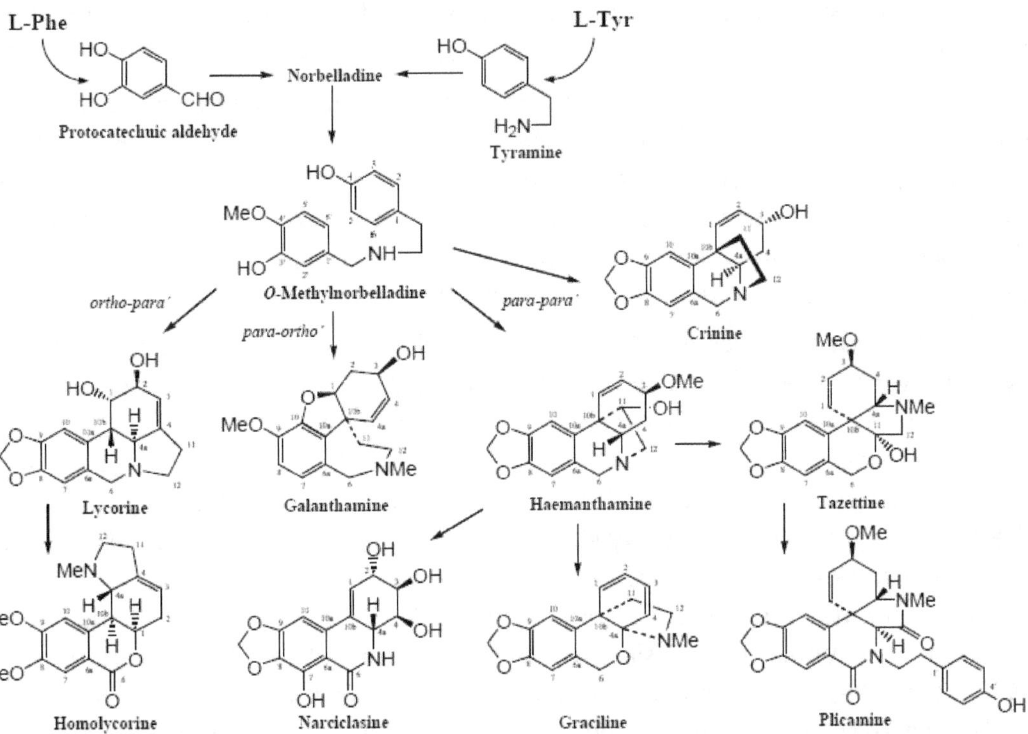

Fig. 1. Biosynthetic pathway of *Galanthus* alkaloids with representative compounds.

The biogenetic pathway of gracilines possibly originates from the 6-hydroxy derivatives of haemanthamine-type species (Noyan *et al.*, 1998), while plicamine-type alkaloids most probably proceed from tazettine-type compounds, considering their structural similarities (Ünver *et al.*, 1999*a*).

4. DISTRIBUTION OF ALKALOIDS IN THE GENUS GALANTHUS

The phytochemical studies of the genus *Galanthus* started in the early fifties of the last century. Two of the first alkaloids reported for the genus were galanthine (Proskurina and Ordzhonikidze, 1953) and galanthamine (Proskurina *et al.,* 1955), which were isolated from *G. voronowii*. To the best of our knowledge, eleven species from the genus *Galanthus* have been phytochemically studied to date and ninety alkaloids have been found and classified in 11 structural types (Table 1, Fig.2).

Until recently, the distribution of alkaloids within the genus has been studied by classical phytochemical approaches. The collected biomass is extracted with alcohol, the neutral compounds removed at low pH and the alkaloids fractionated after basification of the extract. Individual alkaloids have been separated by column chromatography, preparative TLC, prep. HPLC, etc., and identified by spectroscopy, mainly 1D and 2D NMR. The GC-MS technique has proved to be very effective for rapid separation and identification of complex mixtures of Amaryllidaceae alkaloids obtained from low mass samples (Kreh *et al.,* 1995). Thus, the assessment of alkaloid distribution at species, populational and individual levels and the detection of new compounds have become much easier and faster (Berkov *et al.,* 2007*a*, 2009*c*, 2011).

An overview of the literature indicates that the genus *Galanthus* is a very rich source of novel compounds. Thirty-seven alkaloids (namely **12, 22, 26, 29, 34-39, 46-49, 53, 56-58, 62, 67, 69-75, 77-86**) or *ca.* 40% of all identified compounds from the genus have been isolated for the first time from *Galanthus*. What is more, the biochemical evolution of the genus has led to the occurrence of two specific subgroups, namely graciline- and plicamine-type alkaloids.

The most studied species are *G. nivalis* and *G. elwesii*. Due to taxonomical changes over the years, the information on the alkaloids of *G. nivalis* is confusing. Thus, until 1966, only one *Galanthus* species had been recognized in Bulgaria, namely *G. nivalis* L. (Jordanov, 1964). This taxon was subsequently separated into *G. nivalis* L. and *G. elwesii* Hook. (Kozuharov, 1992). At present, it is unclear which plant species the alkaloids isolated in the early sixties from Bulgarian *G. nivalis* can be attributed to (Valkova, 1961; Bubeva-Ivanova and Pavlova, 1965). Kaya *et al.* (2004*b*) have reported five alkaloids for *G. nivalis* L. subsp. *silicicus* (Baker) Guttl.-Tann., a taxon regarded as a synonym of *G. silicicus* Baker by other authors (Davis and Barnett, 1997; Davis, 1999). A recent revelation has substantiated that *G. nivalis* subsp. *cilicicus* is identical to the newly introduced species, G. *trojanus* A. P. Davis and N. Özhatay, a plant species endemic to Northwestern Turkey (Davis and Özhatay, 2001).

Latvala *et al.,* (1995) isolated 18 alkaloids (6 new) from *G. elwesii* in addition to the already reported flexinine, elwesine, tazettine and haemanthamine (Boit and Ehmke, 1955; Boit and Döpke, 1961). The occurrence of elwesine (**26**) in the genus is particularly interesting. This compound displays a β-configuration of its

5,10b-ethano bridge, which is typical of the South African representatives of the family (Viladomat *et al.,* 1997). Although widely accepted that *G. nivalis* was the industrial source of galanthamine (in Bulgaria) during the 1960s (Heinrich and Teoh, 2004), later studies on 32 Bulgarian populations of *G. nivalis* and *G. elwesii* indicate that the distribution of this important compound is limited to a few populations of *G. elwesii*, while just one population of *G. nivalis* has been found to contain galanthamine and only as a minor alkaloid (Sidjimova *et al.,* 2003; Berkov *et al.,* 2011). These studies, however, have also shown a great intra-species diversity of alkaloid synthesis in *G. nivalis* and *G. elwesii.* The populations displayed between 6 and 31 alkaloids in their alkaloid patterns and about 70 compounds have been detected in total. Many of them were left unidentified due to the lack of reference spectra, possibly indicating new structures. This biochemical diversity has led to the isolation of eight more new alkaloids from these well- studied species, after the collection of plant material from populations proven by GC-MS to be a rich source of unknown compounds (Berkov *et al.,* 2007*a*, 2009*c*). Interestingly, many of the *G. elwesii* populations have accumulated the tyramine-type protoalkaloids as major compounds (up to 99 % of all alkaloids). In addition to the tyramine chemotype, homolycorine, lycorine haemanthamine and galanthamine chemotypes have also been found in the studied populations of *G. elwesii.* A galanthamine chemotype population was also found for *G. nivalis*, but in contrast with *G. elwesii*, this *G. nivalis* population accumulated the 4,4a-dihydrogenated derivatives of galanthamine (**12**), lycoramine (**16**) and its isomer (**17**) (Berkov *et al.,* 2011).

As well as a high level of alkaloid diversity and the existence of different chemotypes among the species populations, *G. elwesii* and *G. nivalis* have also shown some important differences in their alkaloid patterns, at least in the studied Bulgarian populations. A study of sympatric populations, and 32 populations from both species showed that the alkaloid pattern of *G. nivalis* is dominated by compounds coming from a *para-para′* oxidative coupling of *O*-methylnorbelladine (haemanthamine- and tazettine-type alkaloids, Fig. 1). The conjugated and free lycorine-type alkaloids proceeding from an *ortho-para′* oxidative coupling were relatively less abundant. Homolycorine-type alkaloids were not detected in this plant species. In contrast to *G. nivalis*, the alkaloid pattern of *G. elwesii* was dominated mainly by compounds coming from *ortho-para′* oxidative coupling: free lycorine- and homolycorine-type alkaloids. The synthesis of *para-para′* oxidative products in *G. elwesii* is relatively weak (only haemanthamine- and no tazettine-type compounds, Berkov *et al.,* 2008, 2011). In total, 46 and 38 alkaloids have been identified in *G. elwesii* and *G. nivalis,* respectively.

Table 1. Alkaloids reported in the genus *Galanthus*

Compound	*G. elwesii*	*G.nivalis*	*G. plicatus*	*G. gracilis*	*G.woronowii*	*G.caucasicus*	*G.ikariae*	*G.krasnovii*	*G. reginae-olgae*	*G.trojanus*	*G. rizehensis*
I. Tyramine type											
Tyramine (**1**)	+1	+1								+25	
Methyltyramine (**2**)	+1	+1									
Hordenine (**3**)	+1,2	+1	+10								
N-feruloyltyramine (**4**)	+2										
II. Norbelladine											
O-Methylnorbelladine (**5**)										+25	
III. Narciclasine type											
Ismine (**6**)	+1	+1,7	+11								
N-Formylismine (**7**)			+12								
Trisphaeridine (**8**)	+1	+1	+13								
5,6-Dihydrobicolorine (**9**)			+13	+15							
Arolycoricidine (**10**)										+25	+27
Narciprimine (**11**)											+27
IV. Galanthamine type											
Galanthamine (**12**)	+1,2	+1			+19	+22	+4	+23	+24		
3-Epigalanthamine (**13**)	+1										
Narwedine (**14**)	+1,2										
N-Demethylgalanthamine (**15**)	+1,2										
Lycoramine (**16**)	+1	+1									
3-Epilycoramine (**17**)	+1	+1									
Sanguinine (**18**)	+2										
N-Formylnorlgalanthamine (**19**)	+1										
Leucotamine (**20**)	+1,2										
O-Methylleucotamine (**21**)	+2										
Nivalidine (**22**)	+3	+8									
V. Haemanthamine type											
Buphanisine (**23**)	+1,4										
Vittatine/crinine (**24**)	+1,5								+24	+26	
Flexinine (**25**)	+6										
Elwesine (**26**)	+6										
Hamayne (**27**, 3-Epihydroxybulbispermine)	+7	+7,9					+4				
11-Hydroxyvittatine (**28**)	+2		+10							+26	
11-Hydroxyvittatine *N*-oxide (**29**)										+25	
Maritidine (**30**)	+1										
8-*O*-Demethylmaritidine (**31**)										+25	

Narcidine (**32**)										+25	
Haemanthamine (**33**)	+1	+1								+25	
11-*O*-(3'-Hydroxybutanoyl)hamayne (**34**)	+1,7	+9									
3,11-*O*-(3',3''-Dihydroxybutanoyl) hamayne (**35**)		+9									
3-*O*-(2''-Butenoyl)-11-*O*-(3'-hydroxybutanoyl)hamayne (**36**)		+9									
3,11,3'-*O*-(3', 3'', 3'''- Trihydroxybutanoyl)-hamayne (**37**)		+9									
3,3'-*O*-(3',3''-Dihydroxybutanoyl) hamayne (**38**)		+7									
11,3'-*O*-(3',3''-Dihydroxybutanoyl) hamayne (**39**)		+7									
VI. Tazettine type 11-Deoxytazettine (**40**)	+1	+1									
6-*O*-Methylpretazettine (**41**)	+1	+1									
Tazettine (**42**)	+1	+7	+11	+15	+20	+22	+4		+24		
Criwelline (**43**)		+6									
Macronine (**44**)		+1									
Epimacronine (**45**)		+7	+11	+15							
3-*O*-Demethyl-3-epimacronine (**46**)			+13								
3-*O*-Demethylmacronine (**47**)				+13							
3-*O*-(3′-Hydroxybutanoyl)tazettinol (**48**)			+12								
Isotazettinol (**49**)				+13							
VII. Lycorine type Anhydrolycorine (**50**)	+1	+1									
11,12-Dehydroanhydrolycorine (**51**)	+1	+1									
Caranine (**52**)		+5									
Galanthine (**53**)	+1,2	+1			+21	+22					
Lycorine (**54**)	+1,2	+1,7	+14	+14	+21	+22			+24	+26	
Incartine (**55**)	+1	+1									
2-*O*-(3'-Hydroxybutanoyl)lycorine (**56**)	+1,7	+1									
2?-*O*-(3'-Hydroxybutanoyl)lycorine isomer (**57**)		+5									
2-*O*-(3'-Acetoxybutanoyl)lycorine (**58**)		+1,9									
Ungeremine (**59**)		+9									
8-*O*-Demethylvasconine (**60**)		+7									
Nartazine (**61**) 8-*O*-Methyldihydrosternbergine *N*-oxide (**62**)		+6								+25	
Dihydrolycorine (**63**)										+25	

VIII. Homolycorine type Homolycorine (**64**)	+1										
8-*O*-Demethylhomolycorine (**65**)	+1,2		+15	+15		+22	+4				
Masonine (**66**)	+5	+6									
2-Methoxy-8-*O*-demethylhomolycorine (**67**)	+1,2										
Hippeastrine (**68**)	+1	+8									
Galwesine (**69**)	+1,2										
8-*O*-Demethylgalwesine (**70**)	+2										
8-*O*-Demethyl-10b-hydroxygalwesine (**71**)	+2										
10b-Hydroxygalwesine (**72**)	+2										
Galasine (**73**)	+2										
2α-Hydroxyhomolycorine (**74**)	+1										
Galanthindole (**75**)			+10								
Neronine (**76**)									+24		
Galanthusine (**77**)						+22					
IX. Graciline type Graciline (**78**)				+16							
11-Acetoxygraciline (**79**)			+16								
3,4-Dihydro-3-hydroxygraciline (**80**)				+12							
3-Epi-3,4-dihydro-3-hydroxygraciline (**81**)				+12							
Digracine (**82**)				+16							
Gracilamine (**83**)				+18							
X. Plicamine type Plicamine (**84**)			+17								
Plicane (**85**)			+12								
Secoplicamine (**86**)			+17								
XI. Other											
Bulbocapnine (**87**)										+26	
Capnoidine (**88**)										+26	
Stylopine (**89**)										+25	
Protopine (**90**)										+25	

1) Berkov *et al.*, (2011); 2) Latvala *et al.*, (1995); 3) Bubeva-Ivanova and Pavlova (1965); 4) Sener *et al.*, (1998); 5) Berkov *et al.*, (2008); 6) Wildman, (1968); 7) Berkov *et al.*, (2009c); 8) Kalashnikov (1970); 9) Berkov *et al.*, (2007a); 10) Ünver *et al.*, (2003); 11) Ak□neri and Günes (1998); 12) Ünver *et al.*, (2001); 13) Ünver *et al.*, (1999a); 14) Kaya *et al.*, (2004*a*); 15) Noyan (1999); 16) Noyan *et al.*, (1998); 17) Ünver *et al.*, (1999b); 18) Ünver and Kaya, (2005); 19) Proskurina *et al.*, (1955); 20) Yakovleva (1963); 21) Proskurina Ordzhonikidze (1953); 22) Tsakadze *et al.*, (1979); 23) Asoeva *et al.*, (1968); 24) Conforti *et al.*, (2010); 25) Kaya *et al.*, (2011); 26) Kaya *et al.*, (2004*b*); 27) Bozkurt *et al.*, (2010).

In a study on sympatric *G. nivalis* and *G. elwesii* populations, it was found that the organs of the plants presented different alkaloid patterns (Berkov *et al.*, 2008). Thus, the predominant alkaloids of *G. nivalis* roots were found to belong to the lycorine and tazettine structural types, bulbs were dominated by tazettine, leaves by lycorine and flowers by haemanthamine-type alkaloids. The predominant alkaloids in *G. elwesii* roots, bulbs and leaves were those of the homolycorine type, whereas the flowers accumulated mainly tyramine-type compounds. To the best of our knowledge, no studies of the dynamics of the alkaloid patterns during ontogenesis have been reported for either of these two species or any other *Galanthus* species. Such studies, however, may contribute to the understanding of the chemoecological role of the alkaloids in the genus *Galanthus* and the Amaryllidaceae as a whole. A remarkably high number of alkaloids conjugated with 3-hydroxybutyryl moieties occur in *G. nivalis*. Co-existence of free and conjugated alkaloids in the plant implies that the latter may have a chemoecological role. Such conjugated alkaloids have rarely been reported for Amaryllidaceae plants.

Another two phytochemically interesting species from which a number of new alkaloids have been isolated are *G. gracilis* and *G. plicatus*. Phytochemical studies on *G. gracilis* resulted in the isolation of three novel monomeric alkaloids (**78, 80, 81**) and a dimeric compound (**82**) bearing a 10b,4a-ethanoiminodibenzo[b,d]pyrane skeleton, which represents a new subgroup of Amaryllidaceae alkaloids named gracilines (Fig. 1, Noyan *et al.*, 1998; Ünver *et al.*, 2001). An unusual pentacyclic dinitrogenous alkaloid, gracilamine (**83**), was also isolated from this species (Ünver and Kaya, 2005). Another new graciline-type alkaloid (**79**, Noyan *et al.*, 1998) has been isolated from *G. plicatus*, together with compounds **84-86** (Ünver et al., 1999*a*, 2001), representing a new subgroup of the Amaryllidaceae alkaloids where the oxygen atom at position 5 of a tazettine molecule is replaced by a nitrogen atom, conjugated with a 4-hydroxyphenethyl moiety. This new subgroup, named after the lead compound plicamine (**84**), was found later in another amaryllidaceous plant, *Cyrtanthus obliquus* (Brine *et al.*, 2002). Apart from plicamines, four new tazettine-type alkaloids (**46-49**) and a compound with a nonfused indole ring (**75**) have also been isolated in *G. plicatus* (Ünver *et al.*, 1999*b*, 2003). In total, 17 and 12 alkaloids have been reported for *G. plicatus* and *G. gracilis,* respectively (Table 1).

The other *Galanthus* species are relatively less studied. Four known alkaloids (**12**, **42**, **53**, and **54**), including galanthamine, have been reported for *G. woronovii* (Proskurina *et al.*, 1955; Proskurina and Ordzhonikidze, 1953; Yakovleva, 1963). A new compound, galanthusine (**78**), has been found in *G. caucasicus*, along with five known alkaloids (**12, 42, 53, 54,** and **65;** Tsakadze *et al.*, 1979). Only galanthamine has been reported for *G. krasnovii* (Asoeva *et al.*, 1968). *G. ikariae* has furnished four known alkaloids (**12, 27, 42,** and **65;** Sener *et al.*, 1998).

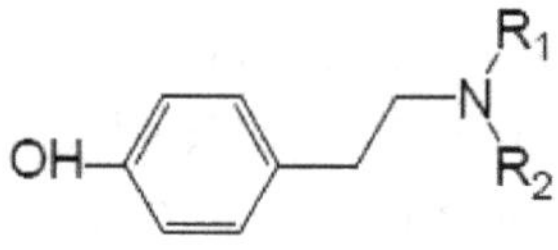

1 $R_1=R_2=H$
2 $R_1=H$, $R_2=Me$
3 $R_1=R_2=Me$
4 $R_1=H$, $R_2=$ feruloyl

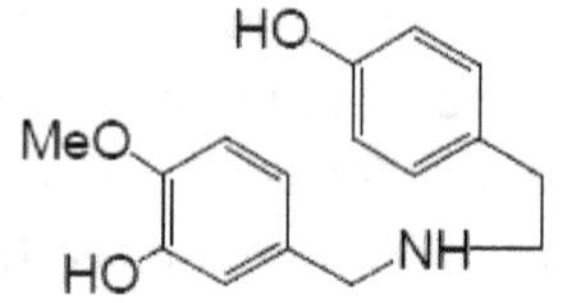

5

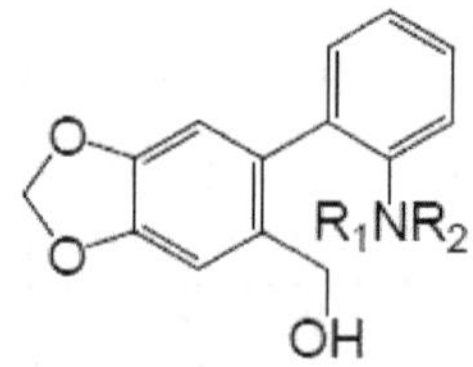

6 $R_1=H$, $R_2=Me$
7 $R_1=CHO$, $R_2=Me$

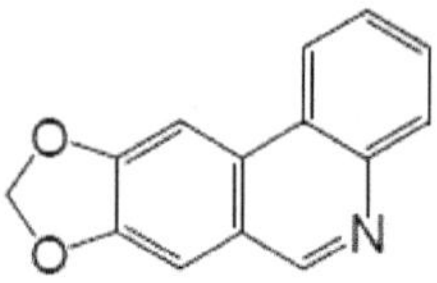

8

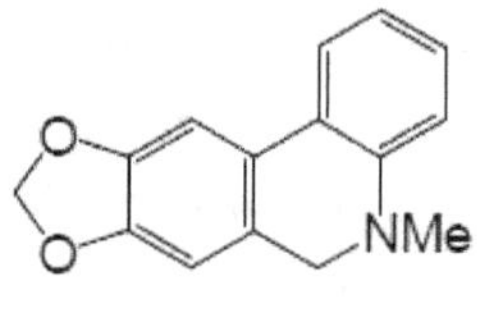

9

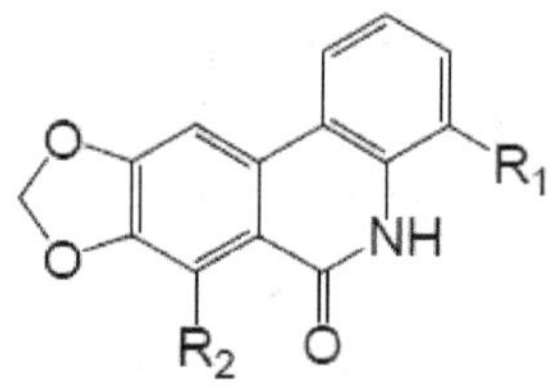

10 $R_1=OH$, $R_2=H$
11 $R_1=R_2=OH$

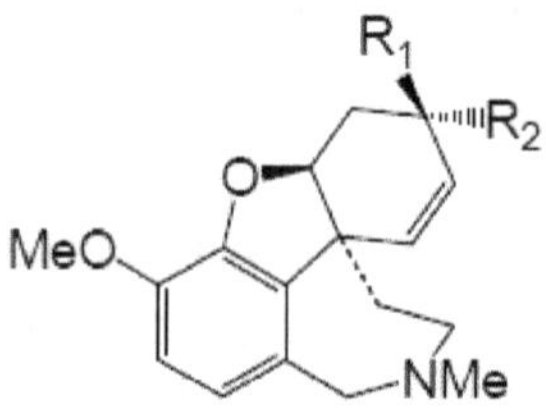

16 $R_1=OH$, $R_2=H$
17 $R_1=H$, $R_2=OH$

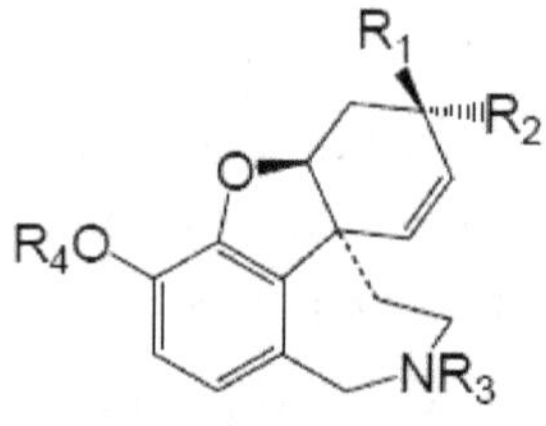

12 $R_1=OH$, $R_2=H$, $R_3=R_4=Me$
13 $R_1=H$, $R_2=OH$, $R_3=R_4=Me$
14 $R_1+R_2=O$, $R_3=R_4=Me$
15 $R_1=OH$, $R_2=R_3=H$, $R_4=Me$
18 $R_1=OH$, $R_2=H$, $R_3=Me$, $R_4=H$
19 $R_1=OH$, $R_2=H$, $R_3=CHO$, $R_4=Me$
20 $R_1=OCOCH_2CHOHMe$, $R_2=H$, $R_3=Me$, $R_4=H$
21 $R_1=OCOCH_2CHOHMe$, $R_2=H$, $R_3=R_4=Me$

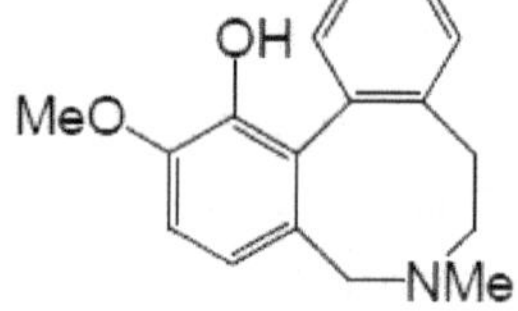

22

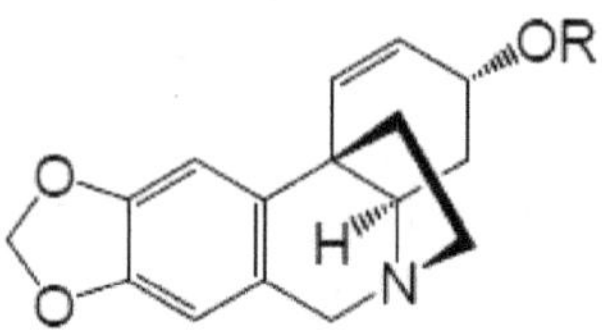

23 R=Me
24 R=H (crinine)

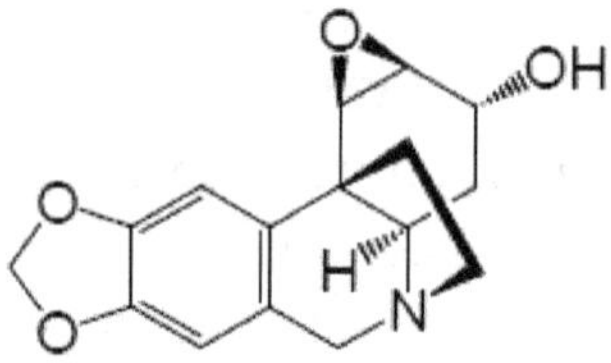

25

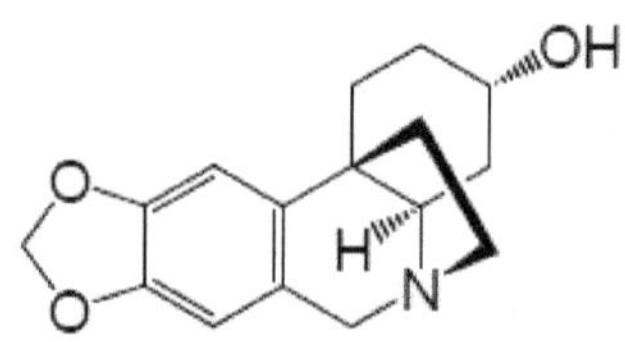

26

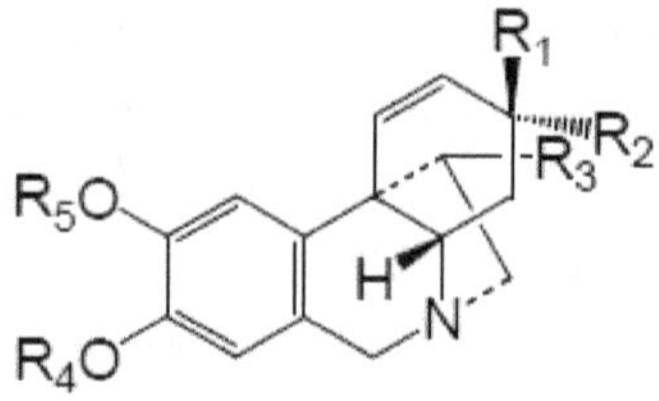

24 $R_1=OH$, $R_2=R_3=H$, $R_4+R_5=CH_2$(vittatine)
27 $R_1=H$, $R_2=R_3=OH$, $R_4+R_5=CH_2$
28 $R_1=OH$, $R_2=H$, $R_3=OH$, $R_4+R_5=CH_2$
30 $R_1=OH$, $R_2=R_3=H$, $R_4=R_5=Me$
31 $R_1=OH$, $R_2=R_3=R_4=H$, $R_5=Me$
32 $R_1=OMe$, $R_2=H$, $R_3=OH$, $R_4=H$, $R_5=Me$
33 $R_1=OMe$, $R_2=H$, $R_3=OH$, $R_4+R_5=CH_2$
34 $R_1=H$, $R_2=OH$, $R_3=OCOCH_2CHOHMe$, $R_4+R_5=CH_2$
35 $R_1=H$, $R_2=R_3=OCOCH_2CHOHMe$, $R_4+R_5=CH_2$
36 $R_1=H$, $R_2=OCOCH=CHMe$, $R_3=OCOCH_2CHOHMe$, $R_4+R_5=CH_2$
37 $R_1=H$, $R_2=OCOCH_2CH(Me)OCOCH_2CHOHMe$, $R_3=OCOCH_2CHOHMe$, $R_4+R_5=CH_2$
38 $R_1=H$, $R_2=OCOCH_2CH(Me)OCOCH_2CHOHMe$, $R_3=OH$, $R_4+R_5=CH_2$
39 $R_1=H$, $R_2=OH$, $R_3=OCOCH_2CH(Me)OCOCH_2CHOHMe$, $R_4+R_5=CH_2$

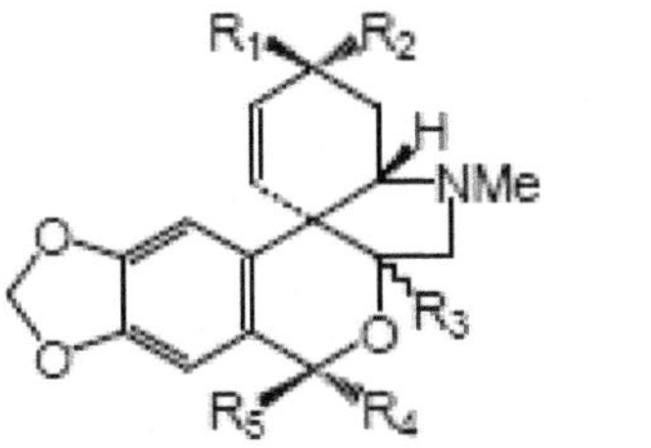

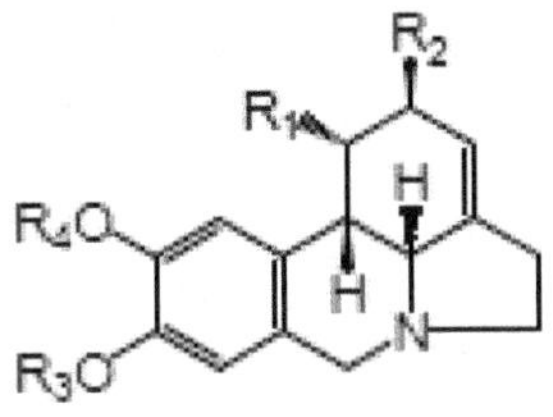

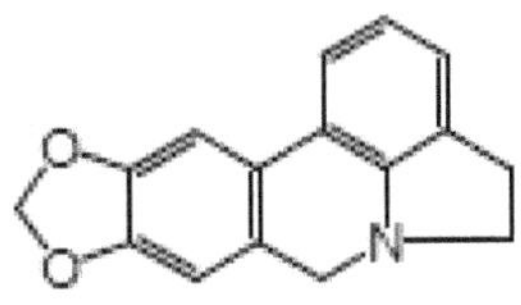

50

40 R_1=OMe, R_2=H, R_3=αH, R_4=R_5=H
41 R_1=OMe, R_2=H, R_3=βH, R_4=H, R_5=OMe
42 R_1=OMe, R_2=H, R_3=αOH, R_4=R_5
43 R_1=H, R_2=OMe, R_3=αOH, R_4=R_5=H
44 R_1=H, R_2=OMe, R_3=βH, R_4+R_5=O
45 R_1=OMe, R_2=H, R_3=βH, R_4+R_5=O
46 R_1=OH, R_2=H, R_3=βH, R_4+R_5=O
47 R_1=H, R_2=OH, R_3=βH, R_4+R_5=O
48 R_1=OCOCH$_2$CHOHMe, R_2=H, R_3=αOH, R_4=R_5=H
49 R_1=H, R_2=OH, R_3=αOH, R_4=R_5=H

52 R_1=OH, R_2=H, R_3+R_4=CH$_2$
53 R_1=OH, R_2=OMe, R_3=R_4=Me
54 R_1=R_2=OH, R_3+R_4=CH$_2$
56 R_1=OH, R_2=OCOCH$_2$CHOHMe, R_3+R_4=CH$_2$
58 R_1=OH, R_2=OCOCH$_2$CHOAcMe, R_3+R_4=CH$_2$

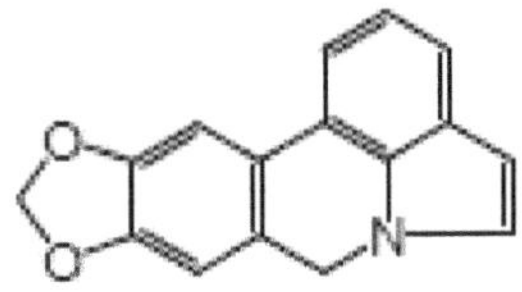

51

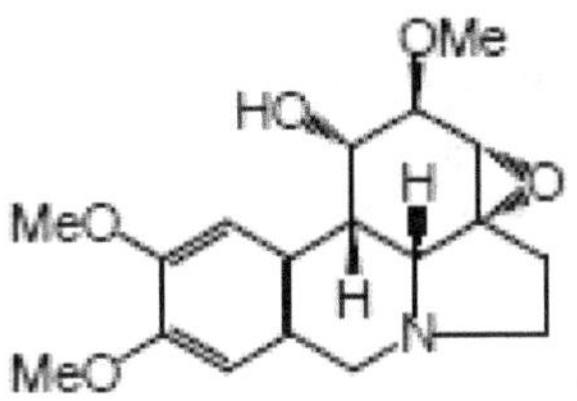

55

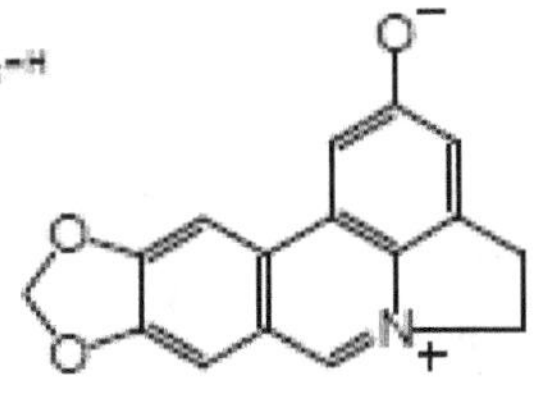

59

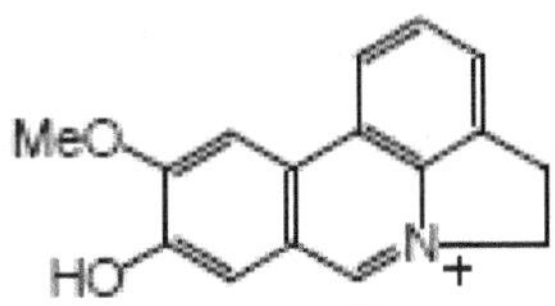

60

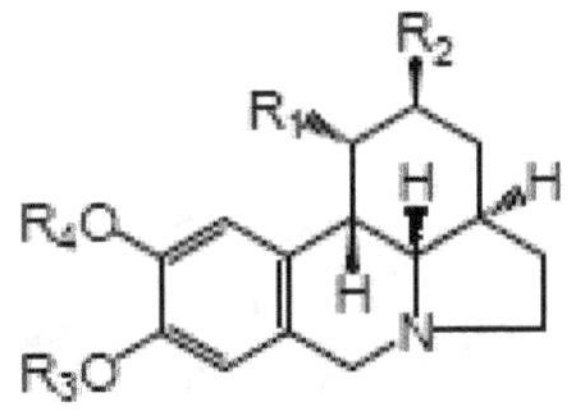

61 R_1=R_2=OAc, R_3+R_4=CH$_2$
63 R_1=R_2=OH, R_3+R_4=CH$_2$

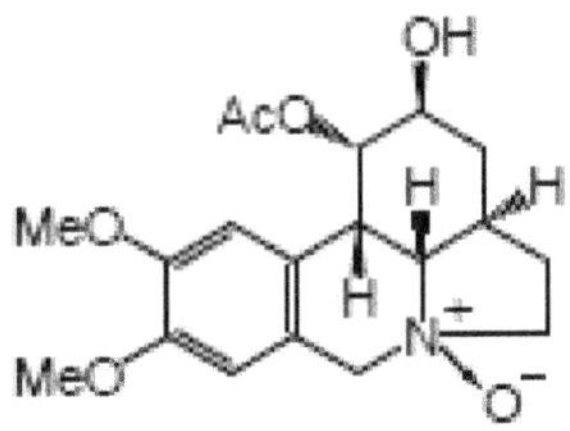

62

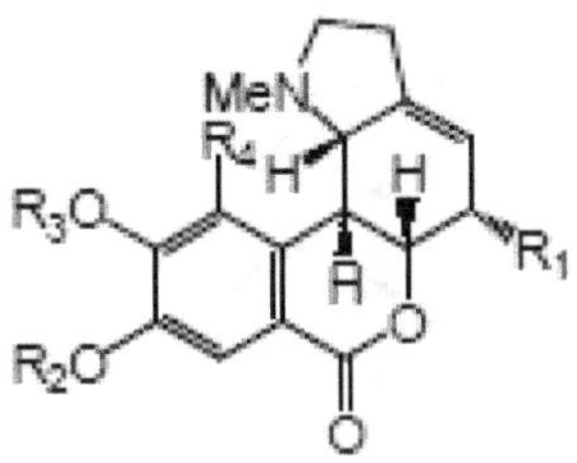

64 R_1=H, R_2=R_3=Me, R_4=H
65 R_1=R_2=H, R_3=Me, R_4=H
66 R_1=H, R_2+R_3=CH$_2$, R_4=H
67 R_1=OMe, R_2=H, R_3=Me, R_4=H
68 R_1=OH, R_2+R_3=CH$_2$, R_4=H
74 R_1=OH, R_2=R_3=Me, R_4=H
76 R_1=OH, R_2+R_3=CH$_2$, R_4=OMe

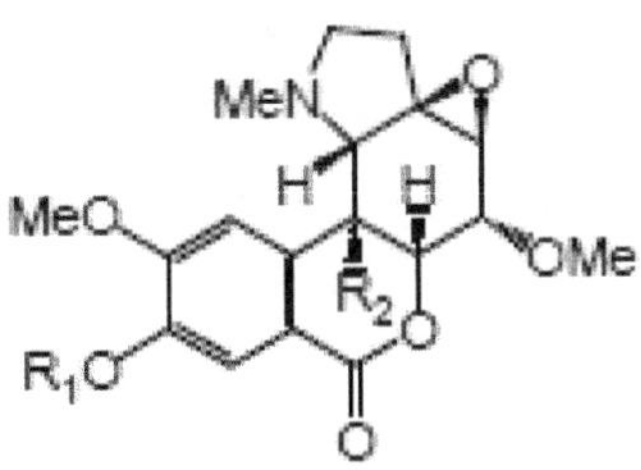

69 R_1=Me, R_2=H
70 R_1=R_2=H
71 R_1=H, R_2=OH
72 R_1=Me, R_2=OH

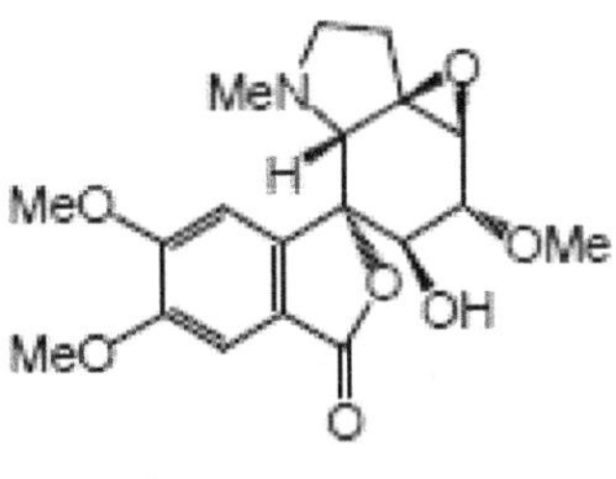

73

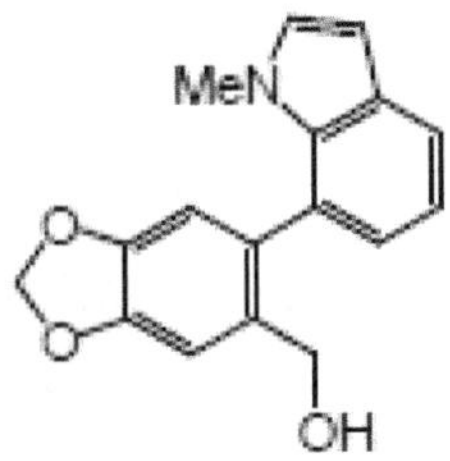

75

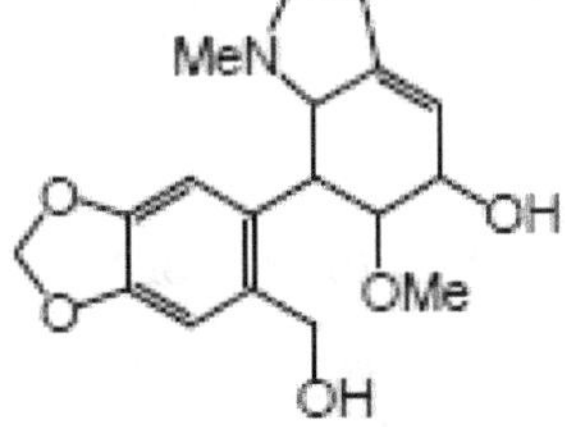

77

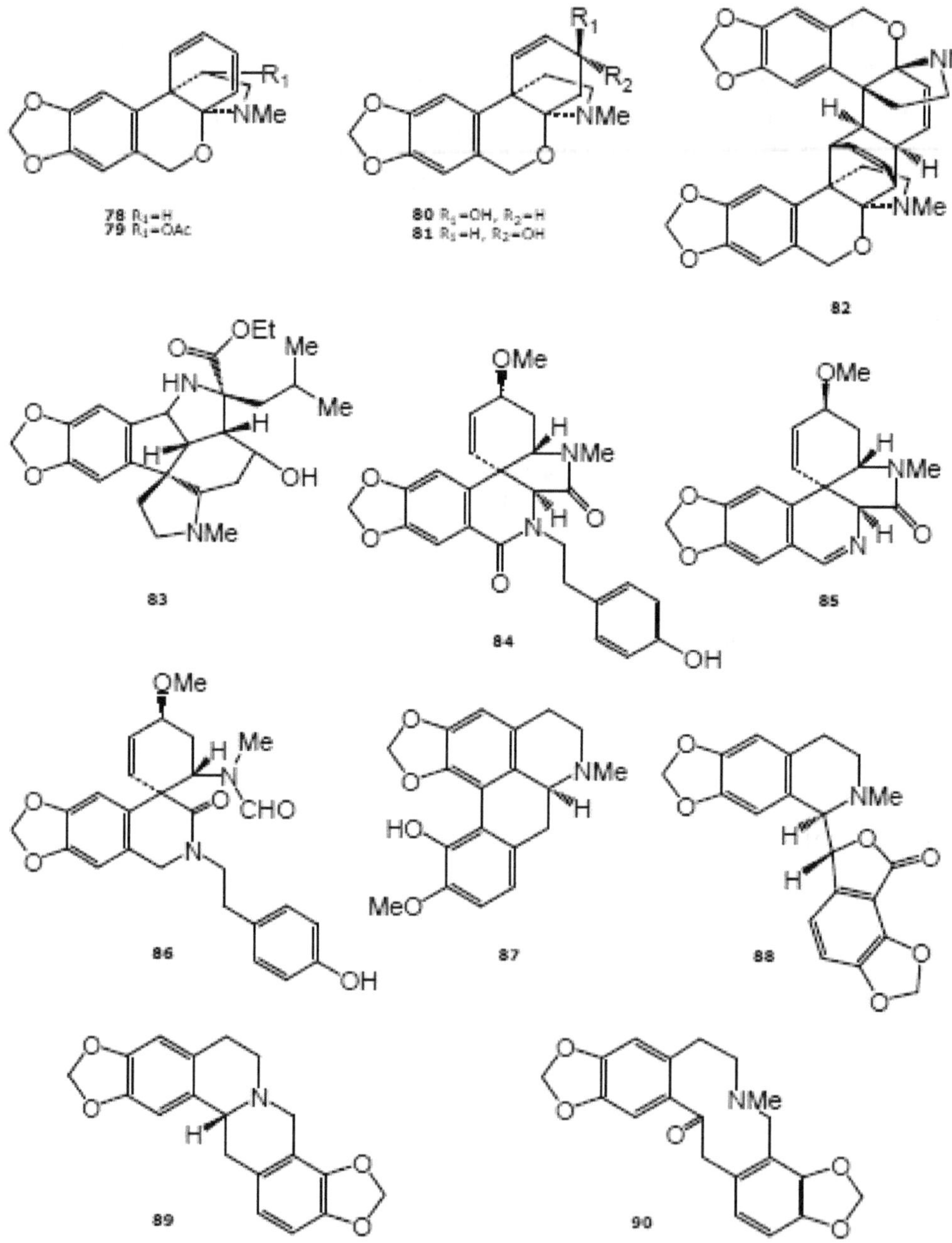

Fig. 2. Structures of the alkaloids found in the genus *Galanthus*.

A recent GC-MS report on *G. reginae-olgae* resulted in the identification of compounds **12, 24, 42**, and **66** (Conforti, *et al.*, 2010). The presence of crinine (with the 5,10b-ethano bridge at the β-position) in this species, as well as in *G. elwesii*, as reported in our earlier GC-MS studies (Berkov *et al.*, 2004), is debatable because the absolute configuration of the 5,10b-ethano bridge cannot be established by GC-MS alone. Later phytochemical studies on *Galanthus* resulted in the isolation of crinane-3-ol derivatives with a α-configuration of their 5,10b-ethano bridges, including vittatine (Kaya *et al.*, 2004*b*), which is the optical isomer of crinine, 11-hydroxyvittatine (Latvala *et al.*, 1995; Kaya *et al.*, 2004*b*); Ünver *et al.*, 2003) and hamayne (Berkov *et al.*, 2007*a*; 2009c). On the other hand, elwesine (**26**, 2,3-dihydrocrinine) and buphanisine (**23**) display a β-configuration of the 5,10b-ethano bridge (Wildman, 1968, Capo and Saa, 1989). Recently initiated phytochemical studies on *G. rizehensis* (Bozkurt *et al.*, 2010) have identified two narciclasine-type compounds, arolycoricidine (**10**) and narciprimine (**11**). An interesting example of biochemical convergence is the presence of bulbocapnine (**87**), capnoidine (**88**), stilopine (**89**) and protopine (**90**) in *G. trojanus* (studied as *G. nivalis* subsp. *silicicus* (Baker) Gottlieb-Tannenhain). Two new alkaloids, the *N*-oxides of 9-*O*- methyldihydrosternbergine (**62**) and 11-hydroxyvittatine (**29**), were also isolated, along with several known alkaloids **2, 5, 10, 24, 28, 29, 31-33, 54, 62** and **63** (Kaya *et al.*, 2004*b*; Ünver 2007). Compounds **84-90** are benzyltetrahydroisoquinoline-, aporphine- and phthalide-type isoquinolines, found in dicotyledonous plants of the Fumariaceae and Papaveraceae families (Kametani and Honda 1985; MacLean, 1985).

5. BIOLOGICAL AND PHARMACOLOGICAL ACTIVITIES OF THE ALKALOID FOUND IN *GALANTHUS*

Alkaloids are important for the well-being of the producing organism. One of their main functions is to provide a chemical defence against herbivores, predators or microorganisms (Wink, 2008). The biological roles of the numerous alkaloids found in the genus *Galanthus* remain largely unknown and only a few have been studied for their pharmacological activities.

Galanthamine-type

The most studied *Galanthus* alkaloid, galanthamine (**12**), is a long-acting, selective, reversible and competitive inhibitor of acetylcholinesterase (AChE) and an allosteric modulator of the neuronal nicotinic receptor for acetylcholine. AChE is responsible for the degradation of acetylcholine at the neuromuscular junction, in peripheral and central cholinergic synapses. Galanthamine has the ability to cross the blood-brain barrier and to act within the central nervous system (Bastida *et al.*, 2006; Heinrich and Teoh, 2006). Owing to its AChE inhibitory activity, galanthamine is used and marketed under the name of Razadine®, formerly Reminyl®, in the USA, for the treatment of certain stages of Alzheimer's Disease (AD). According to data presented by the Alzheimer's Association in 2007, the prevalence of Alzheimer's disease will quadruple by 2050. Galanthamine hydrobromide has superior pharmacological profiles and higher tolerance as compared to the original AChE inhibitors, physostigmine or tacrine (Grutzendler and Morris, 2001).

Epigalanthamine (**13**), with a hydroxylgroup at *α*-position, and narwedine (**14**), with a keto group at C3, are also active AChE inhibitors, but about 130-times less than galanthamine (Thomsen *et al.,* 1998). The loss of the methyl group at the *N* atom, as in *N*- demethylgalanthamine (**15**), decreases the activity 10-fold. On the other hand, sanguinine (**18**), which has a hydroxylgroup at C9 instead of a methoxyl group, is *ca.* 10 times more active than galanthamine. Hydrogenation of the C4-C4a, as in lycoramine (**16**), results in a complete loss of AChE inhibitory activity (López *et al.,* 2002). It is suggested that in plants AChE inhibitors act as pesticides. The synthetic pesticides such as phosphoorganic compounds are non-reversible AChE inhibitors (Hougton *et al.,* 2006).

Tyramine-type

Compounds **1-4** can be attributed to the group of the phenolic amines that impact the hypothalamic-pituitary-adrenal axis (Vera-Avila *et al.,* 1996) due to their structural similarity to adrenaline (epinefrine). The consequent release of adrenocorticotropic hormone and cortisol results in sympathomimetic action with toxic effects in animals (Clement *et al.,* 1998). Hordenine (**3**) possesses diuretic, disinfectant and antihypotensive properties, and acts as a feeding repellent against grasshoppers (Dictionary of Natural Products).

Narciclasine-type

Trisphaeridine (**8**) has a high retroviral activity but a low therapeutic index. Ismine (**6**) shows a significant hypotensive effect on rats and cytotoxicity against Molt 4 lymphoid and LMTK fibroblastic cell lines (Bastida *et al.,* 2006). A recent study revealed that arolycoricidine (**10**) and narciprimine (**11**) were considerably effective in DNA topoisomerase reactions in a dose- dependent manner. Topoisomerase-interfering ability of these alkaloids partially correlated with cytostatic assays, using HeLa (cervix adenocarcinoma), MCF7 (breast adenocarcinoma) and A431 (skin epidermoid carcinoma) cells (Bozkurt *et al.,* 2010). Arolycoricidine showed inhibitory activity against African trypanosomes, (*Trypanosoma brucei rhodesiense)* at micromolar levels (Kaya *et al.,* 2011).

Haemanthamine type

Haemanthamine **(33)** has been shown to be a potent inducer of apoptosis in tumour cells at micromolar concentrations (McNulty *et al.,* 2007). This compound also possesses antimalarial activity against strains of chloroquine-sensitive *Plasmodium falciparum,* hypotensive effects and antiretroviral activity (Bastida *et al.,* 2006; Kaya *et al.,* 2011). Vittatine (**24**) and maritidine (**30)** have shown cytotoxic activity against HT29 colon adenocarcinoma, lung carcinoma and RXF393 renal cell carcinoma (Bastida *et al.,* 2006; Silva *et al.,* 2008). Antibacterial activity against Gram-positive *Staphylococcus aureus* and Gram-negative *E. coli* have been reported for vittatine

(**24**) and 11-hydroxyvittatine (**28**) (Kornienko and Evidente, 2008). Data about the bioactivity of recently isolated compounds **34-39** is still lacking.

Tazettine-type

Moderate cytotoxic activity has been reported for tazettine (**42**), and epimacronine (**45**) (Weniger *et al.*, 1995). Tazettine, however, is an isolation artefact of chemically labile pretazettine, which is indeed present in plants. This compound has shown remarkable cytotoxicity against a number of tumor cell lines, being therapeutically effective against advanced Rauscher leucemia, Ehrlich ascites carcinoma, spontaneous AKR lymphocytic leukaemia, and Lewis lung carcinoma (Bastida *et al.*, 2006).

Lycorine-type

Lycorine (**54**), one of the most frequently occurring alkaloids in Amaryllidaceae plants, possesses a vast array of biological properties. It has been reported as a potent inhibitor of ascorbic acid synthesis, cell growth and division and organogenesis in higher plants, algae, and yeasts, inhibiting the cell cycle during the interphase (Bastida *et al.*, 2006). Additionally, lycorine exhibits antiviral (against poliovirus, vaccine smallpox virus and SARS-associated coronavirus), antifungal (*Saccharomyces cerevisiae, Candida albicans*), and anti-protozoan (*Trypanosoma brucei*) activities (McNulty *et al.*, 2009), and is more potent than indomethacin as an anti-inflammatory agent (Citoglu *et al.*, 1998). Lycorine has also been shown to have insect antifeedant activity (Evidente *et al.*, 1986). As a potential chemotherapeutic drug, this compound has been studied as an antiproliferative agent against a number of cancer cell lines (Likhitwitayawuid *et al.*, 1993). The *in vitro* mode of action in a HL-60 leukemia cell line model is associated with suppressing tumor cell growth and reducing cell survival via cell cycle arrest and induction of apoptosis (Liu *et al.*, 2004). Further investigation showed that it is able to decrease tumor cell growth and increase survival rates with no observable adverse effects in treated animals (Liu *et al.*, 2007), thus being a good candidate for a therapeutic agent against leukaemia (Liu *et al.*, 2009).

Anhydrolycorine (**50**), in contrast to caranine (**52**), has shown a higher ability to inhibit ascorbic acid synthesis than lycorine (Evidente *et al.*, 1986). Analgesic and hypotensive effects have been reported for caranine and galanthine (**53**), the latter also being active against *Tripanosoma brucei rhodesiense* and *Plasmodium falciparum*. Some lycorine-type compounds such as caranine and ungeremine (**59**) have shown acetylcholinesterase inhibitory activity (Bastida *et al.*, 2006). Incartine was found to be cytotoxic and to weakly inhibit AChE (Berkov *et al.*, 2007).

Homolycorine-type

Cytotoxic activity has been demonstrated for homolycorine (**64**), 8-*O*-demethylhomolycorine (**65**), and hippeastrine (**68**). Homolycorine has shown high antiretroviral activity, while hippeastrine is active against *Herpes simplex* type 1. Homolycorine and 8-*O*- demethylhomolycorine have a hypotensive effect on normotensive rats. In addition, hippeastrine shows antifungal activity against *Candida albicans* and also possesses a weak insect antifeedant activity (Bastida *et al.*, 2006).

The bioactivity of the plicamine- and graciline-type alkaloids is largely unknown. Bulbocapline (**87**) and protopine (**90**) have been shown to act as inhibitors of acetylcholinesterase (Kim et al., 1999; Adsersen *et al.*, 2007) and dopamine biosynthesis (Shin *et al.*, 1998). Stylopine (**89**) suppresses the NO and PGE2 production in macrophages by inhibiting iNOS and COX-2 expression (Jang et al., 2004).

6. CONCISE SUMMARY

Despite the limited study of certain species within this phytochemically intriguing genus, it has produced a significant array of novel structures. Furthermore, the significant intraspecies diversity suggests that novel chemicals may be anticipated from previously examined taxa. Only a limited number of the novel alkaloids have undergone screening for their biological and pharmacological properties, perhaps due to the minimal quantities isolated. Thus, their synthesis or computational research will enhance subsequent bioactivity evaluation.

7. REFERENCES

Adsersen, A.; Kjølbye, A.; Dall, O.; Jäger, A.K. (2007). Acetylcholinesterase and butyrylcholinesterase inhibitory compounds from *Corydalis cava* Schweigg. & Kort. *Journal of Ethnopharmacology* 113, 79-82

Alzheimer's Association (2010). Alzheimer's disease facts and figures Alzheimer's & Dementia 6, 158–194

Ak$_1$neri, G.; Gunes, H.S. (1998) *Galanthus plicatus* ssp. *byzantinus* as a new source for some Amaryllidaceae alkaloids. *J Fac Pharm Gazi* 15, 99–106

Asoeva, E.; Murabeva, D.; Molodozhnikov, M.; Rabinovich, I. (1968). *Galanthus krasnovii* – A source for obtaining galanthamine. *Farmacia (Moskow)* 17, 47-49

Bastida, J.; Lavilla, R.; Viladomat, F. (2006). Chemical and biological aspects of *Narcissus* alkaloids. In *The Alkaloids*, Vol. 63, Cordell, G.A. (Ed.), Elsevier Scientific, Amsterdam, pp. 87–179

Berkov, S; Sidjimova, B.; Popov, S.; Evstatieva, L. (2004). Intraspecies variability in alkaloid metabolism in *Galanthus elwesii*. *Phytochemistry* 65, 579-586

Berkov, S.; Bastida, J.; Viladomat, F.; Codina, C. (2007a). Alkaloids from *Galanthus nivalis*. *Phytochemistry* 68, 1791-1798

Berkov, S.; Bastida, J.; Cilpa-Reyes, R.; Viladomat, F.; Codina, C. (2007b). Revized RMN data for incartine: a bioactive compound from *Galanthus elwesii*. *Molecules* 12, 1430-1435

Berkov, S.; Bastida, J.; Sidjimova.; B.; Viladomat, F.; Codina, C. (2008). Phytochemical differentiation of *Galanthus nivalis* and *Galanthus elwesii:* a case study. *Biochemical Systematics and. Ecolology* 36, 638-645

Berkov, S.; Pavlov, A.; Georgiev, V.; Bastida, J.; Burrus, M.; Ilieva, M.; Codina, C. (2009a). *Leucojum aestivum in vitro* cultures: variation in the alkaloid patterns. *Natural Product Communications 4, 359-364*

Berkov, S.; Georgieva, L.; Kondakova, V.; Atanassov, A.; Viladomat, F.; Bastida, J.; Codina, C. (2009b). Plant sources of galanthamine: phytochemical and biotechnological aspects. *Biotechnology and Biotechnological Equipment.* 23, 1170-1176

Berkov, S.; Cuadrado, M.; Osorio, E.; Viladomat, F.; Codina, C.; Bastida, J. (2009c). Three new alkaloids from *Galanthus nivalis* and *Galanthus elwesii. Planta Medica* 75, 1351- 1355

Berkov, S.; Bastida, J.; Sidjimova, B.; Viladomat, F.; Codina, C. (2011). Alkaloid diversity in *Galanthus elwesii* and *Galanthus nivalis. Chemistry and Biodiversity* 8, 115-130

Boit, H. -G.; Ehmke, H. (1955). Alkaloide von *Sprekelia formosissima, Galanthus elwesii, Zephyranthes candida* and *Crinum powellii* (VIII. Mitteil. Über Amaryllidaceen Alkaloide), *Chemische Beichte* 88, 1590-1594

Boit, H, -G.; Döpke, W. (1961). Alkaloide aus *Haemanthus, Zephrantes-, Galanthus-* und *Crinum-* Arten, *Die Naturwissenschaften* 48, 406-407

Bozkurt, B.; Zencir,S.; Ünver, N.; Kaya, G.; Onur, M.; Zupko, I; Topcu, Z; (2010). Biological activity of arolycoricidine and narciprimine on tumor cell killing and topoisomerase enzyme activities. Abstract Book of International Postgraduate Student Meeting on Pharmaceutical Sciences, p. 90, Ismir, Turkey, June 24-27, 2010

Brine, N.; Campbell, W.; Bastida, J.; Herrera M.; Viladomat, F.; Codina, C.; Smith, P. (2002) A dinitrogenous alkaloid from *Cyrtanthus obliquus. Phytochemistry* 61, 443–447

Bubeva-Ivanova, L.; Pavlova, H. (1965). Varhu alkaloidite na *Galanthus nivalis* L. var. *gracilis* (Celak). VIII Saobshtenie. Amaryllidaceae alkaloidi. *Farmacia* 15, 103-105

Capo, M.; Saa, J.M. (1989). Alkaloids from *Leucojum aestivum* sub. *pulchellum* (Amaryllidaceae). *Anales de química. Serie C: Química Orgánica y Bioquímica* 85, 119- 121

Citoglu, G.; Tanker, M.; Gumusel, B. (1998). Antiinfl ammatory effects of lycorine and haemanthidine. *Phytotherapy Research 12,* 205-206

Clement, B.; Goff, C.; Forbes, D. (1998). Toxic amines and alkaloids from *Acacia rigidula. Phytochemistry* 38, 266-279

Conforti, F.; Loizzo, M.; Marrelli, M.; Menichini, F.; Statti, G.; Uzunov, D.; Menichini, F. (2010). Quantitative determination of Amaryllidaceae alkaloids from *Galanthus reginae-olgae* subsp. *vernalis* and *in vitro* activities relevant for neurodegenerative diseases. *Pharmaceutical Biology* 48, 2-9

Davis, A.; Barnett, J. (1997). The leaf anatomy of the genus *Galanthus* L. (Amaryllidaceae J. St.-Hil.). *Botanical Journal of the Linnean Society* 123, 333–352

Davis, A.P. (1999). The genus *Galanthus.* In: Mathew, B. (Ed.) A Botanical Magazine Monograph, Timber Press Inc. Oregon, pp 15–17, 140–155

Davis, A.P.; Özhatay, N. (2001). *Galanthus trojanus*: a new species of *Galanthus* (Amaryllidaceae) from northwestern Turkey. *Botanical Journal of the Linnean Society* 137, 409–412

Dictionary of Natural Products, v17.1, Taylor & Francis Group, http://dnp.chemnetbase.com (accessed on 16.05.2009).

Evidente, A.; Arrigoni, O.; Luso, R.; Calabrese, G.; Randazzo, G. (1986). Further experiments on structure-activity relationships among lycorine alkaloids. *Phytochemistry* 25, 2739–2743

Grutzendler J.; Morris J.C. (2001). Cholinesterase Inhibitors for Alzheimer's Disease, *Drugs* 61, 41-52

Heinrich, M.; Teoh, H.L. (2004). Galanthamine from snowdrop-the development of a modern drug against Alzheimer's disease from local Caucasian knowledge. *Journal of Ethnopharmacology* 92, 147–162

Houghton, P.; Ren, Y.; Howes, M.-J. 2006. Acetylcholinesterase inhibitors from plants and fungi. *Natural Products Reports* 23, 181-199.

Jordanov, D. (1964). Genus *Galanthus* L. In: Jordanov, D. (Ed.), Flora of the People's Republic of Bulgaria, Vol. 2 Izdatelstvo na BAN, Sofia, pp. 318–319

Kalashnikov, I. (1970). Alkaloids from *Galanthus nivalis. Khimija Prirodnix Coedinenii* 6, 380.

Kamari, G. (1981). A biosystematic study of the genus *Galanthus* L. in Greece, part II (Cytology). *Botanika Chronika* 1, 60–98

Kaya, G.; Fillik, A.; Hisil, Y.; Ünver, N. (2004a). High pressure liquid chromatographic analysis of lycorine in four *Galanthus* species grown in Turkey. *Turkish Journal of Pharmaceutical Sciences.* 1, 105-114

Kaya, I.; Ünver, N.; Gözler, B.; Bastida, J. (2004b). (-)-Capnoidine and (+)-bulbocapnine from an Amaryllidaceae species, *Galanthus nivalis* subsp. *cilicicus. Biochemical Systematics and Ecology* 32, 1059-1062

Kaya, G.; Sar□kaya B.; Onur, M.; Unver, N.; Viladomat F.; Codina, C.; Bastida, J.; Lauinger, I.; Kaiser, M.; Tasdemir D. (2011). Antiprotozoal Alkaloids from *Galanthus trojanus. Phytochemistry Letters* 4, 301-305.

Kametani, T.; Honda, T. (1985). Aporphine alkaloids. In: *The alkaloids - chemistry and pharmacology,* Brossi, A.R. (Ed.), Vol. 24, Academic Press Inc., Orlando, pp 153–251

Kim, S.R.; Hwang, S.Y.; Jang, Y.P.; Park, M.J.; Markelonis, G.J.; Oh, T.H.; Kim, Y.C. (1999). Protopine from *Corydalis ternata* has anticholinesterase and antiamnesic activities. *Planta Medica* 65, 218-21

Kornienko, A.; Evidente, A. (2008). Chemistry, biology and medicinal potential of narciclasine and its congeners. *Chemical Reviews* 108, 1982–2014

Kozuharov, S. (1992). Field Guide to the Vascular Plants in Bulgaria. Naouka & Izkustvo, Sofia

Kreh, M.; Matusch, R.; Witte, L. (1995). Capillary gas chromatography–mass spectrometry of Amaryllidaceae alkaloids. *Phytochemistry* 38, 773–776

Latvala, A.; Önür, M.; Gözler, T.; Linden, A.; Kivçak, B.; Hesse, M. (1995). Alkaloids of *Galanthus elwesii. Phytochemistry* 39, 1229-1249

Liu, J.; Hu, W.X.; He, L.F.; Li, Y.; Ye, M. (2004). Effects of lycorine on HL-60 cells via arresting cell cycle and inducing apoptosis, *FEBS Letters* 578, 245–250

Liu, J.; Li, Y.; Tang, L.J.; Zhang, G.P.; Hu, W.X. (2007). Treatment of lycorine on SCID mice model with human APL cells, *Biomedicine and. Pharmacotherapy* 61, 229–234

Liu, X.; Jiang, J.; Jiao, X.; Wu Y.; Lin, J.; Cai, Y. (2009) Lycorine induces apoptosis and down- regulation of Mcl-1 in human leukemia cells. *Cancer Letters* 274 16–24.

Likhitwitayawuid, K.; Angerhofer, C.K.; Chai, H.; Pezzuto, J.M.; Cordell, GA.. (1993). Cytotoxic and antimalarial alkaloids from the bulbs of *Crinum amabile. Journal of Natural Products* 56, 1331-1338

López, S.; Bastida, J.; Viladomat, F.; Codina, C. (2002). Acetylcholinesterase inhibitory activity of some Amaryllidaceae alkaloids and *Narcissus* extracts. *Life Sciences* 71, 2521–2529

Jang, S.; Kim, B.; Lee, W.-Y.; An, S.; Choi, H.;... Jeon, B.; Chung, H-T.; Rho, J.-R.; Kim, Y.-J.; Chai, K.-Y. (2004). Stylopine from *Chelidonium majus* inhibits LPS-induced inflammatory mediators in RAW 264.7 cells. *Archives of Pharmacal Research* 27, 923- 929

Machocho, A.K.; Bastida, J.; Codina, C.; Viladomat, F.; Brun, R.; and Chhabra, S.C. (2004). Augustamine type alkaloids from *Crinum kirkii*, *Phytochemistry* 65, 3143-3149

McNulty, J.; Nair, J.; Codina, C.; Bastida, J.; Pandey, S.; Gerasimoff, J.; Griffin C. (2007). Selective apoptosis-inducing activity of *Crinum*-type Amaryllidaceae alkaloids. *Phytochemistry* 68, 1068-1074

McNulty, J.; Nair, J.; Bastida, J.; Pandey, S.; Griffin, C. (2009). Structure-activity studies on the lycorine pharmacophore: A potent inducer of apoptosis in human leukemia cells. *Phytochemistry* 70, 913–919

MacLean D.B. (1985). Phtalideisoquinoline alkaloids and related compounds. In: *The alkaloids - chemistry and pharmacology,* Brossi, A.R. (Ed.), Vol. 24, Academic Press Inc., Orlando, pp. 253–286

Meerow, A.V.; Snijman, D.A. (1998). Amaryllidaceae, In: The Families and Genera of Vascular Plants, Vol. 3, Kubitzki, K. (Ed.), Springer-Verlag, Berlin Heiderlberg, pp 83-110

Noyan, S.; Rentsch, G.; Önür, M.; Gözler, T.; Gözler, B.; Hesse, M. (1998). The Gracilines: a novel subgroup of the Amaryllidaceae alkaloids. *Heterocycles* 48, 1777-1791

Noyan, S. (1999) Isolation and structural elucidation studies on the alkaloids of *G. gracilis* Célak (Amaryllidaceae) growing wildly in Mount Nif, Kemalpasa, Izmir. Dissertation, Ege University

Proskurina, N.; Ordzhonikidze, S. (1953). Alkaloids of *Galanthus woronovii*. Structure of galanthine. *Dokladi Akademii Nauk SSSR*. 90, 565-567

Proskurina, N.; Yakovleva, A.; Ordzhonikidze, S. (1955). Alkaloids of *Galanthus woronovii* III. Structure of galanthamine. *Zhurnal Obshchei Khimii* 25, 1035-1039

Sener, B.; Koyuncu, M.; Bingöl, F.; Muhtar, F. (1998). Production of bioactive compounds from Turkish geophytes. *Pure and Applied Chemistry* 70, 2131

Shin, J.S.; Kim, K.T.; Lee, M.K. (1998). Inhibitory effects of bulbocapnine on dopamine biosynthesis in PC12 cells. *Neuroscience Letters* 244, 161-164

Sidjimova, B.; Berkov, S.; Popov, S.; Evstatieva, L. (2003). Galanthamine distribution in Bulgarian *Galanthus* species. *Pharmazie* 58, 936-937

Silva, A.; de Andrade J.; Machado K.; Rocha, A.B.;, Apel, M.A.; Sobral, M.E.G.; Henriques, A.T.; Zuanazzi, J.A.S. (2008). Screening for cytotoxic activity of extracts and isolated alkaloids from bulbs of *Hippeastrum vittatum*. *Phytomedicine* 15, 882–885

Thomsen, T.; Bickel, U.; Fischer, J.; Kewitz, H. (1998). Stereoselectivity of cholinesterase inhibition by galanthamine and tolerance in humans. *European Journal of Clinical Pharmacology* 39, 603-605

Tsakadze, D.; Kadirov, K.; Kiparenko, T.; Abdusamatov, A. (1979). New alkaloid from *Galanthus caucasicus*. *Izvestija Akademii Nauk Gruzinskoi SSR, Serija Chimicheskaja* 5, 191-192

Ünver, N.; Noyan, S.; Gözler, T.; Önür, M.; Gözler, B.; Hesse, M. (1999a). Three new tazettine type alkaloids from *Galanthus gracilis* and *Galanthus plicatus* subsp. *byzanthus*. *Planta Medica* 65, 347-350

Ünver, N.; Gözler, T.; Walch, N.; Gözler, B.; Hesse, M. (1999b). Two novel dinitrogenous alkaloids from *Galanthus plicatus* subsp. *byzanthus* (Amaryllidacea). *Phytochemistry* 50, 1255-1261

Ünver, N.; Noyan, S.; Gözler, B.; Gözler, T.; Werner, C.; Hesse, M. (2001). Four new Amaryllidaceae alkaloids from *Galanthus gracilis* and *Galanthus plicatus* subsp. *byzantinus*. *Heterocycles* 55, 641-652

Ünver, N.; Kaya, G.; Werner, C.; Verpoorte, R.; Gozler,B. (2003). Galanthindole:a new indole alkaloid from *Galanthus plicatus ssp. byzantus*. *Planta Medica* 69, 869-871

Ünver, N.; Kaya, G. (2005). An unusual pentacyclic dinitrogenous alkaloid from *Galanthus gracilis*. *Turkish Journal of Chemistry* 29, 547-533

Ünver, N. (2007). New skeletons and new concepts in Amaryllidaceae alkaloids. *Phytochemical Reviews* 6, 125–135

Valkova, A. (1961). Varhu dokazvaneto i opredeljaneto na alkaloidite na *Galanthus nivalis* L. var. *gracilis* i *Leucojum aestivum*, *Farmatsia* 11, 17–22

Vera-Avila, H.; Forbes, T.; Randel, R. (1996). Plant phenolic amines: Potential effects on sympathoadrenal medullary, hypothalamic-pituitary-adrenal, and hypothalamic- pituitary-gonadal function in ruminants. *Domestic Animal Endocrinology 13*, 285-296

Viladomat, F.; Bastida, J.; Codina, C.; Nair, J.; Campbell, W. (1997). Alkaloids of the South African Amaryllidaceae. *Recent Research and Developments in Phytochemistry* 1, 131-171

Weniger, B.; Italiano, L.; Beck, P.; Bastida, J.; Bergoñón, S.; Codina, C.; Lobstein, A. Anton, R. (1995). Cytotoxic Activity of Amaryllidaceae Alkaloids. *Planta Medica* 61, 77-79

Wildman, W.C. (1968). *The Alkaloids Chemistry and Pharmacology*, Vol. 11, Manske, R.H.F.; Holmes, H.L. (Eds.), New York, Academic Press Inc., pp. 307-405.

Willis, J.C. (1988). Amaryllidaceae. In: Shaw, A.H.K. (Ed.), A Dictionary of the Flowering Plants & Ferns, 8th edn. Cambridge University Press, Cambridge

Wink, M. (2008). Ecological Roles of Alkaloids, In: *Modern Alkaloids*, Fattorusso, E.; Taglialatella-Scafati, O., (Eds.),Wiley-VCH Verlag GmbH & Co. KGaA, Weinheim pp. 3-52

World Cheklist of Selected Plant Families, Kew Garden, (accessed May, 2011) http://apps.kew.org/wcsp/qsearch.do;jsessionid=56A58C50E62834507137259ECD D7B0E0

Yakovlea, A. (1963). Alkaloids of *Galanthus woronovii*. VI Isolation of tazettine. *Zhurnal Obshchei Khimii* 33, 1691-1693

Youssef, D.T. (2001). Alkaloids of the flowers of *Hippeastrum vittatum*. *Journal of Natural Products* 64, 839-41.

Zhong, J. (2005). Amaryllidaceae and *Sceletium* alkaloids. *Natural Products Reports* 22, 111- 126

Zonneveld, B.; Grimshaw, J.; Davis, A. (2003). The systematic value of the nuclear DNA content in *Galanthus*. Plant Systematics and Evolution 241, 89–102

Chapter 12

SILYMARIN, NATURAL FLAVONOLIGNANS DERIVED FROM MILK THISTLE

1. INTRODUCTION: AN OVERVIEW

Plants serve as a significant source of medications, food additives, agrochemicals, insecticides, flavors, and pigments. These substances are referred to as secondary metabolites. These compounds are infrequently found in some taxonomic groupings; they are not vital for an organism's life but contribute to the organism's interaction with its environment, hence facilitating its survival within the ecosystem (Verpoorte & Alfermann, 2000).

Milk thistle, often known as St. Mary's thistle [Silybum marianum (L.) Gaertn. (Syn. Cardus marianum), Asteraceae], is an annual or biennial herbaceous plant. The plant originates from the Mediterranean and North African regions (Boulos, 2000). It proliferates naturally across Europe, North Africa, the Americas, and Australia (Hamid et al., 1983). The plant attains heights of 10 feet. The stem measures 20-150 cm in height, is erect, ridged, and branching at the upper section. A notable feature of milk thistle is the white streaks seen in the veins of its dark green leaves (Fig. 1). The expansive leaves are profoundly lobed, measuring 50 cm in length and 25 cm in width. The leaf margins exhibit a yellow coloration and are adorned with woody spines measuring 3-12 mm in length. The leaves are arranged alternately and embrace the stem. Each stem culminates in a solitary composite flower head, approximately 2 inches in diameter, composed of purple disc florets. The flower heads of milk thistle are distinguished from other thistles by the presence of leathery bracts adorned with rigid spines. The fruits (Fig. 2) are hard, flattened achenes, measuring 6-8 mm in length, characterized by a smooth and glossy dark brown hue. The fruits produce 1.5-3% of an isomeric combination of flavonolignans referred to as silymarin (Morazzoni and Bombardelli, 1995). Silymarin predominantly accumulates in the outer layer of the fruits of S. marianum. (Madrid and Corchete, 2010).

2. CHEMISTRY OF FLAVONOLIGNANS

The principal components of silymarin are silybin A, silybin B, isosilybin A, isosilybin B, silychristin A, silychristin B and silydianin (Fig. 3). The first six compounds exist as equimolar mixtures as trans diastereoisomers. These diastereomers have very similar ^{1}H and ^{13}C NMR spectra and have no characteristic signals for facile identification of the individual isomers (Lee and Liu, 2003). A number of other chemically related compounds have been found in the fruits including dehydrosilybin, desoxysilychristin, desoxysilydianin, silandrin, silybinome, silyhermin and neosilymermin. The common feature of these

Fig. 1. Milk thistle with white patches along the veins of dark green leaves.

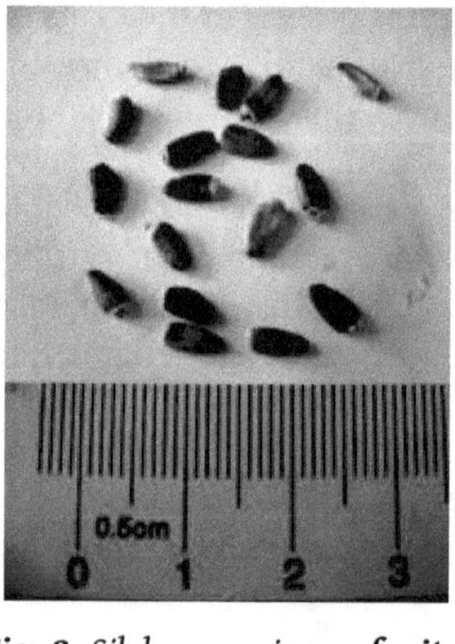

Fig. 2. ***Silybum marianum*** **fruits.**

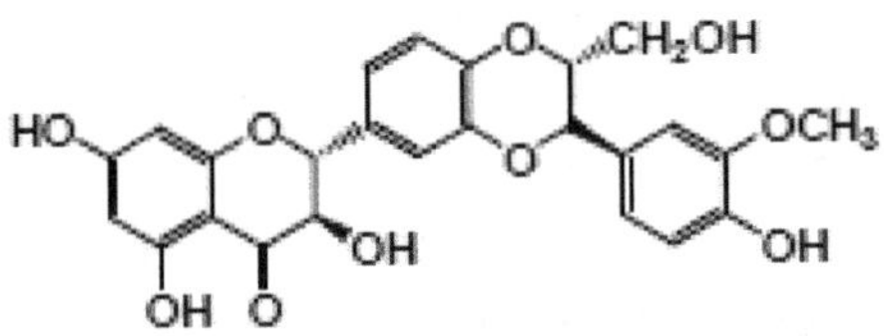

Silybin A

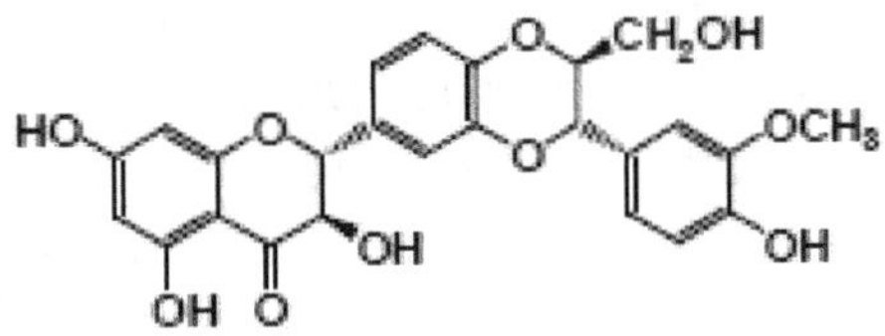

Silybin B

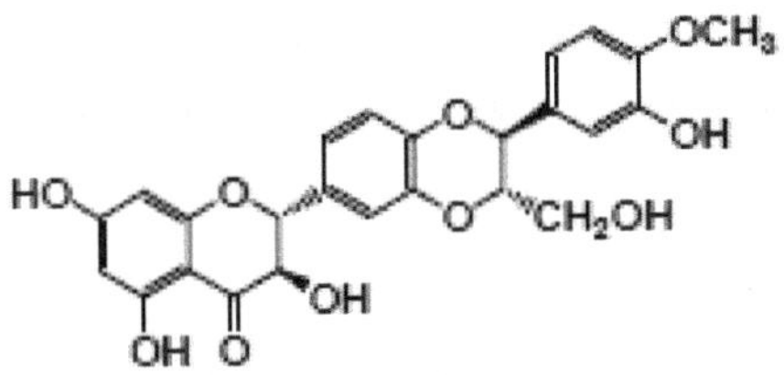

Isosilybin A

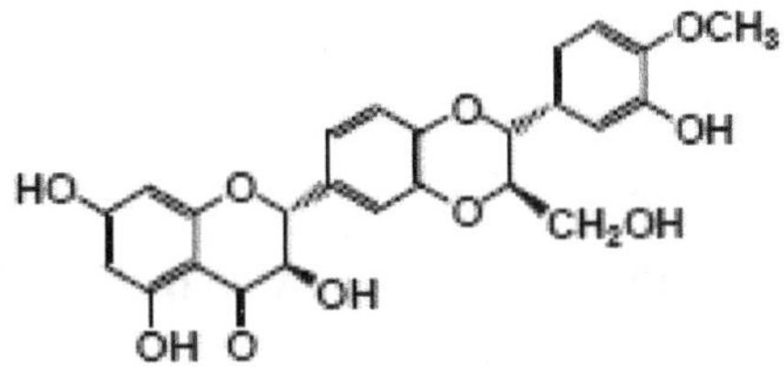

Isosilybin B

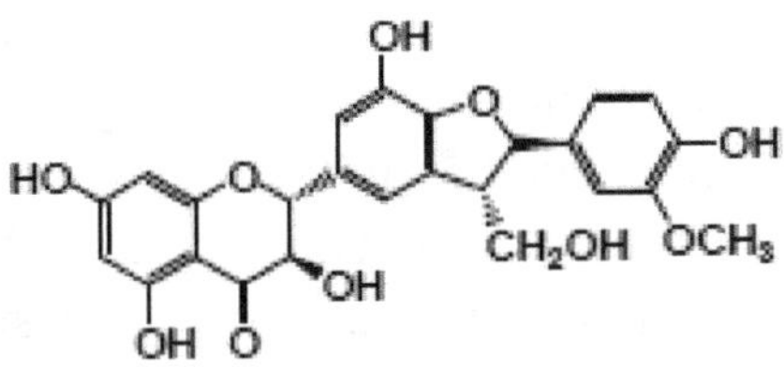

Silychristin A

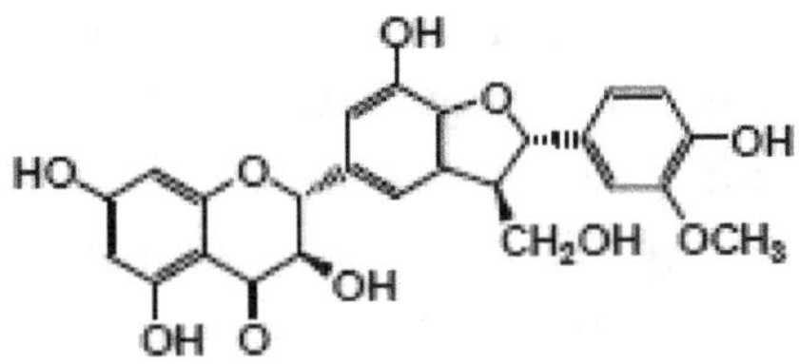

Silychristin B

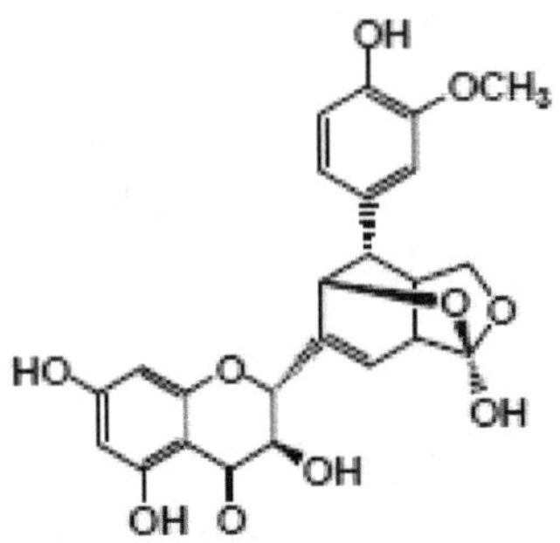

Silydianin

Fig. 3. Chemical structures of silymarin components.

compounds is a flavonolignan skeleton ($C_{25}H_{22}O_{10}$, mol wt 482). Basically, flavonolignan nucleus consists of the dihydroflavanol taxifolin linked to coniferyl alcohol moiety through an oxeran ring. The oxeran ring is responsible for the biological activity of silymarin, and opening of this ring results in loss of activity.

Only silybins and isosilybins contain the 1,4- dioxane ring system in their structure. Silybin and isosilybin have the same trans conformation of C-2, C-3 and C-7', C-8'. Silybin is considered the major and most active component in silymarin (Ligeret et al., 2008; Kim et al., 2009). The chemical structure of silybin has been identified in 1975 using a degradative method (Lee and Liu, 2003). The first trials to synthesize silybin suffered from the problem of giving a product which is a mixture of regioisomers, silybin and isosilybin (57:43). Regioselective synthesis of diastereomeric silybin in 63% overall yield was achieved by synthesizing a key intermediate which was coupled with 2,4,6-trimethoxyacetophenone to form a chalcone intermediate. Epoxidation, deprotection and acidic cyclization were followed (Tanaka et al., 1985).

3. ANALYSIS OF FLAVONOLIGNANS

Extract obtained from the fruits of *S. marianum* is available worldwide in the pharmaceutical market as antihepatotoxic drug under a variety of brand names. There are many products that contain silymarin either as a single component or in a mixture with other active constituents. The extract contains about 80% wt/wt of flavonolignans. Due to its poor water solubility and thus low bioavailability, silymarin is complexed with phosphatidylcholin, β- cyclodextrin or even given as glycosides, which have better water solubility and higher activity. A method for extraction of silymarin from plants on an industrial level has been reported (Madaus et al. 1983). In this method large part of the fruit oil is removed by cold pressing, the compressed mass is broken up, the pressed residue is extracted with ethyl acetate and the ethyl acetate extract is evaporated and processed. There is a need to have a selective and accurate analytical method for qualitative and quantitative determination of silymarin flavonolignan components during standardization of the extract. This is expressed as silymarin percentage and it corresponds to the sum of silybins, isosilybins, and silychristins and silydianin concentrations. It is important that the analytical method characterizes and quantifies each component in silymarin.

3.1 Thin layer chromatography analysis

Flavonolignans were analyzed by Thin Layer Chromatography (TLC) (Wagner et al., 2009). Chloroform-acetone-formic acid (75:16.5:8.5) was used as a solvent system and detection was done using natural products-polyethylene glycol reagent. Silymarin is characterized in UV-365 nm by two intense green-blue fluorescent zones of silybin/isosilybin ($R_f = 0.6$), silychristin ($R_f = 0.35$) and an orange zone of taxifolin ($R_f = 0.4$).

3.2 UV-visible spectrophotometry analysis

UV-visible spectrophotometry was proposed for the quantitative determination of flavonolignans (Famacopea Ufficiale Italiana, 1985). This spectrophotometric method is time consuming, shows a non-satisfactory repeatability and measures total and not individual flavonoligans. A fast, simple and sensitive spectrophotometric method for determination of silymarin in pure form and in pharmaceutical formulations was reported. This method was based on oxidation with potassium permanganate at pH 7. The reaction was followed spectrophotometrically by measuring the decrease in the absorbance at 530 nm (Rahman et al., 2004).

3.3 High performance liquid chromatography analysis

High Performance Liquid Chromatography (HPLC) was proposed as a method for determination of silymarin (Quaglia et al., 1999). Two reversed stationary phases, RP-18 and RP-8, were compared for resolution of all considered flavonolignans. The RP-18 stationary phase showed good separation among silybin and isosilybin, while silydianin and silychristin were not baseline resolved. The increase in water concentration in the mobile phase allowed the separation of two distereomers of silybin. RP-8 stationary phase, a more polar phase, improved the resolution of peaks related to all flavonolignans but did not allow the resolution of the two silybin diastereomers. Among the advantages of this method are precision, sensitivity, ability to measure individual constituents in a mixture, the good separation of all compounds allowed the purity control of each peak, plotting of UV spectra, useful for the peak identification and a more correct quantification. However, time consumption, the need for pre-purification step and availability of pure reference compounds are the main disadvantages of HPLC. Analysis of silymarin components by HPLC on RP-18 in our laboratory only showed separation of the two diastereomers of silybin. However, the two peaks were not base-line resolved (Fig. 4).

3.4 Capillary electrophoresis analysis

Capillary zone electrophoresis has been proposed as a method for separation and determination of silymarin components (Kvasnička et al., 2003). Repeatability, accuracy, linearity and limit of detection were evaluated. The method was comparable to HPLC results. Shorter analysis time and better resolution of silydianin and silychristin from sample constituents were the main advantages of this method. High Performance Capillary Electrophoresis (HPCE) was used for determination of silymarin in the extract of *S. marianum* using borate buffer solution at pH 9. At this pH the flavonolignans having many phenolic groups in their structure were negatively charged (Quaglia et al., 1999). In these conditions isosilybin co-eluted together with silybin. Adding 12 mM dimethyl β–cyclodextrins solution to the running buffer, the separation of silybin from isosilybin was obtained.

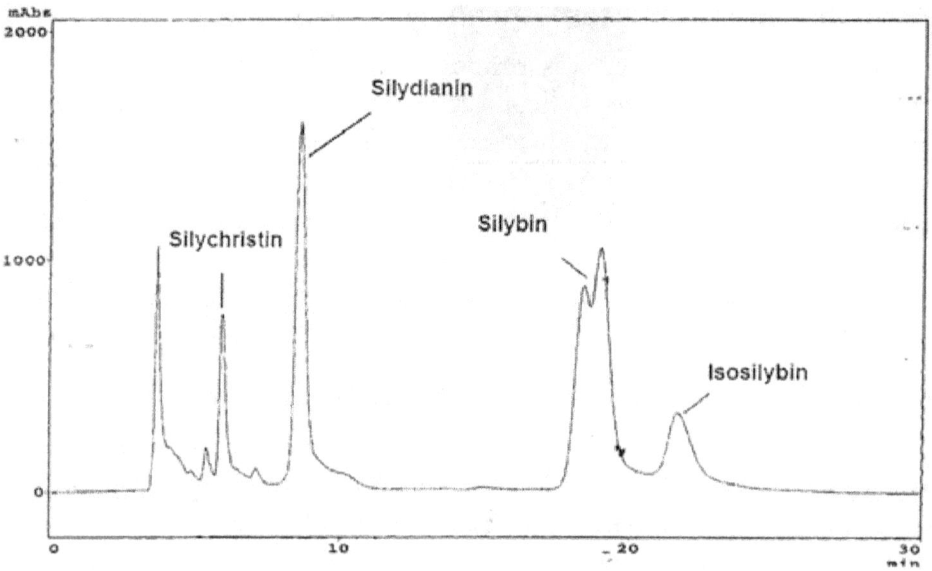

Fig. 5. Analysis of silymarin components by HPLC on RP18 (analysis was carried out in author laboratory).

3.5 Ultra performance liquid chromatography analysis

Ultra-Performance Liquid Chromatography (UPLC) offer many advantages over traditional HPLC for separation and quantification of multicomponent analytes such as silymarin components. Among these advantages are short analysis time, maintaining the resolution and increasing peak capacity and sensitivity. Complete separation of the seven major active flavonolignans of silymarin by UPLC RP18 column was reported (Wang et al., 2010). In this study, the use of electrospray ionization tandem mass spectrometry allowed to obtain detailed analysis of fragmentation and distinguish between the seven flavonolignans for online identification. Advantages and disadvantages of different methods for quantitative analysis of flavonolignan components in silymarin are summarized in table 1.

Table 1. Advantages and disadvantages of different methods for quantitative analysis of flavonolignan components in silymarin.

Method	Advantages	Disadvantages
Spectrophotometric	• Fast and simple • Sensitive	• Individual flavonolignans are not quantified
HPLC	• Precise and sensitive • Individual flavonolignans are quantified • Peak identification • Purity control	• Not all flavonolignans are separated from each other • Time consuming • Needs pre-purification step • Pure reference compounds are needed

HPCE	• Shorter analysis • Less solvent consumption • Individual flavonolignans are quantified including diastereomers	• Needs pre-purification step
UPLC	• Short analysis time • Less solvent consumption • Increased resolution • Increased peak capacity and sensitivity	• Expensive • Needs pre-purification step • Needs calibration curve

4. BIOSYNTHESIS OF FLAVONOLIGNANS IN *SILYBUM MARIANUM*

Flavonolignans are formed by combination of flavonoid and lignan structures. This occurs by oxidative coupling processes between a flavonoid and a phenylpropanoid, usually coniferyl alcohol (Dewich, 2002). Oxidative coupling occurs between free radical generated from the flavanol taxifolin and the free radical generated from coniferyl alcohol. This would lead to an adduct formation. This adduct could cyclize by attachment of the phenol nucleophile on to the quinine methide generated from coniferyl alcohol (Figure 5). The product in this case would be silybin. The fact that silybin exists in *S. marianum* in a mixture of two diastereomers reveals that the radical coupling reaction is not stereospecific. This is also true for isosilybin and silychristin. The latter flavonolignan originate from a mesomer of the taxifolin-derived free radical. Silydianin has a more complex structure and is formed by intramolecular cyclization of the coupling product. This is followed by hemiketal formation.

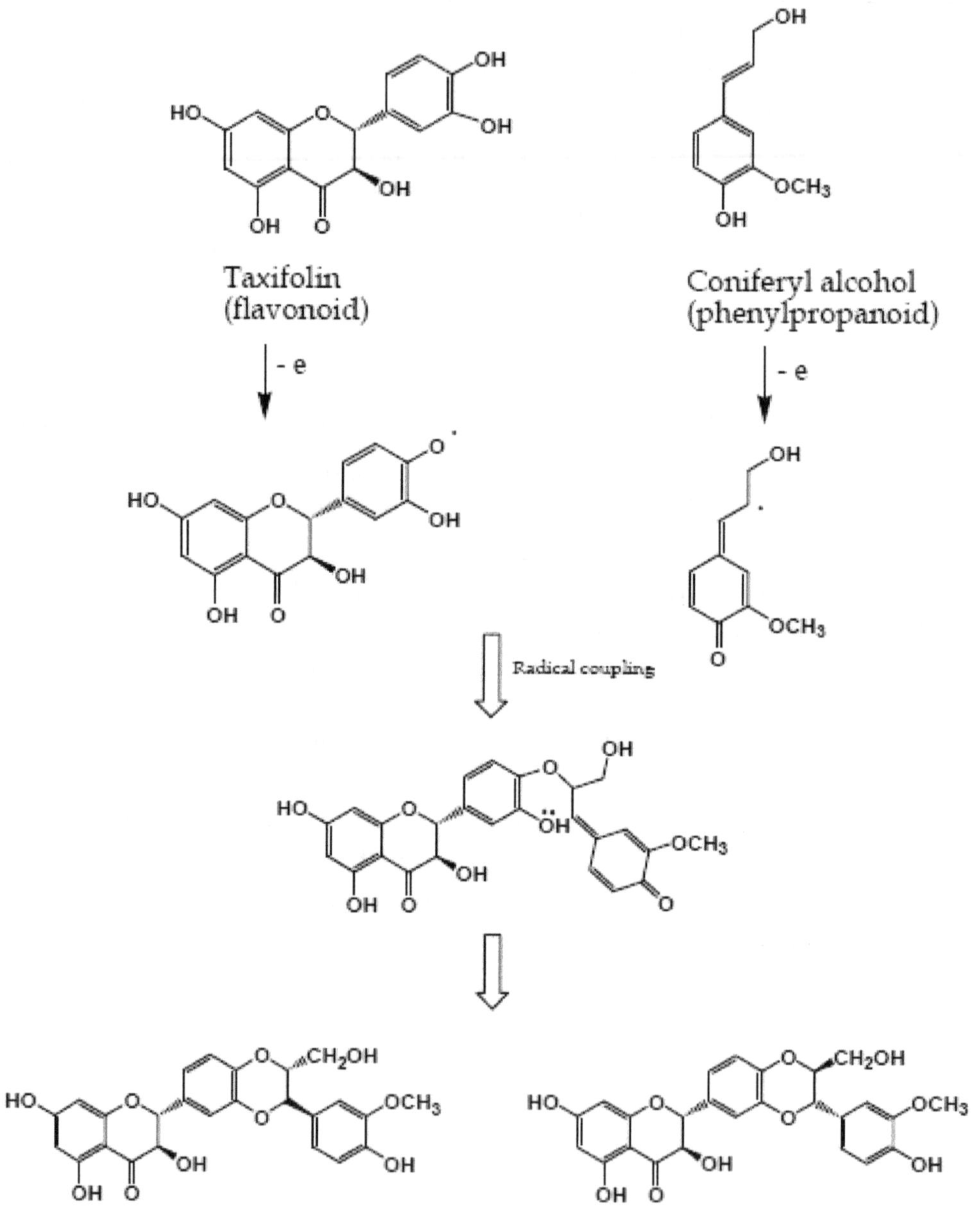

Fig. 5. Proposed biosynthetic pathway to silybin in *Silybum marianum*.

5. BIOLOGICAL ACTIVITY OF FLAVONOLIGNANS

Silymarin has been used for centuries to treat liver, spleen and gall bladder disorders (Shaker et al., 2010). It is known to possess hepatoprotective, antioxidant (Morazzoni and Bombardelli, 1995), anticancer (Zi et al., 1997), anti-inflammatory (De La Puerta, 1996) and anti-diabetic (Maghrani et al., 2004) properties. As a hepatoprotective agent, silymarin is used for oral treatment of toxic liver damage and for the therapy of chronic inflammatory liver diseases (Flora et al., 1998).

5.1 Hepatoprotective activity

Silymarin is one of the most investigated plant extracts with known mechanisms of action for oral treatment of toxic liver damage (Hiroshi et al., 1984). Silymarin is used as a protective treatment in acute and chronic liver diseases (Flora *et al.*, 1998). Silymarin supports the liver cells through multifactor action including binding to cell membrane to suppress toxin penetration into the hepatic cells, increasing superoxide dismutase activity (Feher and Vereckei, 1991), increasing glutathione tissue level (Pietrangelo *et al.*, 1995), inhibition of lipid peroxidation (Bosisio *et al.*, 1992; Carini *et al.*, 1992) and enhancing hepatocyte protein synthesis (Takahara et al., 1986). The hepatoprotective activity of silymarin can be explained based on antioxidant properties due to the phenolic nature of flavonolignans. It also acts through stimulating liver cells regeneration and cell membrane stabilization to prevent hepatotoxic agents from entering hepatocytes (Fraschini et al., 2002). Recently it has been shown that flavonolignans inhibit leucotriene production; this inhibition explains their anti-inflammatory and antifibrotic activity (Dehmlow et al., 1996).

5.2 Anticancer activity of silymarin

Silymarin is also beneficial for reducing the chances for developing certain cancers (Deep et al., 2007; Zhao et al., 1999). The molecular targets of silymarin for cancer prevention have been studied (Ramasamy and Agrawal, 2008). Silymarin interfere with the expressions of cell cycle regulators and proteins involved in apoptosis to modulate the imbalance between cell survival and apoptosis. Sy-Cordero et al., 2010, isolated four key flavonolignan diastereoisomers (silybin A, silybin B, isosilybin A and isosilybin B) from *S. marianum* in gram scale. These compounds and other two related analogues, present in extremely minute quantities, were evaluated for antiproliferative/cytotoxic activity against human prostate cancer cell lines. Isosilybin B showed the most potent activity (Deep et al., 2007; Deep et al., 2008a; Deep et al., 2008b). The isolation of six isomers afforded a preliminary analysis of structure-activity relationship toward prostate cancer prevention. The results suggested that an *ortho* relationship for the hydroxyl and methoxy substituents in silybin A, silybin B, isosilybin A and isosilybin B was more favorable than the *meta* relationship for the same substituents in the minor flavonolignans. Silymarin suppressed UVA-induced oxidative stress that can induce skin damage (Svobodová et al., 2007). Therefore, topical application of silymarin can be a useful strategy for protecting against skin cancer.

5.3 Anti-inflammatory activity

Silymarin seems to possess anti-inflammatory properties by acting through different mechanisms such as its antioxidant action, membrane-stabilizing effect and inhibition of the production or release of inflammatory mediators such as arachidonic acid metabolites (Breschi et al., 2002). Gastric anti-ulcer activity of silymarin has been reported (Alarcon et al., 1992). This action was attributed to the inhibition of enzymatic peroxidation in the lipoxygenase pathway and free radical scavenging activity (Bauman et al., 1980). Silymarin exhibited significant anti-inflammatory and antiarthritic activities in the papaya latex induced model of inflammation and mycobacterial adjuvant induced arthritis in rats (Gupta et al., 2000). This action is mediated through inhibition of 5-lipoxygenase.

5.4 Effect on asthma

Activity of silymarin was examined against bronchial anaphylaxis and against post-anaphylactic, propranolol- or platelet activating factor-induced hyperreactivity in guinea-pigs (Breschi et al., 2002). Silymarin pretreatment reduced the bronchospasm induced by antigen-challenge in sensitized animals. This protective effect was due to indirect mechanism that reduces airway responsiveness to histamine, and consequently the immediate anaphylactic response. Therefore, silymarin can be used as protective agent in the management of asthmatic disorders.

5.5 Immunostimulatory activity

Several studies have reported the immunostimulatory actions of silymarin (Wilarusmee et al., 2002). The effect of treatment with silymarin was studied on glutathione level and proliferation of peripheral blood mononuclear cells of β–thalassemia major patients (Alidoost et al., 2006). In vitro treatment with 10 g/ml silymarin restored glutathione levels and enhanced cellular proliferation. This was explained by its antioxidant activity.

5.6 Treatment of obsessive-compulsive disorder

Many patients cannot tolerate the side effects of pharmaceutical agents available for treatment of obsessive-compulsive disorder, do not respond properly to the treatment or the medications lose their effectiveness after a period of treatment. An 8-week pilot double- blind randomized clinical trial on 35 adult patients was conducted to compare the efficacy of the extract of *S. marianum* with fluoxetine in the treatment of obsessive-compulsive disorder (Sayyah et al., 2010). The results showed that the extract of *S. marianum* has positive effects on obsession and compulsion starting from the fifth week. There were no any serious side effects accompanying *S. marianum* extract administration.

5.7 Hyperprolactinemic effect

S. marianum fruits have been traditionally used by nursing mothers for stimulating milk production (Newall et al., 1996). It was demonstrated that milk thistle increases lactation (Carotenuto and Di Pierro, 2005). The mechanism that led to the increase in lactation has been studied by measuring the concentration of circulating prolactin in female rats treated with silymarin (Capasso et al., 2009). It was shown that silymarin is able to produce a significant increase in circulating prolactin levels after oral administration. The levels of prolactin remains elevated for up to 66 days after silymarin discontinuation. Fig. 6 shows a summary of the wide range of biological activities attributed to silymain.

5.8 Toxicity of silymarin

An average daily dose of silymarin (420 mg/day for 41 months) was found to be non-toxic, relative to placebo, in clinical trials (Tamayo and Diamond, 2007). Drug-drug interaction and liver toxicity by interference with co-drugs by induction or inhibition of cytochrome- P450 is a major concern for the use of silymarin (Izzo and Ernst, 2009). Studies were performed to investigate the potential for hepatotoxicity, cytochrome-P450 isoenzymes induction and inhibition on dry extract from *S. marianum*, as contained in HEPAR-PASC® film-coated tablets (Doehmer et al., 2011). The results indicated that interference or hepatotoxicity of the dry extract from *S. marianum* at the recommended maximum daily dose equivalent to 210 mg silybin is unlikely and is to be considered safe.

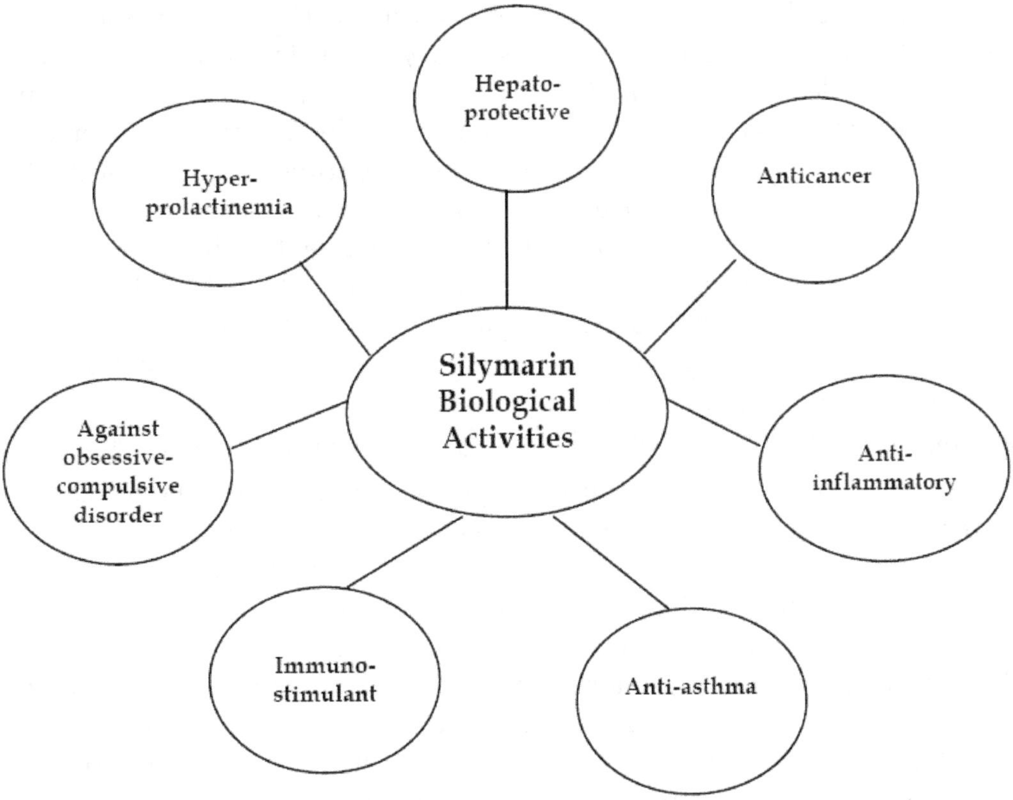

Fig. 6. Biological activities of silymarin.

6. TISSUE CULTURE STUDIES

Plant tissue culture can be a potential source for important secondary metabolites (Misawa, 1994). This is based mainly on using plant cultures in a similar manner to microbial fermentation for factory-type production of pharmaceuticals and food additives. This technology has some advantages over conventional agricultural methods: production is independent of variation in crop quality or failure, yield of secondary metabolites would be constant and geared to demand, there is no difficulty in applying good manufacturing practice to the early stages of production, production would be possible anywhere under strictly controlled conditions, independent of political problems, free from risk of contamination with pesticides, herbicides or fertilizers and new methods of production can be patented (AbouZid et al., 2008). Cell suspension culture and hairy root culture were established from *S. marianum*. The former is established from callus tissue that developed on injured plant surface as a result of wounding or exogenous hormones (Fig. 7). The latter represent an approach to increase the yield of flavonolignans using morphologically differentiated/organized cultures.

6.1 Cell culture

In vitro cultured cells of *S. marianum* may offer an alternative and renewable source for this valuable natural product. However, the yield of silymarin was very low or sometimes not detectable in undifferentiated cultured cells (Becker and Schrall, 1977). In order to obtain silymarin in concentrations high enough for commercial manufacturing, many approaches have been made to stimulate the productivity of silymarin in cultured cells of *S. marianum*. These approaches compromise changes in the media composition (Cacho et al., 1999), treatment with elicitors such as yeast extract and methyl jasmonate (Sánchez-Sampedro et al., 2005a), addition of precursor (Tůmová et al., 2006) and morphological differentiation. Such approaches for improving silymarin production by manipulating plant cell cultures may also help in studying signal transduction pathways, cloning biosynthetic genes, studying metabolic flux and regulation of silymarin production (Zhao et al., 2005).

Becker and Schrall, (1977) cultured cotyledon explants on MS media using different growth hormones for establishment of cell suspension culture. Typical flavonolignans of *S. marianum* were not detected. This was possible after feeding coniferyl alcohol and taxifolin to cell suspension cultures (Schrall and Becker 1977). Feeding the culture medium with precursor of coniferyl alcohol offered enhancement of silydianin production but other components of silymarin were not influenced (Tůmová et al., 2006). Cacho et al. (1999) reported that callus and cell cultures of *S. marianum* could produce silymarin but to a lesser extent than that accumulates in the fruits. They also reported that elimination of calcium ion positively affected silymarin production. This point was further confirmed by

Sánchez- Sampedro et al. (2005a), who also reported that silymarin accumulation was not altered by treatment of cultures with the calcium ionophore A23187.

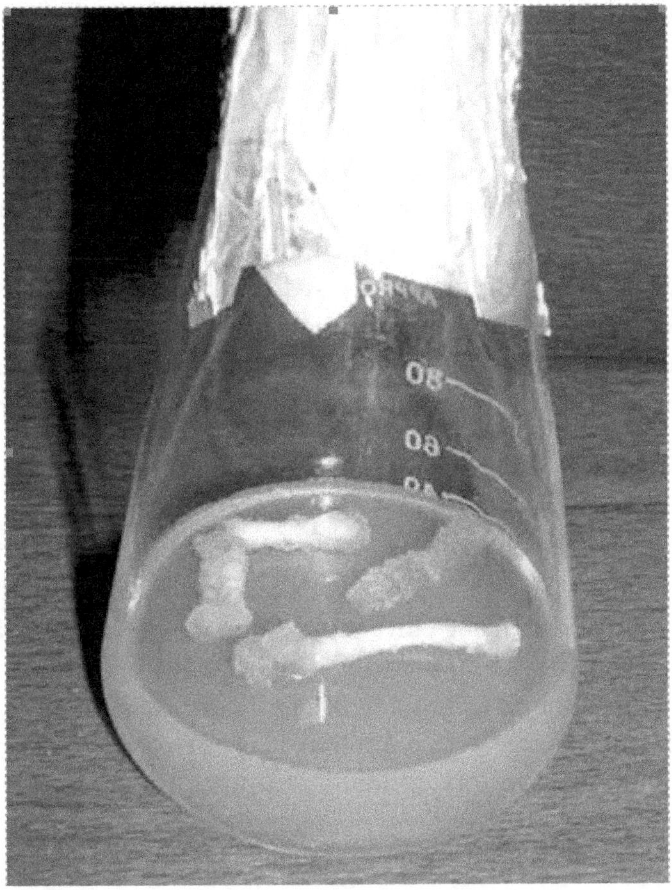

Fig. 7. Callus of *Silybum marianum* developed on explants.

These results suggest that inhibition of external and internal calcium fluxes play a significant role in flavonolignans metabolism in *S. marianum* cell cultures. Sánchez-Sampedro et al. (2005b) reported that yeast extract and methyl jasmonate elicited the production of silymarin. Elicitation is one of the most effective approaches to enhance the yield of secondary metabolites in *in vitro* cultures (Namdeo, 2007). It has been shown that elicitors can affect level of secondary metabolites in medicinal plants by modulating the rates of biosynthesis, accumulation, and/or vacuolar transit, turnover and degradation (Barz et al., 1990). Jasmonic acid and its methyl ester are known to be involved in the plant defense response through altering the gene expression. The mechanism by which jasmonate induces gene expression was studied in *Catharanthus roseus* (van der Fits and Memelink, 2000). In this plant species induction occurs through an ORCA3 transcription factor with a conserved jasmonate-response domain. The use of methyl jasmonate as an elicitor has an advantage of being only one compound of well- defined chemical structure. The effect of elicitation with picloram, jasmonic acid and light on silymarin production was reported (Hasanloo et al., 2008). The greatest silymarin content (0.41 mg/g DW) was obtained with 3 mg/l picloram and 2 mg/l jasmonic acid in the dark after 28 days. The sequence of the signaling processes leading to stimulation of flavonolignan production by methyl jasmonate is not well-known. Madrid and Corchete, 2010, studied the possible involvement of a phospholipase D-mediated lipid signaling in the elicitation of flavonolignans. It was reported that methyl jasmonate increased the activity of phospholipase D. Mastoparan, a phospholipase D activity stimulator, caused a substantial increase in silymarin production. Phosphatidic acid, a product of phospholipase D activity, promoted silymarin accumulation. N-butanol which inhibits phospholipase D activity prevented silymarin elicitation by methyl jasmonate or mastoparan.

6.2 Root culture

Production of flavonolignans from root cultures (Fig. 8) of *S. marianum* was reported before (Alikaridis et al ., 2000). Silybin (1.79×10^{-3} % DW) and silychristin (0.81×10^{-3} % DW) were the major flavonolignans produced by the established root cultures. In the referred study hairy root cultures of *S. marianum* were established. Hairy root cultures are the roots obtained by genetic transformation of plant tissues with the pathogenic soil bacterium *Agrobacterium rhizogenes.* These roots can then be cultured on hormone-free media and have three main advantages: genetic and biochemical stability, cultivation without addition of growth regulators and ability to give high final biomasses from low inocula.

Salicylic acid was effective in increasing the flavonolignan content 2.42 times in hairy root cultures of *S. marianum* higher than control cultures (Khalili et al., 2009). Yeast extract stimulated flavonolignan production in hairy root cultures two-fold higher than the control cultures. Moreover, it was reported that yeast extract treatment induced the activity of lipoxygenase to allow for the production of jasmonate. It was concluded that jasmonate signaling is an integral part of the yeast extract signal transduction for the production of flavonolignans (Hasanloo et al., 2009).

7. FUTURE DIRECTIONS

Plant tissue culture studies have contributed to our understanding of biosynthesis and regulation of silymarin in *S. marianum*. Using elicitation technology may offer an effective approach to improve silymarin production for industrial purpose. However, the possible signaling pathway that may be involved in accumulation of silymarin is still unknown. Understanding the basic components of this pathway is mandatory before these biotechnological methods can replace field crops as the basic source of pharmaceutical raw material. Establishment of plant tissue culture systems able to produce these biologically valuable compounds in high yield will facilitate such studies.

Fig. 8. Root culture of *Silybum marianum* growing in Murashige and Skoog medium.

8. CONCISE SUMMARY

Milk thistle is an annual or biennial herb indigenous to the Mediterranean and North African regions. The plant's fruits contain an isomeric combination of flavonolignans referred to as silymarin. The flavonolignan nucleus comprises a dihydroflavanol taxifolin connected to a coniferyl alcohol moiety via an oxeran ring. There is less knowledge regarding the coupling of coniferyl alcohol to taxifolin. Silymarin is extensively utilized as a hepatoprotective medication for the oral treatment of toxic liver damage and the management of chronic inflammatory liver disorders. Silymarin's hepatoprotective action is attributed to its antioxidant capabilities, the stimulation of liver cell regeneration, and the stability of cell membranes to inhibit the entry of hepatotoxic substances into hepatocytes. Flavonolignans demonstrate a broad spectrum of biological activities, including anticancer, anti-inflammatory, and hyperprolactinemic effects. Diverse methodologies have been established for the investigation of the content and composition of primary silymarin constituents in botanical materials and medicines. The methods include thin layer chromatography, spectrophotometry, high-performance liquid chromatography, capillary zone electrophoresis, and ultra-performance liquid chromatography. In vitro grown cells of S. marianum may provide an alternate and sustainable source for this valuable natural substance. Production of flavonolignans in cell and root cultures of S. marianum has been documented. Numerous strategies have been employed to enhance the yield of flavonolignans in S. marianum tissue culture, including modifications to media composition, the incorporation of precursors, and elicitation.

9. REFERENCES

AbouZid, S., Nasib, A., Khan, S., Qureshi, J., Choudhary, M.I. (2010) Withaferin A production by root cultures of *Withania coagulans*. *International Journal of Applied Research in Natural Products* 2(5), 23-27.

Alarcon, C., Martin, M.J., Marhuenda, E. (1992) Gastric anti-ulcer activity of silymarin, a lipoxygenase inhibitor, in rats. *Journal of Pharmacy and Pharmacology* 44, 929-931.

Alidoost, F., Gharagozloo, M., Bagherpour, B., Jafarian, A., Sajjadi, S.E., Hourfar, H., Moayedi, B. (2006) Effects of silymarin on the proliferation and glutathione levels of peripheral blood mononuclear cells from β-thalassemia major patients. *International Journal of Immunopharmacology* 6, 1305-1310.

Alikaridis, F., Papadakis, D., Pantelia, K., Kephalas, T. (2000) Flavonolignan production from *Silybum marianum* transformed and untransformed root cultures. *Fitoterapia*, 71(4), 379-384.

Barz, W.A., Beimen, B., Drae, U., Jaques, C., Sue, O.E., Upmeier, B. (1990) Turnover and storage of secondary products in cell cultures. In: Charlwood BV, Rhodes MJC, eds. Secondary products from plant tissue culture. Oxford: Clarendon Press, 79-102.

Baumann, J., Wurm, G., Von Bruhhansen, F. (1980) Prostaglandin relation to their oxygen- scavenging properties. *Achieves of Pharmacology* 313, 330-337.

Becker, H., Schrall, R. (1977) Tissue and suspension cultures of *Silybum marianum*: the formation of flavanolignans by flavanoids and coniferyl alcohol. *Planta Medica*, 32(1), 27-32.

Bosisio, E., Benelli, C., Pirola, O. (1992) Effect of the flavolignans of *Silybum marianum* L. on lipid peroxidation in rat liver microsomes and freshly isolated hepatocytes. *Pharmacology Research* 25, 147-154.

Boulos, L. (2000) *Flora of Egypt*, (1st edition), Al Hadara Publishing Inc., Cairo, Egypt.

Breschi, M.C., Martinotti, E., Apostoliti, F., Nieri, P. (2002) Protective effect of silymarin in antigen challenge- and histamin-induced brochoconstriction in in vivo guinea-pigs. *European Journal of Pharmacology* 437, 91-95.

Cacho, M., Moran, M., Corchete, P., Fernandez-Tarrago, J. (1999) Influence of medium composition on the accumulation of flavonolignans in cultured cells of *Silybum marianum* (L.) Gaertn. *Plant Science* 144, 63-68.

Capasso, R., Aviello, G., Capasso, F., Savino, F., Isso, A.A., Lembo, F., Borrelli, F. (2009) Silymarin BIO-C®, and extract from *Silybum marianum* fuits, induces hperprolactinemia in intact female rats. *Phytomedicine* 16, 839-844.

Carini, R., Comoglio, A., Albano, E., Poli, G. (1992) Lipid peroxidation and irreversible damage in the rat hepatocyte model: Protection by the silybin-phospholipid complex IdB 1016. *Biochemical Pharmacology* 43, 10, 2111-2115.

Carotenuto, D., Di Pierro, F. (2005) Studio sulla tollerabilità ed efficacia dela silimarina BIO- C® (Piùlatte®) micronizzata come galattagogo. *Acta Neonatology Pediatrics* 4, 393-400.

De La Puerta, R. (1996) Effect of silymarin on different acute inflammation models and in leukocyte migration. *Journal of Pharmacy and Pharmacology* 48, 9, 968-970.

Deep, G., Oberlies, N.H., Kroll, D.J., Agarwal, R. (2007) Isosilybin B and isosilybin A inhibit growth, induce G1 arrest and cause apoptosis in human prostate cancer LNCaP and 22Rv1 cells. *Carcinogenesis* 28, 1533-1542.

Deep, G., Oberlies, N.H., Kroll, D.J., Agarwal, R. (2008) Isosilybin B causes androgen receptor degradation in human prostate carcinoma cells via P13K-Akt-Mdm2- mediated pathway. *Oncogene* 27, 3986-3998.

Deep, G., Oberlies, N.H., Kroll, D.J., Agarwal, R. (2008) Identifying the differential effects of silymarin constituents on cell growth and cell cycle regulatory molecules in human prostate cancer cells. *International Journal of Cancer* 123, 41-50.

Dehmlow, C., Erhard, J., De Groot, H. (1996) Inhibition of Kupffer cell functions as an explanation for the hepatoprotective properties of silybinin. *Hepatology* 23, 749-754.

Dewick, P.M. (2002) Medicinal natural products. A Biosynthetic Approach, John Wiley & Sons, Ltd, Chichester, UK.

Doehmer, J., Weiss, G., McGregor, G.P., Appel, K. (2011) Assessment of a dry extract from milk thistle (*Silybum marianum*) for interference with human liver cytochrome-P450 activities. *Toxicology in Vitro* 25, 1, 21-27.

Famacopea Ufficiale Italiana, ed. IX, vol. II 1673, Istituto Poligraficao e Zecca cello Stato- Roma, 1985.

Flora, K., Hahn, M., Rosen, H., Benner, K. (1998) Milk thistle (*Silybum marianum*) for the therapy of liver disease. *American Journal of Gastroenterology* 93, 139-143.

Fraschini, F., Demartini, G., Esposti, D. (2002) Pharmacology of silymarin. *Clinical Drug Investigation* 22(1), 51-65.

Gupta, O.P., Sing, S., Bani, S., Sharma, N., Malhotra, S., Gupta, B.D., Banerjee, S.K., Handa, S.S. (2000) Antiinflammatory and antiarithritic activities of silymarin acting through inhibition of 5-lipoxygenase. *Phytomedicine* 7, 21-24.

Hamid, S., Sabir, A., Khan, S., Aziz, P. (1983) Experimental cultivation of *Silybum marianum* and chemical composition of its oil. *Pakistan Journal of Scientific and Industrial Research* 26, 244-246.

Hasanloo, T., Kavari-Nejad, R.A., Majidi, E., Shams, Ardakani, M.R. (2008) Flavonolignan Production in Cell Suspension Culture of *Silybum marianum*. *Pharmaceutical Biology* 46(12), 876-882.

Hiroshi, H., Yoshinobu, K., Wagner, H., Manfred, F. (1984) antihepatotoxic actions of flavonolignans from *Silybum marianum* fruits. *Planta Medica* 51, 248-250.

Kim, S., Choi, J.H., Lim H.I., Lee, S., Kim, W.W., Kim J.S., Kim J., Choe, J., Yang, J., Nam, S.J., Lee, J.E. (2009) Silibinin prevents TPA-induced MMP-9 expression and VEGF secretion by inactivation of the Raf/MEK/ERK pathway in MCF-7 human breast cancer cells. *Phytomedicine* 16, 573-580.

Klassen, C.D., Plaa, G.L. (1969) Comparison of the biochemical alteration elicited in liver of rats treated with carbon tetra chloride and chloroform. *Biochemical Pharmacology* 18(8), 2019- 2027.

Kvasnička, F., Biĩba, B., Ševčík, Voldřich, M., Krátká, J. (2003) Analysis of the active components of silymarin. *Journal of Chromatography A* 990, 239-245.

Izzo, A.A., Ernst, E. (2009) Interactions between herbal medicines and prescribed drugs: and updated systematic review. *Drugs* 69, 1777-1798.

Lee, D.Y.-W., Liu, Y. (2003) Molecular structure and stereochemistry of silybin A, silybin B, isosilybin A, and isosilybin B, isolated from *Silybum marianum* (Milk thistle). *Journal of Natural Products* 66, 1171-1174.

Lee, S.K., Mbwambo, Z.H.Y., Chung, H., Luyengi, L., Gamez, E.J.C., Mehta, R.G., Kinghorn, A.D., Pezzuto, J.M. (1998) Evaluation of the antioxidant potential of natural products. *Combinatorial Chemistry and High Throughput Screening* 1, 35-46.

Ligeret, H., Brault, A., Vallerand, D., Haddad, Y., Haddad, P.S. (2008) Antioxidant and mitochondrial protective effects of silibinin in cold preservation-warm reperfusion liver injury. *Journal of Ethnopharmacology* 115, 507-514.

Madrid, E., Corchete, P. (2010) Silymarin secretion and its elicitation by methyl jasmonate in cell cultures of *Silybum marianum* is mediated by phospholipase D-phosphatidic acid. *Journal of Experimental Botany* 61(3), 747-754.

Maghrani, M., Zeggwagh, N.A., Lemhadri, A., El Amraoui, M., Michael, J.B., Eddouks, M. (2004) Study of the hypoglycaemic activity of *Fraxinus excelsior* and *Silybum marianum* in an animal model of type 1 diabetes mellitus. *Journal of Ethnopharmacology* 91, 309-316.

Madaus, R., Gorler, K., Molls, W. (1983) US patent 4,368,195.

Misawa, M. (1994) Plant tissue culture: an alternative for production of useful metabolites. FAO Agricultural Services Bulletin 108.

Morazzoni, P., Bombardelli, E. (1995) *Silybum marianum (Cardus marianum)*. *Fitoterapia* 66, 3- 42.

Murashige, T., Skoog, F. (1962) A revised medium for rapid growth and bioassays with tobacco tissue culture. *Physiologia Plantarum* 15, 473-497.

Namdeo, A.G. (2007) Plant cell elicitation for production of secondary metabolites: A review. *Pharmacognosy Review* 1, 69-79.

Newall, C.A., Anderson, L.A., Phillipson, J.D. (1996) Herbal Medicines: A Guide for Health- Care Professionals. Pharmaceutical Press, London pp. 46-47.

Quaglia, M.G., Bossù, E., Donati, E., Mazzanti, G., Brandt, A. (1999) Determination of silymarin in the extract from the dried *silybum marianum* fruits by high performance liquid chromatography and capillary electrophoresis. *Journal of Pharmaceutical and Biomedical Analysis* 19, 435-442.

Rahman, N., Khan, N.A., Azmi, S.N.H. (2004) Kinetic spectrophotometric method for the determination of silymarin in pharmaceutical formulations using potassium permanganate as antioxidant. *Pharmazie* 59, 112- 116.

Ramasamy, K., Agrawal, R. (2008) Multitargeted therapy of cancer by silymarin. *Cancer Letters* 269, 352-362.

Pietrangelo, A., Borella, F., Casalgrandi, G. (1995) Antioxidant activity of silybin *in vivo* during long-term iron overload in rats. *Gastroenterology* 109, 1941-1949.

Sánchez-Sampedro, M.A., Fernández-Tárrago, J., Corchete, P. (2005a) Yeast extract and methyl jasmonate-induced silymarin production in cell cultures of *Silybum marianum* (L.) Gaertn. *Journal of Plant Physiology* 162(10), 1177-1182.

Sánchez-Sampedro, M.A., Fernández-Tárrago, J., Corchete, P. (2005b) Some common signal transduction events are not necessary for the elicitor-induced accumulation of silymarin in cell cultures of *Silybum marianum*. *Journal of Plant Physiology* 165(14), 1466-1473.

Sayyah, M., Boostani, H., Pakseresht, S., Malayeri, A. (2010) Comparison of *Silybum marianum* (L.) Gaertn. with fluoxetine in the treatment of Obsessive–Compulsive Disorder. *Progress in Neuro-Psychopharmacology & Biological Psychiatry* 34, 362-365.

Schrall, R., Becker, H. (1977) Produktion von catechinen und oligomeren proanthocyanidinen in callus- und suspensionskulturen von *Cratжgus monogyna, Cratжgus oxyacantha* und *Ginkgo biloba. Planta Medica* 32, 297-307.

Shah, J., Klessig, D.F. (1999) Salicylic acid: signal perception and transduction. In Libbenga, K., Hall. M., Hooykaas, P.J.J., editors. Biochemistry and Molecular Biology of Plant Hormones. Elsevier, Oxford, 513–541.

Shaker, E., Mahmoud, H., Mnaa, S. (2010) Silymarin, the antioxidant component and *Silybum marianum* extracts prevents liver damage. *Food and Chemical Toxicology* 48, 803-806.

Skehan, P., Storeng, R. (1990) New colorimetric cytotoxicity assay for anti cancer drug screening. *Journal of National Cancer Institute* 82, 1107-1112.

Svobodová, A., Zdařilová, A., Walterová, D., Vostálová, J. (2007) Flavonolignans from *Silybum marianum* moderate UVA-induced oxidative damage to HaCaT keratinocytes. *Journal of Dermatological Science* 48, 213-224.

Takahara, E., Ohta, S., Hirobe, M. (1986) Stimulatory effects of silibinin on the DNA synthesis in partially hepatectomized rate livers. *Biochemical Pharmacology* 35, 538- 541.

Tamayo, C., Diamond, S. (2007) Review of clinical trials evaluating safety and efficacy of milk thistle (*Silybum marianum* [L.] Gaertn.). *Integrated Cancer Therapy* 6, 146-157.

Tanaka, H., Shibata, M., Ohira, K., Ito, K. (1985) Total synthesis of (±)-silybin, an antihepatotoxic flavanolignan. *Chemical and Pharmaceutical Bulletin* 33, 1419-1423.

Tůmová, L., Řimáková, J., Tůma, J., Dušek, J. (2006) *Silybum marianum in vitro*-flavonolignan production. *Plant Soil and Environment* 52(10), 454–458.

van der Fits, L., Memelink, J. (2000) ORCA3, a jasmonate responsive transcriptional regulator of plant primary and secondary metabolism. *Science* 289, 295–297.

Verpoorte, R., Alfermann, A.W. (2000) Metabolic Engineering of Plant Secondary Metabolism. Kluwer Academic Publishers, The Netherlands.

Wagner, H., Baldt, S., Rickl, V. (2009) Plant Drug Analysis: A Thin Layer Chromatography Atlas. Second Edition, Springer, Germany.

Wang, K., Zhang, H., Shen, L., Du, Q., Li, J. (2010) Rapid separation and characterization of active flavonolignans of *Silybum marianum* by ultra-performance liquid chromatography coupled with electrospray tandem mass spectrometry. *Journal of Pharmaceutical and Biomedical Analysis* 53, 1053-1057.

Wilasrusmee, C., Kittur, S., Shah, G., Siddiqui, J., Bruch, D., Wilasrusmee, S., Kittur, D.S. (2002) Immunostimulatory effect of *Silybum marianum* (milk thistle) extract. *Medical Science Monitor 8, BR439-434.*

Zhao, B., Wolf, D.M., Agrawal, R. (1999) Inhibition of human carcinoma cell growth and DNA synthesis by silybinin, an active constituent of milk thistle: comparison with silymarin. *Cancer Letters* 147, 77-84.

Zhao, J., Davis, L.C.T., Verpoorte R. (2005) Elicitor signal transduction leading to production of plant secondary metabolites. *Journal of Biotechnology* 23, 283-333.

Zi, X., Mukhtar, H., Agarwal R. (1997) Novel cancer chemopreventative effects of a flavonoid constituent silymarin: inhibition of mRNA expression of an endogenous tumor promoter TNF alpha. *Biochemical and Biophysical Research Communications* 239, 334–339.

13 Chapter | PHYTOCANNABINOIDS

1. INTRODUCTION: AN OVERVIEW

What is cannabis? Marijuana, or Cannabis, is characterized as any formulation of the Cannabis sativa plant utilized for its psychoactive properties, whether for recreational or therapeutic purposes. The 2004 World Drug Report indicates that 3.7% of individuals aged 15-64 utilized marijuana between 2001 and 2003 (World Drug Report, 2004). The consumption of marijuana is linked to several pharmacological effects, the majority of which can be ascribed to tetrahydrocannabinol (THC). The amalgamation of THC and other chemicals from Cannabis sativa may demonstrate distinct pharmacological effects. The isolates derived from Cannabis are referred to as cannabinoids (ElSohly, 2010).

Cannabinoids are a class of C21 terpenophenolic chemicals discovered in Cannabis sativa (Mechoulam & Gaoni, 1967). Phytocannabinoids are the naturally occurring cannabinoids derived from Cannabis species (Pate, 1999). At least 85 cannabinoids have been extracted from Cannabis sativa (El-Alfy et al., 2010). Some of these chemicals are recognized for their medical significance in contemporary society.

To enhance comprehension of the pharmacological effects of phytocannabinoids, human and rodent receptors are employed to assess the binding affinity of these substances to the two cannabinoid receptors documented in literature, CB1 and CB2. CB1 receptors are predominantly situated in the brain, whereas CB2 receptors are generally peripheral, located on mature B cells and macrophages in the tonsils and spleen (Raymon & Walls, 2010). The activation of CB1 receptors produces the euphoric effects associated with Cannabis consumption. Due to the absence of CB1 receptors in the medulla oblongata, the brain stem region governing respiratory and cardiovascular functions, there is no risk of overdose leading to respiratory depression or cardiovascular failure, which may occur with the misuse of other substances, such as opioids. CB2 receptors are purported to provide anti-inflammatory actions.

2. CANNABINOID RECEPTOR FUNCTION

Cannabinoid receptors are G-protein coupled receptors (Figure 1), which are a large family of seven member transmembrane receptors that act in a second messenger fashion. When cannabinoid receptors are activated, they inhibit the enzyme adenylate cyclase.

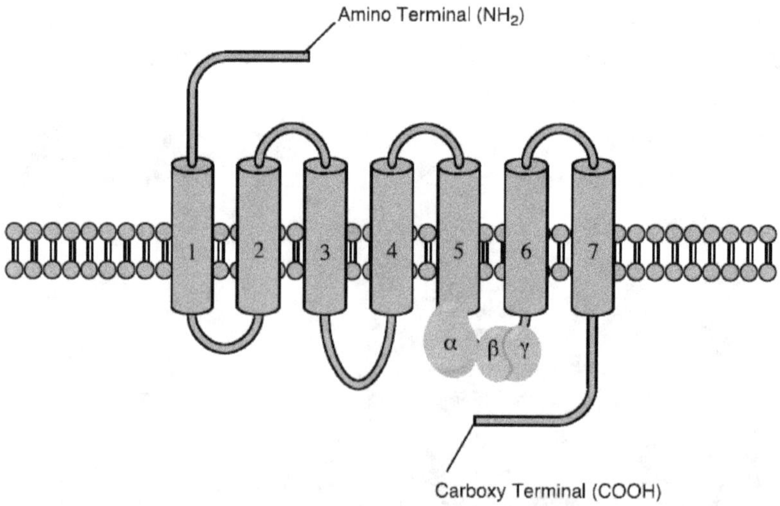

Fig. 1. Example of a G-Protein Coupled Receptor.

Adenylate cyclase is responsible for breaking ATP to form cyclic AMP (cAMP). When a ligand binds to the extracellular surface of cannabinoid receptors, it causes a conformational change of the receptor. This change activates the second messenger by exchanging guanosine diphosphate (GDP) for guanosine triphosphate (GTP). Then, the G- protein's alpha subunit separates from the beta/gamma subunit to cause intracellular proteins to function properly. In CB_1 and CB_2 receptors, cAMP acts as the second messenger. When these receptors are activated, cAMP levels decrease within the cell. Therefore, the result of activating cannabinoid receptors leads to a decrease in cAMP levels, and in turn leads to an inhibition of function.

3. ENDOCANNABINOIDS

Endogenous cannabinoids, or endocannabinoids, are substances produced in the body that activate the cannabinoid receptors. Generally, neurotransmitters are released presynaptically and activate the receptors on a postsynaptic cell. However, unlike most neurotransmitters, the endocannabinoids work in a reverse fashion. Endocannabinoids use retrograde signaling to achieve cannabinoid receptor activation. This means that the ligands are being produced postsynaptically, but acting presynaptically (Lambert, 2009). Another critical point in understanding the function of the endocannabinoids is that the endocannabinoid system can produce endocannabinoids "on demand" in response to an increase in intracellular calcium levels (Sugiura et al., 2006).

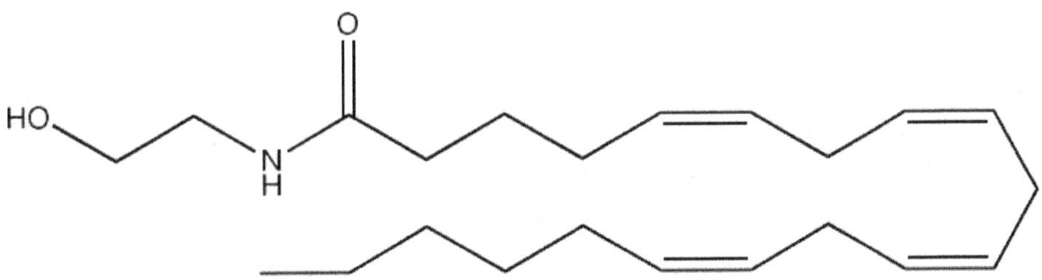

Fig. 2. Chemical structure of anandamide.

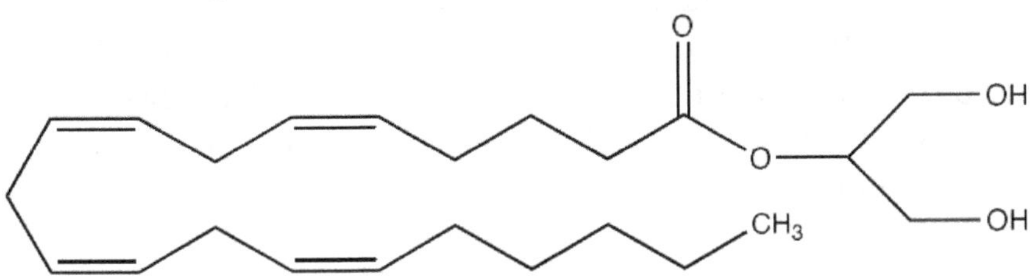

Fig. 3. Chemical structure of 2-AG.

Shortly after the cloning of the cannabinoid receptors, researchers began searching for endogenous ligands that activate these receptors. The first endocannabinoid discovered was anandamide (Figure 2) in 1992 (Devane et al., 1992). Several years after the discovery of anandamide the second endogenous ligand, 2-arachadonoyl-glycerol (2-AG, Figure 3), was discovered (Sugiura et al., 2006). Anandamide and 2-AG act as a partial agonist and full agonist, respectively, at the CB_1 and CB_2 receptors. Although the structure of anandamide differs significantly from THC, both of these ligands have similar pharmacological profiles (Grotenhermen, 2002).

Understanding the mechanism of how cannabinoids produce their effects is in part because of the discovery of the endocannabinoid system.

Although the physiological roles of the endocannabinoids are not fully defined, several pharmacological functions have been described. Studies suggest that these endogenous ligands may aid in pain relief, enhancement of appetite, blood pressure lowering during shock, embryonic development, and blocking of working memory (ElSohly, 2010).

4. PHYTOCANNABINOIDS

The first cannabinoid identified was cannabigerol, and its precursor cannabigeric acid was shown to be the cannabinoid formed in the plant as well as endogenously (Yamauchi, 1975). Today, the most discussed phytocannabinoid is delta-9-tetrahydrocannabinol. In 1964, Gaoni and Mechoulam isolated and elucidated the chemical structure of THC from the leaves of *Cannabis sativa* (Mechoulam & Gaoni, 1964). THC is pharmacologically and toxicologically the best studied constituent of *Cannabis*, responsible for most of the psychoactive effects of natural *Cannabis* preparations (Grotenhermen, 2002). THC and cannabidiol (CBD) are the two most common naturally occurring cannabinoids.

As mentioned earlier, THC (Figure 4) is the main component of *Cannabis* responsible for the psychoactive effects. Other than *Cannabis* being abused to achieve a state of euphoria, it is now being used medicinally to aid in acquired immunodeficiency syndrome (AIDS) patients with wasting syndrome and for pain management, nausea, and vomiting associated with patients receiving cancer chemotherapy. Since THC is responsible for the psychoactive effects of *Cannabis*, people have learned how to genetically increase the concentration of THC within each plant to produce a stronger "high." Since 1980, the concentration of THC within marijuana has increased from less than 1.5% to approximately 20% (ElSohly et al., 2000). THC acts a partial agonist at the CB_1 and CB_2 receptors, but functions via interaction with the CB_1 receptor.

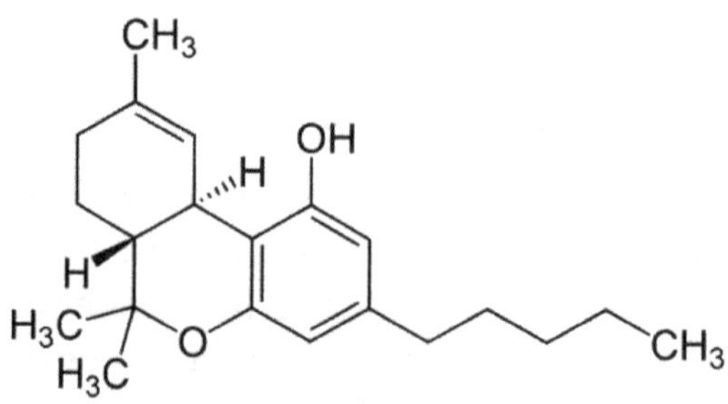

Fig. 4. Chemical structure of delta-9-THC.

The second major constituent of *Cannabis*, cannabidiol (CBD, Figure 5), is responsible for the anti-inflammatory effects due to its interactions with the human CB_2

receptor. CBD was first isolated in 1940 (Adams et al., 1940); however, it was not until 1963 that Mechoulam and Shvo elucidated its correct structure (Mechoulam & Shvo, 1963). At the human CB_2 receptor, CBD's mechanism of action shows inverse agonism activity (Pertwee et al., 2007). In 1995, Benet and colleagues show that cannabidiol is not only responsible for anti-inflammatory effects, but may also aid in reducing unpleasant side effects from THC, including reduced anxiety (Benet et al., 1995). They found that CBD inhibits cytochrome P450 3A11, which causes THC to change into its more potent metabolite 11-hydroxy-THC (Gallily et al., 2002).

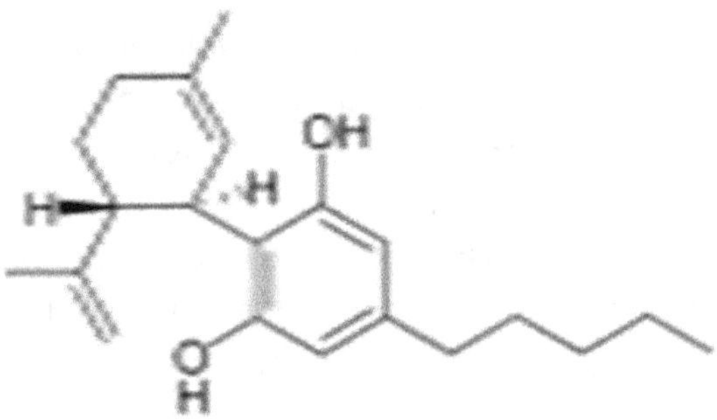

Fig. 5. Chemical structure of cannabidiol.

Tetrahydrocannabinol and cannabidiol are the two most discussed phytocannabinoids, but not the only ones known. ElSohly and co-investigators have divided the phytocannabinoids into ten subclasses: 1) Cannabigerol type – propyl side chains and monomethyl ether derivatives 2) Cannabichromene type – analogs present in the C-5 position 3) Cannabidiol type – analogs varying from C-1 to C-5 positions 4) Delta-9-tetrahydrocannabinol type – double bond in the C-9 position; responsible for psychoactive effects 5) Delta-8- tetrahydrocannabinol type – double bond in the C-8 position; thermodynamically more stable than delta-9-THC , however, 20% less active 6) Cannabicyclol type – five atom ring and C-1 bridge 7) Cannabielsoin type – artifacts formed from CBD 8) Cannabinol and Cannabinodiol types – A ring aromatization 9) Cannabitriol type – additional hydroxyl substitution 10) Miscellaneous types – ex: furano ring, carbonyl function, tetrahydroxy substitution (ElSohly, 2010).

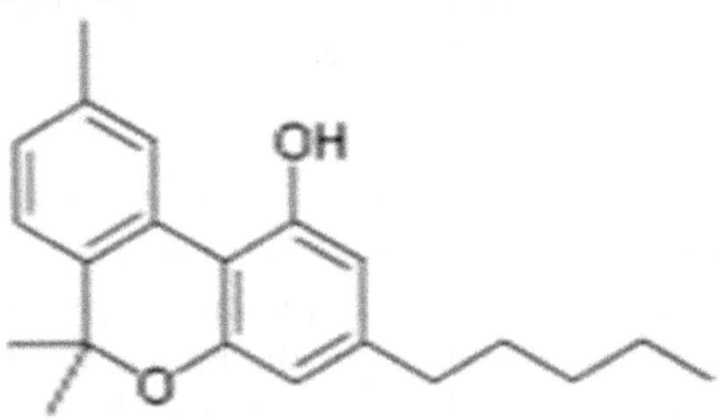

Fig. 6. Chemical structure of cannabinol.

Another phytocannabinoid that shows a significant amount of importance is cannabinol (CBN, Figure 6); it is a metabolite of tetrahydrocannabinol. It was the first cannabinoid identified from *Cannabis sativa.* (Wood et al., 1896). Along with THC, cannabinol is also a psychoactive component of *Cannabis* due to its interaction with CB_1 receptors. Compared to THC, it acts a weak agonist at both the CB_1 and CB_2 receptors.

Extracts that have been isolated from marijuana may be tested to see if they have affinity for each of the CB_1- or CB_2- type receptors. THC remains the best phytocannabinoid in terms of affinity for the cannabinoid receptors with a binding Ki of 14nM (Figure 7). Most of the compounds isolated from *Cannabis* show a sufficient amount of binding activity at both of the cannabinoid receptors. However, not all compounds isolated show interactions with either CB_1 or CB_2. For instance, even though cannabidiol is a major constituent of *Cannabis* and shows pharmacological effects, it has little or no activity for CB_1 or CB_2 receptors (Mechoulam & Rodriguez, 2007). To determine binding affinity and functional activity, in vitro assays are performed.

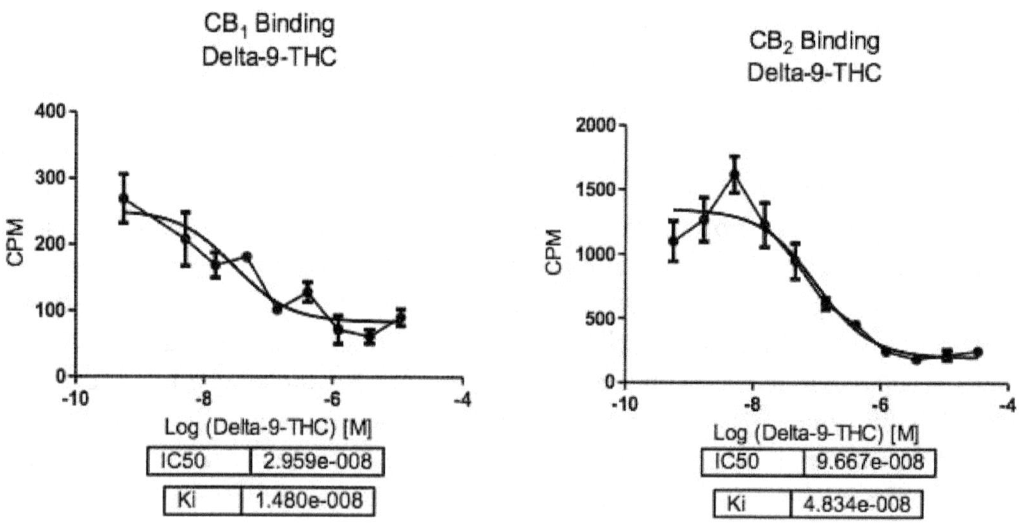

Fig. 7. Binding affinity of delta-9-THC at Cannabinoid Receptor 1 and Cannabinoid Receptor 2.

5. *IN VITRO* BIOASSAYS

In order to have success with in vitro assays, cultured cells containing the specific receptors must be developed. At the University of Mississippi HEK293 cells have been transfected with full length human CB_1 and human CB_2 DNA via electroporation. Once "shocked," the cells open and accept the human CB_1 and CB_2 cDNA with a linked specific antibiotic resistant plasmid. Since not all cells will receive the DNA, a selection process using the specific antibiotic is added to the cultured cells in order to kill off cells without the cDNA. After an allotted time period for growth, a single cell is selected and clonal colonies are grown in cell culture. The replication of a single cell containing either CB_1 or CB_2 DNA allows researchers to guarantee the over expression of cannabinoid receptors on the cell membrane. With this, mass subculture followed by "scraping" of the cells leads to the membrane with the receptors. Once the protein concentration is determined this membrane may be used for in vitro assays.

Phytocannabinoids may be tested for their binding affinity toward each of the cannabinoid receptors. A competitive binding assay is done to determine the binding affinity of each compound. The competition is between the chosen phytocannabinoid and a labeled ligand, such as ^{3}H- CP-55, 940. It is known that the labeled ligand will tightly bind to each of the cannabinoid receptors; therefore, if a test compound shows affinity for the receptors, the amount of labeled ligand bound to the receptor will be low resulting in high binding affinity of the test compound. A compound showing strong binding affinity for either of the cannabinoid receptors, warrants testing to determine the functional activity.

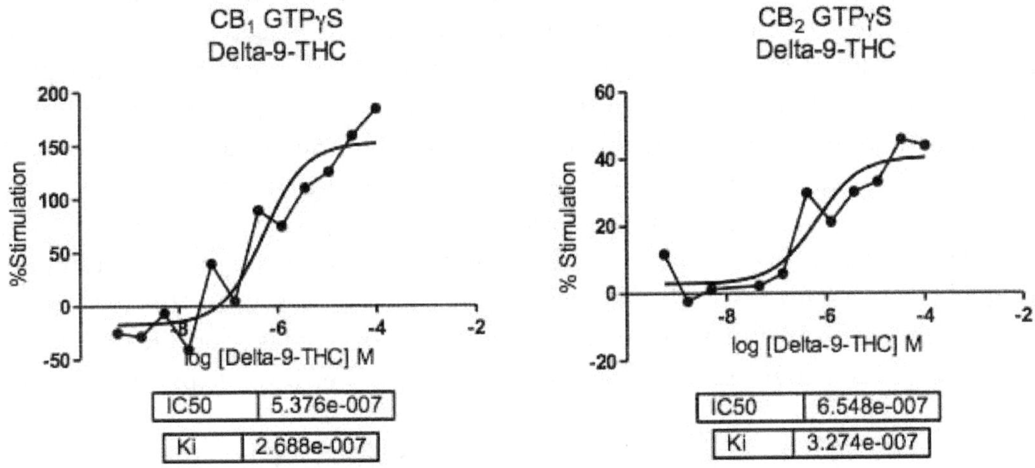

Fig. 8. Functional assay activity of delta-9-THC at Cannabinoid Receptor 1 and Cannabinoid Receptor 2.

A functional assay determines whether the compound is acting as an agonist, antagonist, or inverse agonist. As opposed to the binding assay, an in vitro functional assay is not based upon competitive binding, but rather "tracking" the amounts of guanosine triphosphate (GTP). When the membrane is not stimulated, there is a pool of guanosine diphosphate (GDP) associated with it. Upon stimulation, this pool of GDP is converted into GTP. To monitor this response, ^{35}S labeled GTP is added to the assay to bind to the receptors. Therefore, an increase in GTP is directly proportional to stimulation of the receptor by labeled ligand. An agonist compound is indicated by an increase in GTP. Delta-9-THC has a functional Ki of approximately 300nM, which means it is acting as a partial agonist, yet is still responsible for the psychoactive effects associated with *Cannabis* (Figure 8). To detect an antagonist, the compound must be tested in the presence of a known agonist at that specific receptor. The antagonist blocks the ability of the agonist to fully stimulate the receptor, thus resulting in a right shift of the agonist EC_{50}.

6. *IN VIVO* BIOASSAYS

Cannabinoids that show promising activity in the functional assay, whether acting as an agonist or antagonist, may be tested in vivo using the tetrad assay in mice. In the late 1980s, Little and his colleagues began testing rodents treated with cannabinoids in this tetrad assay. The term tetrad describes a series of four different tests to help evaluate the biological effects of a compound: 1) Locomotor activity 2) Catalepsy 3) Hypothermia and 4) Analgesia. The locomotor activity test allows a researcher to determine if the rodent is acting "lazy." The rodent is placed in a box with perpendicular gridlines, which are beams of light. The test determines the amount of times the beams are broken in an allotted time period, an increase in the number of times broken correlates with a decrease in locomotor activity. To determine if the drug causes cataleptic effects, a rodent is placed on a bar elevated off the ground surface. If the rodent remains immobile, it is considered cataleptic. Hypothermia, also know as a rectal temperature assay, is simply a measure of the rodents rectal temperature after the drug has been administered. For the last part of the tetrad assay, there are two different methods of testing for analgesic effects. One method is the hot plate (Figure 9) assay. In this assay, a rodent is placed upon a hot plate and the time it takes for the rodent to react, usually a small jump, is recorded. The second method is known as the tail-flick assay. In this assay, the rodent is immobilized and a high temperature beam of light is sporadically placed on the tail. If the rodent feels pain, it will move its tail either left or right (Little, 1988).

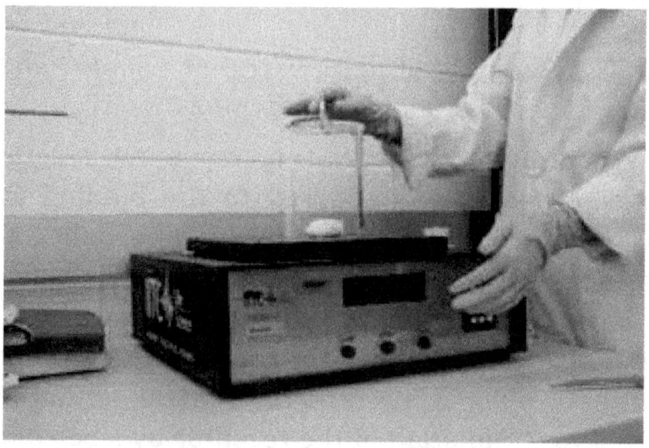

Fig. 9. Analgesic portion of tetrad assay: hot plate test.

7. MEDICINAL USES OF MARIJUANA

According the United Nations, *Cannabis* "is the most widely used illicit substance in the world" (World Drug Report, 2010). There are people who use Cannabis medicinally, and there are others who abuse *Cannabis* in order get "high," or obtain a state of euphoria. Those who use marijuana regularly for medicinal purposes use strict, smaller amounts to control the strength and duration of the "high." However, those who abuse marijuana attempt to smoke or ingest as much as necessary to achieve their own personal state of euphoria. This abuse negatively affects the people who do need *Cannabis* to help with side effects of chemotherapy and AIDS. *Cannabis* is not only used to help those suffering from cancer chemotherapy and AIDS, but it also lowers intraocular eye pressure for those with glaucoma, acts a pain reliever, and more recently has been found to help with symptoms of multiple sclerosis and depression. Therefore, researchers are attempting to formulate a synthetic cannabinoid that resembles the compounds isolated from *Cannabis*, but do not exploit psychotropic properties.

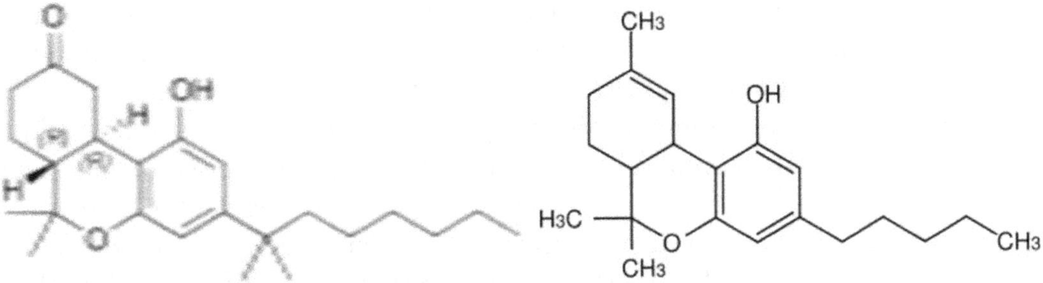

Fig. 10. Chemical structures of Nabilone (left) and Dronabinol (right).

The goal of research in this area is to synthesize a cannabinoid-like compound that warrants a high affinity for either CB_1 or CB_2 receptors, or both, and can help patients without causing some of the unwanted side effects of marijuana, such as the psychotropic effects associated with CB_1. With this said, studies show that *Cannabis* users have fewer psychological side effects than those users administering synthetic THC. There are two synthetic cannabinoid products available on the market in the United States, Nabilone and Dronabinol (Figure 10). Some of these side effects from synthetic cannabinoids include dysphoria, depersonalization, anxiety, and paranoia (Grinsponn & Bakalar, 1997). As previously mentioned, CBD has shown to reduce anxiety and other unpleasant side effects caused by ingestion of pure THC (Zuardi et al., 1982). The preference of whole *Cannabis* over synthetic formulations of THC is due to the lack of extra side effects associated with the whole *Cannabis*. This opens the door for scientists to study what is actually causing all of the side effects associated with synthetic THC. This also shows that some of the compounds associated with *Cannabis sativa* may be working synergistically to alleviate unwanted effects from THC when used alone (McPartland & Russo, 2001).

So, the ultimate goal in cannabinoid drug development would be to mimic the non-psychotropic effects associated with CB_1, mimic the beneficial effects associated with CB_2, and not deal with the negative side effects associated with marijuana or synthetic THC.

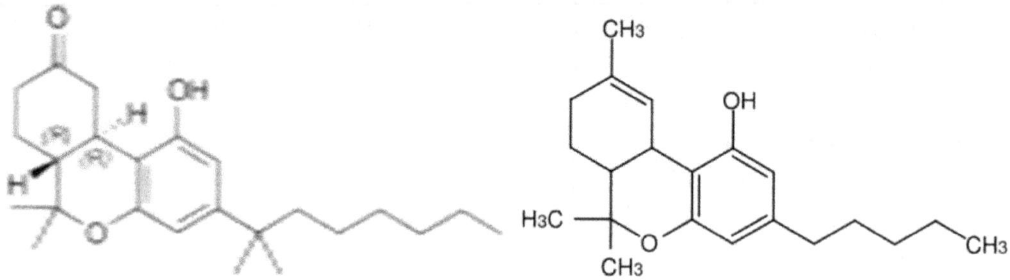

Fig. 10. Chemical structures of Nabilone (left) and Dronabinol (right).

8. PHYTOCANNABINOIDS AND DEPRESSION

Depression may be described as a mood disorder associated with feeling down, sad, angry, or lost that interferes with everyday life. The most commonly associated drug categories for the treatment of depression include monoamine oxidase inhibitors (MAOIs), tricyclic antidepressants (TCAs), selective-serotonin reuptake inhibitors (SSRIs), and serotonin- norepinephrine reuptake inhibitors (SNRIs). A new field of research involving *Cannabis* may be the link to the treatment of depression. However, studies show conflicting data as to whether *cannabis* is beneficial (Grinsponn & Balkar, 1998) or detrimental for the treatment of depression (Bovassa, 2001). Due to the conflicting results of these studies, Witkin switched the focus to the role of the endocannabinoid system and the treatment of depression from exogenously administered cannabinoids (Witkin et al., 2005). Since 2005, it has been concluded that the endocannabinoid system does play a role in the treatment of depression, but differs from minor depression to major depression.

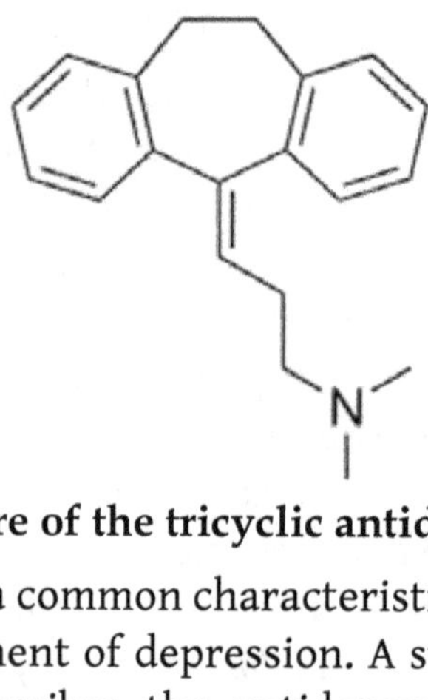

Fig. 11. Chemical structure of the tricyclic antidepressant, Amitriptyline.

New research has found that a common characteristic of *Cannabis*, mood elevation, may be the link to the treatment of depression. A study published by El-Alfy and co-investigators in 2010 describes the antidepressant effects associated with administration of phytocannabinoids. The objective of this study was to isolate the major cannabinoids from *Cannabis* and evaluate the antidepressant effects using the mouse forced swim test (FST), followed by the tail suspension test (TST). Typically in mice, when cannabinoids are administered they exert hypothermia and catalepsy, which means that a psychoactive state is being achieved. For these depression studies, only low dosages of these phytocannabinoids were administered so that the test subjects did not demonstrate psychoactive effects. The cannabinoids isolated and tested were cannabigerol (CBG), cannabinol (CBN), cannabichromene (CBC), cannabidiol (CBD), delta-8- THC, and delta-9-THC (THC) (Figure 12).

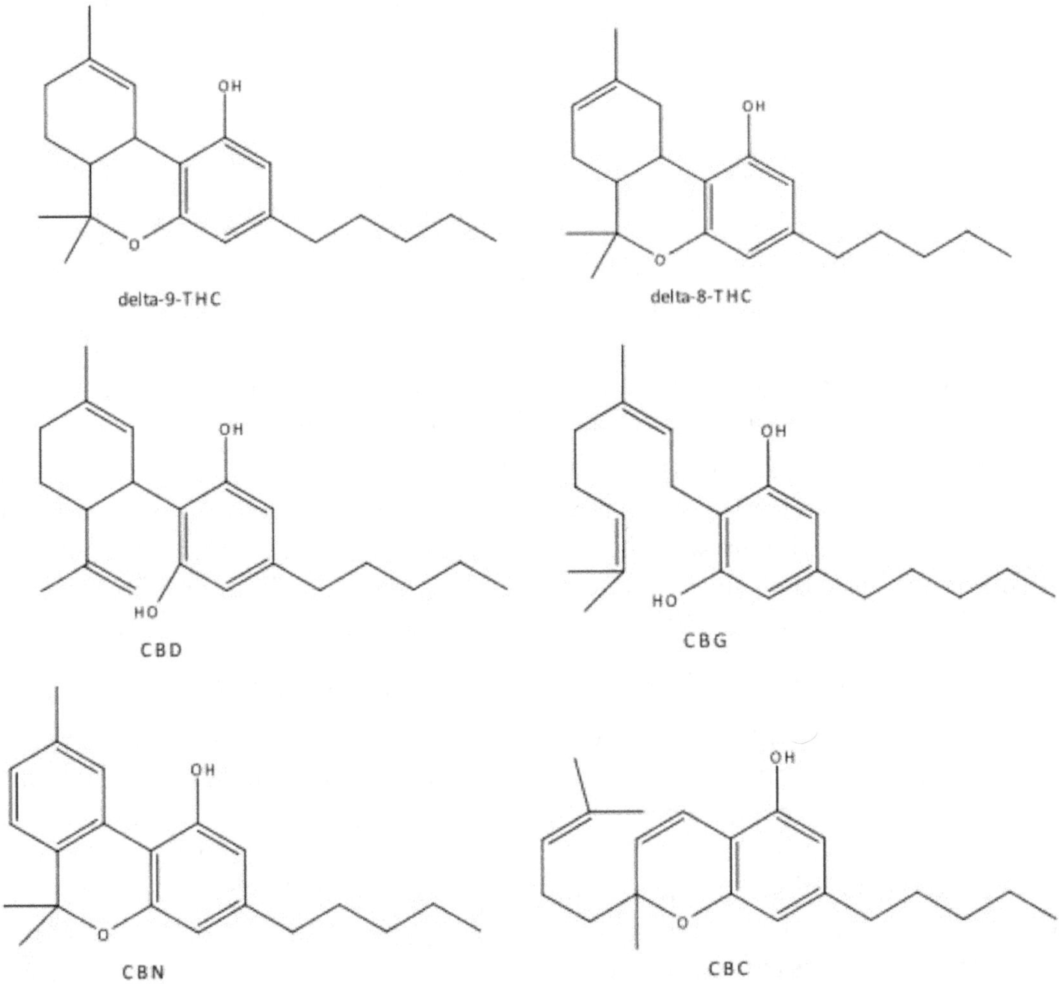

Fig. 12. The six phytocannabinoids tested for antidepressant-like effects (El-Alfy et al., 2010).

To assess that hypothermia and catalepsy were not achieved, the tetrad assay was completed after administration of each cannabinoid. Out of the six cannabinoids tested, only delta-8-THC and delta-9-THC showed a U-shaped dose response in the forced swim test. With this, only delta-9-THC showed significant antidepressant-like effects. Administration of the non-psychoactive components revealed that CBC and CBD displayed antidepressant- like effects in the forced swim test. However, a high dose of CBD was used to display these antidepressant-like effects.

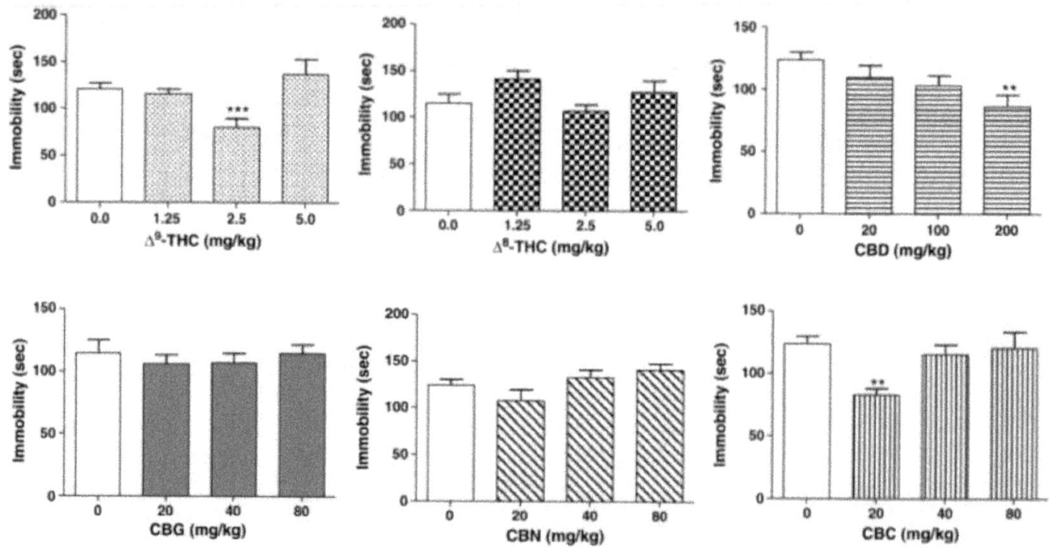

Fig. 13. Effects of each phytocannabinoid on immobility time in the mouse forced swim test (El-Alfy et al., 2010).

To further confirm these tests, delta-9-THC and CBC were evaluated in the tail suspension test. Between these two phytocannabinoids, only delta-9-THC continued to exhibit these antidepressant-like effects at low doses. Therefore, the results of this study show that delta- 9-THC and other phytocannabinoids administered exogenously do indeed aid with the treatment of depression (El-Alfy et al., 2010).

9. PHYTOCANNABINOIDS AND APPETITE STIMULATION

Patients suffering from AIDS are now becoming the main target for the therapeutic use of *Cannabis*. Those with AIDS tend to lose their desire to eat regularly throughout the day. When this occurs, the patient becomes weak, agitated, tired, and anorexic; this occurrence in known as Wasting Syndrome. Research shows that at least 90% of patients who smoked marijuana had the desire to eat immediately after use (Haines & Green, 1970). With the use of *Cannabis* as a therapeutic drug to stimulate appetite, the suffering patients may be able to eat on a regular basis throughout the day, thus improving their quality of life. Several studies have shown that the use of marijuana does increase appetite, which also increases energy in daily life routines.

In a study conducted by Mattes and colleagues, the appetite stimulating effects of cannabinoids, specifically THC, were examined. A major focus in this study, for a means of clarification from previous research, was the route of administration of

THC. The four different ways in which THC was administered includes oral, inhaled, sublingual, and suppository. There are high levels of variability in determining if THC does actually stimulate appetite. Factors such as environment, age, gender, tolerance, dosage, and social influences play a role in the effect of THC on appetite. During one study, the suppository route of administration resulted in the highest energy intake when compared to oral, sublingual, and inhaled administration of THC (Figure 15).

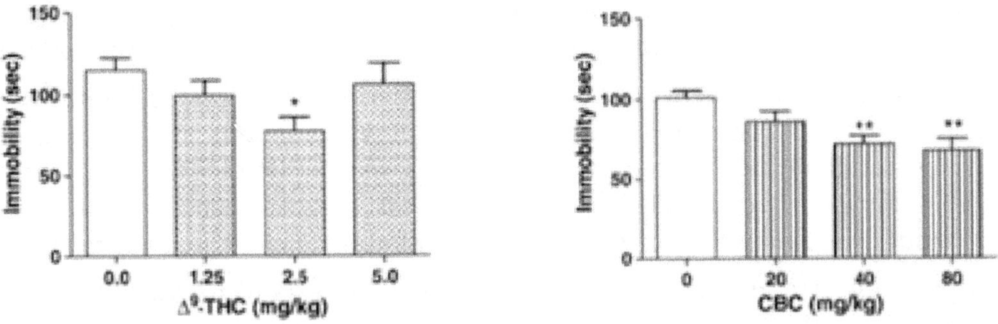

Fig. 14. Effects of THC and CBC on immobility time in the mouse tail suspension test (El- Alfy et al., 2010).

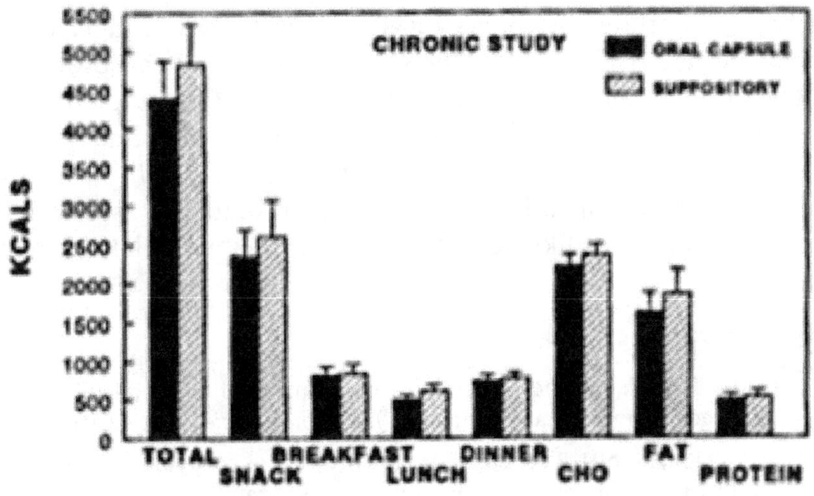

Fig. 15. Mean data from patients dosed orally and via suppository over a 72 hour time period (Mattes et al., 1994).

There is no single outcome on the effect of THC on appetite stimulation no matter the form of administration. The results vary from having no effect to the possibility of having major food cravings. In some circumstances, not only did the food cravings become increased, but during a meal the food seemed to also have an increased taste of delightfulness. The conclusion of this study indicates that THC as an appetite stimulant produces its highest effects on healthy, adult individuals who use low dosage amounts (Mattes et al., 1994).

10. CONCISE SUMMARY

What is cannabis? Marijuana, or Cannabis, is defined as any preparation of the Cannabis sativa plant employed for its euphoric effects, whether for recreational or medicinal use. The 2004 World Drug Report states that 3.7% of individuals aged 15-64 consumed marijuana from 2001 to 2003 (World Drug Report, 2004). The use of marijuana is associated with several pharmacological consequences, primarily attributed to tetrahydrocannabinol (THC). The combination of THC and other compounds from Cannabis sativa may exhibit unique pharmacological effects. The compounds extracted from Cannabis are known as cannabinoids (ElSohly, 2010).

Cannabinoids are a category of C21 terpenophenolic compounds identified in Cannabis sativa (Mechoulam & Gaoni, 1967). Phytocannabinoids are naturally occurring cannabinoids sourced from Cannabis species (Pate, 1999). At least 85 cannabinoids have been isolated from Cannabis sativa (El-Alfy et al., 2010). Certain compounds are acknowledged for their medical importance in modern culture.

To improve understanding of the pharmacological effects of phytocannabinoids, human and rodent receptors are utilized to evaluate the binding affinity of these compounds to the two cannabinoid receptors identified in literature, CB1 and CB2. CB1 receptors are mostly present in the brain, while CB2 receptors are typically found peripherally, specifically on mature B cells and macrophages in the tonsils and spleen (Raymon & Walls, 2010). The stimulation of CB1 receptors induces the euphoric effects linked to Cannabis use. The lack of CB1 receptors in the medulla oblongata, the brainstem area responsible for respiratory and cardiovascular processes, eliminates the possibility of overdose resulting in respiratory depression or cardiovascular failure, a potential consequence associated with the overuse of other compounds like opioids. CB2 receptors are believed to exert anti-inflammatory effects.

11. REFERENCES

Adams, R., Hunt, M., & Clark, J. (1940). Structure of cannabidiol, a product isolated from the marihuana extract of Minnesota wild hemp. I. *J. Am. Chem. Society.* 62, 196-199.

Bornheim, L., Kim, K., Li, J., Perotti, B., & Benet, L. (1995). Effect of cannabidiol pretreatment on the kinetics of Tetrahydrocannabinol metabolites in mouse brain. *Drug Metab. Dispos.* 23, 825-831.

Bovassa, G. (2001). Cannabis abuse as a risk factor for depressive symptoms. *American Journal Psychiatry.* 158: 2033-2037.

Devane, W., Hanus, L., Breur, A., Pertwee, R., Stevenson, L., Griffin, G., Gibson, D., Mandelbaum, A., Etinger, A., & Mechoulam, R. (1992). Isolation and Structure of a Brain Constituent That Binds to the Cannabinoid Receptor. *Science.* 258, 1946-1949.

El-Alfy, A., Ivey, K., Robinson, K., Ahmed, S., Radwan, M., Slade, D., Khan, I., ElSohly, M., & Ross, S. (2010). Antidepressant-like effect of [Delta]9-tetrahydrocannabinol and other cannabinoids isolated from Cannabis sativa L. *Pharmacology Biochemistry and Behavior*, 95, 4, June 2010, 434-442, ISSN 0091-3057.

ElSohly, M., Ross, S., Mehmedic, Z., Arafat, R., Yi, B., & Banahan, B. (2000). Potency trends of Delta-9-THC and other cannabinoids in confiscated marijuana from 1980-1997. *J. Forens. Sci.* 45, 24-30.

ElSohly, M. (2010). *Marijuana and the Cannabinoids*. Humana Press. ISBN: 978-1-61737-581- 1.Totowa, New Jersey.

Gaoni, Y. & Mechoulam, R. (1964). *J. Am. Chem. Soc.* 86, 1646.

Grinsponn, L. & Bakalar, J. (1997). *Marihuana, the forbidden medicine*, revised edition. New Haven, CT: Yale University Press.

Grinsponn, L. & Balkar, J. (1998). The use of cannabis as a mood stabilizer in bipolar disorder: anectodotal evidence and the need for clinical research. *Journal Psychoactive Drugs*. 30, 171-177.

Grotenhermen, F. (2002). Effects of Cannabis and the cannabinoids, in Cannabis and Cannabinoids, In: *Pharmacology, Toxicology, and Therapeutic Potential*, Grotenhermen, F. and Russo, E., pages 55-65, Haworth Press, New York.

Haines, L. & Green, W. (1970). Marijuana use patterns. *British Journal of Addiction*. 65, 347-362.

Lambert, D. (2009). *Cannanbinoids in Nature and Medicine*. Wiley-Verlag Helvetica Chimica Acta. ISBN: 978-3-906390-56-7. Zurich, Switzerland.

Little, P., Compton, D., Johnson, M., Melvin, L., & Martin, B. (1988). Pharmacology and stereoselectivity of structurally novel cannabinoids in mice. *Journal of Pharmacol. Exp. Ther.* 247, 1046-1051.

Mattes, R., Engelman, K., Shaw, L., & ElSohly, M. (1994). Cannabinoids and appetite stimulation. *Pharamcology, Biochemistry, and Behavior*. 49, 187-195

McPartland, J. & Russo, E. (2001). Cannabis and cannabis extracts: greater than the sum of their parts? *J. Cann. Therap.* 1, 103-132.

Mechoulam, R. & Shvo, Y. (1963) Hasish-I. The structure of cannabidiol. *Tetrahedron* 19, 2073-2078.

Mechoulam, R. & Gaoni, Y. (1967). Recent advances in the chemistry of hashish. *Fortschr. Chem. Org. Naturst*, 25, 175-213.

Mechoulam, R., Parker, L., & Gallily, R. (2002) Cannabidiol: an overview of some pharmacological aspects. *J. Clinical Pharmacology*. 42, 11S-19S.

Mechoulam, R. & Peters, M. (2007). Cannabidiol-Recent advances. *Chemistry and Biochemistry* 4, 8, 1678-1692.

Pate, D. (1999) Anandamide structure-activity relationships and mechanisms of action on intraocular pressure in the normotensive rabbit model. *Doctoral dissertation*. Kuopio University Publications, Kuopio.

Pertwee, R. (2005). Pharmacological actions of cannabinoids. *Handbook of Experimental Pharmacology*. 168: 1-51.

Raymon, L. & Walls, H. (2010). Marijuana and the Cannabinoids. *Pharmacology of Cannabinoids*. Page 97-143.

Shoyama, Y., Yagi, M., Nishioka, I., & Yamauchi, T. (1975). Biosynthesis of cannabinoid acids. *Phytochemistry*, 14, 2189-2192.

Sugiura, T., Kishimoto, S., Oka, S., & Gokoh, M. (2006). Biochemistry, pharmacology and physiology of 2-arachadonoylglycerol, and endogenous cannabinoid receptor ligand. *Prog. Lipid Res.* 45, 405-446.

Thomas, A., Baillie, G., Phillips, A., Razdan, R., Ross, R., Pertwee, R. (2007) *British Journal of Pharmacology*. 150, 613.

Witkin, J., Tzavara, E., Davis, R., Li, X., & Nomikos, G. (2005). A therapeutic role for cannabinoid CB1 receptor antagonists in major depressive disorders. *Trends Pharmacol. Sci.* 26, 609-617.

Wood, T., Spivey, W., & Easterfield, T. (1896). *Journal of the Chemical Society*. 69, 539.

Yamauchi, T., Shoyama, Y., Yagi, M., & Nishioka, I. (1975). Biosynthesis of cannabinoid acids. *Phytochemistry*. 14, 10, 2189-2192.

Zuardi, A., Morais, S., Guimaraes, F., & Mechoulam, R. (1995). Antipsychotic effect ofcannabidiol. *J. Clinical Psychiatry*. 56, 485-486.

2004 World Drug Report, United Nations, Office of Drugs and Crime. Oxford University Press, Oxford, United Kingdom.

2010 World Drug Report, United Nations, Office of Drugs and Crime. Oxford University Press, Oxford, United Kingdom.

14 Chapter

ALKALOIDS AND ANTHRAQUINONES DERIVED FROM MALAYSIAN FLORA

1. INTRODUCTION: AN OVERVIEW

The flora of Malaysia is among the most diverse in the world, attributable to its consistently warm and uniformly humid climate. Malaysia is ranked as the 12th most diversified nation in the world (Abd Aziz, 2003) and is predominantly enveloped by tropical rainforests. Tropical rainforests occupy merely 12% of the Earth's terrestrial surface; however, they account for approximately 50% to 90% of global biodiversity. Approximately 25% of contemporary pharmaceuticals are derived from rainforests, despite fewer than 1% of the world's tropical rainforest plant species having been assessed for their pharmacological attributes (Kong, et al., 2003). The great diversity of Malaysian flora, with over 12,000 species of flowering plants, provides significant chemical variations for many biological goals. The flora of Malaysia is a prolific source of various classes of natural compounds, including alkaloids, anthraquinones, and phenolic compounds. Plants are typically studied for their ethnobotanical applications. The phytochemical analysis of various renowned plants in traditional medicine, including Eurycoma longifolia, Labisia pumila, Andrographis paniculata, Morinda citrifolia, and Phyllanthus niruri, revealed numerous bioactive phytochemicals. This paper delineates our research on the alkaloids of Fissistigma latifolium and Meiogyne virgata from the Annonaceae family, as well as the anthraquinones of Renellia and Morinda from the Rubiaceae family.

2. THE FAMILY *ANNONACEAE* AS SOURCE OF ALKALOIDS

Annonaceae, known as *Mempisang* in Malaysia (Kamarudin, 1988) is a family of flowering plants consisiting of trees, shrubs or woody lianas. This family is the largest family in the Magnoliales consisting of more than 130 genera with about 2300 to 2500 species. Plants of the family Annonaceae are well known as source of a variety of alkaloids (Cordell, 1981). Many alkaloids have important physiological effects on human and exhibit marked pharmacological activity which is useful as medicine. For examples, atropine is used widely as an antidote to cholinesterase inhibitors such as physostigmine. Morphine and codeine are narcotic analgesics and antitusive agent while caffeine, which occurs in coffee, tea and cocoa is a central nervous system stimulant. Caffeine is also used as cardiac and respiratory stimulant andbesides as an antidote to barbiturate and morphine poisoning (Parker, 1997). The first report on phytochemical studies of alkaloids from Malaysian Annonaceae plants was on the leaves of *Desmos dasymachalus* which has led to the isolation of new 7- hydroxyaporphine, dasymachaline (Chan & Toh, 1985).

The phytochemical investigation of Malaysian Annoaceous plants for their alkoloidal content continue to flourish. Phytochemical survey of the flora of the Peninsula Malaysia and Sabah, with systematic screening for alkaloids resulted in reports on chemical constituents of several plants from Annonaceae illustrating great interest in this field (Teo, *et al.*, 1990). Lavault *et al.*, (1981) analysed the alkaloid content of three Annonaceae plants; *Disepalum pulchrum*, *Polyalthia macropoda* and *Polyalthia stenopetala* which led to the isolation of several isoquinoline compounds. Isolation of two new 7,7'-bisdehydroaporphine alkaloids; 7,7'-bisdehydro-*O*-methylisopiline and 7-dehydronornuciferine-7'-dehydro-*O*- methylisopiline from bark of *Polyalthia bullata* was reported by Connolly *et al.*, (1996). Kam (1999) reviewed the alkaloids derived from Malaysian flora in a book entitled chemical and biological approach of alkaloids.

In Malaysia, eight species of *Fissistigma* are known. They are *F. mobiforme*, *F. cylindrium*, *F. fulgens*, *F. kingii*, *F. lanuginosum*, *F. latifolium*, *F. munubriatum* and *F. kinabaluensis* (Nik Idris *et al.*, 1994). Not much has been reported on the phytochemical studies of *Fissistigma* species. The studies on the alkaloids from *Fissistigma fulgens* have led to the isolation of aporphine, oxoaporphine and protoberberine alkaloids. Liriodenine, anonaine, argentinine, discretamine and kikemanine were found from this species (Awang, *et al.*, 2000). The phytochemical work on alkaloidal composition of the Malaysian *Fissistigma manubriatum* by Saaid and Awang (2005) yielded two oxoaporphines, lanuginosine and liriodenine together with two tetrahydroprotoberberines, tetrahydropalmatine and discretine. We studied the alkaloids of *Fissistigma latifolium* and reported the isolation of nine alkaloids including a new aporphine compound (Alias *et al.*, 2010).

Meiogyne cylindrocarpa, *Meiogyne monosperma* and *Meiogyne virgata* are the only three *Meiogyne* species found in Malaysia. Only *Meiogyne virgata* was studied by Tadic *et al.* (1987). The sample collected from Mount Kinabalu, Sabah was reported to contain azafluorene alkaloid, kinabaline, together with liriodenine, cleistopholine and other aporphine alkaloids. Our work on *Meiogyne virgata* from Hulu Terengganu yielded nine alkaloids from aporphine, oxoaporphines and azaanthracene groups.

2.1 Alkaloids of Fissistigma latifolium and Meiogyne virgata

Since the last three decades, a large number of alkaloidal compounds have been isolated from some Annonaceae species. Tertiary and quaternary isoquinoline and quinoline alkaloids are pharmacologically important compounds commonly found in Annonaceae plants. Continuing our interest on this family of plants, we pursued phytochemical investigation on *Fissistigma latifolium* and *Meiogyne virgata.*

2.1.1 Alkaloids of Fissistigma latifolium

Fissistigma latifolium (Dunal) Merr. from the genus *Fissistigma* is a climbing shrub found in lowland forest of Malaysia, Sumatra, Borneo and Philippines (Verdout, 1976). The genus *Fissistigma* (Annonaceae) consists of about 80 species and is widely distributed in Asia and Australia (Sinclair, 1955). Several species of the genus *Fissistigma* have been used in Southeast Asia as traditional medicines (Perry, 1980). They have been used for muscular atrophy, hepatomegaly and hepatosplenomegaly (Kan, 1979). In Malaysia, the medicinal uses of *Fissistigma* species was briefly mentioned by Burkill as the treatment for childbirth, malaria, wounds, ulcer and rheumatism (Kamarudin, 1988).

Fig. 1. *Fissistigma latifolium*

Previous studies on *F. fulgens* and *F. manubriatum* have resulted in the isolation of aporphine, oxoaporphine and protoberberine alkaloids. Similarly, the studies on alkaloids from *Fissistigma latifolium* led to the isolation of a new aporphine alkaloid, (-)-*N*- methylguattescidine **1** (Alias, *et al.*, 2010). This alkaloid, together with eight known alkaloids, namely liriodenine **2**, lanuginosine **3**, (-)-asimilobine **4**, dimethyltryptamine **5**, (-)-remerine **6**, (-)-anonaine **7**, columbamine **8** and lysicamine **9**, were obtained from the methanol extract of the bark of the plant. The new compound was characterized by analysis of spectroscopic methods such as NMR (Nuclear Magnetic Resonance), IR (Infrared) and GC-MS (Gas-Chromatography-Mass Spectrometry).

(-)-*N*-Methylguattescidine **1** exhibited a molecular formula of $C_{19}H_{17}O_4N$ based on the HRESIMS spectrum (positive mode), which showed a pseudomolecular ion at *m/z* 324.3581 $[M+H]^+$ (calcd. 324.3595). The UV spectrum showed an absorption band at 310 nm, suggesting the compound was an aporphine alkaloid with substitutions at position 1 and 2. The IR spectrum indicated the presence of C-H aromatic at 3056, C-O at 1266 and OH at 3409 cm^{-1}, respectively. The absorption of methyl group appeared at 2945 and 2833. The ^{13}C- NMR spectrum showed presence of 19 carbons. The signal at δ 198.0 ppm confirmed the presence of the carbonyl group, while the signal at δ 153.1 ppm is evidence for the oxygenated aromatic carbon. The DEPT spectrum revealed three methylene carbons at δ 26.9 ppm, 41.4 ppm and 96.9 ppm. Signal at δ 96.9 ppm is indicative of a methylenedioxy carbon. This is consistent with two doublets at δ 5.99 ppm (*J* = 1.2 Hz) and δ 6.07 ppm (*J* = 1.2 Hz) in the ^{1}H-NMR spectrum for the protons of methylenedioxy group which is typically located at positions 1 and 2. The characteristic ABD aromatic signals of H-11, H-10 and H-8 of aporphine alkaloid were observed at δ 8.24 ppm (*d, J* = 8.7 Hz), δ 7.13 ppm (*dd, J* = 8.7, 2.7 Hz) and δ 7.39 ppm (*d, J* = 2.7 Hz), respectively. The ^{1}H-NMR spectrum also exhibited an *N*- methyl signal at δ 2.34 ppm and another methyl group attached to C-6a gave a singlet at δ1.52 ppm. The assignment of this methyl group at the 6a position is confirmed through its HMBC correlation with C-6a at δ 62.7 ppm, C-1b at δ 118.3 ppm and C-7 at δ 198.0 ppm. HMQC spectrum shows two cross peaks at δ 26.9 ppm (C-4) axis, represented the correlations of C-4 to H-4 (δ 2.55 ppm) and H-4' (δ 3.00 ppm). At δ 41.4 ppm (C-5) axis, two cross peaks showed the correlations between C-5 and H-5 (δ 2.99 ppm) and H-5' (δ 3.01 ppm). The quaternary carbon signals were assigned based on HMBC experiment. C-1a at δ 108.9 ppm, C-7a at δ 126.0 ppm and C-9 at δ 153.1 ppm were assigned based on their correlations with H-11 at δ 8.24 ppm, while C-1b at δ 118.3 ppm and C-2 at δ 143.2 ppm showed correlations with H-3 at δ 6.54 ppm. (-)-*N*-methylguattescidine, is a rare 6a-methylated-7-oxo- aporphine alkaloid, having only been previously reported by Reynald *et al.* in 1982. Presented below are structures and spectroscopic data of the isolated compounds.

Table 1

(-)-N-Methylguattescidine (**1**). yellow amorphous solid; $[\alpha]^{30}_{D}$: -20º (c = 0.1 mg mL^{-1}, $CHCl_3$); MS *m/z*: 324.1242, $C_{19}H_{17}O_4N$; UV λ_{max} nm EtOH: 235, 310; IR υ_{max} cm^{-1}: 3409, 1710, 1266; ^{1}H NMR

(CDCl$_3$, 300 MHz) δ ppm : 8.24 (1H, d, J = 8.7 Hz, H-11), 7.39 (1H, d, J = 2.7 Hz, H-8), 7.13 (1H, dd, J_o = 8.7 Hz; J_m = 2.7 Hz, H-10), 6.54 (1H, s, H-3), 6.07 (1H, d, J = 1.2 Hz, H-2), 5.99 (1H, d, J = 1.2 Hz, H-1), 3.52 (1H, m, H-11a), 3.01 (1H, m, H-5), 3.00 (1H, m, H-4), 2.99 (1H, m, H-5'), 2.55 (1H, m, H-4'); ^{13}C NMR (CDCl$_3$, 75 MHz) δppm : 153.1 (C-9), 143.2 (C-2), 138.8 (C-1), 126.0 (C-7a), 125.3 (C-3a), 123.1 (C-11a), 122.7 (C-11), 122.2 (C-10), 118.3 (C-1b), 110.3 (C-8), 108.9 (C-1a), 103.9 (C-3), 96.9 (O-CH$_2$-O), 62.7 (C-6a), 41.4 (C-5), 34.1 (N-CH$_3$), 26.9 (C- 4), 25.0 (CH$_3$).

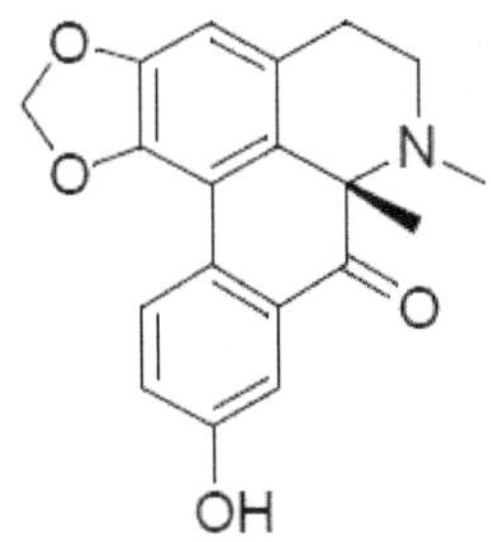

(1)

Liriodenine (**2**), yellow needles; MS m/z : 275, $C_{17}H_9O_3N$; UV λ_{max} nm EtOH : 215, 246, 268, 395, 412; IR υ_{max} cm^{-1} : 3054, 1726, 1421, 1265; ^{1}H NMR (CDCl$_3$, 300 MHz) δ ppm : 8.9 (1H, d, J = 5.1 Hz, H-5), 8.66 (1H, dd, J_o = 7.2 Hz; J_m = 1.2 Hz, H-11), 8.59 (1H, dd, J_o = 7.8 Hz; J_m = 1.2Hz, H-8), 7.79 (1H, d, J = 5.1 Hz, H-4), 7.76 (1H, td, J_o = 7.8 Hz; 7.2 Hz; J_m = 1.5 Hz, H-10), 7.59 (1H, td, J_o = 7.8 Hz;7.2; J_m = 1.2 Hz, H-9), 7.16 (1H, s, H-3), 6.40 (2H, s, O-CH$_2$-O); ^{13}C NMR (CDCl$_3$, 75 MHz) δppm : 151.7 (C-2), 147.9 (C-1), 146 (C-6a), 145.4 (C-3a), 144.9 (C-5), 135.7 (C-1a), 133.9 (C-10), 132.9 (C-7a), 131.3 (C-11a), 128.8(C- 8), 128.6 (C-9), 127.4 (C-11), 124.2 (C-4), 108.2 (C-1b), 103.3 (C- 3), 102.4 (O-CH$_2$-O), 182.4 (C-7)

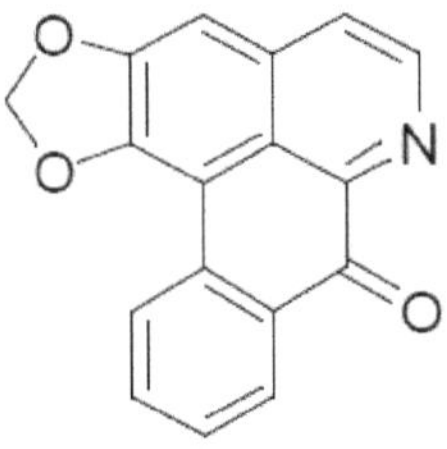

(2)

Lanuginosine (3), yellow needles; MS m/z : 305, $C_{18}H_{11}O_4N$; UV λ_{max} nm EtOH : 246, 271, 315, 258, 283, 334; IR υ_{max} cm^{-1} :3055, 2987, 2306, 1712, 1635, 1363, 1265, 1046, 896; ^{1}H NMR (CDCl$_3$, 300MHz) δ ppm : 8.85 (1H, d, J = 5.4 Hz, H-5), 8.58 (1H, d, J_o = 9.0 Hz, H-11), 8.04 (1H, d, J = 3 Hz, H-8), 7.79 (1H, d, J = 5.4 Hz, H-4), 7.32 (1H, dd, J_o = 9.0 Hz; J_m = 3 Hz, H-10), 7.17 (1H, s, H-3), 6.47 (2H, s, O – CH$_2$ – O); ^{13}C NMR (CDCl$_3$, 75MHz) δ ppm : 158.0 (C-9), 151.0 (C-2), 146.0 (C-1), 144.9 (C- 5), 144.0 (C-6a), 136.0 (C-3a), 133.0 (C-7a), 131.9 (C-1b), 129.1 (C-11), 126.2 (C-11a), 124.3 (C-4), 122.6 (C-10), 110.2 (C-8), 109.0 (C-1a), 102.3 (C-3), 55.8 (OCH$_3$), 102.5 (O – CH$_2$ – O), 182.0 (C-7).

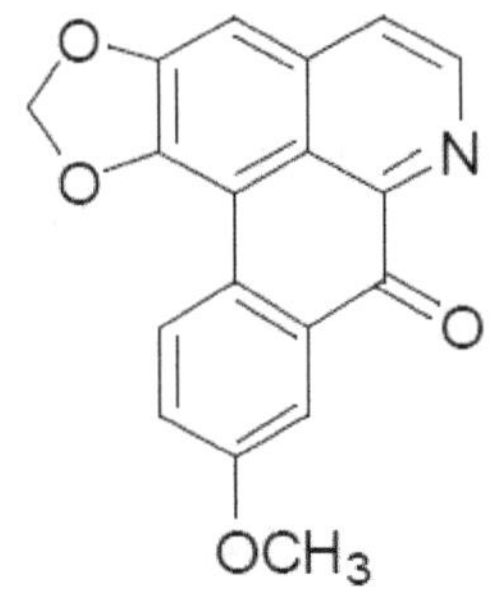

(3)

Asimilobine (**4**), brownish amorphous; MS m/z : 267, $C_{17}H_{17}O_2N$; UV λ_{max} nm EtOH : 274, 308; IR υ_{max} cm^{-1} : 3390, NH 1675, 1600, 1225; ^{1}H NMR (CDCl$_3$, 300MHz) δ ppm : 8.30 (1H, d, J = 7.8 Hz, H-11), 7.36 – 7.25 (3H, m, H-8, H-9, H-10), 6.73 O (1H, s, H-3), 3.92 (1H, m, H-6a), 3.50 (1H, m, H-5'), 3.08 (1H, d, H-4'), 3.04 (1H, d, H-5), 2.99 (1H, m, H7), 2.85 (1H, m, H7), 2.74 (1H, d, H-4), 3.61 (3H, s, OCH$_3$), 2.00 (1H, s, N-H); ^{13}C NMR (CDCl$_3$, 75MHz) δ ppm : 148.6 (C-2), 143.0 (C-1), 135.6 (C-7a), 131.7 (C-11a), 129.4 (C-16), 128.1 (C-3a), 127.7 (C-8), 127.4 (C- 10), 127.3 (C-9), 127.2 (C-11), 125.5 (C-1a), 114.6 (C-3), 53.4 (C- 6a), 42.8 (C-5), 36.7 (C-7), 28.2 (C-4), 60.4 (OCH$_3$).

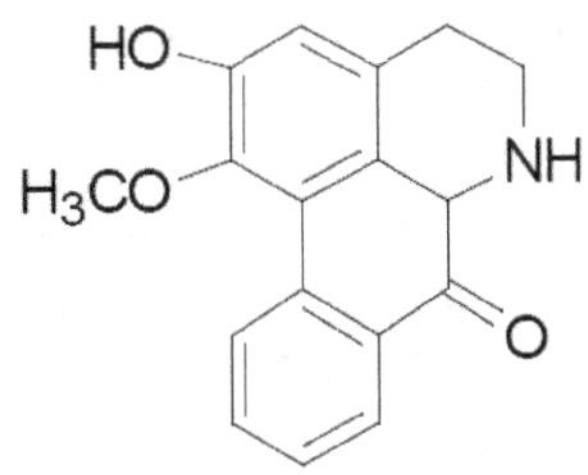

(4)

Dimethyltryptamine (**5**), reddish amorphous; MS m/z : 188, $C_{12}H_{16}N_2$; UV λ_{max} nm EtOH : 240, 252; IR υ_{max} cm^{-1} : 3945, 3055, 2305, 1634, 1422, 1265, 1046, 896; ^{1}H NMR (CDCl$_3$, 300MHz) δ ppm : 7.60 (1H, d, J_o = 5.7 Hz, H-7), 7.38 (1H, d, J_o =7.1 Hz, H-4), 7.20 (1H, td, J_o = 6.9Hz; J_m0.9 Hz, H-5), 7.12 (1H, td, J_o = 6.9 Hz; J_m = 0.9 Hz, H-6), 3.03 (2H, m, H-8), 2.80 (2H, m, H-9), 8.28 (1H, brs, N-H), 2.49 (6H, s, 2(CH$_3$); ^{13}CNMR (CDCl$_3$, 75MHz) δ ppm : 136.0 (C-7), 127.0 (C-3a), 122.0 (C-6), 121.7 (C- 2), 119.2 (C-5), 118.7 (C-4), 59.4 (C-9), 44 (C-3), 22.9 (C-8), 44.9 (2CH$_3$).

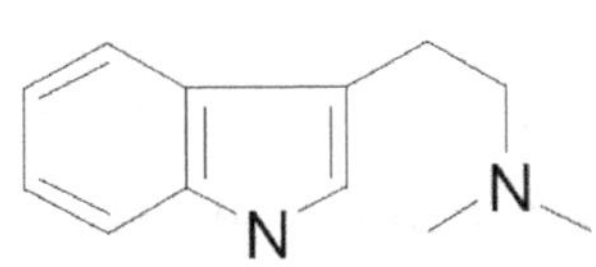

(5)

Remerine (**6**), yellow amorphous; MS m/z : 279, $C_{18}H_{17}O_2N$; UV λ_{max} nm EtOH : 234, 264; IR υ_{max} cm^{-1} : 1401, 1361, 1053, 942; ^{1}H NMR ($CDCl_3$, 300MHz) δ ppm : 8.09 (1H, d, J_o = 7.5 Hz, H-11), 7.34 – 7.24 (3H, m, H-8, H-9, H-10), 6.59 (1H, s, H-3), 4.000 (1H, m, H-6a), 3.4 (1H, m, H-5'), 3.10 (1H, m, H-4'), 3.00 (1H, m, H-5), 2.90 (1H, m, H-7'), 2.80 (2H, m, H-7, H-4), 6.11 (1H, d, J_m = 1.2 Hz, CH-O), 5.96 (1H, d, J_o = 1.2 Hz, CH-O), 2.62 (3H, s, CH_3); ^{13}C NMR ($CDCl_3$, 75MHz) δppm : 146.7 (C- 2), 142.8 (C-1), 136.3 (C-7a), 128.1 (C-8), 128.0 (C-1b), 127.6 (C- 9), 127.0 (C-10), 127.0 (C-11), 125.4 (C-3a), 126.5 (C-1a), 126.0 (C-11a), 125.4 (C-3a), 62.4 (C-6a), 53.3 (C-6a), 43 (C-5), 36.9 (C- 7), 28 (C-4), 100.7 (O – CH_2 – O), 39.0 (CH_3).

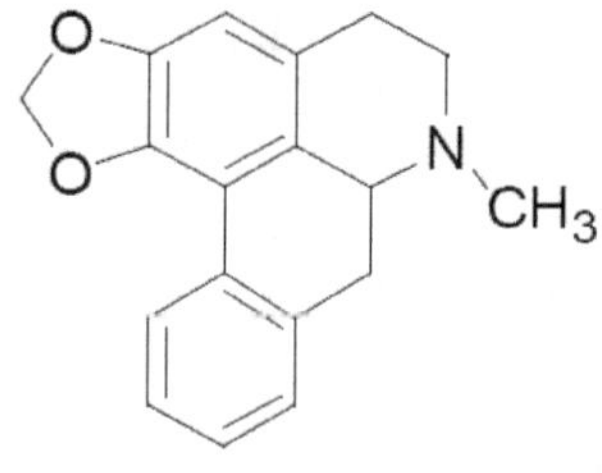

(6)

Anonaine (**7**), yellow amorphous; MS m/z : 265, $C_{17}H_{13}O_2N$; UV λ_{max} nm EtOH : 234, 272, 315; IR υ_{max} cm^{-1} : 1040, 945; ^{1}H NMR ($CDCl_3$, 300MHz) δ ppm : 8.09 (1H, d, J_o = 7.5 Hz, H=11), 7.36 – 7.19 (3H, m, H-8, H-9, H-10), 6.6 (1H, s, H-3), 4.04 (1H, dd, H-6a), 3.48 (1H, m, H-5'), 3.1 (1H, m, H-4'), 3.07 (1H, m, H- 5), 3.02 (1H, m, H-7'), 6.12 (1H, d, J_m = 1.5, CH – O), 5.97 (1H, d, J_m = 1.5, CH – O); ^{13}C NMR ($CDCl_3$, 75MHz) δ ppm : 147.0 (C-2), 143.0 (C-1), 135.4 (C-7a), 131.4 (C-11a), 129.0 (C-1b), 128.0 (C-3a), 127.8 (C-8), 127.7 (C-9), 127.0 (C-10), 126.1 (C-11), 116.3 (C-1a), 53.6 (C-6a), 43.6 (C-5), 37.4 (C-7), 29.6 (C-4), 100.6 (O –CH_2– O).

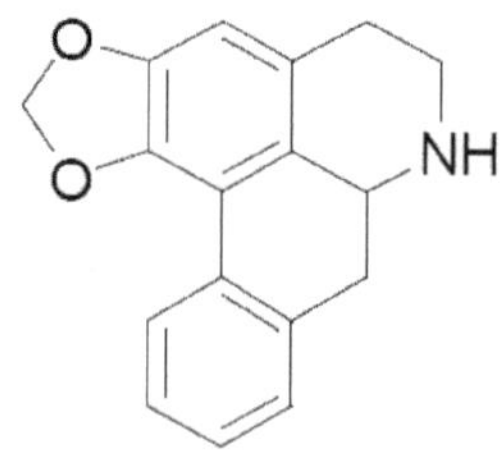

(7)

Columbamine (**8**), red amorphous solid; MS m/z : 338, $C_{20}H_{20}O_4N$; UV λ_{max} nm EtOH : 206, 225, 265, 345; IR υ_{max} cm^{-1}: 3390, 1600; ^{1}H NMR ($CDCl_3$, 300MHz) δ ppm : 9.00 (1H, s, H- 8), 8.08 (1H, s, H-13), 7.65 (1H, d, J_o = 9.0 Hz, H-11), 7.61 (1H, d, J_o = 8.7 Hz, H-12), 7.27 (1H, s, H-1), 6.79 (1H, s, H-4), 4.68 (2H, t, J_o = 6.6 Hz; 6.3 Hz, H-6), 3.18 (2H, t, J_o = 6.0 Hz, H-5), 4.02 (3H, OCH_3), 3.99 (3H, OCH_3), 3.92 (3H, OCH_3). ; ^{13}C NMR ($CDCl_3$, 75MHz) δ ppm : 163.0 (C-10), 151.0 (C-3), 150.0 (C-4a), 149.0 (C-2), 142.0 (C-9), 140.0 (C-8), 139.0 (C-11), 132.0 (C-14), 129.0 (C-12a), 126.0 (C-1a), 123.3 (C-12), 120.0 (C-13), 119.0 (C- 8a), 111.0 (C-4), 108.0 (C-1), 56.0 (C-6), 27.0 (C-5), 60.0 (OCH_3), 58.0 (OCH_3), 57.0 (OCH_3).

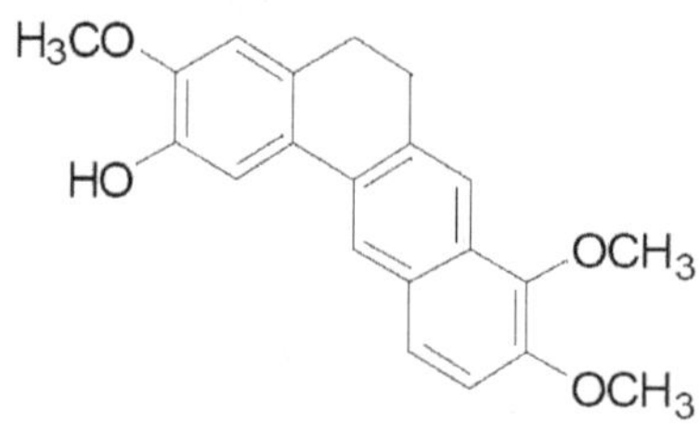

(8)

Lysicamine (**9**), yellow amorphous; MS m/z : 291, $C_{18}H_{13}O_3N$; UV λ_{max} nm EtOH : 214, 250, 255, 261, 319; IR υ_{max} cm^{-1} : 1675, 1600, 1225; ^{1}H NMR ($CDCl_3$, 300MHz) δ ppm : 9.10 (1H, d, J_o = 5.1 Hz, H-11), 8.70 (1H, d, J_o = 6.9 Hz, H-5) 8.48 (1H, dd, J_o = 7.5 Hz; J_m = 1.5 Hz, H-8), 7.76 (1H, td, J_o = 7.2 Hz; J_m = 1.5 Hz, H-10), 7.7 (1H, d, J_o = 6.9 Hz, H-4), 7.55 (1H, td, J_o = 7.2 Hz; J_m = 1.2 Hz, H-9), 7.24 (1H, s, H-3), 4.05 (3H, s, OCH_3), 3.97 (3H, s, OCH_3); ^{13}C NMR ($CDCl_3$, 75MHz) δppm : 156.7 (C-6a), 152.0 (C-2), 145.2 (C-1), 139.0 (C-5), 135.3 (C-3a), 134.7 (C-11a), 132.0 (C-7a), 130.9 (C-10), 125.7 (C-9), 122.0 (C-1b), 119.6 (C-1a), 108.7 (C-3), 65.1 (OCH_3), 56.8 (OCH_3), 182.5 (C=O).

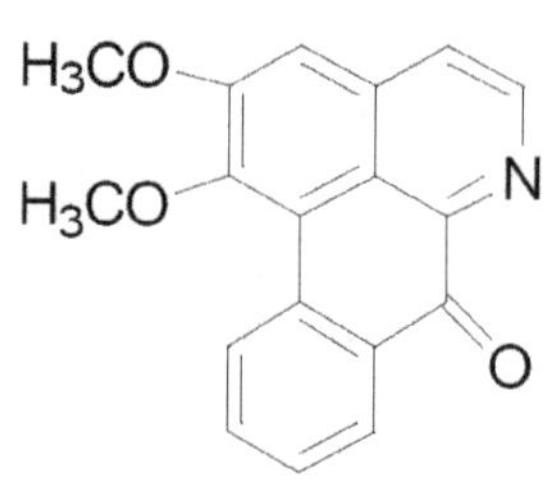

(9)

2.1.1 Alkaloids of *Meiogyne virgata*

Meiogyne virgata is a rainforest tree grows in Peninsular Malaysia, Borneo, Java and Sumatera. The genus *Meiogyne* (Annonaceae) consists of about 24 species and widely distributed in Indo-china, Thailand, Peninsular Malaysia, Sumatra, Java,

Borneo and the Philippines. There is no formal report on the traditional uses of *Meiogyne virgata* in Malaysia. However, being an alkaloid rich species, it could be useful medicinally.

We have conducted phytochemical work on *Meiogyne virgata*. Six of the aporphine alkaloids in *Fissistigma latifolium* were also found in *Meiogyne virgata* collected from the Peninsular Malaysia. Isolation and purification of alkaloids from the bark of *Meiogyne virgata* afforded nine alkaloids; four oxoaporphines, liriodenine **2**, lanuginosine **3**, asimilobine **4** and lysicamine **9**; four aporphines, anonaine **7**, remerine **6**, nornuciferine **10** and norushinsunine **11**; and one azaanthracene alkaloid, cleistopholine **12**.

Most of the compounds are yellowish or colorless hygroscopic liquid at room temperature while impure samples will appear brownish. They have low solubility in water but dissolve well in methanol, chloroform, acetone, dichloromethane and other common organic solvent. They are also soluble in dilute acid as the protonated derivative. The melting point of these thype of compounds in range 100-300 °C.

Most of oxoaporphine and aporphine alkaloids showed IR spectra typified by the 7-oxo group with absorption band in the 1635-1660 cm^{-1} region. The UV spectra data for these type of compounds are quite characteristic for the skeletal type. There is indication that they may also be diagnostic for a particular oxygenation pattern.

Table 2.

Nornuciferine (**10**), Colourless crystalline solid; MS m/z : 281 (M^+); UV λ_{max} nm EtOH : 234, 272, 315; IR υ_{max} cm^{-1} : 1040, 945 ^1H NMR ($CDCl_3$, 300 MHz) δ ppm : 8.39 (1H, *d*,*J*= 7.8 Hz, H-11), 7.33-7.20 (3H, *m*, H-8, H-9, H-10), 6.65 (1H, *s*, H-3), 3.98 (1H, *dd*, *J*= 13.4; 5.2 Hz, H-6a), 3.41 (1H, *dd*, *J*= 12.3; 6.3 Hz, H-5'), 3.90 (3H, *s*, OMe-2), 3.68 (3H, *s*, OMe-1), 3.08 (1H, *dd*, *J*=13.2 Hz, H-4), 3.04 (1H, *td*, *J*= 12.3; 5.1 Hz, H-5), 2.85 (1H, *dd*, *J*= 13.4 ; 5.2 Hz, H-7'), 2.68 (1H, *dd*, *J*= 13.2; 6.0 Hz, H-4'), 2.64 (1H, *t*, *J*= 13.4 Hz, H-7).; ^{13}C NMR ($CDCl_3$, 125 MHz) δ ppm : 152.3 (C- 2), 145.2 (C-1), 135.0 (C-7a), 132.1 (C-1b), 132.1 (C-11a), 131.2 (C-3a), 128.4 (C-8), 127.8 (C-10), 127.4 (C-9), 127.1 (C-11), 126.6 (C-1a), 111.8 (C-3), 60.3 (OMe-1), 55.9 (OMe-2), 53.6 (C-6a), 43.0 (C-5), 37.2 (C-7), 29.7 (C-4).

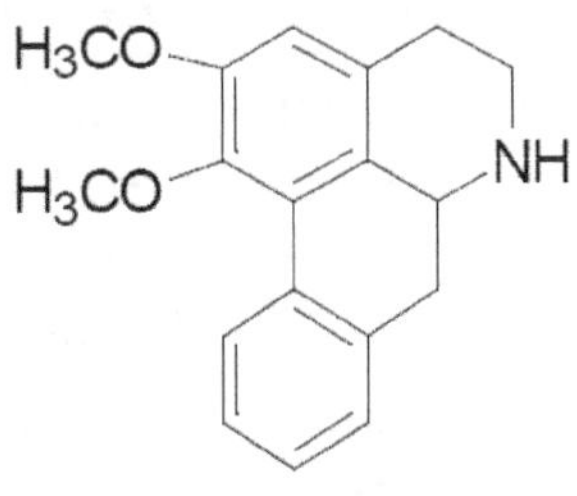

(10)

Norushinsunine (**11**), Colourless crystalline solid; MS m/z : 281; UV λ_{max} nm EtOH : 217, 247, 252, 259, 273, 319; IR υ_{max} cm^{-1} : 3488, 3355, 1574, 1215; ^1H NMR ($CDCl_3$, 300MHz) δ ppm : 8.16 (1H, dd, *J* = 7.2;1.2 Hz, H=11), 7.45 (1H, td, *J* = 8.7;1.2 Hz, H-10), 7.40 (1H, dd, *J* = 8.1;0.9 Hz, H-8), 7.34 (1H, td, *J* = 7.2;1.2 Hz, H-9), 6.59 (1H, s, H-3), 6.11 (1H, d, *J* =1.5 Hz, O – CH_2 – O), 5.95 (1H, d, *J* = 1.2 Hz, O – CH_2 – O), 4.61 (1H, d, *J* = 3.0 Hz, H-7), 4.06 (1H, d, *J* = 3.3 Hz, CH – O). 3.37 (1H, ddd, *J* = 5.0;3.9;1.2 Hz, H-4', 2.68 (1H,dd,

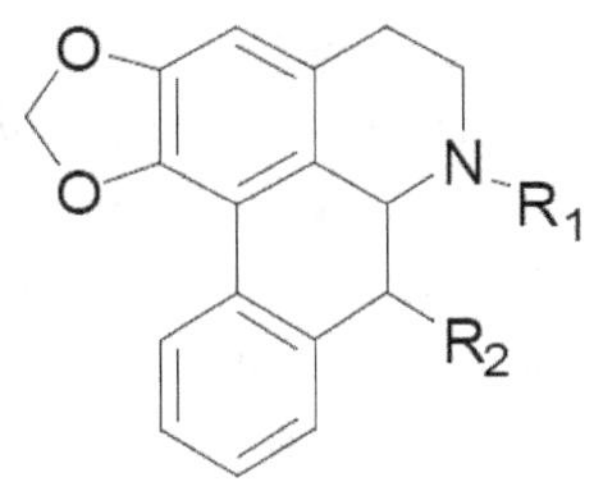

(11)

J=16.2;3.9 Hz, H-4); ^{13}C NMR ($CDCl_3$, 75MHz) δ ppm : 147.1 (C-1), 142.6 (C-2), 135.6 (C-1a), 130.3 (C-7a), 129.4 (C-9), 129.1 (C-3a), 123.6 (C-1b), 115.6 (C-11a), 108.4 (O – CH_2 – O), 71.0 (C-7), 57.2 (C-6a), 43.1 (C-5), 29.2 (C-4).

Cleistopholine (**12**), yellow glassy solid; MS m/z : 281 (M^+); UV λ_{max} nm EtOH : 234, 272, 315; IR ν_{max} cm^{-1} : 1040, 945; ^{1}H NMR ($CDCl_3$, 400 MHz) δ ppm : 8.95 (1H, *d,J*= 4.8 Hz, H-2), 8.31 (1H, *dd,J*= 8.5;2.2 Hz, H-5), 8.21 (1H, *dd,J*= 8.5;2.2 Hz, H-8), 7.79 (1H, m, H-6), 7.79 (1H, m, H-7), 2.89 (1H, s, CH3); ^{13}C NMR ($CDCl_3$, 100.6 MHz) δ ppm : 184.7 (C-9), 181.9 (C-10), 153.4 (C-2), 151.6 (C-4), 150.1 (C-9a), 134.6 (C-7), 134.2 (C-6), 132.6 (C-10a), 131.2 (C-3), 129.1 (C-4a), 127.4 (C-5), 127.2 (C-8), 22.8 (CH3).

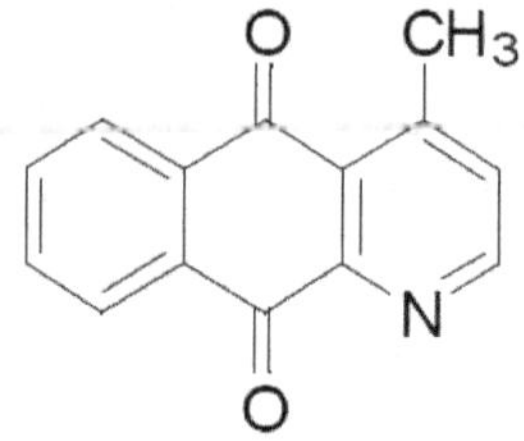

(12)

For example, 1, 2-methylenedioxy derivative in compound **2** gives increase to a bathochromic shift in the 235-250 nm bands on comparison with the corresponding compound **9.** The addition of acid will gives a substantial bathochromic shift of the longest-wavelength band. In oxoaporphine and aporphine, position 1 and 2 are constantly oxygenated. It is frequent to find further oxygen substituent at C-9, C-10 and C-11 and occasionally at C-8. Other than that, H-4 and H-5 will give a characteristic AB system with doublet of doublet at about 7.6 ppm and 8.7 ppm with a coupling constant about 5.4 Hz. The small *J* value is due to the adjacent of electronegative nitrogen atom. The methylenedioxyl group gives singlet peak at about 6.0 ppm due to the inductive effect cause by existence of the neighboring C-7 carbonyl. The C-11 proton usually the most deshielded and the C-3 protons always appeared at a higher field then the aromatic hydrogen (Cordell, 1981). Presented below are structures and spectroscopic data of the isolated compounds.

3. THE FAMILY OF RUBIACEAE AS SOURCE OF ANTHRAQUINONES

Rubiaceae is among the largest flowering plants family comprising of 450 genera and 13,000 species. In Malaysia, 70 genera and 555 species of Rubiaceous plants were reported (Wong, 1989). Most Rubiaceous plants are shrubs or small trees and infrequently herbs (Hutchinson, 1973). Rubiaceous plants are distributed worldwide but they are mainly tropical. They are easily recognized at family level by decussate, entire leaves, presence of stipules, actinomorph flowers and inferior ovary.

Rubiaceous plants are known to accumulate substantial amount of anthraquinones particularly in the roots (Han, *et al.*, 2001). Anthraquinones containing plants are used traditionally for various ailments and health complaints such as diarrhea, loss of appetite, fever, wounds and cancer. The plant extracts are used in form of poultice, lotion and decoction from various plant parts. *Morinda, Hedyotis, Primatomeris* and *Rennellia* are among anthraquinone containing-genera that are widely used in Malaysian traditional medicine (Ismail, *et al.*, 1997; Jasril, *et al.*, 2003; Ahmad, *et al.*, 2005; Lajis, *et al.*, 2006; Osman, *et al.*, 2010).

Morinda comprises of approximately 80 species, distributed worldwide in tropical areas. It is considered to be highly nutritious plant and is used as traditional medicine. In Malaysia *M. citrifolia* and *M. elliptica* are widely used. The roots of *M. elliptica* are used to treat jaundice and gastric complaints and the leaves are used to treat flatulence and fever. *Prismatomeris* and *Hedyotis* species on the other hand are recorded in various traditional medicine systems such as Traditional Chinese Medicine. Several well-known *Prismatomeris species* used in folk medicine in Malaysia are *P. glabra* and *P. malayana*. *P. glabra* is claimed to be aphrodisiac and widely used in the east coast of Malaysia. *P. Malayana* contained the anthraquinones, rubiadin and rubiadin-1-methyl ether (Lee, 1969). *Hedyotis* plants are generally consumed as tonic or febrifuge for treatment of diarrhea and dysentery (Lajis, et al., 2006). Several species of *Hedyotis* native to Malaysia are *H.capitellata*, *H. Herbaceae*, *H.dichtoma*, *H. diffusa* and *H. verticillata*. Besides anthraquinones, *Hedyotis* also contain β-cabolline alkaloids, flavonoids and triterpenes. *Rennellia* is another small genus of Rubiaceae family. Consists of shrubs and small trees, the plants may be found in lowland tropical rainforest of Peninsular Malaysia and Sumatra. *R. elliptica*, is used for general health improvements and dubbed as Malaysian Ginseng most likely due to the appearance of its yellow roots.

Anthraquinones of the Malaysian Rubiaceae are generally of the *Rubia* type. Rings A and B of the anthraquinone skeleton are biosynthetically derived from chorismic acid and *a*- ketoglutarate *via* o-succinylbenzoic acid, whereas ring C is formed from isopentenyl diphosphate *via* the terpenoid pathway (Han, *et al.*, 2001). Chorismate is first converted to isochorismate, and then to *o*-succinoylbenzoic acid (OSB) in the presence of α- ketoglutarate and thiamine diphosphate. OSB is activated at the aliphatic carboxyl group to produce an OSB-CoA ester. It is the ring closure of OSB-CoA which results in the formation of 1,4-dihydroxy-2-naphthoic acid (DHNA) leading to ring A and B. The prenylation of DHNA at C-3, leads to naphthoquinol or naphthoquinone. The ring C formation is a consequence of the cyclization *via* C-C bond between the aromatic ring of the naphthoquinone and an isoprene unit, isopentenyl diphosphate (IPP) or 3,3- dimethylallyl diphosphate (DMAPP).

Of the anthraquinone from Malaysian Rubiaceeae are substituted only on ring C while the remaining are substituted on both ring A and ring C. Anthraquinones from genus *Morinda* are typically substituted at C-1, C-2, and C-5, C-6 or C-7, C-8 and C-1, C-2 and C-3 meanwhile anthraquinones from *Hedyotis* are differed by rare substitution at C-1, C-2 and C- Anthraquinones from *Hedyotis* displayed wide structural variation. *H. capitellata* contains furanoanthraquinones (Ahmad, *et al.*, 2005) and *H. dichotoma* was reported to contain both 9,10- and 1,4-anthraquinone (Hamzah & Lajis, 1998). Genus *Rennellia* is closely related to *Morinda* and anthraquinones reported from *R. elliptica* are similar to those from genus *Morinda* (Osman *et al.*, 2010). One particular difference is the occurrence of anthraquinone with methyl substitution at C-6 which is characteristic to this plant.

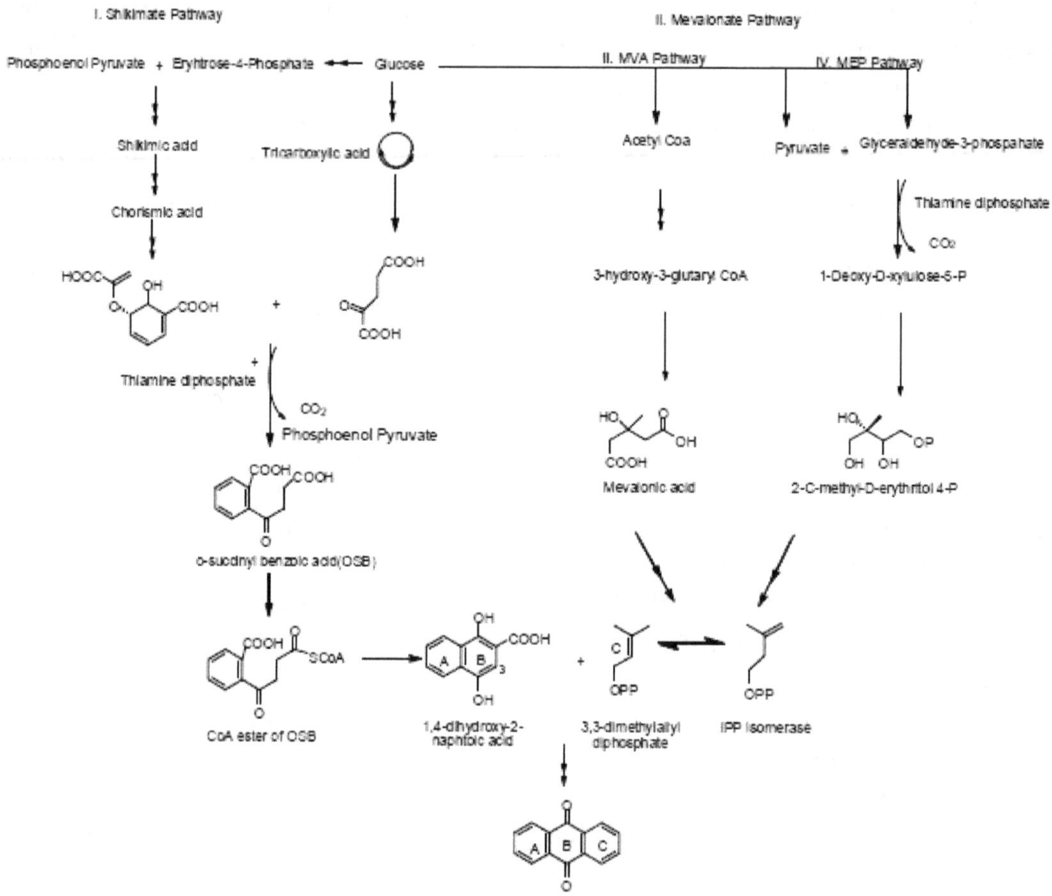

Fig. 2. Biosynthetic Pathway of Anthraquinones

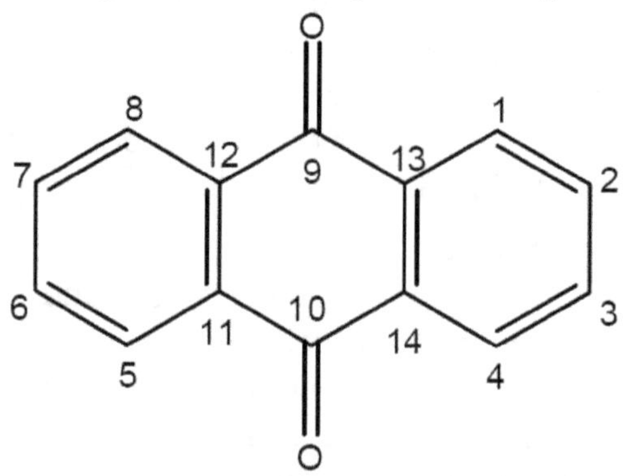

Fig. 3. Basic Skeleton of Anthraquinones

There are several characteristic spectroscopic data that distinguished anthraquinones from other types of compounds. In mass spectra, the major fragmentations are due to two consecutive loss of carbonyls, $[M\text{-}CO]^{+}$ and $[M\text{-}2CO]^{+}$. In the IR spectra, the unchelated carbonyl only viewed as one sharp stretching band at 1670 cm^{-1} due to symmetrical character of 9,10-anthraquinone (Derksen, *et al.*, 2002). Anthraquinones substituted with hydroxyl at *peri* position displayed two carbonyl absorption bands at about 1670 cm^{-1} and 1630 cm^{-1}. Anthraquinones give several chracteristic UV absorptions at 265-280 nm and 285- 290 nm due to electron transfers bonds of benzoid chromophore and at 430-437 nm due local excitation of quinoid carbonyls. The location hydroxyl substituent can be distinguished by observing the absorption maxima in UV spectra. Addition of dilute sodium hydroxide solution caused bathchromic shift of absorption maxima. The shift is useful in distinguishing substitution pattern of polyhydroxyanthraquinones. Proton NMR spectra of 9,10-anthraquinones shows typical A_2B_2 substitution pattern of *ortho*-substituted aromatic ring. An unsubstituted anthraquinone ring can be easily distinguished by the presence of at least two sets of multiplets at ca. δ_H 8.10 and ca. δ_H 7.20 in the aromatic region. Anthraquinones substituted at both rings A and C will give several doublets in the aromatic region. The two carbonyl groups in the molecule can be easily distinguished if hydroxyl substituents present in *para* position. Hydroxyl groups adjacent to carbonyl can be seen as sharp singlets much downfield at δ_H 12-14 due to strong intramolecular hydrogen bonding to the adjacent carbonyl. The presence of hydroxyl adjacent to carbonyl cause significant shift of carbonyl carbon resonance to downfield region at 186-189 ppm.

3.1 Anthraquinones of *Rennellia elliptica* Korth.

R. elliptica Korth. was also previously known as *R. elongata* (King & Gamble) Ridl. It is a shrub of about 2 m tall. This shrub can be found in lowland to hill forest to c. 500m above sea level. *R. elliptica* Korth. is widely distributed from Southern Myanmar to West Malaysia.

Fig. 4. Rennellia elliptica Korth

R. elliptica is used for general health improvements and dubbed as Malaysian Ginseng may be due to the appearance of its yellow roots. Its medicinal uses were documented as treatment of body aches, after-birth tonic and aphrodisiac (Mat Salleh & Latiff, 2002). The root extract of *R. elliptica* was reported to be antimalarial

(Osman, *et al.*, 2010) and antioxidant (Ahmad, *et al.*, 2010). Further study is warranted to investigate the antimalarial potential of roots of *R. elliptca*.

Phytochemical studies of the roots of *R. elliptica* Korth. resulted a new anthraquinone\ 1,2-dimethoxy-6-methyl-9,10-anthraquinone **18**, along with ten known ones. The known anthraquinones were nordamnacanthal **13**, 2-formyl-3-hydroxy 9,10-anthraquinone **14**, damnacanthal **15**, 1-hydroxy-2-methoxy-6-methyl-9,10 anthraquinone **16**, lucidin-ω-methyl ether **17**, 3-hydroxy-2-methoxy-6-methyl 9,10-anthraquinone **19**, rubiadin 2**0**, 3-hydroxy-2- methyl-9,10-anthraquinone **21**, rubiadin-1-methyl ether **22** and 3-hydroxy-2-hydroxymethyl- 9,10-anthraquinone **23**.

Anthraquinone **18**, 1,2-dimethoxy-6-methyl-9,10-anthraquinone, isolated for the first time as bright yellow amorphous solid. The HREIMS of **18** displayed a $[M + H]^+$ peak at 283.0968 [calc 283.3067] suggesting a molecular formula of $C_{17}H_{14}O_4$. The absorption maxima in the UV spectrum were observed at 373, 341 and 257 nm, indicative of an anthraquinone moiety. The IR spectrum did not show presence of chelated carbonyl and hydroxyl groups. The *sp*^{2} C-H stretch for the aromatic ring was observed at 3,081 cm^{-1}. With the exception of the sharp singlet in the downfield region for the hydrogen-bonded hydroxyl group, the ^{1}H NMR spectrum resembles that of compound **16**, suggesting a similar substitution pattern. Splitting pattern of the five aromatic proton signals suggested substitutions on both rings. Two overlapping doublets centered at δ_H 8.17 are due to H-8 (*d*, J = 7.8 Hz) and H-4 (*d*, J = 8.7 Hz), the *peri*-hydrogens. A doublet at δ_H 7.28 (J = 8.7 Hz) is due to H-3, meanwhile H-7 gave another doublet of doublet at δ_H 7.58 (J_o = 7.8 Hz, J_m = 1.7 Hz). These assignments were confirmed by their respective correlations in the COSY spectrum. H-5 resonated as a singlet at 8.06 ppm. In addition, two sharp singlets at δ_H 2.53 (3H, *s*) and 4.02 (6H, *s*) due to a methyl and two methoxy groups, respectively, were also observed. The location of the methoxy groups were established at C-1 and C-2 of ring C based on its NOE correlation with H-3. Thus, the only possible location for the methyl substituent is at C-6. This assignment was confirmed through NOE correlations of the methyl group with H-5 and H-7.

Table 3.

Nordamnacanthal. (**13**) Orange crystals. Mps 216-219 ° [lit. 220° C (Me_2CO) Chang (1984)]. UVλ_{max} EtOH nm421, 295, 259. UVλ_{max} EtOH/ -OH nm: 512, 357, 283. IR v_{max} (KBr) cm^{-1} :3460, 1646, 1627, 1382. MS m/z 268 [M$^+$], 240, 212, 184, 138. ^{1}H NMR ($CDCl_3$, 300MHz): 14.05 (1H, s, 1-OH), 12.70 (1H, s, 3-OH), 10.52 (1H, s, 2-CHO), 8.30 (2H, m, H-5, H-8), 7.88 (2H, m, H-6, H-7), 7.36 (1H, s, H- 4). ^{13}C NMR ($CDCl_3$, 75.5 MHz): 193.9 (2- CHO), 186.8 (C=O, C-9), 181.4 (C=O, C-10), 169.2 (C-OH,C-1), 168. 1 (C-OH, C-3), 139.1 (C-2), 134.8 (C-7), 134.7 (C-6), 133.3 (C-14), 133.2 (C-13), 127. 8 (C-8), 127.0 (H-5), 112.1 (C-14), 109.4 (C-4), 109.1 (C-13)

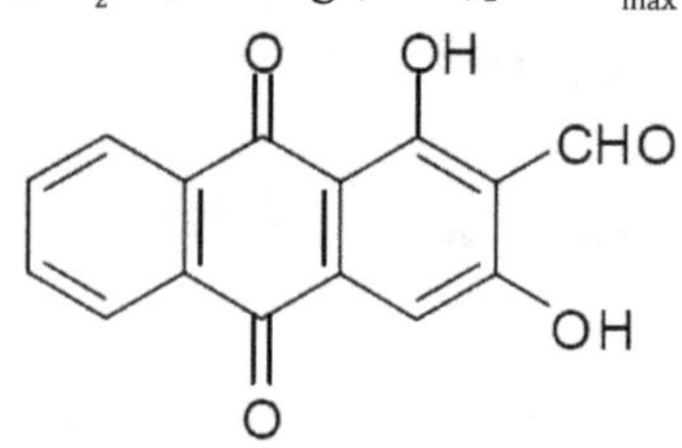

(13)

3-Formyl-2-hydroxy-9,10-anthraquinone (**14**). Bright orange needle crystals. Mps 212-214 °C [259-260 °C, Rath et al.(1995)]. UVλ_{max} EtOH nm: 380, 277, 246. UVλ_{max} EtOH/ -OH nm: 466, 392, 310, 254. IR ν_{max} (KBr) cm^{-1}: 3467, 1655, 1657, 1564. MS m/z 252 [M^{+}], 229, 206, 167, 139. ^{1}H NMR ($CDCl_3$, 300MHz): 11.45 (1H, s, 3-OH), 10.17 (1H, s, 2-CHO), 8.68 (1H, s, H-4), 8.35 (2H, m, H-5, H-8), 7.88 (2H, m, H-6, H-7), 7.86(1H, s, H-1) .^{13}C NMR ($CDCl_3$, 75.5 MHz): 196.8 (2-CHO), 181.0 (C=O, C-9, C- 10), 165.3 (C-OH, C-3), 139.1, 134.8 (C-4), 134.8, 133.3, 127.6, 127.5, 127.3, 126.1, 124.5, 123.4, 116.5 (C-1)

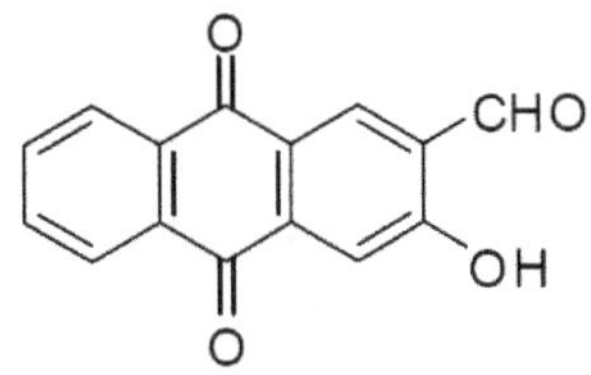

(14)

Damnacanthal (**15**). Yellow crystals. Mps 208-211°C [lit. 218-218.5°C (Me_2CO) Chang (1984)]. UVλ_{max} EtOH nm: 381, 284, 250, 213. UVλ_{max} EtOH/ -OH nm: 460, 379, 315, 262, 250. IR ν_{max} (KBr) cm^{-1}: 3437, 1644, 1561. MS m/z: 282 [M^{+}], 254, 225, 196. ^{1}H NMR ($CDCl_3$, 300MHz): 12.29 (1H, s, 3-OH), 10.49 (1H, s, 2-CHO), 8.25 (2H, m, H-5, H- 8), 7.84 (2H, m, H-6, H-7), 7.68 (1H, s, H-4), 4.14 (3H, s, 1- OCH_3)

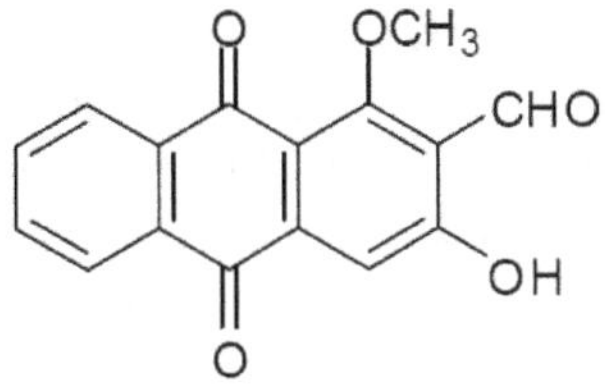

(15)

1-Hydroxy-2-methoxy-6-methyl-9,10-anthraquinone (**16**). Red needle crystals. Mps 220-221 °C. UVλ_{max} EtOH nm: 421, 278, 262, 231. UVλ_{max} EtOH/ -OH nm: 505, 314, 258. IR ν_{max} (KBr) cm^{-1}: 3467, 1653, 1637. MS m/z: 268 [M^{+}], 239, 197, 169,139, 115. ^{1}H NMR (CDCl3, 300MHz): 13.20 (1H, s, 1-OH), 8.23 (1H, d, *J*=8.1, H-8), 8.12 (1H, s, H-5), 7.89 (1H, d, *J*=8.4, H-4), 7.61 (1H, d, *J*=8.1, H-7), 7.19 (1H, d, *J*=8.4, H-3), 4.04 (3H, s, 2-OCH_3), 2.56 (3H, s, 6- CH_3). ^{13}C NMR (CDCl3, 75.5 MHz): 189.1(C=O, C-9), 181.8 (C=O, C-10), 154.0 (C-OH, C-1), 152.7 (C-OCH_3, C2), 146.2 (C-6), 134.6 (C-7), 134.0 (C-11), 131.10 (C-12), 127.8 (C-5), 127.1 (C-8), 125.5 (C-14), 121.0 (C-4), 116.1 (C- 13), 115.6 (C-3), 56.4 (2-OCH_3), 22.0 (6-CH_3)

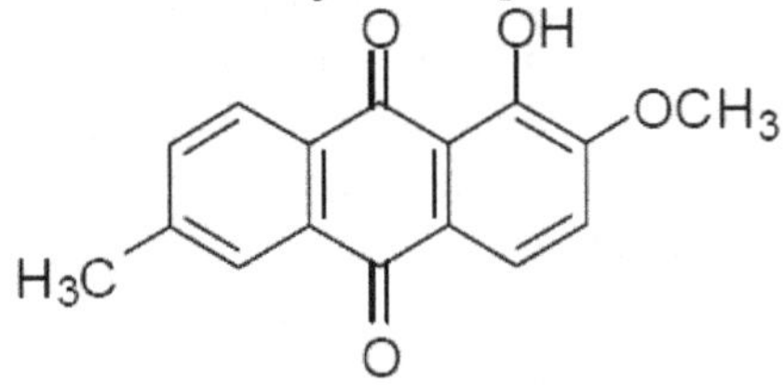

(16)

Lucidin-ω-methyl ether (**17**). Yellow crystals. Mps 175-179°C [lit 170°C, Dictionary of Natural Products (1995); 163-166°C, Leistner (1975)]. UVλ_{max} EtOH nm: 412, 280, UVë$_{max}$ EtOH/ -OH nm: 491, 314, 242 IR ν_{max} (KBr) cm^{-1}: 3428, 2927, 1668, 162. MS m/z: 284 [M^{+}], 263, 241, 213, 185. ^{1}H NMR ($CDCl_3$, 300MHz): 13.29 (1H, s, 1-OH), 9.39 (1H, s br, 3-OH), 8.27 (2H, m, H-5, H-8), 7.80 (2H, m, H-6, H-7), 7.32 (1H, s, H-4), 4.90 (2H, s, 2-**CH_2**OCH_3), 3.59 (3H, s, 2-CH_2**OCH_3**). ^{13}C NMR ($CDCl_3$, 75.5 MHz): 186.9 (C=O, C-9), 182.2 (C=O, C-10), 164.1 (C-1), 161.7 (C- 3), 134.1 (C-11), 134.1 (C-12), 133.6 (C-6), 133.5 (C-7), 126.8 (C-2), 114.4 (C-4), 109.7 (C-8), 109.7 (C-13), 109.6 (C- 5), 109.6 (C-14), 68.9 (2-**CH_2**OCH_3), 59.3 (2-CH_2OCH_3)

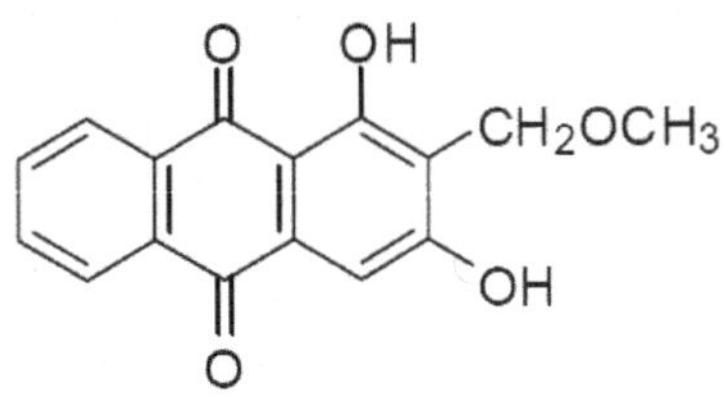

(17)

1,2-Dimethoxy-6-methyl-9,10-anthraquinone (**18**). Bright yellow crystals. Mps 193-196 °C. UVλ_{max} EtOH nm: 373, 341, 257, 222. UVλ_{max} EtOH/ -OH nm: 373, 342, 257, 222. IR ν_{max} (KBr) cm^{-1}:1666, 1601, 1327, 1267MS m/z: 282 [M^{+}], 253, 221, 194, 165, 139. ^{1}H NMR ($CDCl_3$, 300MHz): ^{3} 8.17 (2H, dd, *J*=8.7, 7.8, H-4, H-8), 8.06 (1H, s, H-5), 7.58 (1H, d, *J*=7.8, H-7), 7.28 (1H, d, *J*=8.7, H-3), 4.02 (6H, s, 1- OCH_3, 2-OCH_3), 2.53 (3H, s, 6-CH_3).^{13}C NMR ($CDCl_3$, 75.5 MHz): 182.7 (C=O, C-9), 182.7 (C=O, C-10), 159.1 (C- 1), 149.6 (C-2), 144.6 (C-6), 134.8

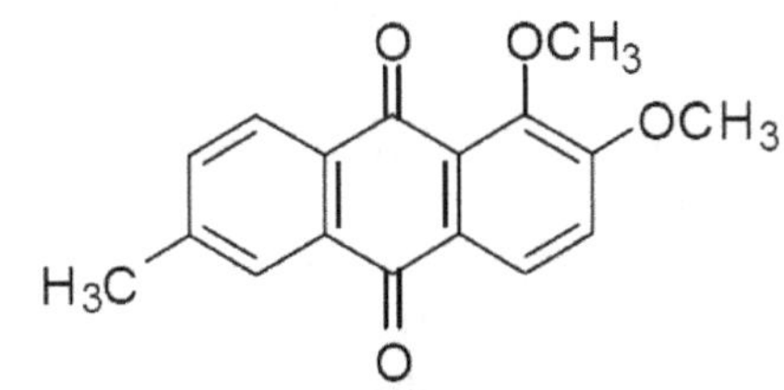

(18)

(C-7), 132. 9 (C-11), 132.9 (C-12), 127.5 (C-14), 127.4 (C-13), 127.1 (C-8), 126.9 (C-5), 125.2 (C-4), 115.9 (C-3), 61.3 (1-OCH_3), 56.3 (2-OCH_3), 21.8 (6-CH_3)

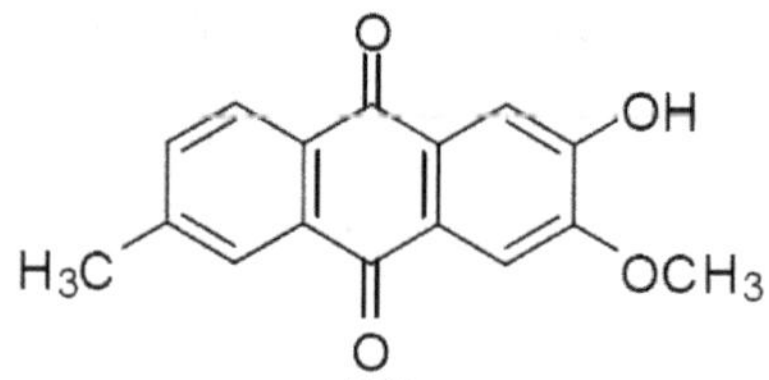

(19)

2-Hydroxy-3-methoxy-6-methyl-9,10-anthraquinone (**19**) Light yellow amorphous solid. Mp 210-215 °C. UVλ_{max} EtOH nm: 393, 286, 244. UVλ_{max} EtOH/ -OH nm: 509, 316, 250. IR ν_{max} (KBr) cm^{-1}: 3203, 2927, 2869, 1666, 1265. MS m/z: 268 (M^+], 239, 207, 169. ^1H NMR ($CDCl_3$, 300MHz): 8.18 (1H, d, *J*=8.1, H-8), 8.08 (1H, s, H-5), 7.79 (1H, s, H-1), 7.76 (1H, s, H-4),7.57 (1H, d, *J*=8.1, H-7), 6.23 (1H, s br, 2-OH), 4.11 (3H, s, 3-OCH_3), 2.54 (3H, s, 6-CH_3). ^{13}C NMR ($CDCl_3$, 75.5 MHz): 182.4 (C=O), 162.8(C-OH, C-2), 151.4 (C-OCH_3, C-3), 144.9, 134.5, 133.6, 127.4, 127.2, 112.6, 108.3, 56.6 (3-OCH_3), 21.9 (6-CH_3)

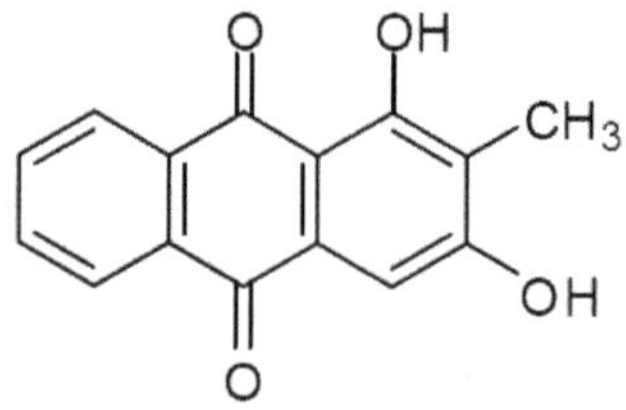

(20)

Rubiadin (**20**). Yellow crystals. Mps 250-258 °C [lit. 280-283 □ C, Leistner (1975)]. UVλ_{max} EtOH nm: 413, 279. UVλ_{max} EtOH/ -OH nm: 496, 314, 241 IR ν_{max} (KBr) cm^{-1}: 3436, 1653, 1626. MS m/z: 254 [M+], 226, 197, 152, 115. ^1H NMR (Acetone$_{-d6}$, 300MHz): 13.20 (1H, s, 1-OH), 8.31 (1H, m, H-8), 8.23 (1H, m, H-5), 7.92 (2H, m, H-6, H-7), 7.38 (1H, s, H), 2.20 (3H, s, 2-CH_3). ^{13}C NMR (Acetone$_{-d6}$, 75.5 MHz): 186.9 (C=O,C9), 181.8 (C=O, C-10), 163.2 (C- OH, C-1), 162.4 (C-OH, C-3), 134.3, 134.2, 133.5, 133.5, 132.4, 126.8, 126.5, 117.9, 107.17, 7.3

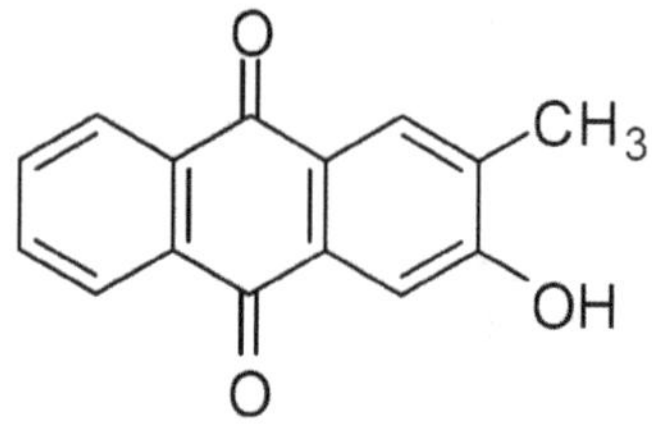

(21)

3-Hydroxy-2-methyl-9,10-anthraquinone (**21**). Yellow crystals. Mps 138- 142 □ C. UVλ_{max} EtOH nm: 379, 329, 274, 245, 239. UVλ_{max} EtOH/ -OH nm: 496, 314, 246 . IR ν_{max} (KBr) cm^{-1}: 3436, 1663, 651. MS m/z 238 [M^+], 238, 210, 181, 152, 105. ^1H NMR (Acetone$_{-d6}$, 300MHz): 8.23 (2H, m, H-5, H-8), 8.05 (1H, s, H-1), 7.89 (2H, m, H-6, H- 7), 7.67 (1H, s, H-4), 2.39 (3H, s, 2-CH_3). ^{13}C NMR (Acetone$_{-d6}$, 75.5 MHz): 182.6 (C=O, C-10), 181.5 (C=O, C-9), 161.0 (C-OH, C-3), 134.1 (C-2), 133.8 (C-14), 133.7 (C-7), 133.6 (C-6), 132.2 (C-13), 130.1 (C-1),126.6 (C-5), 126.5 (C-8), 111.4 (C-4), 15.6 (2-CH_3)

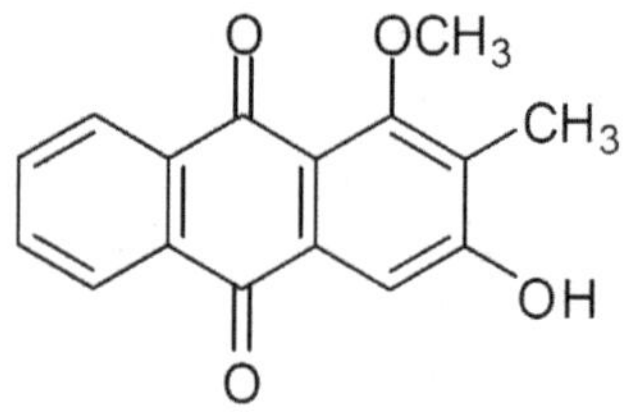

(22)

Rubiadin-1-methyl ether (**22**). Light yellow crystal. Mps 302-304° C [282-284, Briggs (1976); 300°C, Roberts (1977)]. UVλ_{max} EtOH nm: 354, 332, 279. UVλ_{max} EtOH/ - OH nm: 440, 314, 246. IR ν_{max} (KBr) cm^{-1}: 3437, 2913, 2847, 1668, 1651. MS m/z : 268, 239, 207, 181. ^1H NMR (Acetone$_{-d6}$, 300MHz): 9.50 (1H, s, br, 3-OH), 8.20 (2H, m, H-5, H-8), 7.87 (2H, m, H-6, H-7), 7.63 (1H, s, H-4), 3.90 (3H, s, 1-OCH_3), 2.27 (3H, s, 2-CH_3). ^{13}C NMR (Acetone$_{-d6}$, 75.5 MHz): 182.7 (C=O), 180.4 (C=O), 161.2 (C-OH), 140.6, 134.4, 134.2, 133.0, 132.6, 126.8, 126.0, 60.4 (OCH_3), 8.4 (CH_3)

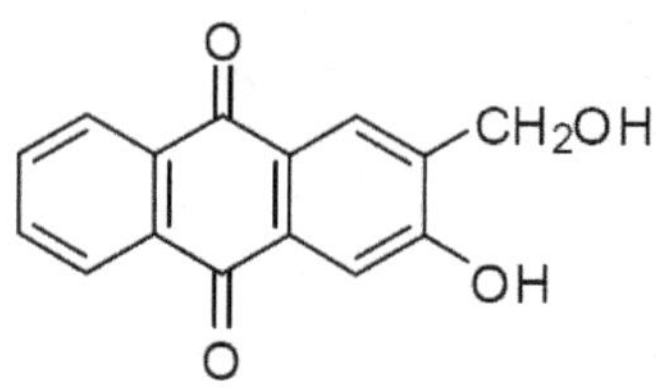

(23)

3-Hydroxy-2-hydroxymethyl-9,10-anthraquinone (**23**). Light yellow solid . UVλ_{max} EtOH nm: 374, 274, 238. UVλ_{max} EtOH/ -OH nm: 481, 311, 246 IR ν_{max} (KBr) cm^{-1}: 3468, 1628. ^1H NMR (Acetone$_{-d6}$, 300MHz): 8.38 (1H, s, H-4), 8.24 (2H, m, H-5, H-8), 7.89 (2H, m, H-6, H-7), 7.63 (1H, s, H-1), 4.803 (2H,s, 2-CH_2OH). ^{13}C NMR (Acetone$_{-d6}$, 75.5 MHz): 182.9 (C=O, C-9), 181.7 (C=O, C-10), 160.1 (C-OH, C-3), 136. 1 (C-2), 134.2 (C-14), 134.1 (C-11),

133.8 (C-6), 133.7 (C-7), 133.6 (C-12), 126.7 (C-4), 126.6 (C-5), 125.9 (C-13), 125.6 (C-8), 111.5 (C-1), 59.0 (2-**CH_2**OH)

The placement of methyl group at C-6 was further confirmed by HMBC experiment which showed a ^3J correlation with H-7. The methine carbons (C-3, C-4, C-5, C-7 and C-8) were assigned through HMQC correlations while the quaternary carbons (C-1, C-2, C-6, C- 11, C-12, C-13 and C-14) were assigned based on careful analysis of HMBC spectrum. Both carbonyl carbons in this compound resonated very closely to each other with only 0.01 ppm difference at δ_C 182.70 and 182.71, which further confirmed the unchelated nature of the carbonyls. Presented below are structures and spectroscopic data of the isolated compounds.

3.2 Anthraquinones of *Morinda elliptica*

Morinda elliptica or locally known as 'mengkudu kecil' is a shrub or small tree and it is very common in wild state of Malay Peninsula and northwards Burma (Burkill, 1966). It can be seen growing wild in newly developed areas, bushes and lowland secondary forest throughout the peninsula. *M. elliptica* is very common and always available and mostly used by the Malays for medicinal purposes. Traditionally, different parts of the plant are used in various ways for a number of health problems and ailments. The leaves may be added to rice for loss of appetite and taken for headache, cholera, diarrhea and wounds. Sometimes a lotion is made and used for hemorrhoid and applied upon body after childbirth (Burkill, 1966). The extracts and anthraquinones isolated from *M. elliptica* were reported to possess wide spectrum of biological activities such as antioxidant (Ismail, *et al.*, 2002; Jasril, *et al.*, 2003), antimicrobial, anti-HIV and anticancer (Ali, *et al.*, 2000).

Fig. 5. *Morinda elliptica*

Five anthraquinones in roots of *M. elliptica* which are nordamnacanthal **13**, damnacanthal **15**, lucidin-ω-methyl ether **17**, rubiadin **20** and rubiadin-l-methyl ether **22** are the same constituents found in *R. elliptica*. The others are 1-hydroxy-2-methylanthraquinone **23**, soranjidiol **25**, morindone **26**, morindone-5- methyl ether **27** and alizarin-1-methyl ether **28.** In addition, 2-formyl-1-

hydroxyanthraquinone **24** was reported as a new naturally occuring anthraquinone from roots of *M. elliptica.* HR-MS of **24** showed molecular ion peak at 252.0414 consistent with molecular formula of $C_{15}H_{14}O_4$. A bathchromic shift (407 to 531 nm) upon adding NaOH suggested the presence of OH at C-1 of the anthraquinone skeleton. The presence of hydroxyl group was evident from the broad stretching band observed at 3448 cm^{-1}. Two sharp stretching vibrations due to chelated and unchelated carbonyls were observed at 1638 and 1676 cm^{-1}, respectively. In the proton NMR, the signal for chelated hydroxyl group is at δ_H 13.26. The splitting pattern of ^{1}H NMR suggest substitution pattern on ring C only.

TaBLE

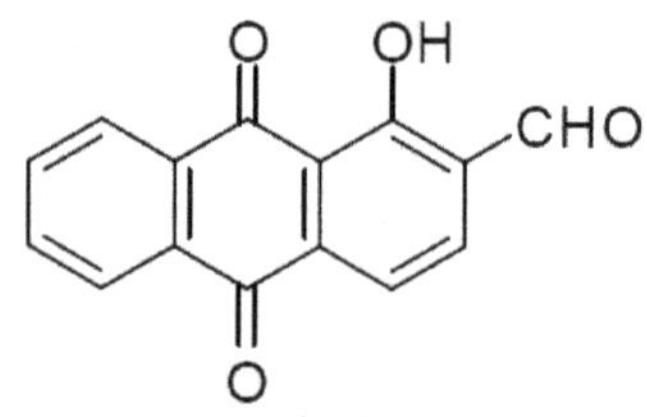

(24)

2-Formyl-1-hydroxy-9,10-anthraquinone (**24**). Mps 183-185 ⍰ C [lit. 259-260 ⍰ C, Rath et al. (1995)]. UVλmax EtOH nm: 229, 278, 331, 407. UVλmax EtOH/ -OH nm: 229, 280, 308, 531. IR vmax (KBr) cm-1: 3448 (OH), 1696 (aldehyde), 1676 (C=O unchelated), 1638 (C=O chelated), 1592 (C=C aromatic). MS m/z 252 (M+), 2224, 196, 168. ^{1}H NMR ($CDCl_3$, 500MHz): 13.26 (1H,s, 1-OH), 10.63 (1H, s, CHO), 8.35 (1H, m, H-8), 8.32 (1H, m, H-5), 8.23 (1H, d, J= 8.0 Hz, H-3), 7.89 (1H, d, J= 8.0 Hz, H-4), 7.88 (2H, m, H-6, H-7). 13C NMR(CDCl3, 125 MHz): 164.5 (C-1), 128.4 (C-2), 135.4 (C-3), 118.7 (C-4), 127.7 (C-5), 134.7 (C-6), 135.3 (C-7), 127.2 (C-8), 188.9 (C-9), 181.8 (C-10), 117.4 (C-11), 137.2 (C-12), 134.8 (C-13), 133.3 (C-14) and 188.0 (C-15)

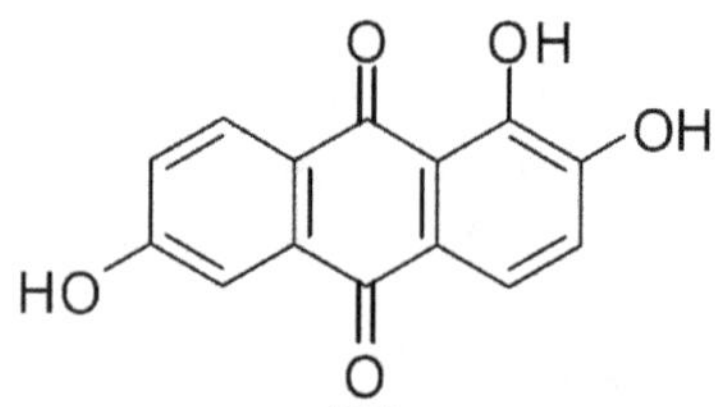

(25)

Soranjidiol. Yellow-orange neddles (**25**) Mps 276-273 ⍰ C [lit. 271-272 ⍰ C, Adesogan (1973)]. UVλ_{max} EtOH nm: 265, 409. UVλ_{max} EtOH/ -OH nm: 308, 489. IR v_{max} (KBr) cm^{-1}: 3401 (OH), 1667 (C=O unchelated), 1635 (C=O chelated), 1593 (C=C aromatic). MS m/z 254 (M^+), 226, 197, 115. ^{1}H NMR ($DMSO_{-d6}$, 500 MHz): 13.10 (1H,s, 1-OH), 11.21 (1H, s, 6- OH), 7.63 (1H, d, J= 7.57 Hz, H-3), 7.57 (1H, d, J=7.57 Hz, H-4), 7.25 (1H, dd, $J_{7,8}$ = 8.55 Hz, $J_{7,5}$ = 2.69 Hz, H-7), 7.45 (1H, d, J= 2.69 Hz, H-5), 2.27 (3H, s, CH3). ^{13}C NMR ($DMSO_{-d6}$, 125 MHz): 187.6 (C=O), 181.8 (C=O), 163.8 (C- OH), 160.0 (C-OH), 136.9, 135.6, 134.2, 131.1, 129.8, 124.5, 121.4, 118.6, 114.7, 112.5, 15.8 (CH_3)

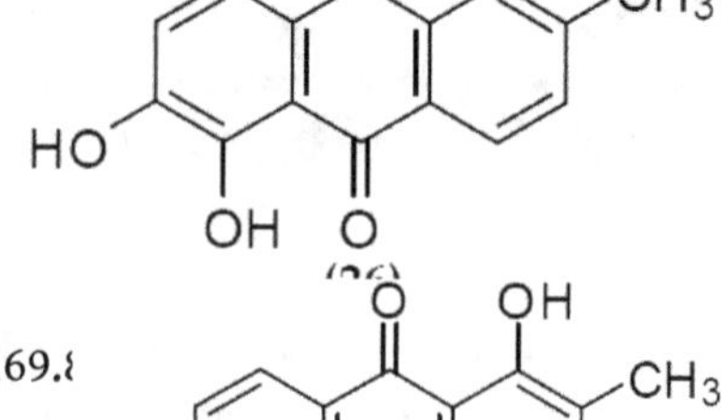

(26)

Morindone (**26**). Orange needles. Mps 240-241 ⍰ C (CHCl3) [lit. 248-249.5 ⍰ C, Leistner (1975)]. UVλ_{max} EtOH nm: 260, 299, 448. UVλ_{max} EtOH/ -OH nm: 260, 302, 338, 558. IR v_{max} (KBr) cm^{-1}: 3462 (OH), 1628 (C=O chelated). MS m/z 270 (M^+), 242, 135.^{1}H NMR ($CDCl_3$, 500MHz): 13.21 (1H, s, 1-OH), 12.95 (1H, s, 5-OH), 7.85 (1H, d- J=8.2 Hz, H-8), 7.75 (1H, d, J= 7.7 Hz, H-4), 7.52 (1H, d, J=7.6Hz, H-3), 7.26 (1H, d, J=8.2 Hz, H-7), 6.32 (1H, s, 6-OH), 2.39 (3H, s, CH_3). ^{13}C NMR ($CDCl_3$, 125 MHz): 186.6 (C=O),179.9 (C=O), 170.4 (C-OH), 169. 149.4, 136.6, 130.9, 121.3, 119.8, 118.9, 115.3, 16.3 (CH_3)

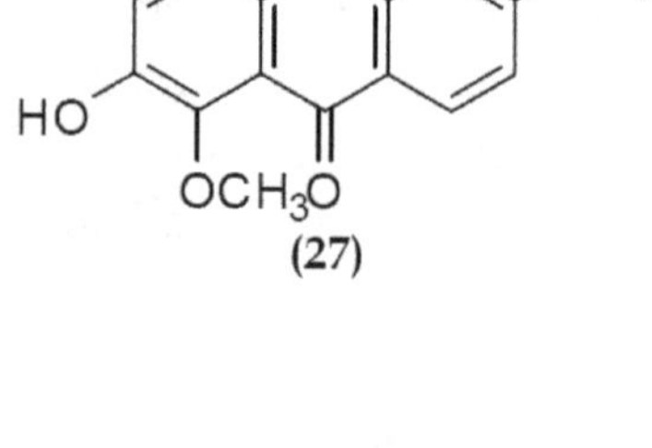

(27)

Morindone-5-methyl ether (**27**). Orange cystals. Mp 232 ⍰ C [lit. 223 ⍰ C, Chang & Lee (1984)]. UVλ_{max} EtOH nm: 410, 497. UVλ_{max} EtOH/ -OH nm: 314, 388, 498. IR v_{max} (KBr)

cm^{-1}: 3389 (OH), 2926, 1672 (C=O unchelated), 1630 (C=O chelated), 1581 (C=C aromatic). MS m/z 284 (M^+), 266, 238, 197. ^1H NMR ($CDCl_3$, 500MHz): 13.02 (1H, s, 1-OH),8.14 (1H, d, J=8.55 Hz, H-8), 7.70 (1H, d, J= 8.06 Hz, H-4), 7.51 (1H, d, J=7.81 Hz, H-3), 7.35 (1H, d, J=8.54 Hz, H-7), 6.73 (1H, s, 6-OH), 4.03 (3H, s, OCH_3), 2.37 (3H, s, CH_3). ^{13}C NMR ($CDCl_3$, 125 MHz): 187.8 (C=O), 182.0 (C=O), 160.6 (C-OH), 155.9, 146.8, 136.9, 134.5, 132.3, 127.1, 125.9, 125.5, 112.0, 118.9, 114.7, 62.3 (OCH_3), 16.1 (CH_3)

Alizarin-1-methyl ether. Yellow-orange crystals (**28**). Mp 164 [lit 178-179 ⍰ C, Chang & Lee (1984)]. UVλ_{max} EtOH nm: 313, 378, 485. UVλ_{max} EtOH/ -OH nm: 315, 333, 493. IR v_{max} (KBr) cm^{-1}: 3443 (OH), 2926, 1671 (C=O unchelated), 1589 (C=C) aromatic. MS m/z 254 (M^+), 236, 208, 183. ^1H NMR ($DMSO_{-d6}$, 500MHz): 8.28 (2H, m, H-5, H-8), 8.15 (1H, d, J= 8.55 Hz, H-4), 7.78 (2H, m, H-7, H-6), 7.37 (1H, d, J = 8.54 Hz, H-3), 6.70 (1H, s, 2-OH), 4.04 (3H, s, OCH_3). ^{13}C NMR ($DMSO_{-d6}$ 125 MHz): 182.7 (C=O), 182.1 (C=O), 155.5, 146.6, 131.4, 133.9, 132.9, 127.5, 127.1, 126.8, 125.8, 125.6, 120.2, 62.3 (OCH_3)

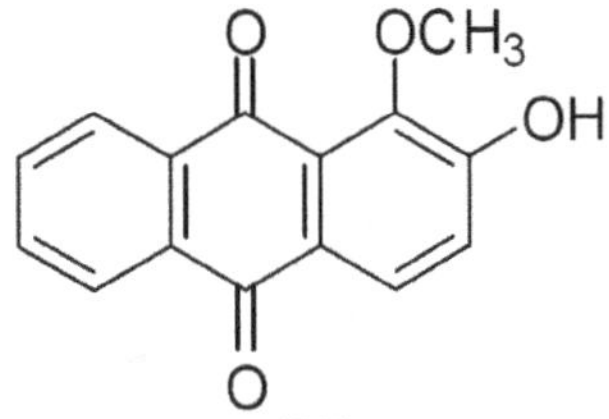

(28)

H-3 and H-4 appeared as doublets at δ_H 8.23 and 7.89 respectively. A formyl group (δ_H 10.63) is attached to C-2. HMBC correlations of C-10 with H-3 and H-5 confirmed the assignment of the protons at their respective positions and supported by their respective COSY correlations. ^{13}C NMR showed fifteen carbons peaks as expected. One of the chelated carbonyl carbon was further downfield at δ_C 188.9 (C-9), confirming the chelated nature of this carbonyl. The assignment of carbons were accomplished using FGHMQC and FGHMBC experiment. Presented below are structures and spectroscopic data of the isolated compounds.

4. CONCISE SUMMARY

The phytochemical investigation of Fissistigma latifolium and Meiogyne virgata (Annonaceae) identified twelve alkaloids: (-)-N-methylguattescidine 1, liriodenine 2, lanuginosine 3, (-)-asimilobine 4, dimethyltryptamine 5, (-)-remerine 6, (-)-anonaine 7, columbamine 8, lysicamine 9, nornuciferine 10, norushinsunine 11, and cleistopholine 12. Tryptamine alkaloids have not been documented in Fissistigma species, but (-)-N-methylguattescidine 1 is a rare example of a naturally occurring 6a-methylated-7-oxo-aporphine alkaloid. Alkaloids 3, 6, 9, and 10 have not been documented in Meiogyne species.

Rennellia and Morinda are frequently conflated owing to their analogous traditional applications. Both plants are conventionally employed for the treatment of fever, postpartum recovery, and bodily pain. Our phytochemical analysis of the root extract of R. elliptica revealed notable parallels in major anthraquinones with those present in Morinda species. The primary components of R. elliptica, including nordamnacanthal, damnacanthal, rubiadian, rubiadin methyl ether, and lucidin-ω-methyl ether, are also found in M. elliptica and M. citrifolia.

5. REFERENCES

Abd Aziz, R. (2003). *Siri Syarahan Perdana Professor*. Skudai: Universiti Teknologi Malaysia.

Adesogan, E.K. (1973). Anthraquinones and anthraquinols from *Morinda lucida*. The biological significance of oruwal and oruwalol, *Tetrahedron* 29 (1973), p. 4099.

Ahmad, R., Mahbob, E. N. M., Noor, Z. M., Ismail, N. H., Lajis, N. H., & Shaari, K. (2010). Evaluation of antioxidant potential of medicinal plants from Malaysian Rubiaceae (subfamily Rubioideae). *African Journal of Biotechnology, 9*(46), 7948- 7954.

Ahmad, R., Shaari, K., Lajis, N. H., Hamzah, A. S., Ismail, N. H., & Kitajima, M. (2005). Anthraquinones from *Hedyotis capitellata. Phytochemistry, 66*(10), 1141-1147.

Ali, A. M., Ismail, N. H., Mackeen, M. M., Yazan, L. S., Mohamed, S. M., Ho, A. S. H., et al. (2000). Antiviral, Cytotoxic and Antimicrobial Activities of Anthraquinones Isolated from the Root of *Morinda elliptica. Pharmaceutical Biology, 38*, 298-301.

Alias, A., Hazni, H., Mohd Jaafar, F., Awang, K. and Ismail, N. H. (2010). Alkaloids from *Fissistigma latifolium* Dunal Merr. *Molecules*, 15, 4583-4588.

Awang, K.; Hamid, A.; Hadi, A. (2000). Protoberberine Alkaloids From *Fissistigma fulgens* Merr. (Annonaceae). *Malaysian J. Sci., 19*, 41–44.

Briggs, L.H., Beachen, J.F., Cambie, R.C., Dudman, N.P.B., Steggles, A.W. & Rutledge, P.S. (1976) Chemistry of *Coprosma* genus. Part XIV. Constituents of five New Zealand species. *J. Chem. Soc. Perkin Trans. I*, 1789–1792

Burkill, I. H. (1966). *A Dictionary of the Economic Product of the Malay Peninsular*. Volumes I &II. Ministry of Agriculture and Cooperatives, Kuala Lumpur.

Chan, K. C. & Toh, H. T. (1985). A new aporphinoid from *Desmos dasymachallus*. In: *Proceedings of the 2nd Meeting of the Natural Products Group*, edited by Said, I.M. & Zakaria, Z. 17-20. Jabatan Kimia, Fakulti Fizis dan Gunaan, Universiti Kebangsaan Malaysia, Bangi.

Chang, P. and Lee, K., Cytotoxic antileukemic anthraquinones from *Morinda parvifolia. Phytochemistry*, 23, 1733-1736. (1984).

Connolly, J. D., Haque, M. D. E., Kadir, A. A. (1996): Two 7,7-bisdehydroaporphine alkaloids from *Polyalthia bullata. Phytochemistry* 43: 295–297.

Cordell, G.A. (1981). *Introduction to Alkaloid: A Biogenetic Approach;* John Wiley & Sons: New York, NY, USA, 6–19.

Derksen, G. C. H., Van Beek, T. A., & Atta ur, R. (2002). Rubia tinctorum L *Studies in Natural Products Chemistry* (Vol. Volume 26, Part 7, pp. 629-684): Elsevier.

Hamzah, A. S., & Lajis, N. H. (1998). Chemical Constituents of Hedyotis herbaceae. *ARBEC, Article II*, 1-6.

Han, Y.-S., der Heiden, R. V., & Verpoorte, R. (2001). Biosynthesis of Anthraquinones in Cell Cultures of the Rubiaceae. *Plant Cell, Tissue and Organ Culture, 67*, 201-220.

Hutchinson, J. (1973). *The Families of Flowering Plants* (3rd ed.): Oxford University Press.

Ismail, N. H., Ali, A. M., Aimi, N., Kitajima, M., Takayama, H., & Lajis, N. H. (1997). Anthraquinones from *Morinda elliptica. Phytochemistry, 45*(8), 1723-1725.

Ismail, N. H., Mohamad, H., Mohidin, A., & Lajis, N. H. (2002). Antioxidant Activity of Anthraquinones from *Morinda elliptica. Natural Product Science, 8*(2), 48-51.

Jasril, Lajis, N. H., Lim, Y. M., Abdullah, M. A., Sukari, M. A., & Ali, A. M. (2003). Antitumor Promoting and Antioxidant Activities of Anthraquinones Isolated from the Cell Suspension Culture of *Morinda elliptica. Asia Pacific Journal of Molecular Biology and Biotechnology, 11*(1), 3-7.

Kan, W. S. (1979). In *Pharmaceutical Botany*; National Research Institute of Chinese Medicine: Taiwan, 268.

Kam, T. S. (1999). Alkaloids from Malaysian Flora, in Alkaloids: Chemical and Biological Perspectives, S. W. Pelletier (Ed.), Pergamon, Amsterdam, Volume 14, Chapter 2, 285-435.

Kamaruddin, M. S. (1998). Proceedings Malaysian Traditional Medicine, 10-11 June 1998, Universiti Malaya, Kuala Lumpur, 80.

Kong, J.-M., Goh, N.-K., & Chia, T.-F. (2003). Recent Advances in Traditional Plant Drugs and Orchids. *Acta Pharmacologica Sinica, 24*(1), 17-21.

Lajis, N. H., Ahmad, R., & Atta-ur, R. (2006). Phytochemical studies and pharmacological activities of plants in genus Hedyotis/oldenlandia *Studies in Natural Products Chemistry* (Vol. Volume 33, Part 13, pp. 1057-1090): Elsevier.

Lavault, M., Cabalion, P. and Bruneton, J. (1981). Alkaloids of Uncaria guianensis. *Planta Med.* 42, 50.

Lee, H. H. (1969). Colouring matters from Prismatomeris malayana. *Phytochemistry, 8*(2), 501-503.

Leistner,E.,1975. Isolation,identifica tion and biosynthesis of anthraquinones in cell suspension cultures of Morinda citrifolia. Planta Med. (Suppl.) 214–224.

Mat Salleh, K., & Latiff, A. (2002). *Tumbuhan Ubatan Malaysia*: Universiti Kebangsaan Malaysia & Kem, Sains, Teknologi dan Alam Sekitar.

Nik Idris, Y., Lim, S. Y., Zaemah, J. and Ikram, M. S. (1994). *Alkaloid daripada Batang Fissistigma latifolium (Annonaceae) dan Potensinya Sebagai Dadah Anti-Leukaemia*, Laporan Teknik FSFG 4 : 199-204

Osman, C. P., Ismail, N. H., Ahmad, R., Ahmat, N., Awang, K., & Jaafar, F. M. (2010). Anthraquinones with Antiplasmodial Activity from the Roots of Rennellia elliptica Korth. (Rubiaceae). *Molecules, 15*(10), 7218-7226.

Parker, Sybil P. (1997). *Chemistry; Dictionaries*. McGraw-Hill (New York). Perry, L. M. (1980). *Medicinal Plants of Southeast Asia.* MIT: Cambridge, 19.

Rath, G., Ndonzao, M., & Hostettmann, K. (1995). Antifungal Anthraquinones from *Morinda lucida. Int J. Pharmacogn.*, 33, 107-114.

Roberts, J. L., Rutledge, P. S., and Trebilcock, M. J. (1977). Experiments Directed Towards the Synthesis of Anthracyclinones. I Synthesis of 2-Formylmethoxyanthraquinones *Aust. J. Chem.*, , 30, 1553.

Saaid M., and Awang, K. (2005) Alkaloids of *Fissistigma manubriatum*. Malaysian *Journal of Science*. 24 (1), 41-45.

Sinclair, J., (1955) A revision of the Malayan Annonaceae. *The Gardens' Bulletin Singapore,* 14, 149-69.

Tadic, D., Cassels, B. K., Leboeuf, M. and Cavé, A. (1987). Kinabaline and the aporphinoid biogenesis azaanthracene and azafluorene alkaloids. *Phytochemistry*, 26, 537–541.

Teo, L.E., Pachiaper, G., Chan, K.C., Hadi, H.A., Weber, J.F., Deverre, J.R., David, B. & Sevenet, T. (1990). A new phytochemical survey of Malaysia V. Preliminary screening and plant chemical studies. *J. Ethnopharmacol.* 28(1) : 63-101.

Verdout, B. (1976). *Annonaceae, Flora of Tropical East Africa*; Crown Agents for Oversea Government and Administrations: London, UK, pp. 101–102.

Wong, K. M. (1989). Rubiaceae (from the genus Rubia). In F. S. P. Ng (Ed.), *Tree Flora of Malaya; A Manual for Foresters* (Vol. 4, pp. 324-337, 404-405): Longman Malaysia.

15 Chapter

PHYTOCHEMISTRY, PHARMACOLOGY, AND AGRONOMY OF MEDICINAL PLANTS: AN INTERDISCIPLINARY STUDY OF AMBURANA CEARENSIS

1. INTRODUCTION: AN OVERVIEW

Plants serve as a significant source of physiologically active compounds and have been utilized for therapeutic purposes since antiquity. Plant materials are utilized as home cures, in over-the-counter pharmaceuticals, dietary supplements, and as raw materials for the extraction of phytochemicals. The utilization of medicinal plants is typically grounded in traditional knowledge, from which their therapeutic effects are frequently validated through pharmacological tests.

A significant proportion of prescribed medications continues to derive from botanical sources, exhibiting various pharmacological activities, including morphine (I) (analgesic), scopolamine (II), atropine (III) (anticholinergics), galantamine (IV) (Alzheimer's disease), quinine (V) (antimalarial), paclitaxel (VI), vincristine (VII), and vinblastine (VIII) (anticancer agents), in addition to digitalis glycosides (IX) (heart failure) (Fig. 1). The flexibility of biological functions is due to the extensive quantity and diverse range of secondary metabolites in plant species, which belong to many chemical classes such as alkaloids, coumarins, flavonoids, tannins, terpenoids, and xanthones.

The significant use of herbal medications, despite the efficacy of synthetic pharmaceuticals, stems from the perception that natural products are non-toxic and/or possess fewer adverse effects, the inclination for alternative therapies, and their comparatively cheaper expenses. In poor nations, herbal medicine constitutes the primary mode of healthcare. In Brazil, which boasts one of the highest biodiversities of plants globally, pharmaceutical aid programs, such as "Living Pharmacies," play a significant role in promoting the sensible use of medicinal plants, particularly for impoverished populations, as acknowledged by the World Health Organization (WHO). Moreover, herbal medications constitute a substantial segment of the pharmaceutical market in several industrialized nations, such as Germany.

Conversely, herbal medications are often dismissed by the majority of health experts due to insufficient scientific studies validating their efficacy and safety. Physicians generally experience insecurity in prescribing herbal remedies, as they do not undergo clinical studies, phytochemical research, and their active constituents remain undetermined. Consequently, herbal medicines lack a specified dosage, details regarding their chemical composition, and cautions about potential hazards. The inadequate quality control of herbal medications, susceptible to adulteration and inherent features associated with the raw materials utilized, results in variability and unpredictable outcomes. Moreover, the majority of herbal pharmaceuticals are derived from wild sources, constraining industrial production and endangering the species utilized.

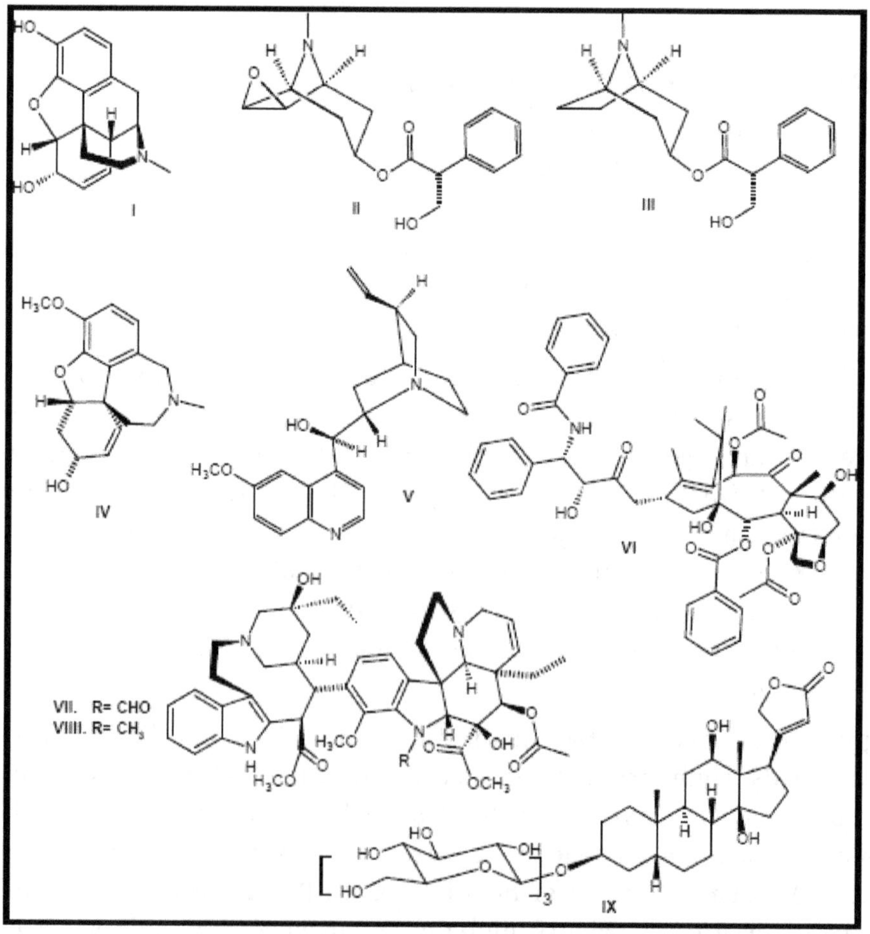

Fig. 1. Chemical structures of plant-derived drugs

In light of the aforementioned limitations and drawbacks of herbal medicine, we aim to provide a comprehensive review of the chemical, pharmacological, and agronomic studies of Amburana cearensis, exemplifying successful scientific research on wild plants and serving as a model for the sustainable economic utilization of medicinal plants.

2.HERBAL DRUGS AND PHYTOPHARMACEUTICALS

According to the WHO definition, herbal drugs are preparations containing plant parts (leaves, roots, seeds, stem bark, etc.) or whole plant materials in the crude or processed form, as active ingredients, besides some excipients. Herbal preparations can be found under different forms: oral tablets, capsules, gel caps, syrup, extracts and infusions. In general, combinations with chemically defined active substances or isolated constituents, are not considered to be herbal medicines. (Calixto, 2000).

The information about the therapeutical properties and usage of medicinal plants are commonly based on the empirical knowledge of ancient people, which was passed over several generations and originated the traditional medicine systems, utilized all over the world (Traditional Chinese Medicine, Ayurvedic system, Western and African Herbalisms). An estimated quantity of 50 000 plant species are used for medicinal purposes, from which the stand out species of the following families, such as Apocynaceae, Araliaceae, Apiaceae, Asclepiadaceae, Canellaceae, Clusiaceae and Menispermaceae (Schippmann et al., 2002). From the total of 252 drugs in the WHO's essential medicine list, 11% are exclusively derived from plant origins (Sahoo et al., 2010)

Herbal drugs are consumed by three-quarters of the world's population in the treatment of mainly chronic diseases, particularly headache, rheumatological disorders and asthma (Inamdar et al., 2010). In the developing countries, the population relies basically on medicinal plants for primary health care, since modern medicine is expensive and not easily accessible. However, the consumption of herbal drugs is also large in developed countries. Phytotherapy is popular in many countries of Western Europe (Germany, France, Italy, etc.), since people believe that either herbal drugs are devoid of side-effects or seek a healthier life style. Americans usually buy herbal products as a dietary supplement in the United States, aiming at preventing aging and diseases like cancer, as well as diabetes (Calixto, 2000).

Herbal drugs have some features which distinguish themselves considerably from synthetic drugs. Herbal medicines are always formed from a complex mixture of chemical compounds (eg. *Scutellaria baicalensis* has over 2000 components), and they may be constituted by many plants, therefore herbal drugs show an ample therapeutic usage. It is quite common to find a medicinal plant with several therapeutic properties (Sahoo et al., 2010; Calixto, 2000). The combination of either many plants, containing diverse bioactive substances or a pool of structural analogs, can produce a synergistic action that results in a stronger effect, therefore, permitting a reduction of dosage, which implies in lower risks of intoxication and undesirable side effects. As some diseases (e.g. AIDS or various types of cancer) possess a multi-causal etiology and a complex pathophysiology, a medical treatment may be more effective through well-chosen drug combinations than a single drug. Ginkgolides A and B, isolated from *Ginkgo biloba*, duly demonstrated a greater effect on the thrombocyte aggregation inhibition, when used as a mixture as opposed to what would be expected from the sum of the two compounds separately (Wagner, 2011).

On the other hand, plant-based products do not possess a well-defined chemical composition, due partially to chemical complexity stated above. Hence, the active principles of herbal drugs are frequently unknown, in addition to their standardization and quality control, being hardly achieved (Calixto, 2000) owing to mainly chemical variability in raw material. Secondary metabolites are the bioactive components from herbal drugs and their contents are strongly influenced by several factors: genetic (genotypes, chemotypes), physiologic (circadian rhythm, phenology, age), environmental (climate, sunlight exposure, water availability, soil, agronomic conditions) and manufacturing conditions (harvesting, storage and processing) (Tab. 1), (Sahoo et al., 2010; Gobbo-Beto & Lopes, 2007).

Table 1. Effects on the production of secondary metabolites

Evaluated Effect	Plant	Compound (s)	Result
Seasonality	*Hypericum perforatum*	Hypericin	300x content larger in summer than winter. (Southwell & Bourke, 2001)
Harvesting time	*Ocimum gratissimum*	Eugenol	98 % at 17h, but it is 11 % at midday (Silva et al. 1999)
Age	*Papaver somniferum*	Morphine	6x content larger on the 75th day after germination than on the 50th day (Williams & Ellis, 1989)
Phenology	*Gentiana lutea*	Mangiferin/ isoorientin	Before flowering-↑ [Mangiferin]; during flowering- ↑ [Isoorientin] (Menković et al., 2000)
Temperature	*Nicotiana tabacum*	Scopolamine	[Scopolamine] 4x larger after freezing (Koeppe et al, 1970)
Water availability	*H. perforatum*	Hyperforin/ Hypericin	Under water stress: ↑ 2x [Hyperforin] and ↓ [Hypericin] (Zobayed et al, 2007)

Furthermore, most herbal drugs are utilized and commercialized without having a proven efficacy and safety through well-controlled double-blind clinical and toxicological trials, as pharmaceuticals are usually tested prior to being marketed. The safety and efficacy of herbal drugs are supported by their long historical use. Nevertheless, it is known that various herbal drugs fail, after testing in clinical trials and there are numerous reports on intoxication cases associated with their consumption. The WHO database has over sixteen thousand suspected case reports, related to intoxication by herbal drugs. The most frequent adverse reactions are hypertension, hepatitis, convulsions, thrombocytopenia and allergic reactions. Cardiovascular problems with the use of ephedra, hepatotoxicity caused by the consumption of kava-kava and comfrey, as well as licorice-related water retention, are some side effects claimed by the pharmacovigilance authorities. In addition to intrinsic factors mentioned above, the herbal drugs efficacy and safety, may also be seriously affected due to botanical misidentification or intentional usage of fake plants, contamination with pesticide residue, toxic heavy metals, pathogens and mycotoxins, as well as adulterants added to increase potency (synthetic substances) or the weight of herbal products in order to reduce costs (Sahoo et al., 2010; Calixto, 2000).

For the purpose of overcoming or mitigating the aforesaid inconvenient issues, WHO has developed a series of technical guidelines and documents in relation to the safety and the quality assurance of medicinal plants and herbal drugs preparations, such as "Quality Control Methods for Medicinal Plant Materials" (a collection of recommended test procedures for assessing the identity, purity and content of medicinal plant materials), "Guidelines on good agricultural and collection practices for medicinal plants", as well as "WHO guidelines for assessing quality of herbal medicines with reference to contaminants and residues". In turn, pharmaceutical laboratories have been investing in the enhancement of the quality for herbal products, aiming at the approval by governmental regulatory agencies, as a strategy to offer more reliable products, therefore conquering the confidence of health care professionals and consumers (Sahoo et al., 2010).

The interest in herbal drugs is continuously growing and they account for a significant share in the pharmaceutical market. The global herbal pharmaceutical industry (including drugs from herbal precursors and registered herbal medicines) invoices approximately US$ 50 billion/year (2008). In 2006, the best selling herbal products were: Ginseng (>U$ 1 billion global sales), Ginkgo (U$ 1 billion), Noni (U$ 1 billion), Saw Palmetto (U$ 600 millions) Echinacea (U$ 500 millions), Valerian (U$ 450 millions), and Green Tea (U $ 450 millions) (Gruenwald, 2008- Entrepreneur). The United States, China, Japan, Germany, South Korea and India, are the largest market. Medicinal plants have also been utilized as source of phytochemicals for the pharmaceutical, cosmetic and agrochemical industries. The most successful examples are paclitaxel (an anticancer drug from *Taxus baccata*), artemisinin (an antimalarial agent from *Artemisia annua*), vincristine/vinblastine (anticancer substances from *Catharanthus roseus*). The Pharmaceutical industry is interested in phytochemicals, however, the availability of quantities of pure chemical substances is normally a limiting factor, since the market demand for phytochemicals, usually reaches a scale from hundreds to thousands of kilograms per annum. (McChesney, 2007)

3. MEDICINAL PLANTS THREATENED BY EXTINCTION

The increasing demand for medicinal plants has endangered several species, since the main source of herbal drugs is the wild plant and the amount required from plant materials invariably exceeds the supply available from its natural source. Although the Convention on Biological Diversity (CBD), held in 1992, has established as goals, the conservation of biological diversity, the sustainable use of its components as well as the fair and equitable sharing of the benefits from the usage of genetic resources, it is still estimated that slightly more than 4000 medicinal plant species are under threat of extinction. The Convention on International Trade of Endangered Species of Wild Fauna and Flora (CITES), being the principal tool for monitoring or restricting the international trade of species threatened by over-exploitation, has published a biannual list of medicinal species like: *Taxus wallichiana*, *Panax quinquefolius*, *Dioscorea deltoidea*, *Hydrastis canadensis*, *Prunus africana* , *Rauvolfia serpentina* and *Pterocarpus santalinus* (Schippmann et al., 2002).

The overexploitation of a certain wild medicinal plant and consequent depletion of its raw material affect inevitably the economic feasibility of any phytopharmaceutical business in medium or long-term, since the production cost tends to be higher and the product supply become discontinuous. Furthermore, the extractivism provokes loss of genetic diversity, becoming the remaining plant population more vulnerable to diseases/pests and diminishing the variability of genotypes with features of interest such as yielding, bioactive substance content and resistance to biotic and abiotic factors (Rao et al., 2004). Hence, agencies concerned with conservation policies are recommending that wild species be brought into cultivation systems in order to assure the economic and environmental sustainability of herbal medicines trade (Schippmann et al., 2002). *Ginkgo biloba* and *Hypericum perforatum* are some of the top- selling medicinal plants, however they are not endangered, because their plant materials have been obtained by cultivation for a long time (Canter et al., 2005).

4. CULTIVATION OF MEDICINAL PLANTS

The cultivation of medicinal plants is advocated as a means for meeting current and future demands for large quantities of herbal drugs, but also as a way to relieve the pressure of harvesting on wild populations (Schippmann et al., 2002). In China, one of the largest markets of herbal medicine, 380,000 ha of lands are utilized for farming of medicinal plants.

Medicinal plants are also cultivated for supplying phytochemicals. Bristol-Myers Squibb developed a system of production based upon isolation of a precursor of Taxol from the leaves or needles of cultivated *Taxus baccata* or *T. wallichiana* that provide the hundreds of kilograms of Taxol required per year for the treatment of cancer patients. (McChesney, 2007) From the perspective of the market, domestication and cultivation provide a number of advantages over wild harvest for production of herbal drugs: (1) reliable botanical identification; (2) uniform and high quality raw material. As wild plants are dependent on many factors that cannot be controlled and the irregularity of supply is a common feature, the cultivation assures a steady source of raw material; (3) price and volume between farmer and pharmaceutical companies can be more easily negotiable, since the production forecast is more precise; (4) genetic breeding and biotechnology tools can lead to the development of plant materials with agronomically and commercially desirable features, permitting to optimize yield and to meet regulations as well as consumer preferences, respectively; (5) cultivated material can be easily certified as "organic product" (Schippmann et al., 2002; Canter et al, 2005).

Cultivated plants account for 60-90 % in terms of amount of plant material employed by Herbal medicines companies, but the number of wild species still is larger. Although the cultivation is apparently more advantageous than wild harvesting, only 130-140 species are cultivated in Europe, while just 20 out of 400 medicinal plants marketed in India are grown in field. Likewise, amongst 1000 plants more commonly used with medicinal purposes in China, only 100-250 species are sourced from cultivation. There are some reasons that can explain this low utilization of cultivated plants:

Belief of that wild specimens are more potent than cultivated plants. Chinese believe that the physical appearance of wild roots to the human body symbolizes vitality and this feature is crucial for the potency of the ginseng roots, nevertheless cultivated roots do not exhibit this characteristic shape. Furthermore, some scientific studies support partly this hypothesis saying that secondary metabolites, the main responsible for therapeutic properties of herbal medicines, are biosynthesized by plants under particular conditions of stress and competition in their natural environments. Hence, perhaps the secondary metabolites would not be so expressed in monoculture conditions, therefore the active ingredient levels can be much lower in cultivated plant.

Domestication of wild plant is not always technically possible. Many species are difficult to cultivate because of certain biological features or ecological requirements (slow growth rate, special soil requirements, low germination rates, susceptibility to pests, etc.).

Economical feasibility. Domestication requires a long time of agronomical studies and high financial investment for the plantation. Generally, production costs through cultivation are higher than wild harvesting, thus few species can be marketed at a high sufficient price to make cultivation profitable, for instance *Garcinia afzelii, Panax quinquefolius, Saussurea costus* and *Warburgia salutaris.* Hence, many endangered medicinal plants only will bring into cultivation, if exists governmental incentive (Schippmann et al., 2002).

However, the cultivation of medicinal plant in agroforestry system can be a good alternative for more viable and environmentally sustainable farming. In China, ginseng (*Panax ginseng*) and other medicinal plants are grown in pine (*Pinus* spp.), *Paulownia tomentosa* and spruce (*Picea* spp.) forests; besides some medicinal herbs are often planted with bamboo (*Bambusa* spp.). In New Zealand, American ginseng showed better growth under *Pinus radiate.* The shade offered by forest species seems to favor the growth of medicinal plants. Likewise, quinine yields of *Cinchona ledgeriana* increase when it is protected by shade of other species, such as *Crotalaria anagyroides* and *Tephrosia candida*. In India, some medicinal plants that have also been successfully intercropped with fuel wood trees (e.g., *Acacia auriculiformis* and *Eucalyptus tereticornis*) and coconut. Intercropping gives some income to farmers during the period when the main trees have not started production. (Rao et al., 2004).

Application of traditional and biotechnological plant-breeding techniques can become the cultivation of medicinal plants a trade more attractive (eg. increasing the yielding) as well as it can improve features of the plant that affect the efficacy and safety of a herbal drug (eg. levels of bioactive compounds or presence of potentially toxic substances). *Mentha* spp (mints) have been engineered to modify essential oil production and to enhance the resistance of the plant to fungal infection and abiotic stresses. Genetic engineering allowed the enhancement of scopolamine and artemisinin in *Atropa belladonna* and *Artemisia annua*, respectively. (Canter et al., 2005)

5. AMBURANA CEARENSIS

A. cearensis (Fabaceae) is a native tree from "Caatinga" (a kind of vegetation found in the Brazilian semi-arid region), where it is popularly known as "cumaru" or "imburana-de- cheiro" (Fig. 2). Because of these said popular names, *A. cearensis* is usually misidentified as *Dipteryx odorata* (Fabaceae) and *Commiphora leptophloeos*

(Burseraceae). *A. cearensis* occurs widely in South America (from Peru to Argentina), along with another species of this taxon, *Amburana acreana*, which is found chiefly in the southwestern region of the Amazon Forest. *A. cearensis* can reach 15 m of height and 50 cm of diameter, but it is characterized by white flowers and dark pods containing only one seed each, besides its stem bark possessing reddish stains and a vanilla-like aroma of coumarin (**1**). At the early stage of development (seedlings), *A. cearensis* displays a hypertrophied and subterraneous tube-like structure, called xylopodium, which acts as a storage of water and nutrients, therefore it is considered an adaptive strategy for arid habitats (Lima, 1989; Cunha & Ferreira, 2003).

Fig. 2. A wild specimen of *Amburana cearensis* **in its natural habitat.**

Given the various applications, *A. cearensis* has a great commercial importance in Northeastern region of Brazil. Its wood is used in the carpentry for the manufacturing of furniture, doors and crates, owing to its recognized durability, whereas the seeds are used as flavoring and insect repellents. The wood powder from it can be added to alcoholic beverage barrels for accelerating the aging process of sugar cane distilled spirits (cachaça) (Aquino et al, 2005). The seeds and stem bark are traditionally utilized for treating respiratory diseases, such as influenza, asthma and bronchitis due to anti-inflammatory, analgesic and bronchodilator properties. As far as folk medicine is concerned, *A. cearensis* is consumed as a homemade medication called "lambedô (a sugary drink), however in an industrial scale, the syrup is a pharmaceutical form, which is produced by the government and private laboratories (Fig. 3).

The medicinal use of *A. cearensis* is based on scientific studies, which demonstrated that this plant possesses therapeutic properties that justify its recommendation for the treatment of respiratory illnesses. Preclinical tests demonstrated bronchodilator, analgesic and anti- inflammatory activities for the hydro-alcohol extract from the stem bark of the *A. cearensis*, which also showed to be free of toxicity in therapeutic doses. The chemical composition of the stem bark and seeds from it, consists basically of coumarin, flavonoids, phenol acids and phenol glucosides. Some of them were tested individually and showed pharmacological activities similar to the extract, hence they were considered the active principles of the *A. cearensis.*

However, the intense commercial use of *A. cearensis* has led to the threat of extinction for this specie. In order to ensure the conservation and the economic utilization of *A. cearensis*, we proposed the replacement of its stem bark of a wild adult plant for a young specimen, cultivated under controlled agronomic parameters. In an interdisciplinary study, ethanol extracts of cultivated plants were compared to the extracts of this wild plant through preclinical trials and phytochemical analysis.

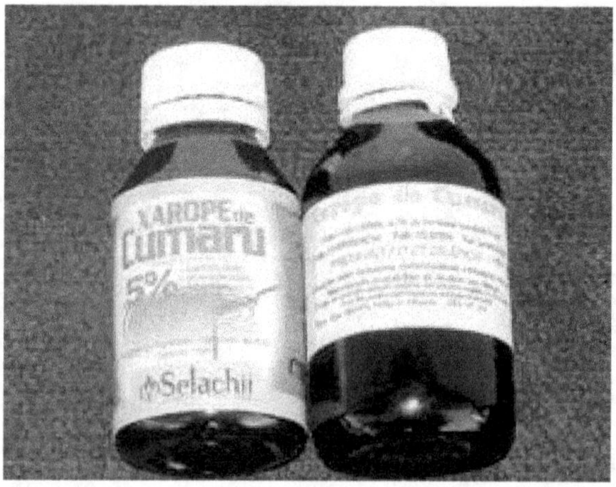

Fig. 3. Syrups made from the trunk bark of *Amburana cearensis*.

5.1 Agronomical study of A. cearensis

The agronomical study of *A. cearensis* was carried out with seedlings obtained by seed germination. Each plot consisted of six regularly spaced rows of 20 cm, whose sowing density was 50 seeds/row. The seedlings were transplanted to four garden beds (1.2m×10 m), fertilized prior to an organic fertilizer (2.8 kg m^{-2}), containing each of them 20 young plants.

Fig. 4. Seedlings of *A. cearensis* **harvested in 8th month of growth**

Eight harvestings were performed monthly, starting on the 2nd month until the 9th month after the sowing. The plants harvested were evaluated with the following parameters: fresh plant weight, plant height, xylopodium diameter, root size, ethanol extract yield from the aerial part and xylopodium (Fig. 4). The fresh biomass production of *A. cearensis* seedlings increased almost eight-fold, during the 2nd through 9th month after the sowing. With reference to ethanol extract yield, there was a tendency of decrease for the extract weigh/xylopodium weight ratio over a period of time, while an oscillatory behavior was observed for yield of the ethanol extract from its aerial part, achieving a plateau on the 3rd and 7th month (Leal et al. 2011)

5.2 Phytochemistry of A. cearensis

5.2.1 Wild plant

While the pharmacological research about *A. cearensis* advanced, it was necessary to have a better understanding of the chemistry in this specie, previously limited to coumarin and a few phenol compounds, in order to discover its active principles. Bastos (1983), in her master thesis, found coumarin (**1**) as the most abundant component and described the isolation of isokaempferide (**2**), methyl 3,4-dimethoxy-cinnamate (**3**), afrormosin (**4**), 7-hidroxy-8,4'-dimethoxy-isoflavone (**5**), 24-methylenecycloartanol (**6**), β-sitosterol (**7**) and 6- hidroxy-coumarin (**8**). Additionally, the seeds presented a high oil content (23 %), which is composed of triglycerides of the following fatty acids: oleic acid (53 %), palmitic acid (19 %), stearic acid (8 %) and linoleic acid (7 %). Bravo et al. (1999) isolated amburosides A [(4-O-β- D-glucopyranosylbenzyl) protocatechuate] (**9**) and B [(4-O-β-D-glucopyranosylbenzyl) vanillate] (**10**) from the ethyl acetate extract of the trunk bark, utilizing just silica gel preparative Thin-Layer Chromatography (TLC) (Fig. 5) Canuto & Silveira (2006) carried out a phytochemical investigation of the ethanol extract from the stem bark. The ethanol extract was partitioned with water and ethyl acetate. The aqueous phase showed to be very rich in sucrose (**11**), whereas the organic phase was dried and submitted to successive chromatographic columns on silica gel and dextran gels. These chromatographic separations led to the isolation of coumarin (**1**), two phenol acids [vanillic acid (**12**) and protocatechuic acid (**13**)], five flavonoids [isokaempferide (**2**), afrormosin (**4**), kaempferol (**14**), quercetin (**15**) and 4'-methoxy-fisetin (**16**)], amburoside A (**9**) and a mixture of β-sitosterol and stigmasterol glycosides (**17-18**). Later on, this same methodology was applied to isolate an isoflavone formononetin (**19**) and a novel coumarin 6-coumaryl protocatechuate (**20**) (Canuto et al., 2010) (Fig. 5). Continuing with the phytochemical study of *A. cearensis*, the seeds revealed high presence of phenol glucosides. Liquid-liquid partitioning from the ethanol extracts followed by chromatography on Sephadex LH-20 and a reversed-phase HPLC chromatography of the ethyl acetate fraction, resulted in the isolation of six new amburosides (C-H). 4-*O*-β-D-(6′′-*O*-galloylglucopyranosyl)-benzyl protocatechuate (**21**), 4-*O*-β-D-(6′′-*O*-acetylglucopyranosyl)-benzyl protocatechuate (**22**), 4- *O*-β-D-(6′′-*O*-protocatechuoylglucopyranosyl)-benzyl protocatechuate (**23**), 4-*O*-β-D-(6′′-*O*- feruloylglucopyranosyl)-benzyl protocatechuate (**24**), 4-*O*-β-D-(6′′-*O*- vanilloylglucopyranosyl)-benzyl protocatechuate (**25**) and 4-*O*-β-D-(6′′-*O*- sinapoylglucopyranosyl)-benzyl protocatechuate (**26**). Additionally, amburoside A (**9**), isokaempferide (**2**), vanillic acid (**12**), 6-hydroxycoumarin (**8**) and (*E*)-*o*-coumaric acid (**27**) were isolated from this same extract. (Canuto et al., 2010) (Fig. 5).

The isolation of 6-coumaryl protocatechuate (**20**) (trunk bark) and 6-hydroxycoumarin (**8**) (seeds) from *A. cearensis* presents an intriguing finding for biosynthesis of coumarins, since monoxygenated-coumarins are preferentially substituted at C-7 position (umbelliferone and its derivatives), according to biogenetic rules. This substitution pattern is due to the usual precursor of coumarins, *p*-coumaric acid, which is biosynthentized by the shikimate pathway from either tyrosine-deamination or *p*-hydroxylation of cinnamic acid (Dewick, 2002). Nevertheless, despite a large occurrence of simple coumarins oxygenated at the C-7 position, 6-hydroxycoumarin (**8**) was also found in some species like *Bidens parviflora* (Asteraceae), *Paeonia suffruticosa* (Paeoniaceae) and *Hydrangea chinensis* (Hydrangeaceae) (Tommasi et al., 1992; Wu et al., 2002; Khalil et al., 2003)

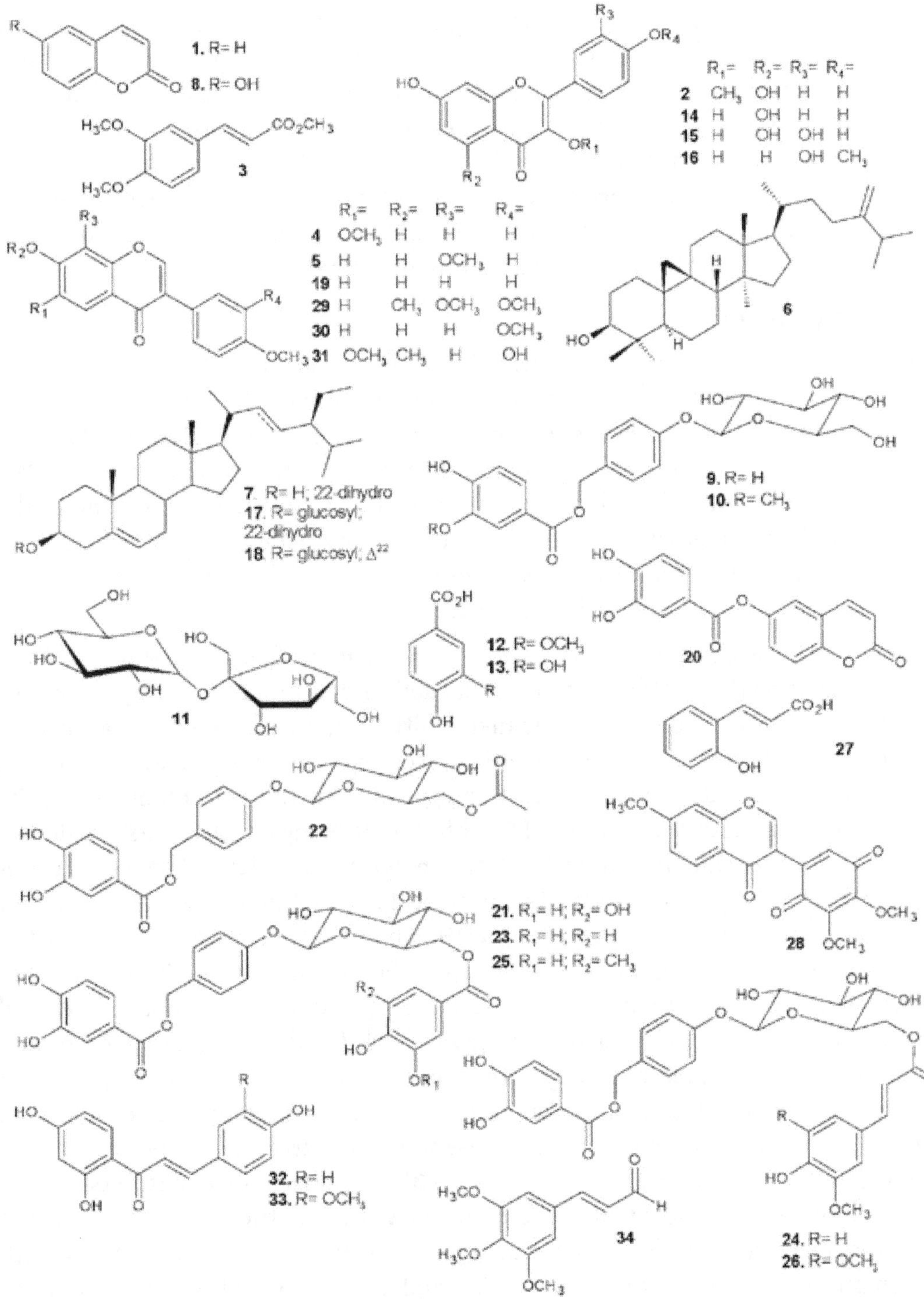

Fig. 5. Chemical structures of constituents isolated from wild *A. cearensis*.

Recently, Bandeira et al (2011) studied a resin exuded from the trunk of *A. cearensis* and found a flavonoid-rich material. The resin ethanol extract was partitioned with water and organic solvents, yielding an ethyl acetate fraction, which was chromatographed on silica gel. From this chromatography, a chloroform fraction was separated by a Sephadex LH-20, resulting in the isolation of a novel compound 3',4'-dimethoxy-1'-(7-methoxy-4-oxo-4H-cromen-3-yl)-benzo-2',5'-quinone (**28**), along with six known compounds:, 7,8,3',4'- tetramethoxyisoflavone (**29**), 3',4'-dimethoxy-7-hydroxyisoflavone (**30**) and 6,7,4'-trimethoxy-3'-hydroxyisoflavone (**31**), 4,2',4'-trihydroxychalcone (**32**), 4,2',4'-trihydroxy-3-methoxychalcone (**33**), 3,4,5-trimethoxycinnamaldehyde (**34**).

5.2.2 Cultivated plant

In order to seek a sustainable alternative for an economic utilization of *A. cearensis*, our research became focused on this *A. cearensis* cultivated plant. The chemical study of the cultivated *A. cearensis* was divided into two parts: (1) a Nuclear Magnetic Resonance (NMR) and the HPLC profiling of ethanol extracts obtained from the aerial part (EEAP) and xylopodium (EEX) of specimens cultivated according to the growing conditions described above; (2) A refined phytochemical analysis of EEAP and EEX extracts produced from specimens in 7 months of growth, where was chosen with basis on pharmacological results.

NMR profiling was performed with extracts of specimens from the 2nd through the 9th month of growth. ^{1}H NMR spectra, recorded in deuterated dimethylsulfoxide, revealed that the extracts from specimens with 2, 4, 7 and 9 months of growth, presented significantly different profiles, requiring further analysis. Hence, these extracts were duly analyzed comparatively with the wild plant extract by Photodiode Array detector (PDA)-HPLC profiling, utilizing constituents previously isolated from the wild *A. cearensis* as an analytical standard. The separations were performed on a C18 analytical column and the mobile phase was a gradient composed of H_2O (pH 3, H_3PO_4-Et_3N)/MeOH. The run time was 40 min and the chromatograms were observed at 254 nm. The qualitative analysis consisted of identification from analytical standards in the chromatograms of ethanol extracts, which were derived from the trunk bark (wild plant), xylopodium and the aerial part (cultivated plant) by retention time and UV spectra. Only 8 out of the 13 standards injected were detected in the samples. Coumarin (**1**) and vanillic acid (**12**) were the only substances present in all extracts of the *A. cearensis.* Amburoside B (**10**) and protocatechuic acid (**13**) were found in all extracts, except in the xylopodium extracts from specimens harvested in the 2nd and 4th month of growth, respectively. (*E*)-*o*-coumaric acid (**27**) and its glucoside (**35**), along with isokaempferide (**2**), afrormorsin (**4**) and kaempferol (**14**) were not detected in any extracts. Ayapin (**36**) and (*Z*)-*o*-coumaric acid glucoside (**37**) were found only in cultivated plants (Table 2).

A quantitative analysis was carried out for four major constituents of the *A. cearensis* [coumarin (**1**), amburoside A (**9**), vanillic acid (**12**), protocatechuic acid (**13**)] in ethanol extracts from trunk bark (EETB) and seeds (EES) of the said wild plant, as well as the whole plant (EEWP), xylopodium (EEX), as well as the aerial part (EEAP) of cultivated specimens in the four following selected months (2, 4, 7 and 9), accounting for 10 samples. The HPLC method was developed in chromatographic condition similar to the once described above and validated according to analytical parameters, defined by the Brazilian Health Surveillance Agency and the Brazilian Institute of Metrology, Standardization and Industrial Quality:

Table 2. Distribution of some constituents in *A. cearensis* **(+, presence; - ,not detected)**

Substances	Retention time (min.)	Trunk Bark	Xylopodium (month)				Aerial Part (month)			
			2	4	7	9	2	4	7	9
(Z)-Coumaric acid	3,85	-	+	+	+	+	-	-	-	-
p-Hidroxi-benzoic	4,99	+	+	+	+	+	-	-	-	-
Protocatechuic acid	7,36	+	+	-	+	+	+	+	+	+
Vanillic acid	10,87	+	+	+	+	+	+	+	+	+
Coumarin	13,98	+	+	+	+	+	+	+	+	+
Amburoside A	14,65	+	+	-	+	+	-	+	+	+
Ayapin	14,85	-	+	+	+	-	+	+	+	+
Amburoside B	17,17	+	-	+	+	+	+	+	+	+

linearity, selectivity, accuracy, precision as well as the limit of detection and the limit of quantification. As can be noticed on Table 3, amburoside A (**9**) was the most abundant component in EETB, followed by coumarin (**1**), protocatechuic acid (**13**) and vanillic acid (**12**). However, coumarin (**1**) was the only component detected in EES. Among cultivated plants extracts, vanillic acid (**12**) was the principal component in 3 out of 4 periods analyzed through EEW, while coumarin (**1**) appeared as the major compound in the 7th month. Protocatechuic acid (**13**) and amburoside A (**9**) were below the limit of quantification in all extracts, except in the 9th month, whereby amburoside A (**9**) had a considerable content (Leal et al., 2011). In EEAP, vanillic acid (**12**) was the main constituent evaluated in all seasons, reaching the highest concentration in the 7th month (8520 mg/100g ext). On the other hand, coumarin (**1**) was the major component in xylopodium (4th month: 3760 mg/100g ext), except in the last month, when amburoside A (**9**) was the most abundant (680 mg/100g ext). Protocatechuic acid (**13**) presented measurable levels only in the EEAP of 4 months (360 mg/100g ext). Coumarin (**1**) and vanillic acid (**12**) were found preferentially in the aerial part, while amburoside A (**9**) was present mainly in the xylopodium. In comparison with wild plants, EEX of 9 months was the only cultivated plant extract which had amburoside A as the major component like EETB, even so at a concentration being three-fold lower than in the latter.

Table 3. Concentrations of four major compounds of *A.cearensis* in different extracts.

Extracts	Concentration (mg/100g extract)- CV (%)			
	Protocatechuic acid	Vanillic acid	Coumarin	Amburoside
Wild Plant	320			2180
Trunk bark	(6,3)	200(5, 6)	1340 (6, 8)	(5,5)
Seeds	ND	ND	23520 (3,9)	ND
Cultivated Plant				
Whole				
2 months	ND	1520 (1,3)	1020 (5,7)	ND
4 months	ND	2680 (13,4)	2000 (7,0)	ND
7 months	ND	3440 (4,3)	4060 (6,5)	ND
9 months	ND	1520 (6,1)	660 (1,8)	400 (5,7)
Aerial Part				
2 months	ND	4780 (10,6)	1540 (3,0)	ND
4 months	360 (3,3)	6120 (4,8)	1660 (2,0)	260 (8,7)
7 months	ND	8520 (1,0)	6060 (7,9)	ND
9 months	ND	3460 (6,3)	1500 (3,3)	ND
Xylopodium				
2 months	ND	780 (5,3)	1320 (1,0)	380 (5,3)
4 months	ND	760 (7,5)	3760 (0,9)	ND
7 months	ND	1380 (2,9)	2500 (4,2)	300 (10,4)
9 months	ND	540 (9,8)	420 (10,8)	680 (9,4)

EEAP and EEX extracts harvested in the 7th month of growth were submitted to partition H_2O/EtOAc, yielding aqueous and ethyl acetate fractions from each extract. Isokaempferide (**2**), amburoside B (**10**), vanillic acid (**12**), *p*-hidroxy-benzoic acid (**38**) and the coumarin ayapin (**36**) were isolated from the ethyl acetate fraction derived from EEAP after being chromatographed on silica and dextran gels, while the aqueous fraction of EEAP yielded (*E*)-melilotoside (**35**) and amburoside B (**10**), again, through C18 solid-phase extraction (SPE) and C18 HPLC. On the flip side, adsorption and exclusion chromatography of the ethyl acetate fraction derived of EEX, afforded the isolation of amburoside A (**9**) and protocatechuic acid (**13**), whereas (Z)-melilotoside (**37**) was isolated by Sephadex LH-20 followed with the purification through C18 HPLC (Fig. 6). (*E/Z*)- melilotoside (**35** and **37**) are trivial names for *o*-coumaric acid glucoside in allusion to the genus *Melilotus*, where these compounds were firstly identified. Interestingly, the *E* and *Z*-melilotosides were found in different parts of the *A. cearensis*: (*E*)-stereoisomer exclusively in the aerial part, while (Z)- stereoisomer was present only in xylopodium. In *Melilotus alba* (a legume), the melilotosides are considered the precursors of coumarin, being one of the major constituents of *A. cearensis*. (*E*)-melilotoside (**35**) exposed to UV radiation (sunlight) may be converted to the less stable stereoisomer (*Z*), which do undergo enzyme-catalyzed lactonization to yield coumarin (**1**) (Dewick et al., 2002).

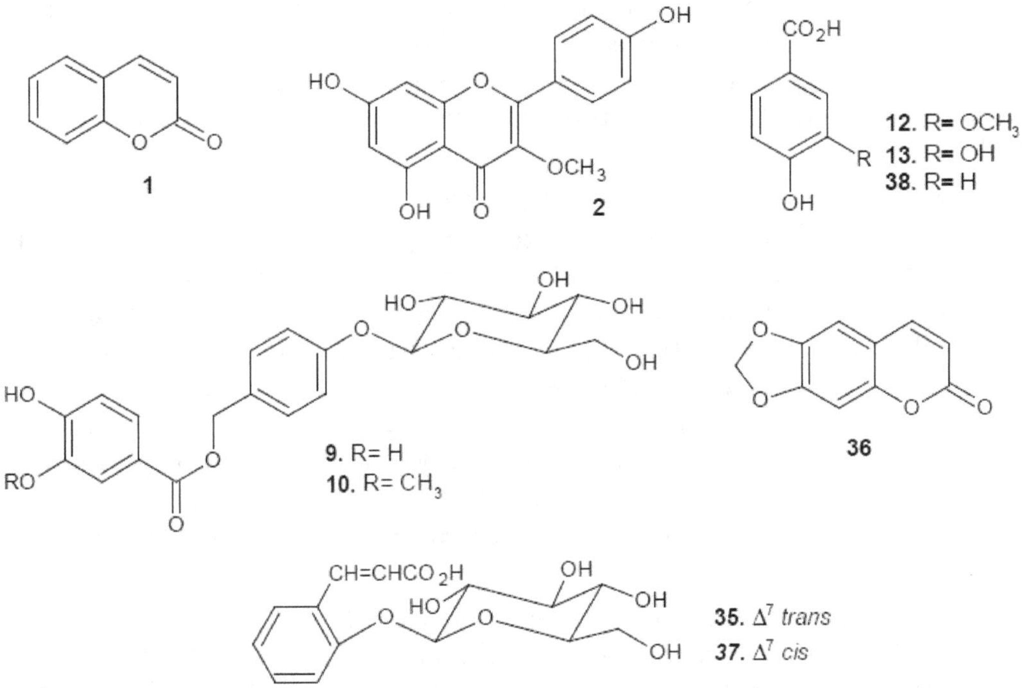

Fig. 6. Chemical structures of constituents isolated from cultivated *A. cearensis*.

As part of our effort for finding out which substances are responsible by medicinal properties of *A. cearensis*, isokaempferide (**2**), afrormosin (**4**), amburoside A (**9**), vanillic acid (**12**) and protocatechuic acid (**13**) obtained from this work were assayed in diverse pharmacological tests, which will be discussed briefly. The chemical structures of the new compounds were elucidated by means of spectroscopic techniques such as IR, HRMS, 1D and 2D NMR (COSY, HSQC, HMBC and NOESY).

5.3 Pharmacology of A. cearensis

The literature reports several toxicological and pharmacological studies carried out with the extracts and substances isolated from wild and cultivated *A. cearensis*. The focus of them is on the anti-inflammatory, antioxidant, smooth muscle relaxant, antinociceptive, neuroprotector and platelet antiaggregant effects (Leal, 1995; 2006; Leal et al., 1997; 2000; 2001; 2003ab; 2005; 2006ab; 2008).

A toxicological study carried out with the hydroalcoholic extract (HAE) from the trunk bark of the *A. cearensis* administered to rats by the oral route did not show any toxic effects (Leal et al., 2003). Further studies demonstrated that the HAE administered to rats daily for 50 days did not interfere with the pregnancy rate and development during the 1st as well as the 2nd generation of animals (Leal et al., 2003a, Leal et al., 2006a). The cytotoxicity of isokaempferide (**2**), kaempferol (**14**), amburoside A (**9**) and protocatechuic acid (**13**) from the *A. cearensis,* were evaluated on tumor cell lines and on the sea urchin egg development, as well as their lytic properties on mouse erythrocytes. The results showed that isokaempferide (**2**) and kaempferol (**14**), but not amburoside A (**9**) and protocatechuic acid (**13**), inhibited the sea urchin egg development, as well as tumor cell lines. However, only protocatechuic acid (**13**) induced lysis on mouse erythrocytes (Costa-Lotufo et al., 2003).

Previous studies (Leal et al., 1997; 2000) reported the antinociceptive, antiedematogenic and smooth muscle relaxant properties of HAE, coumarin (**1**), and the flavonoid fraction, from wild *A. cearensis.* The antiedematogenic activity was manifested in inflammatory process dependents on polimorphonuclear cells, while the antinociceptive effect of coumarin (**1**) and HAE seems to occur by a mechanism at least in part dependent on the opioid system. Nevertheless, the nitridergic system has also an important role in the coumarin nociception. Additional studies about the pharmacological potential of the HAE, coumarin (**1**) and the flavonoid fraction emphasized the anti-inflammatory potential of these species, which seems to be related to the presence of coumarin (**1**) in the plant (Leal et al., 2003).

Like other medicinal plants containing coumarin (**1**) such as *Justicia pectoralis, Pterodon polygaliflorus, Hybanthus ipecacuanha* and *Eclipta alba, A. cearensis,* also has a relaxing activity on isolated guinea pig tracheal muscles (Leal et al., 2000). Confirming this as particular effect, it was recently (Leal et al., 2006) demonstrated the relaxant action of the isokaempferide (**2**). The relaxation of the guinea-pig isolated trachea, induced by isokaempferide (**2**), was a direct and an epithelium-independent phenomenon, resulting from several intracellular actions through a common pathway e.g., the opening of Ca^{2+} and ATP-sensitive K^{+} channels.

Previous studies (Leal et al., 2003; Leal, 2006) showed that the anti-inflammatory activity of HAE, coumarin (**1**), isokaempferide (**2**) and amburoside A (**9**) from *A. cearensis,* seems to occur by an inhibitory action on the release of inflammatory mediators, and/or alternatively by interfering with a certain phase of the neutrophil migration into the inflammatory focus. Other data (Leal et al., 2008) corroborated this hypothesis showing that both the isokaempferide (**2**) and amburoside A (**9**) exert their anti-inflammatory activities mainly by inhibiting the lipopolysaccharide-induced release of TNF-α, although the involvement of other inflammatory mediators cannot be excluded. Furthermore, inhibitions of some biological functions of neutrophils, namely, accumulation of cells and activity of hydrolytic enzymes, as myeloperoxidase, may also play a role.

Amburoside A (**9**) showed a hepatoprotective property in the CCl_4-induced liver toxicity model in rats. This effect may be due to its capacity to modulate the oxidative stress, especially by reducing of the lipid peroxidation, as well as by a significant restoration to normal levels of the catalase activity and GSH contents as observed in CCl_4-intoxicated rats after the amburoside A (**9**) treatment (Leal et al., 2008).

The large-scale usage and demand for the wild *A. cearensis,* as a medicinal plant by communities in the Northeastern of Brazil, governmental programs of phytotherapy as well as the pharmaceutical industry, are contributing to decrease availability on these species, presently considered as endangered ones. In this sense, our laboratory has conducted comparative studies on the pharmacological profile of the ethanolic extract (EtOHE) or vanillic acid (**12**) from the wild and cultivated *A. cearensis*, by evaluating their antinociceptive and antiedematogenic activities in several experimental models, such as the formalin test, carrageenan or dextran-induced edema and carragenan-induced neutrophil migration into the rat peritoneal cavity (Leal et al., 2010).

The acute treatment with both the EtOHE prepared from all parts of the cultivated *A. cearensis* (4, 7 or 9 months) or the wild *A. cearensis,* present antinociceptive and anti- inflammatory activities (Fig. 7). In addition, vanillic acid (**12**), which together with coumarin (**1**) are the major compounds present in cultivated *A. cearensis*, also showed an antinociceptive activity by inhibiting both phases of the formalin test in mice, and this effect was partially blocked by naloxone. Thus, the data suggest that antinociceptive effect of vanillic acid (**12**) occur by a mechanism at least in part dependent on the opioid system (Leal et al., 2010).

Coumarin (**1**) has been found in several Brazilian medicinal plants including *J. pectoralis*, *M. glomerata* and *A. cearensis*. It has been reported that the antinociceptive and the anti- inflammatory activities of these species seems to be related at least in part to the presence of coumarin (**1**) (Leal et al., 1997; 2003; Leal et al., 1997; Lino et al., 1997; Leal et al., 2000; Freitas et al., 2008). The biological effects of coumarin (**1**) include antibacterial, antiviral, antiedematogenic, antioxidant, lipoxygenase inhibition, lipid peroxidation inhibition, and scavenging of superoxide hydroxyl radicals (Hoult & Paya, 1996; Chang et al., 1996; Casley- Smith et al., 1993; Rajarajeswari & Pari, 2011).

Recently (Leal et al., 2010), it was also determined that the anti-inflammatory effect of vanillic acid (**12**) is isolated from the cultivated *A. cearensis.*

This compound orally administered to rats, was shown to significantly inhibit the carrageenan, but not the dextran-induced edema. It also reduced the accumulation of PMN into the peritoneal cavity of rats and this effect was comparable to that observed with dexamethasone, used as a standard drug (Fig. 7).

Vanillic acid (**12**) is a benzoic acid derivative that is used as a flavoring agent. It is an intermediate in the production of vanillin from ferulic acid (Prince et al., 2011; Kim et al., 2011). Previous studies have shown antifilarial, antibacterial, antioxidant, hepatoprotective and anti-inflammatory (Kim et al., 2011; Prince et al., 2011; Itoh et al., 2009) effects of the vanillic acid. This compound exerts its anti-inflammatory effect by suppressing the production of prostaglandin E2, nitric oxide and cytokines. Furthermore, it also suppressed the activation of nuclear-factor-kappa B and caspase (Kim et al., 2011; Itoh et al., 2009). These findings confirm the anti-inflammatory activity of vanillic acid (**12**) as demonstrated by our laboratory.

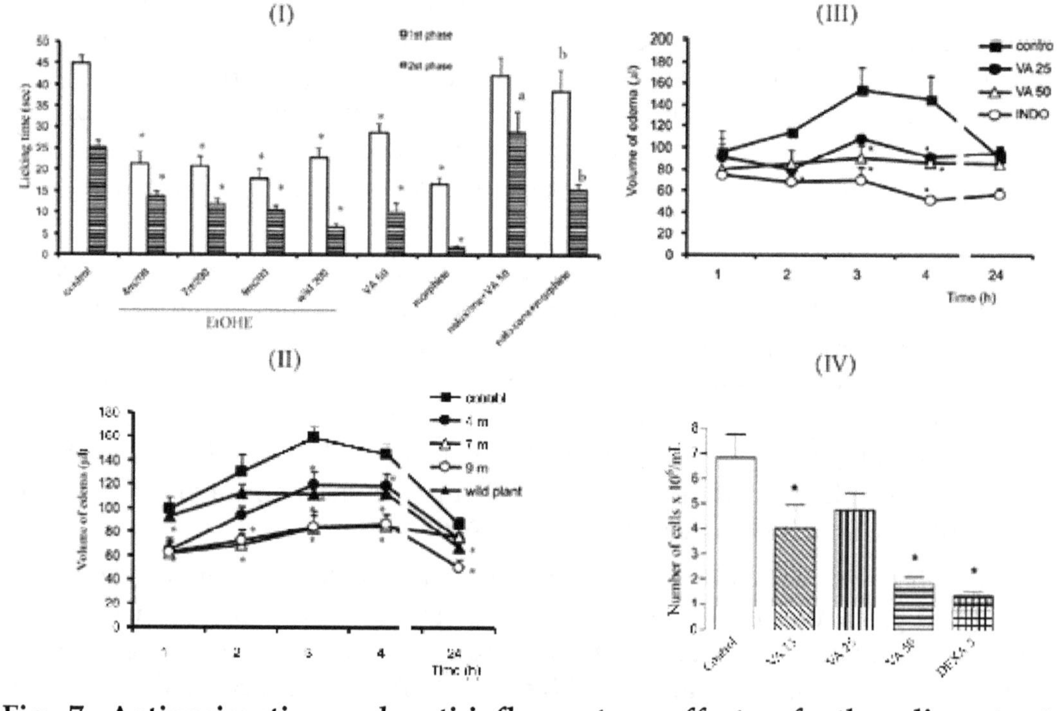

Fig. 7. Antinociceptive and anti-inflammatory effects of ethanolic extracts (EtOHE) and vanillic acid (VA) from *Amburana cearensis* in rodents.[1]

A growing body of evidence suggests that the extract and chemical constituents from the wild *A. cearensis* have pharmacological properties which justify at least in part

1 I : effects of EtOHE (cultivated, (4, 7 and 9 months(m) or wild plants: 200 mg/kg, p.o.), VA (50 mg/kg, p.o.) or morphine (MP, 5 mg/kg, s.c.) on the formalin-induced nociception in mice (6-18 animals/group). II, III and IV: anti-inflammatory effects of EtOHE and VA on the carrageenan (Cg)- induced mice paw edema and Cg-inuced rat peritonitis.

its traditional use in the treatment of asthma. Among others, the anti-inflammatory activity is possible due to their capacity to modular several responses, especially those related to oxidative stress, the production of inflammatory mediators, and the accumulation and/or activation of inflammatory cells as neutrophils.

The preliminary pharmacological study of the cultivated *A. cearensis* (Leal et al., 2010) showed that both cultivated and wild plants have antinociceptive and anti-inflammatory activities in rodents. Coumarin (**1**) and vanillic acid (**12**) are possibly responsible for the pharmacological activities of the cultivated *A. cearensis* extracts, however the pharmacological importance of other chemical constituents present in the cultivated species cannot be ruled out.

6. CONCISE SUMMARY

The interdisciplinary examination of A. cearensis demonstrated that its ethanol extracts from both cultivated and wild sources have comparable phytochemical profiles; thus, both extracts show analogous pharmacological activity. Consequently, these findings endorse the use of cultivated plants for the production of herbal drug formulations by pharmaceutical laboratories, ensuring a consistent and high-quality supply of raw materials while preserving wild specimens in their native ecosystems. This research suggests favorable opportunities for the judicious utilization of A. cearensis; nonetheless, further advancement is required in agronomic, phytochemical, and pharmacological understanding of these species. The impact of several agronomic characteristics (such as plant spacing, shade or sunlight exposure, and water supply) on chemical composition will be assessed. A metabolomic method or chemical markers will be established to assess the impact of the previously described agronomic parameters on chemical composition. Pharmacological testing utilizing several animal models of inflammation and clinical trials will be conducted to explain the mechanisms of action of the active principles of A. cearensis and to assess their efficacy in humans. An economic analysis should be conducted to assess the economic viability of producing A. cearensis herbal medicine formulations from cultivated sources.

7. REFERENCES

Aquino, F.W.B., Rodrigues, S., Nascimento, R.F. & Casimiro, A.R.S. (2005). Phenolic compounds in imburana (*Amburana cearensis*) powder extracts, *European Food Research and Technology* Vol. 221: 739-745.

Bandeira, P.N., Farias, S.S., Lemos, T.L.G., Braz-Filho, R., Santos, H.S., Albuquerque, M.R.J.R. & Costa, S.M.O. (2011). New isoflavone derivative and other flavonoids from the resin of *Amburana cearensis*, *Journal of the Brazilian Chemical Society* Vol. 22: 372-375.

Bastos, C.R.V. (1983). *Contribuição ao conhecimento químico de Torresea cearensis* (Fr. All.) (Master Thesis), Universidade Federal do Ceará, Fortaleza, pp.

Bravo, J.A., Sauvain, M., Gimenez, A., Muñoz, V., Callapa, J., Le Men-Olivier, L., Massiot, G., & Lavaud, C. (1999). Bioactive phenolic glycosides from *Amburana cearensis*, *Phytochemistry* Vol. 50: 71-74.

Calixto, J.B. (2000). Efficacy, safety, quality control, marketing and regulatory guidelines for herbal medicines (phytotherapeutic agents), *Brazilian Journal of Medical and Biological Research.* Vol. 33: 179-189.

Canter, P.H., Thomas, H. & Ernst, E. (2005). Bringing medicinal plants into cultivation: opportunities and challenges for biotechnology, *TRENDS in Biotechnology* Vol. 23: 180-185.

Canuto, K.M., Lima, M.A.S. & Silveira, E.R., (2010). Amburosides C-H and 6-*O*- protocatechuoyl coumarin from *Amburana cearensis, Journal of the Brazilian Chemical Society* Vol. 21: 1746-1753.

Canuto, K.M., Silveira, E.R. & Bezerra, A.M.E., (2010). Estudo Fitoquímico de espécimens cultivados de cumaru (*Amburana cearensis* A. C. Smith), *Quimica Nova* Vol. 33: 662- 666.

Canuto, K.M. & Silveira, E.R., (2006). Chemical constituents of trunk bark of *Amburana cearensis* A.C. Smith., *Quimica Nova* Vol. 29: 1241-1243.

Carvalho, P.E.R. (1994). *Espécies florestais brasileiras: recomendações silviculturais, potencialidades e uso da madeira,* EMBRAPA, Brasília, pp.

Casley-Smith, J.R., Morgan, R.G. & Piller, N.B. (1993). Treatment of lymphedema of the arms and legs with 5,6-benzo-[α]pyrone. The *New England Journal of Medicine* Vol. 329: 1158–1163

Chang, W.S., Lin, C.C., Chuang, S.C. & Chiang, H.C. (1996). Superoxide anion scavenging effect of coumarins. *The American Journal of Chinese Medicine* Vol. 24: 11-17

Costa-Lotufo, L.V., Jimenez, P. C., Wilke, D.V., Leal, L.K.A.M., Cunha, G.M.A., Silveira, E.R.,

Canuto, K.M., Viana, G.S.B., Moraes, M.E.A., Moraes, M.O. & Pessoa, C., (2003). Antiproliferative effects of several compounds isolated from *Amburana cearensis* A.C. Smith. *Zeitschrift fur Naturforschung C* Vol. 58c: p. 675-680.

Cunha, M.C.L. Ferreira, R.A. (2003). Aspectos morfológicos da semente e do desenvolvimento da planta jovem de *Amburana cearensis* A.C. Smith- Leguminosae Papilionoideae. Revista Brasileira de Sementes Vol. 25: 89-96.

Dewick, P.M. (2002). *Medicinal Natural Products: a biosynthetic approach,* 2nd ed., John Wiley & Sons, New York, 515 p.

Freitas, T.P., Silveira, P.C., Rocha, L.G., Rezin, G.T., Rocha, J., Citadini-Zanette, V., Romao, P. T., Dal-Pizzol, F., Pinho, R.A., Andrade, V.M. & Streck, E.L. (2008). Effects of *Mikania glomerata* Spreng. and *Mikania laevigata* Schultz Bip. ex Baker (Asteraceae) extracts on pulmonary inflammation and oxidative stress caused by acute coal dust exposure. *Journal of Medicinal Food* Vol. 11: 761-766.

Gobbo-Neto, L., Lopes, N.P. (2007). Medicinal plants: factors of influence on the content of secondary metabolites. *Quimica Nova,* Vol. 30: 374-381, 2007

Gruenwald, J. (2008). The global herbs & botanicals market; Herbs and botanicals are currently showing the most potential in functional foods and cosmetics. *Entrepreneur* URL: http://www.entrepreneur.com/tradejournals/article/181916708_1.html

Guedes, R.S., Alves, E.U., Gonçalves, E.P., Viana, J.S., Moura, M.F. & Costa, E.G. (2010). Emergence and vigor of *Amburana cearensis* (Allemão) A.C. Smith seedling in function of the sowing position and depth, *Semina: Ciências Agrárias* Vol. 31: 843- 850.

Guedes, R.S., Alves, E.U., Gonçalves, E.P., Viana, J.S., França, P.R.C. & Santos, S.S. (2010b). Physiological quality of *Amburana cearensis* (Allemão) A.C. Smith seeds stored, *Semina: Ciências Agrárias* Vol. 31: 331-342.

Hoult, J.R.S. and Paya, M. (1996). Pharmacological and biochemical actions of simple coumarins: natural products with therapeutic potential. *General Pharmacology* Vol. 27: 713-722.

Itoh, A., Isoda, K., Kondoh, M., Kawase, M., Kobayashi, M., Tamesada, M. & Yagi, K., (2009). Hepatoprotective effect of syringic acid and vanillic acid on concanavalin a- induced liver injury. *Biological & Pharmaceutical Bulletin* Vol. 32: 1215–1219.

Khalil, A.T., Chang, F.R., Lee, Y. H., Chen, C.Y., Liaw, C.C., Ramesh, P., Yuan, S.S.F., & Wu, Y.C. (2003). Chemical constituents from the *Hydrangea chinensis. Archives Pharmacal Research* Vol. 31: 15-20.

Kim, M.C., Kim, S.J., Kim, D.S., Jeon, Y.D., Park, S.J., Lee, H.S., Um, J.Y., & Hong, S.H. (2011). Vanillic acid inhibits inflammatory mediators by suppressing NF-κB in lipopolysaccharide-stimulated mouse peritoneal macrophages *Immunopharmacology and Immunotoxicology* Vol. 33:1-8

Koeppe, D.E., Rohrbaugh, L.M., Rice, E.L. & Wender, S.H. (1970). Effect of age and chilling temperature on the concentration of scopolin and caffeoylquinic acids in tobacco, *Physiologia Plantarum* Vol. 23: 258-266.

Leal, L.K.A.M., Pierdoná, T.M., Góes, J.G.S., Fonsêca, K.S., Canuto, K.M., Silveira, E.R., Bezerra, A.M.E., Viana, G. S. B. (2011). A comparative chemical and pharmacological study of standardized extracts and vanillic acid from wild and cultivated *Amburana cearensis* A.C. Smith. *Phytomedicine* Vol. 18: 230-233.

Leal, L.K.A.M., Canuto, K.M., Costa, K.C.S., Nobre-Júnior, H.V., Vasconcelos, S.M., Silveira, E.R., Ferreira, M.V.P., Fontenele, J.B., Andrade, G. M., Viana, G. S. B. (2009). Effects of amburoside A and isokaempferide, polyphenols from *Amburana cearensis*, on rodent inflammatory processes and myeloperoxidase activity in human neutrophils. *Basic & Clinical Pharmacology & Toxicology* Vol. 104: 198.

Leal, L.K.A.M., Fonseca, F.N., Pereira, F.A., Canuto, K.M., Felipe, C.F.B., Fontenele, J. B., Pitombeira, M.V., Silveira, E.R., Viana, G.S.B., (2008). Protective effects of amburoside A, a phenol glucoside from *Amburana cearensis*, against CCl_4-induced hepatotoxicity in rats. *Planta Medica* Vol. 74: 497.

Leal, L.K.A.M., Costa, M.F., Pitombeira, M., Barroso, V.M., Silveira, E.R., Canuto, K.M., Viana, G.S.B., (2006). Mechanisms underlying the relaxation induced by isokaempferide from *Amburana cearensis* in the guinea-pig isolated trachea. *Life Sciences* Vol. 79: 98-104.

Leal, L.K.A.M., Nobre, H.V., Cunha, G.M.A., Moraes, M.O., Pessoa, C., Oliveira, R.A., Silveira, E.R., Canuto, K.M., Viana, G.S.B., (2005). Amburoside A, a glucoside from *Amburana cearensis*, protects mesencephalic cells against 6-hydroxydopamine- induced neurotoxicity. *Neuroscience Letters* Vol. 388, 86-90.

Leal, L.K.A.M., Nechio, M., Silveira, E.R., Canuto, K.M., Fontenele, J.B., Ribeiro, R.A., Viana, G.S.B., (2003a). Anti-inflammatory and smooth muscle relaxant activities of the hydroalcoholic extract and chemical constituents from *Amburana cearensis* A. C. Smith. *Phytotherapy Research* Vol. 17: 335-340.

Leal, L.K.A.M; Oliveira, F.G.; Fontenele, J.B.; Ferreira, M.A.D. & Viana, G.S.B. (2003b). Toxicological study of hydroalcoholic extract from *Amburana cearensis* in rats. *Pharmaceutical Biology* Vol. 41: p. 308-314.

Leal, L.K.A.M., Ferreira, A.A.G. & Viana G.S.B. (2000). Antinociceptive, anti-inflammatory and bronchodilador activities of Brazilian medicinal plants containing coumarin: a comparative study. *Journal of Ethnopharmacology* Vol. 70: 151–9.

Leal, L.K.A.M, Matos, M.E.; Matos, F.J.A., Ribeiro, R.A., Ferreira, F.V. & Viana, G.S.B. (1997). Antinociceptive and antiedematogenic effects of the hydroalcoholic extract and coumarin from *Torresea cearensis* Fr. All. *Phytomedicine* Vol. 4: p. 221-227.

Lima, D.A. (1989) *Plantas das caatingas*, Academia Brasileira de Ciências Rio de Janeiro, 243 p. Lino, C.S., Taveira, M.L., Viana, G.S.B. & Matos, F.J.A. (1997). Analgesic and antiinflammatory activities of *Justicia pectoralis* Jacq and its main constituents: coumarin and umbelliferone. *Phytotherapy Research* Vol. 11: 211-215.

McChesney, J.D., Venkataraman, S.K., & Henri, J.T. (2007). Plant natural products: Back to the future or into extinction? *Phytochemistry* Vol. 68: 2015–2022

Menković, N., Šavikin-Fodulović, K. & Savin, K. (2000). Chemical composition and seasonal variations in the amount of secondary compounds in *Gentiana lutea* leaves and flowers, *Planta Medica* Vol. 66: 178-180

Prince, P.S.M., Dhanasekar, K. & Rajakumar, S. (2011). Preventive effects of vanillic acid on lipids, Bax, Bcl-2 and myocardial infarct size on isoproterenol-induced myocardial infarcted rats: a biochemical and in vitro study, *Cardiovascular Toxicology* Vol. 11: 58–66.

Rajarajeswari, N & Pari, L. (2011). Antioxidant Role of Coumarin on Streptozotocin– Nicotinamide-Induced Type 2 Diabetic Rats, *Journal of Biochemical and Molecular Toxicology* Vol. 25: 1-7.

Rao, M.R., Palada, M.C. & Becker, B.N. (2004). Medicinal and aromatic plants in agroforestry systems, *Agroforestry Systems* Vol. 61: 107–122.

Sahoo, N., Manchikanti, P. & Dey, S. (2010). Herbal drugs: Standards and regulation *Fitoterapia* Vol. 82: 462–471.

Schippmann, U., Leaman, D.J. & Cunnningham, A.B. (2002). *Impact of cultivation and gathering of medicinal plants on biodiversity: global trends and issues.* FAO, Rome, pp. 1– 21.

Silva, M.G.V., Craveiro, A.A.; Matos, F.J.A., Machado, M.I.L. & Alencar, J.W. (1999). Chemical variation during daytime of constituents of the essential oil of *Ocimum gratissimum* leaves, *Fitoterapia* Vol. 70: 32-34.

Southwell, I.A. & Bourke, C.A. (2001). Seasonal variation in hypericin content of *Hypericum perforatum* L. (St. John's Wort), *Phytochemistry* Vol. 56: 437-.441

Tommasi, N., Feo, V., Pizza, C. & Zhou, Z.I. (1992). Constituents of *Bidens parviflora. Fitoterapia* Vol. 63: 470.

Wagner, H. (2011). Synergy research: Approaching a new generation of phytopharmaceuticals. *Fitoterapia* Vol. 82: 34–37.

Williams, R.D. & Ellis, B.E. (1989). Age and tissue distribution of alkaloids in *Papaver somniferum, Phytochemistry* Vol. 28: 2085-2088.

Wu, S.; Ma, Y., Luo, X., Hao, X. & Wu, D. (2002). Studies on chemical constituents in root back of *Paeonia suffruticosa. Zhongcaoyao*, Vol. 33: 679-680.

Zobayed, S.M.A., Afreen, F., Kozai, T. (2007). Phytochemical and physiological changes in the leaves of St. John's wort plants under a water stress condition. *Environmental and Experimental Botany* Vol. 59: 109–116

INDEX

www.ingramcontent.com/pod-product-compliance
Lightning Source LLC
LaVergne TN
LVHW081939200926
848484LV00009B/2476